Nonequilibrium Molecular Dynamics

Theory, Algorithms and Applications

This book describes the growing field of nonequilibrium molecular dynamics (NEMD), written in a form that will appeal to the general practitioner in molecular simulation. It introduces the theory fundamental to the field, namely nonequilibrium statistical mechanics and nonequilibrium thermodynamics, provides state-of-the-art algorithms and advice for designing reliable NEMD code, and examines applications for both atomic and molecular fluids. It discusses homogenous and inhomogeneous flows and pays considerable attention to studies of highly confined fluids. In addition to statistical mechanics and thermodynamics, the book covers such themes as temperature, thermodynamic fluxes and their computation, the theory and algorithms for homogeneous shear and elongational flows, response theory and its applications, heat and mass transport algorithms, applications in molecular rheology, highly confined fluids (nanofluidics), the phenomenon of slip and generalised hydrodynamics.

Billy D. Todd undertook his bachelor and doctoral studies in physics at the University of Western Australia and Murdoch University in Perth, Australia. He then completed postdoctoral appointments at the University of Cambridge and the Australian National University, before moving to CSIRO in Melbourne in 1996. In 1999 he took up an academic appointment at Swinburne University of Technology, where he is currently Professor and Chair of the Department of Mathematics. His research focus is on statistical mechanics, nonequilibrium molecular dynamics and computational nanofluidics. He is a Fellow of the Australian Institute of Physics and a former President of the Australian Society of Rheology.

Peter J. Daivis holds Bachelor and Master's degrees in Applied Physics from Royal Melbourne Institute of Technology, a Graduate Diploma in Applied Colloid Science from Swinburne University of Technology, and a PhD from Massey University. After completing his PhD he worked on computational and theoretical investigations of transport processes at the Australian National University. He joined Royal Melbourne Institute of Technology in 1995 as a lecturer and has held the position of Professor since 2011. His research interests include applications of thermodynamics, statistical mechanics and computational physics to nonequilibrium phenomena. He is a member of the Australian Institute of Physics, the Institute of Physics UK and is currently President of the Australian Society of Rheology.

Nonequilibrium Molecular Dynamics

Theory, Algorithms and Applications

BILLY D. TODD
Swinburne University of Technology

PETER J. DAIVIS
Royal Melbourne Institute of Technology

CAMBRIDGE
UNIVERSITY PRESS

CAMBRIDGE
UNIVERSITY PRESS

University Printing House, Cambridge CB2 8BS, United Kingdom

One Liberty Plaza, 20th Floor, New York, NY 10006, USA

477 Williamstown Road, Port Melbourne, VIC 3207, Australia

314-321, 3rd Floor, Plot 3, Splendor Forum, Jasola District Centre, New Delhi - 110025, India

79 Anson Road, #06-04/06, Singapore 079906

Cambridge University Press is part of the University of Cambridge.

It furthers the University's mission by disseminating knowledge in the pursuit of education, learning and research at the highest international levels of excellence.

www.cambridge.org
Information on this title: www.cambridge.org/9780521190091
10.1017/9781139017848

First published 2017

A catalogue record for this publication is available from the British Library

Library of Congress Cataloging in Publication data
Names: Todd, Billy D., author. | Daivis, Peter J., author.
Title: Nonequilibrium molecular dynamics : theory, algorithms and applications / Billy D. Todd (Swinburne University of Technology), Peter J. Daivis (Royal Melbourne Institute of Technology).
Description: Cambridge, United Kingdom ; New York, NY : Cambridge University Press, 2017. |
Includes
bibliographical references and index.
Identifiers: LCCN 2016044167 | ISBN 9780521190091 (alk. paper) | ISBN 0521190096 (alk. paper)
Subjects: LCSH: Nonequilibrium statistical mechanics. | Nonequilibrium thermodynamics. |
Molecular dynamics.
Classification: LCC QC174.86.N65 T63 2017 | DDC 541/.394 – dc23
LC record available at https://lccn.loc.gov/2016044167

ISBN 978-0-521-19009-1 Hardback

Contents

Preface

In 2007 we wrote a review, entitled 'Homogeneous nonequilibrium molecular dynamics simulations of viscous flow: techniques and applications' [1]. Our aim then was to write a comprehensive review of the current state of the field. Though limited only to homogeneous fluids, it was clear to us then that such a review was necessary because of the growing popularity of nonequilibrium molecular dynamics (NEMD) as a powerful tool to study the transport of molecular fluids far from equilibrium. While NEMD is powerful, it is also subtle and it is often quite easy to make fundamental errors in the design and implementation of algorithms and, hence, generate results that are not what the researcher actually intends.

There are several books that deal with NEMD methods, but only the one by Evans and Morriss is entirely devoted to the field [2]. However, this influential book concentrates more on the theoretical foundations of the field, rather than providing broad algorithmic guidance for those interested in writing NEMD programs. Furthermore, the mathematical depth of the treatment presents the subject in a way that may not be readily absorbed or implemented by graduate students or nonspecialist scientists or engineers who wish to make use of reliable NEMD algorithms in their research.

It is with this point central in our thoughts that we felt it timely to write a book that could appeal to the general practitioner in the broader field of molecular simulation: not only one which builds upon previous knowledge, but also one that provides a more general overview of the field – its motivations and theoretical foundations – introduces state-of-the-art algorithms, and provides guidance on how to design reliable NEMD code for both atomic and molecular fluids. Furthermore, this book now addresses the shortcoming of our 2007 review, in that we discuss techniques to simulate highly confined fluids, thus enabling researchers to apply these methods to the realm of nanofluidics. In this realm, traditional concepts of local transport coefficients must be questioned, and the principles of generalised hydrodynamics embraced. Not every NEMD method is discussed, and at the outset we acknowledge this limitation; time and space restrictions make it impossible to condense all methods into a single book. But we have hopefully discussed many of the important methods used to simulate liquids far from equilibrium, as well as their strengths and limitations.

The book is largely self-contained; however, it is assumed that readers have a basic knowledge of statistical mechanics, thermodynamics and are familiar with the underlying principles of molecular dynamics. The books by McQuarrie [3] and de Groot and Mazur [4] are excellent references for statistical mechanics and nonequilibrium thermodynamics, respectively, while those of Allen and Tildesley [5], Rapaport [6], Frenkel and Smit [7] and Sadus [8] provide solid background for molecular dynamics methods implemented in Fortran, C or C++.

Additional resources including code examples in Fortran 90 are available under the Resources tab at http://www.cambridge.org/9780521190091.

The authors would like to thank Professor Denis Evans for introducing us to nonequilibrium molecular dynamics and nonequilibrium statistical mechanics while we were both postdoctoral researchers in the Research School of Chemistry, ANU. His inspiring leadership of the group at ANU has led to our enduring interest in this field of research.

Finally, we would like to thank our families for their patience, understanding and support over the years it has taken us to write this book, as well as a number of our colleagues and graduate students who have provided an enormous amount of intellectual stimulation, encouragement, good humour and advice over an extended period of time. In particular Jesper Hansen, Federico Frascoli, Adrian Menzel and Stephen Hannam are thanked for their careful reading of our manuscript and valuable suggestions for improvement.

Acknowledgement

This book contains extracts from two articles we published in Molecular Simulation http://dx.doi.org/10.1080/08927020412331332730 and http://dx.doi.org/10.1080/08927020601026629, reproduced here with the permission of the publisher, Taylor & Francis.

1 Introduction

We live in a world out of equilibrium – a nonequilibrium world. We are surrounded by phenomena occurring in nature, in industrial and technological processes and in controlled experiments that we can only understand with the aid of a theoretical framework that encompasses nonequilibrium processes. Our understanding of these phenomena is largely based on a macroscopic theory that starts with the balance equations for the densities of mass, momentum, energy and other macroscopic quantities. To solve these equations, it is necessary to introduce relationships based on experiments that relate the observable properties of materials to the variables that define their macroscopic state. These relationships may describe equilibrium or locally equilibrium states of the material and in this case they are called equations of state. But we also need other relationships that relate the fluxes of properties to the property gradients that drive them. These are called constitutive or transport equations. The main subject of this book is the study of these transport equations and the material properties, such as the transport coefficients that account for the differences in the behaviour of different substances, using molecular dynamics simulation methods.

The molecular dynamics (MD) simulation method was developed soon after the Monte Carlo (MC) method, for the purpose of studying relaxation and transport phenomena [9]. Both MC and MD employed periodic boundary conditions, in which the system of interest is assumed to be replicated periodically in all directions, to limit (but not totally eliminate) the effects of the finite system size. At first, applications of this new technique focused on the structure, dynamics and equations of state of equilibrium systems [10–12]. The development in the 1950s of the Green–Kubo formalism, relating linear transport coefficients to equilibrium fluctuations in the corresponding fluxes [13, 14], made it possible to use equilibrium simulations to study nonequilibrium properties. However these methods, based on the computation of time correlation functions, were difficult to apply to all of the transport properties except self-diffusion due to their large computational requirements in comparison to the computing power available at that time. In addition, they could only address transport processes in the linear regime, i.e. where the flux is directly proportional to the thermodynamic driving force. These factors motivated the development of nonequilibrium molecular dynamics (NEMD) methods.

Early nonequilibrium simulations of shear flow by Lees and Edwards [15], Gosling, McDonald and Singer [16] and Ashurst and Hoover [17] set the stage for decades of innovation and advancement through the interplay of theory, experiment and simulation.

There is no better example of this than the extraordinary growth of nonequilibrium statistical mechanics and molecular simulation as they both matured during the 1980s. For interested readers, much of the historical development of nonequilibrium molecular dynamics methods can be traced through Hoover's review and the accompanying original research articles in a special issue of *Physica* in 1983 [18], Hoover's 1993 review [19] and the book by Evans and Morriss [2].

Throughout the development of nonequilibrium simulation methodology there have emerged three distinct ways of performing nonequilibrium simulations. The first class of methods tries to model a real physical system as closely as possible and applies perturbations that are physically realistic. The Ashurst and Hoover [17] paper represents a prototype of this approach. This method has the great advantage that there is never any uncertainty regarding the relationship between the thermodynamic force applied in the simulation and the natural one. The disadvantages are that the simulated systems must always represent a tiny portion of any realistic system (unless the system that we want to mimic is actually nanoscopic), and the perturbations that are applied must be enormous compared to those existing in nature, due to the large thermal noise in the quantities being computed for small systems. This type of simulation is often called a 'boundary-driven' nonequilibrium simulation because the thermodynamic force arises due to the conditions of the momentum, heat or chemical species reservoirs attached to the system of interest. In these simulations, heat that is generated by the nonequilibrium processes occurring in the system of interest is transferred by natural thermal conduction to the boundary regions where it is removed by velocity rescaling [17] or one of the modern synthetic thermostats discussed in the following chapters. Even though the Lees and Edwards method employs periodic boundary conditions, it can be seen as a member of this class because it is the motion of the periodic images above and below the main simulation box that drives the shear flow. They therefore act as momentum reservoirs. Whereas nonperiodic reservoir methods result in spatially inhomogeneous properties, the Lees-Edwards method results in spatially homogeneous properties. This is a huge advantage for the accurate determination of transport properties.

A second class of nonequilibrium simulation method is one in which a spatially periodic (usually sinusoidal) transverse perturbation is applied through an explicit external force to drive shear flow, so that it is compatible with standard periodic boundary conditions. The method devised by Gosling, McDonald and Singer [16] is the prototype of this class. In this method, there are no reservoirs to absorb the dissipated heat. Gosling, McDonald and Singer allowed the temperature of the system to rise with time, yielding the temperature dependence of the properties during the course of the simulation. Later variants of this sinusoidal transverse force (STF) technique applied a synthetic thermostat to the fluid to allow the development of a true steady state. This method results in spatially inhomogeneous properties, because the strain rate, density and temperature all follow a sinusoidal spatial dependence.

The third method uses explicit external forces in the equations of motion, combined with periodic boundary conditions, to drive homogeneous fluxes of the desired type. This method was pioneered by Hoover, Evans and others in the early 1980s [20, 21]. Like the STF method, this method has the advantage that the perturbation is explicit in the equations of motion, making it possible to apply response theory to the system. Because the perturbation is spatially homogeneous, the response of the system is also homogeneous so the properties are computed at a specific, known thermodynamic state. Finally, the absence of reservoirs means that surface effects can be eliminated, without requiring excessively large systems. The transport properties can be obtained by direct averaging of the fluxes [20], by using the 'subtraction method' to reduce noise [22] or by forming their nonequilibrium transient time correlation functions and applying nonlinear response theory to determine their average values (the TTCF method) [23].

By the early 1990s when the first edition of the book by Evans and Morriss [2] appeared, many of the basic technical aspects of nonequilibrium molecular dynamics simulation methods were already settled. These techniques had been developed and tested using simple atomic liquids as a testbed. With the growth of computer power and increasing sophistication of algorithms and software, it became possible to embark on more ambitious investigations of nonequilibrium phenomena, and the range of applications of NEMD grew rapidly.

In this book, we have decided to restrict our attention to homogeneous NEMD methods based on the Hoover-Evans approach, STF methods based on the Gosling McDonald and Singer method and inhomogeneous NEMD methods where the inhomogeneity (i.e. the presence of interfaces in confined flows) is itself of interest. We have not discussed the wide range of very useful techniques based on reservoir methods superficially similar to the original Ashurst and Hoover simulations, but differing in important ways. This is not due to a lack of confidence in these methods on our part, but rather the simple lack of time it would take us to do justice to them, the limited space available to us, and our own lack of personal experience with them. These techniques have been successfully used to study shear flow [24], thermal conductivity and thermal diffusion [25] and we encourage the curious reader to seek further information in the vast ocean of scientific literature.

The remainder of this book is summarised as follows. In Chapter 2, we introduce the theory of nonequilibrium thermodynamics, which provides the macroscopic foundations of our description of nonequilibrium phenomena and helps us to decide which properties are worth computing. The nonequilibrium thermodynamics of the transport of mass and heat in multicomponent fluids, and the transport of spin angular momentum in molecular fluids are treated in detail.

Chapter 3 outlines the elements of nonequilibrium statistical mechanics that are necessary for an understanding of the algorithms and methods to be described later. Particular attention is paid to the development of response theory in a form that is suitable for the analysis of homogeneous nonequilibrium molecular dynamics algorithms and the TTCF method.

The statistical mechanical derivations of expressions for the temperature and fluxes are the focus of Chapter 4. Derivations of expressions for the temperature, pressure tensor and heat flux vector are carried out in detail for different circumstances in order to display to those who are new to the field some of the finer points that are often omitted from original research articles.

The underlying theory of the SLLOD algorithm for simulating homogeneous flows of atomic fluids is discussed in Chapter 5, with a focus on shear and elongational flows. The intricate details of homogeneous thermostats for nonequilibrium systems and periodic boundary conditions that are compatible with the various flows are discussed.

Applications of the SLLOD equations of motion for atomic fluids are presented in Chapter 6. Steady and oscillatory flows are discussed for different types of deformation including shear and elongation.

Algorithms for heat and mass transport are discussed in Chapter 7. Single component and multicomponent systems are considered and the connections between the phenomenological coefficients for heat flow, diffusion and the cross-effects and their corresponding practical transport coefficients are discussed.

Chapter 8 considers nonequilibrium flows of molecular liquids. Some simple models for molecular liquids and the molecular version of the SLLOD algorithm are described. Simulation methods for different ensembles are discussed and results for a few cases of molecular liquids are described.

Simulations of inhomogeneous systems including STF, planar Couette and Poiseuille flow simulations are covered in Chapter 9. Chapter 10 extends this discussion to the increasingly important case of confined molecular liquids.

In the final chapter, we consider the implications of the breakdown of standard Navier-Stokes-Fourier hydrodynamics in the analysis of NEMD simulations of highly inhomogeneous fluids. This requires the consideration of generalised hydrodynamics, in which the stress becomes a nonlocal function of the strain rate. A proper consideration of slip, which is usually negligible in macroscopic hydrodynamics, is seen to be vitally important for nanofluidic flows.

2 Nonequilibrium Thermodynamics and Continuum Mechanics

Molecular dynamics simulations provide us with a numerical solution to the equations of motion for all of the particles in a system. For a system of N particles, this gives us $3N$ positions and $3N$ momenta, which embody the full microscopic description of the system. By itself, this is too much information. What we require is a way to compute measurable properties from this microscopic description. Experimental studies of thermal and mechanical processes in fluids at the macroscopic and mesoscopic scales are based on coarse-grained measurements of field variables that obey the laws of continuum thermodynamics and mechanics. In this chapter, we will review some of the basic results of continuum thermodynamics and mechanics that we need to make the link between the microscopic mechanical variables and the variables of continuum thermodynamics and mechanics. Fortunately, a well-developed formalism for the unified treatment of continuum thermodynamics and mechanics already exists, namely nonequilibrium thermodynamics. Many excellent books and review articles on nonequilibrium thermodynamics are available, ranging from classical treatments of linear nonequilibrium thermodynamics [4] to more advanced treatments of nonequilibrium thermodynamics for systems that are far from equilibrium [26, 27]. Here, we will provide a simplified introduction to this vast and growing field, with an emphasis on the ideas that are pertinent to our discussion of nonequilibrium molecular dynamics simulations.

2.1 Thermodynamics

The first law of thermodynamics for a macroscopic system is expressed as

$$\Delta U = Q - W, \tag{2.1}$$

where ΔU is the internal energy change, Q is the heat absorbed by the system and W is the thermodynamic work done by the system. The second law of thermodynamics, which introduces the entropy S, expresses the entropy difference between two equilibrium states A and B in the form

$$\int_A^B \frac{dQ}{T} \leq S(B) - S(A), \tag{2.2}$$

where dQ is an infinitesimal heat transfer into the system along some arbitrary thermodynamic path and T is the temperature of the heat bath from which the heat is absorbed.

If the path is a reversible or "quasi-static" one, then the infinitesimal heat absorbed for any segment of that path is denoted as dQ_R, and the definition of an entropy change in terms of the reversible heat transfer is obtained

$$dQ_R/T = dS. \tag{2.3}$$

In classical thermodynamics, the second law is expressed as an equality for a reversible process, but for an arbitrary process it is an inequality. This is clearly unsatisfactory, for a number of different reasons. First, it means that the second law of thermodynamics becomes purely qualitative and it loses its ability to predict the progress of the entropy through processes that are not "quasi-static". Secondly, it results in a paradox, because common statements of the second law imply that the entropy of an arbitrary nonequilibrium state for an isolated system is defined and it increases until the system reaches its equilibrium state. Both of these issues are resolved if we adopt the local equilibrium hypothesis. When the local equilibrium hypothesis is satisfied, the local entropy density remains well defined, and the total entropy of the system is the volume integral of the local entropy density. When the local equilibrium hypothesis is not satisfied, even the existence of the entropy and the uniqueness of the temperature remain contentious.

One of the primary goals of nonequilibrium thermodynamics is to obtain an equality for the entropy change for certain types of irreversible processes. Here, we will restrict our attention to processes for which the local equilibrium hypothesis is satisfied and this goal can be achieved.

In general, the properties of a material undergoing a thermal or mechanical change will not necessarily be homogeneous in space or constant in time. Therefore, it is useful to introduce local, time-dependent quantities in place of the extensive thermodynamic variables if we want a description of the processes occurring inside the system's boundaries. For example, the volume is replaced by the specific volume, defined as the limit as the mass goes to zero of the volume per unit mass of a small but macroscopic mass element of the system,

$$v(\mathbf{r},t) = \frac{1}{\rho(\mathbf{r},t)} = \lim_{\delta m \to 0} \frac{\delta V}{\delta m}, \tag{2.4}$$

where we have also introduced the more commonly used mass density, $\rho(\mathbf{r},t)$. All of the other extensive thermodynamic variables can similarly be converted into local specific (per unit mass) field variables.

Let us consider a small but macroscopic mass element δm, undergoing a reversible thermal process. The first law of thermodynamics can be applied to find the infinitesimal internal energy change $d(\delta U)$ due to any combination of a reversible infinitesimal heat transfer where the differential of the heat absorbed is given by $d(\delta Q_R) = Td(\delta S)$, and an infinitesmal reversible compression or expansion where the differential of the work done by the material is $d(\delta W_R) = pd(\delta V)$, as

$$d(\delta U) = Td(\delta S) - pd(\delta V). \tag{2.5}$$

Now we can substitute $\delta U = u(\mathbf{r},t)\,\delta m$, $\delta S = s(\mathbf{r},t)\,\delta m$ and $\delta V = \delta m/\rho(\mathbf{r},t)$, take the limit as the mass approaches zero, and rearrange the result to obtain the reversible part of the change in the entropy due to changes in the thermodynamic fields,

$$T(\mathbf{r},t)ds(\mathbf{r},t) = du(\mathbf{r},t) - \frac{p(\mathbf{r},t)}{\rho^2(\mathbf{r},t)}d\rho(\mathbf{r},t). \tag{2.6}$$

This is the local form of the Gibbs equation. The Gibbs equation and its generalisations are central to classical and extended treatments of nonequilibrium thermodynamics. Note that although the specific internal energy, entropy and volume are very useful thermodynamic field variables, it is often more convenient to work with their densities. For example, the internal energy density (internal energy per unit volume) is equal to the specific internal energy multiplied by the mass density, $\rho(\mathbf{r},t)\,u(\mathbf{r},t)$.

2.2 Continuum Mechanics

The basic principles of continuum mechanics can be found in any one of a large number of excellent books. Treatments with an emphasis on fluids include those by Bird *et al.* [28, 29], Tanner [30], Phan-Thien and Huilgol [31] and Truesdell [32]. Our aim here is to introduce some basic concepts that we will use later in our development of nonequilibrium thermodynamics and our interpretations of molecular dynamics simulations.

2.2.1 Pressure Tensor

The forces acting on a small element of fluid can be classified as either short-ranged, in which case they are regarded as contact forces acting on the surface of the element, or long-ranged, where they are regarded as acting throughout the entire element. Continuum mechanics treats contact forces as contributions to the pressure tensor, whereas forces that act throughout a fluid are treated as body forces. Of course, the distinction between these two types of force is not always clear, particularly if we consider a very small volume element with a size approaching the range of the "short-ranged" forces, but it is usually the case that intermolecular forces fall into the short-ranged category and external gravitational and electric fields can be regarded as long-ranged.

Consider an element of surface area on a closed surface in the fluid. The element of surface area has an orientation defined by the outward pointing normal vector at that point on the surface. The total force $d\mathbf{F}$ acting on the oriented surface element $d\mathbf{A}$ due to the pressure tensor arising from molecular motion and intermolecular interactions within the material is given by

$$d\mathbf{F} = -\mathsf{P}^T \cdot d\mathbf{A}. \tag{2.7}$$

This is the mechanical definition of the pressure tensor. In an isotropic equilibrium fluid, the pressure tensor is isotropic, $\mathsf{P} = p\mathbf{1}$, where $\mathbf{1}$ is the second-rank isotropic tensor (the unit tensor) and p is the scalar pressure of the equilibrium fluid, $p = (1/3)\,\mathrm{Tr}(\mathsf{P})$. In the context of fluid mechanics, it is often preferred to separate the equilibrium and nonequilibrium parts of the pressure tensor and define the nonequilibrium part of the pressure tensor as $\mathsf{\Pi} = \mathsf{P} - p\mathbf{1}$ and the stress as $\boldsymbol{\sigma} = -\mathsf{\Pi}^T$. Note that the transpose in this relationship is required because stress is defined in terms of the force exerted by the surrounding medium on a fluid element, $d\mathbf{F} = \boldsymbol{\sigma} \cdot d\mathbf{A}$ whereas the pressure tensor is defined as the diffusive transport of momentum through the surface. This distinction is discussed in more detail by Bird *et al.* [28]. In many cases, the pressure tensor is symmetric, but in situations where the intrinsic angular momentum density of the material plays a role, the pressure tensor may possess an antisymmetric component.

2.2.2 Deformation

In continuum mechanics, we can characterise the kinematics of material deformation in terms of the motion of a set of material points which are defined by their positions at a certain time. For example, a point $\mathbf{r}'$ could represent the position of a certain material element at time $t = 0$, and the label $\mathbf{r}'$ can be used to identify this point at all future times, regardless of the motion that the material may undergo. The trajectory of a material point is given by its path function $\mathbf{r} = \mathbf{M}(\mathbf{r}', t)$ [31] and the deformation gradient tensor is defined as $\mathsf{F} = (\partial \mathbf{r}/\partial \mathbf{r}')^T$, where the superscript T denotes the transpose. The relative deformation gradient $\mathsf{F}_t(\tau) = (\partial \mathbf{r}''/\partial \mathbf{r})$ is defined as the derivative of the position at some intermediate time τ with respect to its position at another time t. In general, neither of these times need be equal to $t = 0$. Deformation can be defined in terms of the "right relative Cauchy-Green strain tensor", which is given by

$$\mathsf{C}_t(\tau) = \mathsf{F}_t(\tau)^T \mathsf{F}_t(\tau). \tag{2.8}$$

As the name implies, this measure of strain is not unique. In particular, for finite strains, as distinct from infinitesimal strains, different strain measures are not identical. The differences between these different measures of strain arise from different possible choices of the reference state. Alternative definitions of strain can be found in many of the previously mentioned books on theoretical rheology or continuum mechanics [28, 30–32]. For our current purposes, it is sufficient to choose a commonly used definition and develop our treatment consistently with this definition.

Standard treatments of rheology discuss fluid deformation in terms of the "rheologically simple fluid" model. A "rheologically simple fluid" is defined as one for which the stress is given by a spatially local functional of the strain history. For a "rheologically simple fluid" the deformation can also be expressed in terms of the Rivlin-Ericksen tensors, which are defined as the material (streaming) derivatives of the strain tensor, given by

$$\left.\frac{d^n}{d\tau^n}\mathsf{C}_t(\tau)\right|_{\tau=t} = \mathbf{A}_n(t), \tag{2.9}$$

where $n = 0, 1, 2, \ldots$. This can also be expressed in terms of the velocity gradient tensor $\mathbf{L} = \nabla \mathbf{v}$ and its transpose. Using the results for the material derivative of the relative deformation gradient and its transpose, $\dot{\mathsf{F}} = \mathbf{L}\mathsf{F}$ and $\dot{\mathsf{F}}^T = \mathsf{F}^T \mathbf{L}^T$, the rate of change of the strain is found to be

$$\frac{d}{d\tau}\mathsf{C}_t(\tau) = \frac{d}{d\tau}\left(\mathsf{F}_t(\tau)^T \mathsf{F}_t(\tau)\right)$$
$$= \mathsf{F}_t^T \cdot \mathbf{L} \cdot \mathsf{F}_t + \mathsf{F}_t^T \cdot \mathbf{L}^T \cdot \mathsf{F}_t. \tag{2.10}$$

Evaluating this at time $t = \tau$ results in $\mathsf{F}_t^T = \mathsf{F}_t = \mathbf{1}$ which then gives

$$\mathbf{A}_1(t) = \frac{d}{d\tau}\mathsf{C}_t(\tau)\bigg|_{\tau=t} = \mathbf{L} + \mathbf{L}^T \tag{2.11}$$

for the rate of strain. This reflects the fact that the rate of strain is related to only the symmetric part of the velocity gradient tensor. In the standard continuum mechanics treatment, the higher-order time derivatives are defined in terms of the gradients of the particle accelerations and higher derivatives as

$$\mathbf{L}_n = \frac{d^n}{d\tau^n}\mathsf{F}_t(\tau)\bigg|_{\tau=t} = \nabla\frac{d^n\mathbf{r}}{dt^n} = \nabla\frac{d^{n-1}\mathbf{v}}{dt^{n-1}} \tag{2.12}$$

and the Rivlin-Eriksen tensors are then given by

$$\mathbf{A}_n(t) = \frac{d^n}{d\tau^n}\mathsf{C}_t(\tau)\bigg|_{\tau=t} = \sum_{r=0}^{n}\binom{n}{r}\mathbf{L}_{n-r}^T\mathbf{L}_r, \tag{2.13}$$

where $\binom{n}{r} = \frac{n!}{r!(n-r)!}$, $\mathbf{L}_1 = \mathbf{L}$ and $\mathbf{L}_0 = \mathbf{1}$. A recursion relation due to Oldroyd [31] can be used to generate the $(n+1)^{th}$ Rivlin-Eriksen tensor from the nth,

$$\mathbf{A}_{n+1} = \frac{d}{dt}\mathbf{A}_n + \mathbf{A}_n \cdot \mathbf{L} + \mathbf{L}^T \cdot \mathbf{A}_n. \tag{2.14}$$

These results will be used later when simulation results for the stress tensor in shear and planar elongational flows are described in terms of some standard continuum mechanical constitutive equations.

2.3 Nonequilibrium Thermodynamics

The equations of change for the thermal and mechanical fields, i.e. the density, the streaming velocity and the internal energy, are derived from basic physical principles – conservation of mass, Newton's second law of motion and conservation of the total energy. In keeping with our simplified approach, which we hope will most clearly display the underlying physical ideas, we will apply these principles to the motion of a mass element and derive the balance equations for the density, velocity and specific internal energy fields.

2.3.1 Mass Balance

We begin our derivation of the balance equation for the mass with the assumption that our small but macroscopic subsystem is defined as a cuboidal element of the continuum that maintains a constant mass while it undergoes thermal and mechanical changes.[1] Thus, we can write

$$\frac{d(\delta m)}{dt} = \frac{d}{dt}(\rho \delta V) = \rho \frac{d(\delta V)}{dt} + \delta V \frac{d\rho}{dt} = 0. \tag{2.15}$$

The volume of the mass element can only change if the velocities of the fluid at opposing faces of the cuboid are different, so the term containing the rate of change of the volume can be written as

$$\frac{1}{\delta V}\frac{d(\delta V)}{dt} = \frac{1}{\delta x \delta y \delta z}\frac{d}{dt}(\delta x \delta y \delta z) = \frac{1}{\delta x}(\delta v_x) + \frac{1}{\delta y}\left(\delta v_y\right) + \frac{1}{\delta z}(\delta v_z), \tag{2.16}$$

which becomes $\nabla \cdot \mathbf{v}$ in the limit as the size of the mass element approaches zero. Thus, we obtain the equation of change for the local fluid density,

$$\frac{d\rho}{dt} = -\rho \nabla \cdot \mathbf{v}. \tag{2.17}$$

2.3.2 Momentum Balance

Applying Newton's second law of motion to the small mass element, we obtain

$$\delta m \frac{d\mathbf{v}}{dt} = \delta \mathbf{F}^t, \tag{2.18}$$

where $\delta \mathbf{F}^t$ is the total force acting on the material element. We will consider two types of force that may act on the mass element. The first is the force due to interactions with the surrounding fluid (via the pressure tensor, Equation (2.7)), and the second is the body force due to an external field $\mathbf{F}^e$, which we will write as $\delta \mathbf{F} = \delta m \mathbf{F}^e$. This explicitly nominates the external field as a gravitational field, since it couples to the mass, but the body forces due to electric and other fields can be written in a similar way, by altering the variable to which thc field couples. The explicit form of the x-component of the total force due to the stresses on all six faces of the cuboid and the external body force is

$$\begin{aligned} \delta F_x^t = & -\delta y \delta z \left[P_{xx}(x+\delta x) - P_{xx}(x)\right] - \delta z \delta x \left[P_{yx}(y+\delta y) - P_{yx}(y)\right] \\ & - \delta x \delta y \left[P_{zx}(z+\delta z) - P_{zx}(z)\right] + \delta m F_x^e \end{aligned} \tag{2.19}$$

and the other components are similar. Figure 2.1 shows some of these forces schematically. Substituting Equation (2.19) and the corresponding equations for the y and z components of the force into Equation (2.18) and then dividing both sides by the volume of the mass element $\delta V = \delta x \delta y \delta z$ and taking the continuum limit, we find

$$\rho \frac{d\mathbf{v}}{dt} = -\nabla \cdot \mathsf{P} + \rho \mathbf{F}^e. \tag{2.20}$$

[1] A cuboid is a parallelepiped with rectangular faces.

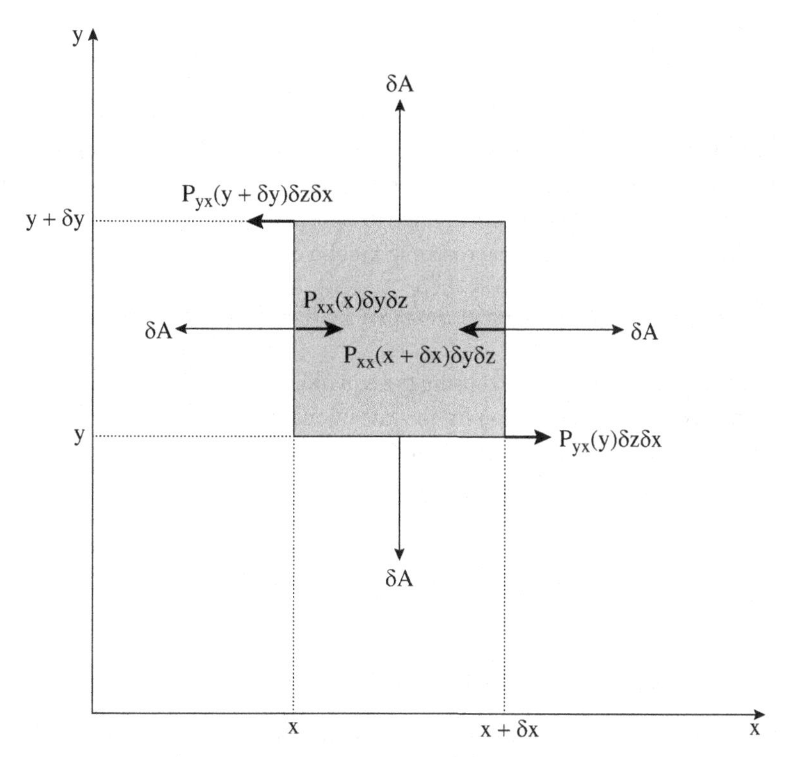

Figure 2.1 Contributions to the x-component of the force on a mass element of fluid (heavy arrows). Only those forces due to components of the pressure tensor that act on the x and y facing sides of the element are shown. For example, the force on the top face of the fluid mass element due to the shear pressure is $\delta \mathbf{F}_x = -(\delta \mathbf{A} \cdot \mathbf{P})_x = -\delta z \delta x[P_{yx}(y + \delta y)]$ because $\delta \mathbf{A}$ points in the positive y-direction for this face. If P_{yx} is negative (e.g. for shear flow with $\partial v_x / \partial y > 0$), then this force will be in the positive x-direction.

2.3.3 Internal Energy Balance

The total energy of the mass element, $e(\mathbf{r}, t)\delta m$, consists of the macroscopic kinetic energy of its centre of mass, the macroscopic potential energy of the mass element due to its interactions with external conservative fields $\delta m \psi$, the microscopic kinetic energy due to atomic motion relative to the centre of mass and the microscopic potential energy due to interatomic forces. The last two are combined into the internal energy, $u\delta m$, and from the macroscopic point of view, this provides us with a definition of the internal energy – it is the total energy minus the macroscopic kinetic and potential energies, that is,

$$e\delta m = \frac{1}{2}\delta m v^2 + \psi \delta m + u\delta m. \tag{2.21}$$

Taking the time derivative of this and then dividing by the volume of the mass element and taking the infinitesimal limit, we find that the rate of change of the specific energy is given by

$$\rho\frac{de}{dt} = \frac{1}{2}\rho\frac{dv^2}{dt} + \rho\frac{d\psi}{dt} + \rho\frac{du}{dt}. \tag{2.22}$$

We will ignore chemical, electronic, nuclear and other subatomic contributions to the internal energy and assume that the internal energy consists entirely of the thermal part of the kinetic energy plus the energy due to intermolecular forces.

The rate of change of the macroscopic kinetic energy of a mass element is

$$\frac{d}{dt}\left(\frac{1}{2}\delta m v^2\right) = \frac{1}{2}\delta m\left(2\mathbf{v}\cdot\frac{d\mathbf{v}}{dt}\right) = \delta m\mathbf{v}\cdot\frac{d\mathbf{v}}{dt}. \tag{2.23}$$

Taking the continuum limit, and using the equation for the rate of change of the momentum, this becomes an expression for the rate of change of the specific kinetic energy due to changes in the streaming velocity of the fluid

$$\frac{1}{2}\rho\frac{dv^2}{dt} = -\mathbf{v}\cdot(\nabla\cdot\mathbf{P}) + \rho\mathbf{v}\cdot\mathbf{F}^e. \tag{2.24}$$

Similarly, the rate of change of the specific potential energy (for time-independent fields) is

$$\rho\frac{d\psi}{dt} = \rho\mathbf{v}\cdot\nabla\psi = -\rho\mathbf{v}\cdot\mathbf{F}^e, \tag{2.25}$$

showing that conservative external forces do not change the total energy of a one component system, since their effect on the potential energy is exactly cancelled by their effect on the kinetic energy. The total energy of the mass element can only change by two mechanisms; through the work done on it by stresses and nonconservative external forces, or by the diffusive transport of internal energy through its boundaries (heat flow). This is expressed as

$$\frac{d}{dt}(e\delta m) = \delta m\frac{de}{dt} = \frac{d\left(\delta W^t\right)}{dt} + \frac{d\left(\delta Q\right)}{dt}. \tag{2.26}$$

Dividing by the volume of the mass element and taking the infinitesimal limit, we obtain

$$\rho\frac{de}{dt} = \rho\frac{dw^t}{dt} + \rho\frac{dq}{dt}, \tag{2.27}$$

where w^t is the specific mechanical work done by stresses and nonconservative external forces and q is the specific heat absorption. The work done by stresses on one face of the cuboid is given by

$$d\left(\delta W\right) = -\delta\mathbf{A}\cdot\mathbf{P}\cdot d\mathbf{r}. \tag{2.28}$$

The total work done by stresses is a sum over all six faces of the cuboid. The sum of the work done by the x-components of the forces acting on the mass element is

$$\begin{aligned}\sum\delta F_{xi}dx_i = &-\delta y\delta z\left[P_{xx}(x+\delta x)d(x+\delta x) - P_{xx}(x)dx\right]\\ &-\delta z\delta x\left[P_{yx}(y+\delta y)d(x+\delta x) - P_{yx}(y)dx\right]\\ &-\delta x\delta y\left[P_{zx}(z+\delta z)d(x+\delta x) - P_{zx}(z)dx\right]\end{aligned} \tag{2.29}$$

and the y and z components are similar. Dividing by the volume and forming the derivative of the position with respect to time and then adding the three components together to find the density of the total rate at which work is done on the fluid, we obtain

$$\frac{1}{\delta V}\frac{d\left(\delta W^t\right)}{dt} = \sum \frac{\delta F_{xi}}{\delta V}\frac{dx_i}{dt} + \frac{\delta F_{yi}}{\delta V}\frac{dy_i}{dt} + \frac{\delta F_{zi}}{\delta V}\frac{dz_i}{dt}. \tag{2.30}$$

In the limit as the mass element becomes infinitesimally small, this becomes

$$\rho\frac{dw^t}{dt} = -\nabla\cdot(\mathbf{P}\cdot\mathbf{v}). \tag{2.31}$$

In general the heat absorbed through the surface of a specified volume of material is given by

$$\frac{dQ}{dt} = -\oint_S \mathbf{J}_q\cdot d\mathbf{A}, \tag{2.32}$$

where $\mathbf{J}_q$ is the heat flux vector. This can be applied to the finite element of fluid, and the net absorption of heat by the element is the sum of the contributions from all six faces. Considering only the two faces with their normal vectors along the x-axis, we find

$$\frac{1}{\delta V}\frac{d\left(\delta Q_x\right)}{dt} = -\frac{1}{\delta V}\left(-J_{qx}(x)\delta y\delta z + J_{qx}(x+\delta x)\delta y\delta z\right), \tag{2.33}$$

which shows that the total rate of heat absorption depends on the difference in the value of the heat flux vector on opposite faces of the element. In the continuum limit, this again corresponds to a divergence, so we find that the density of the rate of heat absorption is

$$\rho\frac{dq}{dt} = -\nabla\cdot\mathbf{J}_q. \tag{2.34}$$

Now we can combine all of the terms in the rate of change of the total energy, and then rearrange to obtain the internal energy balance equation

$$\rho\frac{du}{dt} = \rho\frac{dw^t}{dt} + \rho\frac{dq}{dt} - \frac{1}{2}\rho\frac{dv^2}{dt} - \rho\frac{d\psi}{dt}. \tag{2.35}$$

Substituting the individual terms, we obtain

$$\rho\frac{du}{dt} = -\nabla\cdot(\mathbf{P}\cdot\mathbf{v}) - \nabla\cdot\mathbf{J}_q + \mathbf{v}\cdot(\nabla\cdot\mathbf{P}). \tag{2.36}$$

Using the result $\nabla\cdot(\mathbf{P}\cdot\mathbf{v}) = \nabla\cdot\mathbf{P}\cdot\mathbf{v} + \mathbf{P}^T : \nabla\mathbf{v}$ we finally obtain the internal energy balance equation

$$\rho\frac{du}{dt} = -\nabla\cdot\mathbf{J}_q - \mathbf{P}^T : \nabla\mathbf{v}. \tag{2.37}$$

In this equation, the left-hand side represents the rate of change of the specific internal energy multiplied by the density at some point moving with the fluid, the first term on the right represents the diffusive heat transfer at that point, and the last term represents the thermodynamic work done by stresses. If the material is a purely viscous one, all of the work done by the stresses is dissipated and the rate at which the stresses do work is equal to the rate of dissipation of energy. If the material is viscoelastic, the work done by the stresses may be partially elastic (stored) and partially viscous (dissipated), depending on the deformation history.

We now have the balance equations for the mass density, the specific momentum and the specific internal energy in a single-component fluid, which are collected below for future reference:

$$\frac{d\rho}{dt} = -\rho\nabla\cdot\mathbf{v} \tag{2.38a}$$

$$\rho\frac{d\mathbf{v}}{dt} = -\nabla\cdot\mathbf{P} + \rho\mathbf{F}^e \tag{2.38b}$$

$$\rho\frac{du}{dt} = -\nabla\cdot\mathbf{J}_q - \mathbf{P}^T : \nabla\mathbf{v}. \tag{2.38c}$$

These equations are written in the Lagrangian form, which means that the derivatives are a special form of the total derivative, where the rate of change of a quantity is evaluated for a point that is moving with the streaming velocity, $\mathbf{v}$. In other words, a derivative evaluated in this way includes the rate of change of a quantity due to its explicit time dependence and also the change in the quantity due to its position dependence, consistent with the flow of the fluid. Another way that the rates of change of thermal and mechanical properties can be evaluated is to consider the rate of change at a fixed point in space due only to explicit time dependence, represented by a partial derivative. The Eulerian form of the balance equations can be obtained from the Lagrangian form by applying the following transformation rule. The rate of change of the local density $\rho(\mathbf{r},t)a(\mathbf{r},t)$ of a property $a(\mathbf{r},t)$ is given by

$$\rho\frac{da}{dt} = \frac{\partial(\rho a)}{\partial t} + \nabla\cdot(\rho a\mathbf{v}). \tag{2.39}$$

Thus, the Eulerian form of the balance equations is

$$\frac{\partial\rho}{\partial t} = -\nabla\cdot(\rho\mathbf{v}) \tag{2.40a}$$

$$\frac{\partial(\rho\mathbf{v})}{\partial t} = -\nabla\cdot\mathbf{P} - \nabla\cdot(\rho\mathbf{v}\mathbf{v}) + \rho\mathbf{F}^e \tag{2.40b}$$

$$\frac{\partial(\rho u)}{\partial t} = -\nabla\cdot\mathbf{J}_q - \nabla\cdot(\rho u\mathbf{v}) - \mathbf{P}^T : \nabla\mathbf{v}. \tag{2.40c}$$

These equations can be integrated over an arbitrary volume V which we may regard as either instantaneously at rest (the Eulerian picture) or co-moving with the fluid (the Lagrangian picture). In the first instance we denote the volume as V and in the second, we denote it as $V(t)$. Since the volume may be of arbitrary size, we should more correctly say that in the Lagrangian picture each point on its surface is moving with the local streaming velocity, so its shape and volume may be changing. The Reynolds transport theorem provides us with a link between the Lagrangian and Eulerian pictures. Consider the time derivative of the total amount of some quantity A contained in a co-moving volume $V(t)$,

$$\frac{dA(V(t))}{dt} = \frac{d}{dt}\int_{V(t)} \rho a dV. \tag{2.41}$$

The Reynolds transport theorem relates this derivative to the rate of change of the amount of A in the same volume V, but with V kept fixed,

$$\frac{d}{dt}\int_{V(t)} \rho a dV = \int_V \frac{\partial(\rho a)}{\partial t} dV + \int_S \rho a \mathbf{v} \cdot d\mathbf{A} \tag{2.42a}$$

$$= \int_V \left[\frac{\partial(\rho a)}{\partial t} + \nabla \cdot (\rho a \mathbf{v})\right] dV \tag{2.42b}$$

$$= \int_V \rho \frac{da}{dt} dV \tag{2.42c}$$

$$= \int_V [-\nabla \cdot \mathbf{J}_A + \sigma_A] dV, \tag{2.42d}$$

where the divergence theorem has been used to obtain the second line and the last line represents the diffusion of the quantity A into the volume V plus the total production of A in the volume V at time t. If A is conserved, its production is zero and it can only change by the diffusive flux of A through the surface bounding V. These equalities apply to any arbitrarily chosen volume, so they are often used to derive the balance laws.

2.3.4 Entropy Production

When the local equilibrium hypothesis is valid, the entropy of an isolated system is defined and it has its maximum value at equilibrium. That means that the entropy of an isolated system that is not at equilibrium must increase as it progresses towards equilibrium. But the second law of thermodynamics says that the entropy change for a reversible process is given by $dS = dQ_R/T$. Any change in an isolated system must have $dQ = 0$, so the entropy change inside the system cannot occur by heat transfer with the environment. A postulate of nonequilibrium thermodynamics is that the entropy change inside an isolated system occurs instead through internal entropy production. In a system that is not isolated, it is assumed that entropy changes can occur by a combination of reversible and irreversible processes. The derivation of expressions for the entropy production is one of the central features of the subject of nonequilibrium thermodynamics. In this section, we will outline a derivation of the expression for the entropy production in the simple case of a one component fluid with heat flow and viscous flow.

We begin by postulating that the change in the local specific entropy can be written in a similar form to the change in the other thermodynamic fields, as a flux term plus a source term:

$$\rho \frac{ds}{dt} = -\nabla \cdot \mathbf{J}_s + \sigma(\mathbf{r}, t). \tag{2.43}$$

The entropy flux is assumed to be reversible, so it is given by the field analogue of the equation for reversible entropy transfer,

$$\mathbf{J}_s(\mathbf{r}, t) = \frac{\mathbf{J}_q(\mathbf{r}, t)}{T(\mathbf{r}, t)}. \tag{2.44}$$

The entropy production is therefore

$$\sigma(\mathbf{r}, t) = \rho \frac{ds}{dt} + \nabla \cdot \left(\frac{\mathbf{J}_q(\mathbf{r}, t)}{T(\mathbf{r}, t)} \right). \tag{2.45}$$

We now introduce an expression for the rate of change of the entropy in terms of the thermodynamic variables, using the Gibbs equation. By using the Gibbs equation, we are assuming that the specific entropy can be expressed as the same function of the specific internal energy and density as it is at equilibrium. Consequently, its time dependence is also given by the derivatives of these variables,

$$T(\mathbf{r}, t) \frac{ds(\mathbf{r}, t)}{dt} = \frac{du(\mathbf{r}, t)}{dt} - \frac{p(\mathbf{r}, t)}{\rho^2(\mathbf{r}, t)} \frac{d\rho(\mathbf{r}, t)}{dt}. \tag{2.46}$$

This is known as the local equilibrium approximation. We can now use this in the equation for the entropy production to obtain

$$\begin{aligned} \sigma(\mathbf{r}, t) &= \frac{1}{T} \left(\rho \frac{du}{dt} - \frac{p}{\rho} \frac{d\rho}{dt} \right) + \nabla \cdot \mathbf{J}_s \\ &= -\frac{1}{T} \mathbf{\Pi}^T : \nabla \mathbf{v} - \frac{1}{T^2} \mathbf{J}_q \cdot \nabla T, \end{aligned} \tag{2.47}$$

where we have used the internal energy and mass balance equations and we define the nonequilibrium part of the local pressure tensor as

$$\mathbf{\Pi} = \mathbf{P} - p\mathbf{1}. \tag{2.48}$$

Thus, the last line of Equation (2.47) shows that the entropy production contains terms that are products of the fluxes (to be discussed in greater detail in the following chapter) and their corresponding thermodynamic forces [4]. If we express each thermodynamic flux and force in the entropy production in terms of their irreducible components, the entropy production naturally separates into a scalar part, a vector part and a traceless symmetric second-rank tensor part. The heat flux is an ordinary first-rank polar tensor and is already in irreducible form. Using the irreducible representation of a second-rank tensor for the viscous pressure tensor and the velocity gradient tensor, we find

$$\mathbf{\Pi} = \Pi \mathbf{1} + \mathbf{\Pi}^{ts} + \mathbf{\Pi}^a \tag{2.49}$$

$$\nabla \mathbf{v} = \tfrac{1}{3} \nabla \cdot \mathbf{v} \mathbf{1} + (\nabla \mathbf{v})^{ts} + (\nabla \mathbf{v})^a, \tag{2.50}$$

where $\Pi = \mathrm{Tr}(\mathbf{\Pi})/3$ is the isotropic part of the pressure tensor, $\mathbf{\Pi}^{ts}$ is the traceless symmetric part of the pressure tensor and $\mathbf{\Pi}^a$ is its antisymmetric part. We can further decompose the entropy production associated with the viscous pressure tensor into three distinct terms

$$\mathbf{\Pi}^T : \nabla \mathbf{v} = \Pi \nabla \cdot \mathbf{v} + \mathbf{\Pi}^{ts} : (\nabla \mathbf{v})^{ts} - \mathbf{\Pi}^a : (\nabla \mathbf{v})^a. \tag{2.51}$$

In this section, we will assume that the pressure tensor is symmetric, so the third term is zero. The final form of the entropy production in terms of the irreducible components of the thermodynamics fluxes and forces is

$$\sigma(\mathbf{r},t) = -\Pi\frac{\nabla\cdot\mathbf{v}}{T} - \mathbf{J}_q\cdot\frac{\nabla T}{T^2} - \mathbf{\Pi} : \frac{(\nabla\mathbf{v})^{ts}}{T}. \tag{2.52}$$

2.3.5 Constitutive Equations

The formulation of constitutive relations between the thermodynamic forces and their corresponding fluxes has a long history, encompassing Newton's law of viscosity, Fick's law of diffusion, Ohm's law of electrical conduction and Fourier's law of thermal conduction. The modern theory of nonequilibrium thermodynamics can be seen as a systematisation of these apparently diverse transport phenomena within a common framework, with the entropy production as the unifying concept. Our description is necessarily brief, but for more detailed information, the reader should consult one of the many excellent references on this subject (e.g. reference [4]).

In general, we can expect each of the thermodynamic fluxes to be a nonlinear, nonlocal functional of all of the thermodynamic forces. However, if we consider only isotropic materials and limit these dependencies to simple linear, local functions, symmetry requirements place restrictions on the possible dependencies. The principle that symmetry restricts the possible linear dependencies of the thermodynamic fluxes on the forces in isotropic materials is known as Curie's principle. For a more detailed discussion of the effect of symmetry on the properties of isotropic materials, we refer the reader to de Groot and Mazur [4] or Evans and Morriss [2], and for a more comprehensive discussion encompassing anisotropic materials such as crystalline solids and liquid crystals, specialised texts such as Juretchke [33] and de Gennes [34] should be consulted.

After taking these symmetry restrictions into account and limiting our consideration to a linear and local response in both space and time, the linear phenomenological relations between the fluxes and thermodynamic forces are obtained. The only scalar flux is the isotropic nonequilibrium pressure. It can only couple to the scalar thermodynamic force $\nabla\cdot\mathbf{v}/T$. Similarly, the heat flux is the only vector flux and it can only couple to the temperature gradient. Lastly, the traceless symmetric pressure tensor can only couple to the traceless symmetric velocity gradient tensor, so the linear phenomenological equations are

$$\Pi = -L_\Pi\frac{\nabla\cdot\mathbf{v}}{T} \tag{2.53}$$

$$\mathbf{J}_q = -L_q\frac{\nabla T}{T^2} \tag{2.54}$$

$$\mathbf{\Pi} = -L_{\mathbf{\Pi}}\frac{(\nabla\mathbf{v})^{ts}}{T}. \tag{2.55}$$

The linear constitutive equations are usually expressed in terms of the practical transport coefficients defined by empirical relations determined from experiments

$$\Pi = -\eta_v \nabla \cdot \mathbf{v} \tag{2.56}$$

$$\mathbf{J}_q = -\lambda \nabla T \tag{2.57}$$

$$\mathbf{\Pi}^{ts} = -2\eta \left(\nabla \mathbf{v}\right)^{ts}, \tag{2.58}$$

which are Stokes' law of bulk viscosity, Fourier's law of thermal conduction and Newton's law of shear viscosity. These equations were already well known before the development of nonequilibrium thermodynamics, but their derivation via the entropy production illustrates that classical nonequilibrium thermodynamics encompasses a wide range of phenomena within a unified and consistent framework. Comparison of the two sets of equations leads to the straightforward identification of the linear phenomenological coefficients with their corresponding practical transport coefficients as $\eta_v = L_\Pi / T$, $\lambda = L_q / T^2$ and $2\eta = L_{\mathbf{\Pi}} / T$.

2.3.6 Differential Equations

The Navier-Stokes equations are obtained by substituting the linear constitutive equations into the momentum balance equation, giving

$$\rho \frac{d\mathbf{v}}{dt} = -\nabla p + \eta \nabla^2 \mathbf{v} + \left(\frac{\eta}{3} + \eta_v\right) \nabla \left(\nabla \cdot \mathbf{v}\right) + \rho \mathbf{F}^e. \tag{2.59}$$

Similarly, the linear constitutive equations for the heat flux vector and the pressure tensor can be substituted into the internal energy equation, giving

$$\rho \frac{du}{dt} = \lambda \nabla^2 T + \frac{p}{\rho}\frac{d\rho}{dt} + \eta_v \left(\nabla \cdot \mathbf{v}\right)^2 + 2\eta \left(\nabla \mathbf{v}\right)^{ts} : \left(\nabla \mathbf{v}\right)^{ts}. \tag{2.60}$$

As it is written, this equation is not in a form that can directly be solved, because it has the internal energy on the left-hand side, and the temperature on the right. To recast this in a form that is more useful, we must express the internal energy in terms of the temperature (or the converse). We have a choice of two different thermodynamic representations for the internal energy. We could express it as a function of temperature and pressure, or as a function of temperature and density. The distinction between these two representations is important for compressible substances. Let us begin by introducing the specific enthalpy though $u = h - p/\rho$. The specific enthalpy is then regarded as a function of the temperature and pressure, $h = h\left(p, T\right)$ and expressed as

$$\begin{aligned} dh &= \left(\frac{\partial h}{\partial p}\right)_T dp + \left(\frac{\partial h}{\partial T}\right)_p dT \\ &= \frac{1}{\rho}\left(\rho L_p + 1\right) dp + c_p dT, \end{aligned} \tag{2.61}$$

where c_p is the specific heat capacity at constant pressure and L_p is termed the latent heat of pressure rise. Together, these results give the desired relationship between the derivatives of the internal energy and temperature fields as

$$\rho \frac{du}{dt} = \rho L_p \frac{dp}{dt} + \rho c_p \frac{dT}{dt} + \frac{p}{\rho} \frac{d\rho}{dt}. \tag{2.62}$$

When this is used in the internal energy balance equation, we find

$$\rho c_p \frac{dT}{dt} = \lambda \nabla^2 T + \eta_v (\nabla \cdot \mathbf{v})^2 + 2\eta (\nabla \mathbf{v})^{ts} : (\nabla \mathbf{v})^{ts} - \rho L_p \frac{dp}{dt}. \tag{2.63}$$

In most situations, mechanical equilibrium is established very quickly and the term involving the time derivative of the pressure will be negligible.

2.4 Multicomponent Fluids and Coupled Transport Processes

We will now consider the linear nonequilibrium thermodynamics of a multicomponent solution, allowing for the diffusion of mass, heat and momentum, but in the absence of chemical reactions.

2.4.1 Balance Equations

The mass, momentum and internal energy balance equations must be supplemented by equations for the concentration of each component when a multicomponent system is studied. It is also found that the internal energy balance equation must be modified to include the thermodynamic work done by an external body force. Taking these points into consideration, we find that the balance equations for a multicomponent fluid are

$$\frac{d\rho}{dt} = -\rho \nabla \cdot \mathbf{v} \tag{2.64a}$$

$$\rho \frac{d\mathbf{v}}{dt} = -\nabla \cdot \mathbf{P} + \sum_k \rho_k \mathbf{F}_k^e \tag{2.64b}$$

$$\rho \frac{du}{dt} = -\nabla \cdot \mathbf{J}_q - \mathbf{P}^T : \nabla \mathbf{v} + \sum_k \mathbf{J}_k \cdot \mathbf{F}_k^e \tag{2.64c}$$

$$\rho \frac{dc_k}{dt} = -\nabla \cdot \mathbf{J}_k, \tag{2.64d}$$

where

$$\rho = \sum_k \rho_k \tag{2.65}$$

is the total density expressed as the sum of the concentrations with dimensions of mass/volume, and

$$\mathbf{J}_k = \rho_k (\mathbf{v}_k - \mathbf{v}) \tag{2.66}$$

is the diffusion flux of component k. This definition of the diffusion flux uses the streaming velocity or barycentric velocity as the reference velocity. Other choices are possible, and a full discussion of the different choices and their relationships to the definitions of various diffusion coefficients is given in de Groot and Mazur [4].

2.4.2 Entropy Production

The expression for the entropy production for a multicomponent fluid is derived by a method very similar to that used to obtain the entropy production for a single component fluid. However, there are some important differences. For example, Equation (2.43) remains unchanged, but the equation for the entropy flux through a co-moving surface changes when there are two or more components.

As usual, the foundation of the local equilibrium treatment is the Gibbs equation, this time generalised to a multicomponent solution,

$$Tds = du - \frac{p}{\rho^2}d\rho - \sum_{k=1}^{n} \mu_k dc_k, \tag{2.67}$$

where μ_k is the chemical potential (or the partial specific Gibbs free energy) of component k and c_k is the mass fraction of component k. The mass fraction is related to the density of component k and the total density $\rho = \sum_{k=1}^{n} \rho_k$ by $c_k = \rho_k/\rho$. The local equilibrium hypothesis allows us to obtain the time derivative of the local specific entropy by asserting that the entropy only depends implicitly on time through the time dependence of the internal energy, density and mass fractions, thus

$$T\frac{ds}{dt} = \frac{du}{dt} - \frac{p}{\rho^2}\frac{d\rho}{dt} - \sum_{k=1}^{n} \mu_k \frac{dc_k}{dt}. \tag{2.68}$$

We now substitute the time derivatives on the right-hand side of Equation (2.68) using the balance equations for the internal energy, density and concentration, to find

$$\rho\frac{ds}{dt} = -\frac{1}{T}\left(\nabla \cdot \mathbf{J}_q + \mathbf{P}^T : \nabla\mathbf{v}\right) + \frac{p}{T}\nabla \cdot \mathbf{v} + \frac{1}{T}\sum_{k=1}^{n} \mu_k \nabla \cdot \mathbf{J}_k. \tag{2.69}$$

The divergence of the heat flux and diffusive flux can be written in terms of the divergence of the flux divided by the temperature, and the definition of the nonequilibrium pressure tensor is introduced to give

$$\begin{aligned} \rho\frac{ds}{dt} = &-\nabla \cdot \left[\frac{1}{T}\left(\mathbf{J}_q - \sum_{k=1}^{n} \mu_k \mathbf{J}_k\right)\right] - \frac{1}{T^2}\mathbf{J}_q \cdot \nabla T - \frac{1}{T}\mathbf{\Pi}^T : \nabla\mathbf{v} \\ &- \frac{1}{T}\sum_{k=1}^{n} \mathbf{J}_k \cdot \left(T\nabla\left(\frac{\mu_k}{T}\right) - \mathbf{F}_k^e\right). \end{aligned} \tag{2.70}$$

When this is compared with the expression for the entropy balance given by Equation (2.43), we see that the entropy flux can be identified as the term inside the divergence of Equation (2.69),

$$\mathbf{J}_s = \frac{1}{T}\left(\mathbf{J}_q - \sum_{k=1}^{n} \mu_k \mathbf{J}_k\right) \tag{2.71}$$

and the entropy production can be identified as the remainder of the right-hand side of Equation (2.69),

$$\sigma = -\frac{1}{T^2}\mathbf{J}_q \cdot \nabla T - \frac{1}{T}\mathbf{\Pi}^T : \nabla \mathbf{v} - \frac{1}{T}\sum_{k=1}^{n} \mathbf{J}_k \cdot \left(T\nabla\left(\frac{\mu_k}{T}\right) - \mathbf{F}_k^e\right) \tag{2.72}$$

in which each term again appears as a flux multiplied by a thermodynamic force. This form of the entropy production is not unique. By choosing a different combination of terms, it is possible to define an alternative but equivalent expression for the entropy production that employs a different heat flux vector. The alternative form may be more convenient in some situations. The temperature gradient appears in the expressions for the thermodynamic forces corresponding to both the heat flux vector and the diffusive flux. If we prefer to group the terms in such a way that the thermodynamic forces are distinct terms exclusively involving the temperature gradient, the velocity gradient and the purely concentration dependent part of the chemical potential gradient, we can write

$$\nabla \mu_k(T, p, c_i) = \left(\frac{\partial \mu_k}{\partial T}\right)_{p,c_i} \nabla T + \left(\frac{\partial \mu_k}{\partial p}\right)_{T,c_i} \nabla p + \left(\frac{\partial \mu_k}{\partial c_k}\right)_{T,p,c_{i\neq k}} \nabla c_k. \tag{2.73}$$

If the pressure gradient is zero, we can write this as

$$\begin{aligned}\nabla \mu_k(T, p, c_i) &= \left(\frac{\partial \mu_k}{\partial T}\right)_{p,c_i} \nabla T + (\nabla \mu_k)_{T,p,c_{i\neq k}} \\ &= -s_k \nabla T + (\nabla \mu_k)_{T,p,c_{i\neq k}}\end{aligned} \tag{2.74}$$

and using the fact that $\mu_k = h_k - Ts_k$, we can rewrite the entropy production as

$$\sigma = -\frac{1}{T^2}\mathbf{J}'_q \cdot \nabla T - \frac{1}{T}\mathbf{\Pi}^T : \nabla \mathbf{v} - \frac{1}{T}\sum_{k=1}^{n} \mathbf{J}_k \cdot \left[(\nabla \mu_k)_{T,p,c_{i\neq k}} - \mathbf{F}_k^e\right], \tag{2.75}$$

where $(\nabla \mu_k)_{T,p,c_{i\neq k}}$ is the chemical potential gradient of component k due to its concentration gradient at constant temperature, pressure and concentration of the other components, and we have introduced a new heat flux vector which we call the primed heat flux vector, defined as

$$\mathbf{J}'_q = \mathbf{J}_q - \sum_{k=1}^{n} h_k \mathbf{J}_k. \tag{2.76}$$

This leads to differential equations that are expressed in terms of the temperature, concentration and velocity fields – all of which are directly measurable (unlike the chemical potential and internal energy). For these reasons, the primed heat flux is often called the measurable heat flux.

2.4.3 Constitutive Equations

The linear constitutive equations that result from our derivation of the entropy production for a multicomponent solution share some similarities with their one component fluid analogues. The terms involving the bulk and shear viscosities, corresponding to viscous flow are identical. Therefore, we only need to consider the vectorial terms, i.e. the expressions for the heat flux and the diffusive fluxes. In an isotropic fluid, the heat flux can depend on both the temperature gradient and the part of the chemical potential gradient due to the concentration gradient, since these thermodynamic forces both have the same vector character as the fluxes. To simplify the analysis, we will at this stage only consider a binary liquid without external body forces and we will obtain the constitutive equations for the unprimed heat flux vector and the diffusive fluxes. The extension to multicomponent systems is straightforward.

Following on from Equation (2.72), the entropy production for a binary solution in the absence of viscous flow and external forces can be written as

$$\sigma = -\frac{1}{T^2}\mathbf{J}_q \cdot \nabla T - \sum_{k=1}^{2} \mathbf{J}_k \cdot \left(\frac{1}{T} (\nabla \mu_k)_{T,p,c_{i\neq k}} - \frac{h_k}{T^2} \nabla T \right). \tag{2.77}$$

The fluxes and thermodynamic forces are not independent, due to certain relationships between them. The diffusive fluxes must obey

$$\sum_{k=1}^{2} \mathbf{J}_k = 0 \tag{2.78}$$

and the chemical potential gradients under constant temperature and pressure conditions must obey the Gibbs-Duhem equation,

$$\sum_{k=1}^{2} c_k (\nabla \mu_k)_{T,p} = 0 \tag{2.79}$$

giving

$$\sum_{k=1}^{2} \mathbf{J}_k \cdot (\nabla \mu_k)_{T,p} = \left(1 + \frac{c_1}{c_2} \right) \mathbf{J}_1 \cdot (\nabla \mu_1)_{T,p} . \tag{2.80}$$

Note that the notation for the chemical potential gradient has been simplified here, but it must be kept in mind that it is taken while keeping the concentration of components other than k constant. Then the entropy production can be expressed in terms of the independent fluxes and thermodynamic forces as

$$\sigma = -\mathbf{J}_q \cdot \mathbf{X}_q - \mathbf{J}_1 \cdot \mathbf{X}_1, \tag{2.81}$$

with

$$\mathbf{X}_q = \left(\frac{\nabla T}{T^2} \right) \tag{2.82a}$$

$$\mathbf{X}_1 = \frac{1}{T} \left[\left(1 + \frac{c_1}{c_2} \right) (\nabla \mu_1)_{T,p} - \frac{1}{T} (h_1 - h_2) \nabla T \right]. \tag{2.82b}$$

The linear phenomenological equations for this form of the entropy production are

$$\mathbf{J}_q = -L_{qq}\mathbf{X}_q - L_{q1}\mathbf{X}_1 \tag{2.83a}$$
$$\mathbf{J}_1 = -L_{11}\mathbf{X}_1 - L_{1q}\mathbf{X}_q \tag{2.83b}$$

and if the primed heat flux is used instead, the entropy production is

$$\sigma = -\mathbf{J}'_q \cdot \mathbf{X}_q - \mathbf{J}_1 \cdot \mathbf{X}'_1 \tag{2.84}$$

with the primed thermodynamic force given by

$$\mathbf{X}_1 = \frac{1}{T}\left[\left(1 + \frac{c_1}{c_2}\right)(\nabla\mu_1)_{T,p} - \frac{1}{T}(h_1 - h_2)\nabla T\right] = \mathbf{X}'_1 - (h_1 - h_2)\mathbf{X}_q. \tag{2.85}$$

The corresponding linear phenomenological equations are

$$\mathbf{J}'_q = -L'_{qq}\mathbf{X}_q - L'_{q1}\mathbf{X}'_1 \tag{2.86a}$$
$$\mathbf{J}_1 = -L'_{11}\mathbf{X}'_1 - L'_{1q}\mathbf{X}_q. \tag{2.86b}$$

The entropy production must be invariant with respect to the redefinition of the heat flux and the thermodynamic force for diffusion, so if we substitute the linear phenomenological equations into the expressions for the entropy production and then express the entropy production for the unprimed variables in terms of the primed thermodynamic forces, we find the following relationships between the primed and unprimed phenomenological coefficients

$$L'_{qq} = L_{qq} - L_{q1}(h_1 - h_2) + L_{11}(h_1 - h_2)^2 - L_{1q}(h_1 - h_2) \tag{2.87a}$$
$$L'_{q1} = L_{q1} - L_{11}(h_1 - h_2) \tag{2.87b}$$
$$L'_{11} = L_{11} \tag{2.87c}$$
$$L'_{1q} = L_{1q} - L_{11}(h_1 - h_2). \tag{2.87d}$$

The Onsager reciprocal relations require that $L_{q1} = L_{1q}$, which also leads to $L'_{q1} = L'_{1q}$. Equation (2.86) is an example of coupled transport processes, because the heat flux depends on both the temperature gradient and the concentration gradient, and the diffusive flux also depends on both thermodynamic forces.

The practical transport coefficients are defined by

$$\mathbf{J}_1 = -\rho D\nabla c_1 - \rho c_1 c_2 D'\nabla T \tag{2.88a}$$
$$\mathbf{J}'_q = -\lambda\nabla T - \rho_1 \frac{\partial\mu_1}{\partial c_1} T D''\nabla c_1, \tag{2.88b}$$

where D is the mutual diffusion coefficient, D' is the thermal diffusion coefficient, D'' is the Dufour coefficient and λ is the thermal conductivity. These practical transport coefficients are related to the phenomenological coefficients defined earlier by

$$D = \frac{L'_{11}}{\rho c_2 T}\left(\frac{\partial \mu_1}{\partial c_1}\right)_{T,p} \tag{2.89a}$$

$$D' = \frac{L'_{1q}}{\rho c_1 c_2 T^2} \tag{2.89b}$$

$$D'' = \frac{L'_{q1}}{\rho c_1 c_2 T^2} \tag{2.89c}$$

$$\lambda = \frac{L'_{qq}}{T^2}. \tag{2.89d}$$

As we will see later, the distinction between the primed and unprimed phenomenological coefficients becomes quite important when the transport coefficients such as the thermal conductivity, thermal diffusion coefficient and Dufour coefficient are computed for fluids of more than one component from the Green–Kubo relations. It turns out that the unprimed coefficients are easiest to evaluate from the Green–Kubo relations (because the unprimed heat flux vector has a readily computed microscopic expression) while the primed ones are most useful experimentally.

2.4.4 Differential Equations

The Navier-Stokes equations for the velocity field are unchanged when we go from a single component to a multicomponent system, but the partial differential equation for the temperature field is modified and the equation for the concentration field is new. To obtain the partial differential equation for the temperature field, we begin with the internal energy balance equation given in Equation (2.64). We again introduce the relationship between the specific internal energy and the specific enthalpy, $u = h - p/\rho$ and assume that the specific enthalpy can be expressed as a function of the temperature, pressure, and concentration, $h = h(p, T, c_k)$. This leads to

$$\begin{aligned} dh &= \left(\frac{\partial h}{\partial p}\right)_{T,c_k} dp + \left(\frac{\partial h}{\partial T}\right)_{T,c_k} dT + \sum_k \left(\frac{\partial h}{\partial c_k}\right)_{p,T,c_{i\neq k}} dc_k \\ &= \frac{1}{\rho}\left(\rho L_p + 1\right) dp + c_p dT + \sum_k h_k dc_k, \end{aligned} \tag{2.90}$$

which then gives the left-hand side of the internal energy balance equation as

$$\begin{aligned} \rho\frac{du}{dt} &= \rho\frac{dh}{dt} - \frac{dp}{dt} + \frac{p}{\rho}\frac{d\rho}{dt} \\ &= \rho L_p\frac{dp}{dt} + \rho c_p\frac{dT}{dt} + \sum_k h_k\frac{dc_k}{dt} + \frac{p}{\rho}\frac{d\rho}{dt}. \end{aligned} \tag{2.91}$$

This can then be combined with the concentration balance equation and the relationship $\nabla \cdot \sum_k h_k \mathbf{J}_k = \sum_k h_k \nabla \cdot \mathbf{J}_k + \sum_k \mathbf{J}_k \cdot \nabla h_k$ to give

$$\rho c_p \frac{dT}{dt} = -\nabla \cdot \mathbf{J}'_{\mathrm{q}} - \sum_k \mathbf{J}_k \cdot \nabla h_k - \rho L_p \frac{dp}{dt} - \mathbf{\Pi}^{\mathrm{T}} : \nabla \mathbf{v} + \sum_k \mathbf{J}_k \cdot \mathbf{F}_k^e. \tag{2.92}$$

Substitution of the constitutive equation (2.88b) for the heat flux into this equation, and assuming that the velocity gradient and external force are zero, pressure is time-independent and partial specific enthalpies are spatially uniform, gives us the standard heat equation.

The differential equation for the concentration is much simpler. It follows directly when the constitutive equation (2.88a) is inserted into the balance equation for the concentration, Equation (2.64d).

2.5 Spin Angular Momentum

In the classical macroscopic treatment of momentum balance, angular momentum is neglected, and the pressure tensor is assumed to be symmetric at all times. To allow for the possibility that molecules can possess intrinsic (spin) angular momentum, the spin angular momentum density is introduced and, along with this, it becomes necessary to allow for an antisymmetric component of the pressure tensor. This is actually easier to understand from the microscopic point of view, as we will see later.

2.5.1 Balance Equations

The balance equation for the translational part of the momentum density is, as before,

$$\rho \frac{d\mathbf{v}}{dt} = -\nabla \cdot \mathbf{P} + \rho \mathbf{F}^e. \tag{2.93}$$

In the absence of external forces, the total translational momentum is conserved. Similarly, the balance equation for the total angular momentum $\mathbf{M}$ is

$$\rho \frac{d\mathbf{M}}{dt} = -\nabla \cdot \mathbf{Q}_M + \rho \mathbf{G}_M^e, \tag{2.94}$$

where $\mathbf{Q}_M$ is the diffusive flux of total angular momentum, $\mathbf{G}_M^e$ is the sum of the orbital and spin external torques and the total angular momentum $\mathbf{M}$ is the sum of the orbital and spin angular momenta,

$$\mathbf{M} = \mathbf{L} + \mathbf{S}. \tag{2.95}$$

In the absence of external torques, the total angular momentum is conserved.

The translational momentum balance equation and the total angular momentum balance equation are independently satisfied, but they are also linked by the fact that the orbital component of the angular momentum satisfies a balance equation. The balance equation for the orbital angular momentum is found by forming the cross product of

Equation (2.93) with the radius vector $\mathbf{r}$ from the origin of the coordinate system:

$$\rho\frac{d\mathbf{L}}{dt} = -\mathbf{r}\times\nabla\cdot\mathbf{P} + \rho\mathbf{r}\times\mathbf{F}^e, \tag{2.96}$$

where $\mathbf{L} = \mathbf{r}\times\mathbf{v}$ is the specific orbital angular momentum. Using the identity

$$\nabla\cdot\left(\mathbf{r}\times\mathbf{P}^T\right)^T = \mathbf{r}\times\nabla\cdot\mathbf{P} - \boldsymbol{\varepsilon}:\mathbf{P} \tag{2.97}$$

and the definition of the pseudovector dual of an antisymmetric tensor, we find

$$\rho\frac{d\mathbf{L}}{dt} = \nabla\cdot(\mathbf{P}\times\mathbf{r}) + 2\mathbf{P}^a + \rho\mathbf{r}\times\mathbf{F}^e. \tag{2.98}$$

Note that $\boldsymbol{\varepsilon}$ is the Levi-Civita isotropic third-rank tensor, which is equal to $+1$ when ijk is a cyclic permutation of 123, -1 for all other permutations of these indices and 0 when two or more of the indices are equal.

This confirms that the orbital angular momentum is not conserved, even when there is no orbital torque. Instead, we see an additional term that accounts for the conversion of orbital to spin angular momentum. Subtracting this equation from the total angular momentum balance equation, we obtain the balance equation for the spin angular momentum,

$$\rho\frac{d\mathbf{S}}{dt} = -\nabla\cdot\mathbf{Q} - 2\mathbf{P}^a + \rho\boldsymbol{\Gamma}^e \tag{2.99}$$

where $\mathbf{Q}$ is now the intrinsic couple tensor describing the diffusive flux of spin angular momentum, defined as the total couple tensor minus its orbital component

$$\mathbf{Q} = \mathbf{Q}_M - \mathbf{P}\times\mathbf{r}. \tag{2.100}$$

$\boldsymbol{\Gamma}^e$ is the specific spin torque (which can be similarly defined), and $\mathbf{P}^a$ is the pseudovector dual of the antisymmetric part of the pressure tensor. The explicit introduction of spin angular momentum also changes the internal energy balance equation. The specific internal energy must now be defined as the energy per unit mass e minus the sum of the streaming translational kinetic energy per unit mass e_t and the streaming rotational kinetic energy per unit mass e_r

$$u = e - e_t - e_r = e - \frac{1}{2}\mathbf{v}^2 - \frac{1}{2}\boldsymbol{\omega}\cdot\boldsymbol{\Theta}\cdot\boldsymbol{\omega} \tag{2.101}$$

where $\boldsymbol{\Theta}$ is the specific average moment of inertia tensor and $\boldsymbol{\omega}$ is the streaming component of the angular velocity. The rate of change of the translational kinetic energy is, as found previously, given by

$$\rho\frac{de_t}{dt} = -\mathbf{v}\cdot(\nabla\cdot\mathbf{P}). \tag{2.102}$$

For an isotropic fluid, the averaged moment of inertia tensor is isotropic, so it can be written as

$$\boldsymbol{\Theta} = \Theta\mathbf{1} \tag{2.103}$$

where **1** is the second-rank isotropic tensor. Then the rate of change of streaming angular kinetic energy for an isotropic fluid is

$$\rho \frac{de_r}{dt} = -\boldsymbol{\omega} \cdot (\nabla \cdot \mathbf{Q}) - 2\boldsymbol{\omega} \cdot \mathbf{P}^a. \tag{2.104}$$

The rate of change of the total energy is

$$\rho \frac{de}{dt} = -\nabla \cdot (\mathbf{P} \cdot \mathbf{v}) - \nabla \cdot (\mathbf{Q} \cdot \boldsymbol{\omega}) - \nabla \cdot \mathbf{J}_q, \tag{2.105}$$

where the terms on the right-hand side are the total work done by stresses and couples respectively, and the heat absorbed. The rate of change of the internal energy is then the rate of change of the total energy minus the rate of change of the sum of the streaming translational and rotational energies,

$$\rho \frac{du}{dt} = -\nabla \cdot \mathbf{J}_q - \mathbf{P}^T : \nabla \mathbf{v} - \mathbf{Q}^T : \nabla \boldsymbol{\omega} + 2\boldsymbol{\omega} \cdot \mathbf{P}^a. \tag{2.106}$$

The work terms on the right-hand side of this equation represent the thermodynamic work done by stresses and couples, i.e. the parts of the total work that can change the internal energy. Separating the thermodynamic work due to stresses into the products of its three irreducible tensor components, we have

$$\mathbf{P}^T : \nabla \mathbf{v} = (p + \Pi) \nabla \cdot \mathbf{v} + \mathbf{P}^{ts} : (\nabla \mathbf{v})^{ts} - \mathbf{P}^a : (\nabla \mathbf{v})^a, \tag{2.107}$$

where the sign of the last term follows from the fact that taking the transpose of an antisymmetric tensor changes its sign. p is the local equilibrium pressure and Π is the nonequilibrium part of the isotropic pressure, so that

$$p + \Pi = \frac{1}{3} \mathrm{Tr}(\mathbf{P}). \tag{2.108}$$

Similarly, the thermodynamic work due to couples can be decomposed into three irreducible tensor components

$$\mathbf{Q}^T : \nabla \boldsymbol{\omega} = Q \nabla \cdot \boldsymbol{\omega} + \mathbf{Q}^{ts} : (\nabla \boldsymbol{\omega})^{ts} - \mathbf{Q}^a : (\nabla \boldsymbol{\omega})^a. \tag{2.109}$$

The product of two antisymmetric tensors can be written in terms of the scalar product of the two corresponding pseudovectors, e.g.

$$\mathbf{P}^a : (\nabla \mathbf{v})^a = -\mathbf{P}^a \cdot \nabla \times \mathbf{v}, \tag{2.110}$$

so the internal energy balance equation becomes

$$\begin{aligned} \rho \frac{du}{dt} = &- \nabla \cdot \mathbf{J}_q - (p + \Pi) \nabla \cdot \mathbf{v} - \mathbf{P}^{ts} : (\nabla \mathbf{v})^{ts} - \mathbf{P}^a \cdot (\nabla \times \mathbf{v} - 2\boldsymbol{\omega}) \\ &- Q \nabla \cdot \boldsymbol{\omega} - \mathbf{Q}^{ts} \cdot (\nabla \boldsymbol{\omega})^{ts} - \mathbf{Q}^a \cdot (\nabla \times \boldsymbol{\omega}). \end{aligned} \tag{2.111}$$

2.5.2 Entropy Production

The local equilibrium approximation allows us to substitute the Gibbs equation for the rate of change of the internal energy on the left-hand side

$$\rho \frac{du}{dt} = \rho T \frac{ds}{dt} + \frac{p}{\rho} \frac{d\rho}{dt}. \tag{2.112}$$

Using the mass balance equation, we eliminate the term involving the local equilibrium pressure from both sides. Then we can write

$$\rho \frac{ds}{dt} = -\nabla \cdot \mathbf{J}_s + \sigma \tag{2.113}$$

with the entropy flux given by

$$\mathbf{J}_s = \mathbf{J}_q / T \tag{2.114}$$

and the entropy production

$$\begin{aligned} \sigma = -\frac{1}{T} \Bigg[& \frac{1}{T} \mathbf{J}_q \cdot \nabla T + \Pi \nabla \cdot \mathbf{v} + \mathbf{P}^{ts} : (\nabla \mathbf{v})^{ts} + \mathbf{P}^a \cdot (\nabla \times \mathbf{v} - 2\boldsymbol{\omega}) \\ & + Q \nabla \cdot \boldsymbol{\omega} + \mathbf{Q}^{ts} : (\nabla \boldsymbol{\omega})^{ts} + \mathbf{Q}^a \cdot (\nabla \times \boldsymbol{\omega}) \Bigg]. \end{aligned} \tag{2.115}$$

2.5.3 Constitutive Equations

The linear constitutive relations resulting from this form of the entropy production are:

$$\mathbf{J}_q = -\lambda \nabla T - \zeta_{qr} \nabla \times \boldsymbol{\omega} \tag{2.116a}$$

$$\Pi = -\eta_v \nabla \cdot \mathbf{v} \tag{2.116b}$$

$$\mathbf{P}^{ts} = -2\eta (\nabla \mathbf{v})^{ts} \tag{2.116c}$$

$$\mathbf{P}^a = -\eta_r (\nabla \times \mathbf{v} - 2\boldsymbol{\omega}) \tag{2.116d}$$

$$Q = -\zeta_v \nabla \cdot \boldsymbol{\omega} \tag{2.116e}$$

$$\mathbf{Q}^{ts} = -2\zeta (\nabla \boldsymbol{\omega})^{ts} \tag{2.116f}$$

$$\mathbf{Q}^a = -\zeta_{rr} (\nabla \times \boldsymbol{\omega}) - \zeta_{rq} \nabla T, \tag{2.116g}$$

where η_r is the vortex viscosity and the ζ coefficients are the spin viscosities and coupling coefficients. The only cross-couplings that can occur are those between the heat flux and the antisymmetric part of the spin angular velocity gradient, and its Onsager reciprocal effect, the antisymmetric part of the couple tensor and the temperature gradient. This is because all other pairs of a flux and force of the same tensor rank have different parities. For example, the scalar (isotropic) component of the nonequilibrium pressure tensor is polar and the thermodynamic force corresponding to the isotropic part of the spin angular velocity gradient is axial [35, 36].

2.5.4 Differential Equations

When these constitutive relations are substituted into the momentum and spin angular momentum balance equations, we arrive at the extended Navier-Stokes equations

$$\begin{aligned}\rho\frac{d\mathbf{v}}{dt} &= -\nabla p + \eta_v \nabla(\nabla\cdot\mathbf{v}) + \eta\left(\frac{1}{3}\nabla(\nabla\cdot\mathbf{v}) + \nabla^2\mathbf{v}\right)\\ &\quad - \eta_r\left(\nabla(\nabla\cdot\mathbf{v}) - \nabla^2\mathbf{v} - 2\nabla\times\boldsymbol{\omega}\right) + \rho\mathbf{F}^e\\ &= -\nabla p + \left(\eta_v + \frac{1}{3}\eta - \eta_r\right)\nabla(\nabla\cdot\mathbf{v})\\ &\quad + (\eta+\eta_r)\nabla^2\mathbf{v} + 2\eta_r\nabla\times\boldsymbol{\omega} + \rho\mathbf{F}^e,\end{aligned} \tag{2.117}$$

where we have used

$$\begin{aligned}\nabla\cdot\mathsf{P}^a &= \nabla\cdot\boldsymbol{\varepsilon}\cdot\mathbf{P}^a\\ &= -\eta_r\nabla\cdot\boldsymbol{\varepsilon}\cdot(\nabla\times\mathbf{v} - 2\boldsymbol{\omega})\\ &= \eta_r(\nabla\times\nabla\times\mathbf{v} - 2\nabla\times\boldsymbol{\omega})\\ &= \eta_r\left(\nabla(\nabla\cdot\mathbf{v}) - \nabla^2\mathbf{v} - 2\nabla\times\boldsymbol{\omega}\right)\end{aligned} \tag{2.118}$$

and

$$\begin{aligned}\nabla\cdot\mathsf{P}^{ts} &= -2\eta\nabla\cdot(\nabla\mathbf{v})^{ts}\\ &= -\eta\left(\nabla^2\mathbf{v} + \frac{1}{3}\nabla(\nabla\cdot\mathbf{v})\right).\end{aligned} \tag{2.119}$$

The corresponding equation for the spin angular velocity is found by beginning with the balance equation for the spin angular velocity

$$\rho\frac{d\mathbf{S}}{dt} = -\nabla\cdot\mathsf{Q} - 2\mathbf{P}^a + \rho\boldsymbol{\Gamma}^e. \tag{2.120}$$

Now we specialise our treatment to homogeneous isotropic fluids, so that the specific inertia tensor is a constant isotropic second-rank tensor and the specific spin angular momentum is

$$\mathbf{S} = \boldsymbol{\Theta}\cdot\boldsymbol{\omega} = \Theta\boldsymbol{\omega}. \tag{2.121}$$

Then the differential equation for the streaming angular momentum is

$$\begin{aligned}\rho\Theta\frac{d\boldsymbol{\omega}}{dt} &= \zeta_v\nabla(\nabla\cdot\boldsymbol{\omega}) + \zeta\left(\frac{1}{3}\nabla(\nabla\cdot\boldsymbol{\omega}) + \nabla^2\boldsymbol{\omega}\right)\\ &\quad - \zeta_{rr}\left(\nabla(\nabla\cdot\boldsymbol{\omega}) - \nabla^2\boldsymbol{\omega}\right) + 2\eta_r(\nabla\times\mathbf{v} - 2\boldsymbol{\omega}) + \rho\boldsymbol{\Gamma}^e\\ &= \left(\zeta_v + \frac{1}{3}\zeta - \zeta_{rr}\right)\nabla(\nabla\cdot\boldsymbol{\omega}) + (\zeta+\zeta_{rr})\nabla^2\boldsymbol{\omega}\\ &\quad + 2\eta_r(\nabla\times\mathbf{v} - 2\boldsymbol{\omega}) + \rho\boldsymbol{\Gamma}^e.\end{aligned} \tag{2.122}$$

The equations of motion for the translational and angular streaming velocities are coupled by the appearance of the angular velocity in the equation for the translational velocity and vice-versa. The coupling parameter is the rotational viscosity. If this parameter is zero, the equations are independent and the two velocity fields evolve separately.

In principle, the coupling will always be present as long as the rotational viscosity is nonzero, but we now demonstrate how the coupling occurs in practice under specific conditions. Consider a planar Poiseuille flow between two parallel walls, as represented in Figure 9.6 (Poiseuille flow will be discussed in greater detail in Chapter 9, but for now we use it as an illustrative example). We will assume that the flow is driven by a gravity-like field in the x-direction, and that the walls confine the fluid to a channel of width L in the y-direction. The system is assumed to be periodic and infinite in the x- and z-directions. In this geometry, the translational and angular velocity fields are divergenceless for stable low Reynolds number flows, and the extended Navier-Stokes equations reduce to

$$\rho \frac{\partial v_x}{\partial t} = (\eta + \eta_r) \frac{\partial^2 v_x}{\partial y^2} + 2\eta_r \frac{\partial \omega_z}{\partial y} + \rho F^e, \tag{2.123a}$$

$$\rho \Theta \frac{\partial \omega_z}{\partial t} = (\zeta + \zeta_{rr}) \frac{\partial^2 \omega_z}{\partial y^2} - 2\eta_r \left(\frac{\partial v_x}{\partial y} + 2\omega_z \right) + \rho \Gamma_z^e. \tag{2.123b}$$

We see that the differential equation for the translational streaming velocity depends on the angular velocity and the differential equation for the angular velocity depends on the translational streaming velocity, so these equations are interdependent and must be solved simultaneously. We will return to these coupled differential equations again in Chapter 10.

3 Statistical Mechanical Foundations

A detailed microscopic description of nonequilibrium classical mechanical systems can be achieved through the methods of nonequilibrium statistical mechanics. Beginning with the development of the Green–Kubo relations in the 1950s, modern nonequilibrium statistical mechanics has now reached a high level of sophistication, culminating in time-dependent nonlinear response theory and the derivation of fluctuation theorems. Several different approaches to nonequilibrium statistical mechanics that have slightly different fundamental perspectives have been developed. Some examples of these differing approaches can be seen in the books by McLennan [37], Eu [38], Zwanzig [39], Zubarev [40] and Gaspard [41]. The methodology we will use closely follows Evans and Morriss [2]. It evolved in close connection with the development of many of the nonequilibrium molecular dynamics simulation algorithms discussed in this book, and it is particularly suitable for our discussion. Readers seeking a broader introduction to nonequilibrium statistical mechanics are advised to consult the references provided.

3.1 Fundamentals of Classical Mechanics

3.1.1 Equations of Motion

Before proceeding with our discussion of nonequilibrium statistical mechanics, we will introduce the main concepts of classical mechanics. This will allow us to establish our notation and some basic ideas before moving on to the more complex material that follows. The simplest and most familiar formulation of classical mechanics for a single particle is based on Newton's second law of mechanics,

$$m\frac{d^2\mathbf{r}}{dt^2} = \mathbf{F}_T\left(\mathbf{r}, \dot{\mathbf{r}}, t\right), \tag{3.1}$$

where m is the particle's mass, $\mathbf{r}$ is its position and $\dot{\mathbf{r}}$ is its velocity. The total force $\mathbf{F}_T$ may be a function of the particle's position, velocity and may also be an explicit function of time. For a system of interacting particles, the equation of motion for particle i may be a function of the positions and velocities of all other particles in the system, as well as an explicit function of time. The Lagrangian and Hamiltonian formalisms are richer and more elegant formulations of classical mechanics. In particular, the concepts of constrained motion and generalised coordinates (such as the Euler angles for rotational motion of a rigid body) are easily accommodated by these more sophisticated formulations. Detailed descriptions of the different formulations of classical mechanics can be found in many excellent books (e.g. Goldstein [42] and Tolman, [43]). Here we will only summarise the main points. For problems involving constrained motion (for example, motion constrained to a line or surface, or where there are mathematical relationships between the positions or velocities of sets of particles, as in rigidly bonded atoms within a molecular model) the concept of generalised coordinates is very useful. In the case where a relationship between the positions of certain particles makes it possible to eliminate a coordinate, the constraint is called a holonomic constraint. In such a case, the number of coordinates, and hence the number of equations of motion, is reduced and the problem is sometimes significantly simplified. The remaining independent coordinates may be quite different from the usual physical positions and velocities in Cartesian coordinates. For this reason, the notion of generalised coordinates, denoted by q, was introduced. The Lagrangian for a system of particles is defined as $L = T(q, \dot{q}) - V(q, t)$ where T is the kinetic energy and V is the total potential energy due to internal and external interactions and q represents the set of generalised coordinates of the particles in the system. The kinetic energy may in general be a function of the generalised coordinates and their derivatives. By minimising the integral of the Lagrangian with respect to the independent variables q, $\dot{q}$ and t between two times along a trajectory of the system, the Lagrangian equations of motion are obtained,

$$\frac{d}{dt}\left(\frac{\partial L}{\partial \dot{q}}\right) - \frac{\partial L}{\partial q} = 0. \tag{3.2}$$

These equations can easily be used to describe the dynamics of systems such as a set of interacting particles constrained to move on a surface, or a system of particles subject to rigid bond constraints between various particles. By performing a Legendre transformation, we can perform a change in the independent variables in the equations of motion from the generalised coordinates and their times derivatives (velocities) to the generalised coordinates and the generalised momenta,

$$p = \frac{\partial L(q, \dot{q}, t)}{\partial \dot{q}}. \tag{3.3}$$

First, we observe that

$$dL(q, \dot{q}, t) = \frac{\partial L}{\partial q}dq + \frac{\partial L}{\partial \dot{q}}d\dot{q} + \frac{\partial L}{\partial t}dt. \tag{3.4}$$

Then define $H = \frac{\partial L}{\partial \dot{q}}\dot{q} - L$ and take the differential of this to obtain

$$\begin{aligned} dH &= pd\dot{q} + \dot{q}dp - \frac{\partial L}{\partial q}dq - pd\dot{q} - \frac{\partial L}{\partial t}dt \\ &= \dot{q}dp - \frac{\partial L}{\partial q}dq - \frac{\partial L}{\partial t}dt. \end{aligned} \tag{3.5}$$

Now use Lagrange's equation of motion to give

$$dH = \dot{q}dp - \dot{p}dq - \frac{\partial L}{\partial t}dt. \tag{3.6}$$

From the definition of the differential of H, we see that the equations of motion in terms of the Hamiltonian must be

$$\dot{q} = \frac{\partial H}{\partial p}, \quad \dot{p} = -\frac{\partial H}{\partial q}. \tag{3.7}$$

Hamilton's equations of motion, expressed as two first-order differential equations, are equivalent to the second-order Lagrangian and Newtonian equations of motion. For a conservative system, the Hamiltonian and the Lagrangian are not explicitly dependent on time. When constraints are present, the conjugate momentum is not necessarily equal to the usual momentum of Newtonian mechanics. Although the Hamiltonian is often identified with the total energy of the system, this need not always be the case. The conditions under which this is so are discussed by Goldstein [42]. For a system with any forces that cannot be expressed as the gradient of a potential, the right-hand side of Lagrange's equation is nonzero and the idea of generalised forces must be introduced.

3.1.2 Phase Space

The instantaneous microscopic state of a classical N-particle system without holonomic constraints is given by the values of all $6N$ generalised coordinates and momenta of the particles comprising the system at that instant in time. This corresponds to a point in $6N$ dimensional phase space represented by the phase vector $\mathbf{\Gamma} = (q_1, q_2, \ldots, q_{3N}, p_1, p_2, \ldots, p_{3N})$. The time evolution of the system is represented by the line in phase space traced out by the phase point $\mathbf{\Gamma}(t) = \mathbf{\Gamma}(\mathbf{\Gamma}_0, t)$ which can be computed from the initial conditions and the elapsed time, if the equations of motion are given. Classical statistical mechanical systems of more than two particles are typically chaotic, and the dynamics exhibit Lyapunov instability. One consequence of this Lyapunov instability is that the numerical solution of the equations of motion and the finite precision floating point representation of numbers in computers leads to an exponential deviation of the computed trajectory from the "true" trajectory. However, this is usually of no practical consequence when we only wish to compute the statistical properties of the system, such as the mean values of phase variables and their fluctuations. Chaotic dynamics and Lyapunov instability and their relationship to nonequilibrium statistical mechanics are discussed in detail by Evans and Morriss [2] and Gaspard [41].

The relationship between the statistical properties of a system of particles and the equations of motion governing the system's dynamics is one of the central themes of statistical mechanics. The most successful method of determining the statistical properties of a statistical mechanical system from its equations of motion is Gibbs's ensemble method, in which the average value of a phase variable is evaluated from the probability of finding the system at a particular phase point (regardless of the way it got there). This probability can be evaluated by considering the behaviour of a representative ensemble of systems. A representative ensemble of systems is a set of phase points distributed over all of the accessible phase space. To evaluate an equilibrium ensemble average, we could use the representative ensemble of systems and their probabilities at any time, since the results of the equilibrium ensemble average are time-independent. The situation for nonequilibrium systems is more complicated. The time at which the ensemble average of a property is calculated is then vitally important.

3.2 The Liouville Equation

Consider a continuum of points in phase space with a number density given by

$$f(\mathbf{\Gamma}, t) = \sum_{i=1}^{\infty} \delta(\mathbf{\Gamma} - \mathbf{\Gamma}_i(t)). \tag{3.8}$$

The density of ensemble members $f(\mathbf{\Gamma}, t)$ is proportional to the phase space probability density (also known as the N-particle distribution function). The proportionality constant is unimportant for our current purposes, and has been omitted. Each point in phase space represents an ensemble member. The rate of change of the probability density at a fixed point in phase space is given by

$$\begin{aligned}
\frac{\partial f(\mathbf{\Gamma}, t)}{\partial t} &= \sum_{i=1}^{\infty} \frac{\partial}{\partial t} \delta(\mathbf{\Gamma} - \mathbf{\Gamma}_i(t)) \\
&= \sum_{i=1}^{\infty} \frac{\partial \mathbf{\Gamma}_i}{\partial t} \cdot \frac{\partial}{\partial \mathbf{\Gamma}_i} \delta(\mathbf{\Gamma} - \mathbf{\Gamma}_i(t)) \\
&= -\frac{\partial}{\partial \mathbf{\Gamma}} \cdot \sum_{i=1}^{\infty} \dot{\mathbf{\Gamma}}_i \delta(\mathbf{\Gamma} - \mathbf{\Gamma}_i(t)) \\
&= -\frac{\partial}{\partial \mathbf{\Gamma}} \cdot (\dot{\mathbf{\Gamma}} f),
\end{aligned} \tag{3.9}$$

where a generalisation of the identity $\partial \delta(x - y)/\partial x = -\partial \delta(x - y)/\partial y$ has been used to obtain the third line from the second. This equation and its variants are commonly known as the Liouville equation. It is virtually identical in form to the mass density balance equation encountered previously, but it describes the evolution of a probability density in abstract phase space rather than the mass density in physical space. The analogy with the mass density balance equation is useful, because we can write the Liouville equation in terms of the streaming or hydrodynamic derivative as

$$\begin{aligned}
\frac{df(\mathbf{\Gamma}, t)}{dt} &= \dot{\mathbf{\Gamma}} \cdot \frac{\partial f}{\partial \mathbf{\Gamma}} + \frac{\partial f}{\partial t} \\
&= \dot{\mathbf{\Gamma}} \cdot \frac{\partial f}{\partial \mathbf{\Gamma}} - \dot{\mathbf{\Gamma}} \cdot \frac{\partial f}{\partial \mathbf{\Gamma}} - f \frac{\partial}{\partial \mathbf{\Gamma}} \cdot \dot{\mathbf{\Gamma}} \\
&= -f \frac{\partial}{\partial \mathbf{\Gamma}} \cdot \dot{\mathbf{\Gamma}} \\
&= -f \Lambda(\mathbf{\Gamma}).
\end{aligned} \tag{3.10}$$

$\Lambda(\mathbf{\Gamma})$ is the phase space probability density compression factor, usually called the phase space compression factor. Loosely speaking, it represents the divergence of the phase space velocity field, and it physically represents the rate of compression of the flow of trajectories in phase space. For systems with equations of motion that can be expressed in canonical Hamiltonian form, it is identically zero, but for thermostatted systems it is not. In most introductory discussions of statistical mechanics, canonical Hamiltonian equations of motion are assumed to apply so the flow of phase space trajectories is then incompressible and the Liouville equation takes the form

$$\frac{df(\mathbf{\Gamma},t)}{dt}=0. \tag{3.11}$$

3.3 Time Evolution

3.3.1 Time Evolution of the Probability Density

The Liouville equation provides us with an equation for the time dependence of the phase space probability density. This equation can be written in terms of a differential operator that operates on the phase space probability density as

$$\begin{aligned}\frac{\partial f(\mathbf{\Gamma},t)}{\partial t}&=-\frac{\partial}{\partial \mathbf{\Gamma}}\cdot\left(\dot{\mathbf{\Gamma}}f\right)\\&=-f\frac{\partial}{\partial \mathbf{\Gamma}}\cdot\left(\dot{\mathbf{\Gamma}}\right)-\dot{\mathbf{\Gamma}}\cdot\frac{\partial f}{\partial \mathbf{\Gamma}}\\&=-\left(\Lambda f+\dot{\mathbf{\Gamma}}\cdot\frac{\partial f}{\partial \mathbf{\Gamma}}\right)\\&=-\left(\Lambda+\dot{\mathbf{\Gamma}}\cdot\frac{\partial}{\partial \mathbf{\Gamma}}\right)f\\&=-iL_f f,\end{aligned} \tag{3.12}$$

which defines iL_f, the Liouville operator or f-Liouvillean. We now assume that the equations of motion are not explicitly time-dependent, so that the Liouville operator is also not explicitly time-dependent, which allows us to write the formal solution of the Liouville equation as

$$f(\mathbf{\Gamma},t)=\exp\left(-iL_f t\right)f(\mathbf{\Gamma},0). \tag{3.13}$$

The exponential operator, known as the propagator, is a symbolic representation of the infinite series expansion

$$\exp\left(-iL_f t\right)=\sum_{n=0}^{\infty}\frac{(-t)^n}{n!}\left(iL_f\right)^n. \tag{3.14}$$

This type of operator notation is very similar to that used in quantum nonequilibrium statistical mechanics, where the equivalent of the Liouville operator is then a super-operator, because the observables are already operators themselves (e.g. Zwanzig [39]).

3.3.2 Time Evolution of Phase Variables

A phase variable is a property that depends on the phase space vector of the system, i.e. the instantaneous microscopic state of the system, so it can be written as $A(\mathbf{\Gamma})$. Let us at this point specialise our discussion to a system of N atoms with no holonomic constraints, so that $\mathbf{\Gamma} = (\mathbf{r}_1, \mathbf{r}_2, \ldots, \mathbf{r}_N, \mathbf{p}_1, \mathbf{p}_2 \cdots \mathbf{p}_N)$. Typical phase variables include the internal energy, pressure, temperature, heat flux vector, etc. The time dependence of a phase variable is implicit because it is entirely due to the time dependence of the phase vector, so a normal phase variable cannot be *explicitly* time-dependent. Using the chain rule, we can write

$$\begin{aligned} \frac{dA}{dt} &= \frac{d\mathbf{\Gamma}}{dt} \cdot \frac{\partial A}{\partial \mathbf{\Gamma}} \\ &= \sum_{i=1}^{N} \left(\dot{\mathbf{r}}_i \cdot \frac{\partial}{\partial \mathbf{r}_i} + \dot{\mathbf{p}}_i \cdot \frac{\partial}{\partial \mathbf{p}_i} \right) A(\mathbf{\Gamma}) \\ &= iL_p(\mathbf{\Gamma}) A(\mathbf{\Gamma}), \end{aligned} \tag{3.15}$$

which defines the p-Liouvillean L_p. The formal solution of this equation can again be written in terms of an exponential operator as

$$A(t) = \exp\left(iL_p t\right) A(0). \tag{3.16}$$

The phase variable and phase space density Liouvilleans are adjoint operators. To prove this, we write the ensemble average of the time derivative of a phase variable as

$$\begin{aligned} \langle \dot{A}(t) \rangle &= \int d\mathbf{\Gamma} f(0)\, iL_p A(\mathbf{\Gamma}) \\ &= \int d\mathbf{\Gamma} f(0)\, \dot{\mathbf{\Gamma}} \cdot \frac{\partial}{\partial \mathbf{\Gamma}} A(\mathbf{\Gamma}). \end{aligned} \tag{3.17}$$

Integrating by parts gives

$$\begin{aligned} \langle \dot{A}(t) \rangle &= \left[f(\mathbf{\Gamma}) \dot{\mathbf{\Gamma}} A(\mathbf{\Gamma}) \right]_S - \int d\mathbf{\Gamma} A(\mathbf{\Gamma}) \frac{\partial}{\partial \mathbf{\Gamma}} \cdot \left(f(\mathbf{\Gamma}) \dot{\mathbf{\Gamma}} \right) \\ &= -\int d\mathbf{\Gamma} A(\mathbf{\Gamma})\, iL_f f(\mathbf{\Gamma}), \end{aligned} \tag{3.18}$$

where the boundary term is zero because f approaches zero when the momenta go to plus or minus infinity. Since these two expressions must be equal, we have

$$\int d\mathbf{\Gamma} f(\mathbf{\Gamma})\, iL_p A(\mathbf{\Gamma}) = -\int d\mathbf{\Gamma} A(\mathbf{\Gamma})\, iL_f f(\mathbf{\Gamma}), \tag{3.19}$$

which is the relation satisfied by adjoint operators. If the phase space compression factor is zero, the f and p Liouvilleans are self-adjoint, or Hermitian.

3.3.3 Schrödinger and Heisenberg Representations of Ensemble Averages

If we want to compute the ensemble average of a phase variable A, we could sample the initial state from a distribution function $f(\mathbf{\Gamma}, 0)$ and then propagate the value of A

from the initial state to the state at time t. This is called the Heisenberg picture and it is written as

$$\langle A(t)\rangle = \int d\mathbf{\Gamma} f(\mathbf{\Gamma}, 0) A(t). \tag{3.20}$$

However, this is not the only way to formulate the ensemble average of A at t. We could alternatively evaluate $A(\mathbf{\Gamma})$ at time 0, and then average over the distribution function after it is propagated from 0 to t. This called the Schrödinger picture and it is written as

$$\langle A(t)\rangle = \int d\mathbf{\Gamma} f(\mathbf{\Gamma}, t) A(\mathbf{\Gamma}). \tag{3.21}$$

To prove the equality of these two pictures, we expand the exponential operator that generates $A(t)$ from $A(0)$ and use the adjoint relationship between the f and p Liouville operators to shift the Liouville operators from the phase variable to the distribution function. First, we write the Heisenberg picture of the ensemble average as

$$\langle A(t)\rangle = \int d\mathbf{\Gamma} f(\mathbf{\Gamma}, 0) \exp\left(iL_p t\right) A(0). \tag{3.22}$$

Then, expand the exponential operator, giving

$$\langle A(t)\rangle = \int d\mathbf{\Gamma} f(\mathbf{\Gamma}, 0) \sum_{n=0}^{\infty} \frac{(t)^n}{n!} \left(iL_p\right)^n A(0). \tag{3.23}$$

The $n = 0$ term of the exponential operator is equal to 1. Using the fact that L_f and L_p are adjoint operators, the $n = 1$ term is

$$\int d\mathbf{\Gamma} f(\mathbf{\Gamma}, 0) \frac{(t)}{1!} iL_p A(0) = \int d\mathbf{\Gamma} \frac{(-t)}{1!} iL_f f(\mathbf{\Gamma}, 0) A(0). \tag{3.24}$$

Repeatedly using the adjoint operator relationship for the higher-order terms, we find

$$\int d\mathbf{\Gamma} f(\mathbf{\Gamma}, 0) \frac{(t)^n}{n!} \left(iL_p\right)^n A(0) = \int d\mathbf{\Gamma} \frac{(-t)^n}{n!} \left(iL_f\right)^n f(\mathbf{\Gamma}, 0) A(0). \tag{3.25}$$

The summation of all terms in the expansion generates the f-propagator, so the ensemble average can then be written as

$$\begin{aligned} \langle A(t)\rangle &= \int d\mathbf{\Gamma} \sum_{n=0}^{\infty} \frac{(-t)^n}{n!} \left(iL_f\right)^n f(\mathbf{\Gamma}, 0) A(0) \\ &= \int d\mathbf{\Gamma} \exp\left(-iL_f t\right) f(\mathbf{\Gamma}, 0) A(0) \\ &= \int d\mathbf{\Gamma} f(\mathbf{\Gamma}, t) A(0). \end{aligned} \tag{3.26}$$

The Schrödinger and Heisenberg formulations of the ensemble average are equivalent, and they are strongly analogous to the Lagrangian and Eulerian pictures of fluid dynamics. These are not the only two choices of times at which we can evaluate A and f. Intermediate times are also possible.

3.4 Response Theory

Linear response theory is one of the foundations on which nonequilibrium statistical mechanics and nonequilibrium molecular dynamics algorithms are built. Excellent derivations and discussions of linear response theory exist in the literature so it would be somewhat redundant to repeat them here. Instead, we will take the route suggested by Evans and Morriss [2], and consider nonlinear response theory first. The linear response is then obtained as a limiting case. This approach has the added advantage that a recent derivation of nonlinear, time-dependent response theory that is very direct and illuminating [44] can be used at the outset.

3.4.1 Nonlinear Response to the SLLOD Equations of Motion

In many situations, we are interested in the steady state nonlinear response of a dissipative system. In this case, it is essential to include a thermostat in the equations of motion because it would be impossible to achieve a steady state without removing the dissipated heat. To make our discussion of the nonlinear response more concrete, we will consider a specific example: the nonlinear response of a system that is simulated using the thermostatted SLLOD equations of motion. The SLLOD equations of motion, which we will discuss in more detail in Chapter 5, are used to apply a velocity gradient $\nabla\mathbf{v}$, in order to produce shear, elongational and other types of flow. These equations of motion are often used in conjunction with a Gaussian isokinetic thermostat, which keeps the thermal kinetic energy of the system constant. Thermostats are discussed in more detail in Chapter 6. Combining the SLLOD algorithm with the Gaussian isokinetic thermostat, we obtain the following equations of motion,

$$\dot{\mathbf{r}}_i = \frac{\mathbf{p}_i}{m_i} + \mathbf{r}_i \cdot \nabla\mathbf{v} \tag{3.27a}$$

$$\dot{\mathbf{p}}_i = \mathbf{F}_i - \mathbf{p}_i \cdot \nabla\mathbf{v} - \alpha\mathbf{p}_i, \tag{3.27b}$$

where

$$\alpha = \frac{\sum_{i=1}^{N} \mathbf{p}_i \cdot (\mathbf{F}_i - \mathbf{p}_i \cdot \nabla\mathbf{v}) \big/ m_i}{\sum_{i=1}^{N} \mathbf{p}_i^2 \big/ m_i} \tag{3.28}$$

is the thermostat multiplier and the kinetic temperature is defined as

$$T_K = \frac{1}{(3N - N_C)\,k_B} \sum_{i=1}^{N} \frac{\mathbf{p}_i^2}{m_i}, \tag{3.29}$$

where N_C is the number of kinetic constraints. Let us now consider the f and p Liouville operators for these equations of motion with an isokinetic thermostat. The f-Liouville operator is defined as

$$iL_f = \Lambda + \dot{\mathbf{\Gamma}} \cdot \frac{\partial}{\partial \mathbf{\Gamma}} = \Lambda + iL_p, \tag{3.30}$$

where the phase space density compression factor and the Liouvillean were defined earlier. Explicitly using the isokinetic SLLOD equations of motion, we find that the phase space probability density compression factor Λ is given by

$$\frac{\partial}{\partial \mathbf{\Gamma}} \cdot \dot{\mathbf{\Gamma}} = \sum_i \left(\frac{\partial}{\partial \mathbf{r}_i} \cdot \dot{\mathbf{r}}_i + \frac{\partial}{\partial \mathbf{p}_i} \cdot \dot{\mathbf{p}}_i \right) = 3N\left(\nabla \cdot \mathbf{v} - \nabla \cdot \mathbf{v} - \alpha\right) = -3N\alpha, \quad (3.31)$$

where terms of order 1 have been neglected. This assumes that the applied velocity gradient is spatially homogeneous (i.e. uniform). The phase space compression factor can be related to the total internal energy balance by considering the rate of change of the internal energy. The total internal energy is

$$U = K + \Phi, \quad (3.32)$$

where K is the thermal kinetic energy and Φ is the internal potential energy due to (pairwise) interatomic interactions

$$K = \sum_{i=1}^{N} \frac{\mathbf{p}_i \cdot \mathbf{p}_i}{2m_i} \quad (3.33a)$$

$$\Phi = \frac{1}{2} \sum_{i=1}^{N} \sum_{j \neq i}^{N} \phi_{ij}\left(r_{ij}\right). \quad (3.33b)$$

The rate of change of the internal energy for a simple atomic system is

$$\frac{dU}{dt} = \frac{dK}{dt} + \frac{d\Phi}{dt}. \quad (3.34)$$

Taking the time derivative of each term separately and using the equations of motion to eliminate the position and peculiar momentum derivatives, we find

$$\frac{dK}{dt} = \sum_{i=1}^{N} \frac{\mathbf{p}_i}{m_i} \cdot \mathbf{F}_i - \sum_{i=1}^{N} \frac{\mathbf{p}_i \mathbf{p}_i}{m_i} : \nabla \mathbf{v} - \alpha \sum_{i=1}^{N} \frac{\mathbf{p}_i \cdot \mathbf{p}_i}{m_i} \quad (3.35)$$

and

$$\frac{d\Phi}{dt} = -\sum_{i=1}^{N} \frac{\mathbf{p}_i}{m_i} \cdot \mathbf{F}_i + \frac{1}{2} \sum_{i=1}^{N} \sum_{i=1}^{N} \left(\mathbf{r}_{ij} \mathbf{F}_{ij}\right)^T : \nabla \mathbf{v}. \quad (3.36)$$

The first term of each of these equations represents the conversion of kinetic into potential energy and vice versa. The second term in each equation represents a portion of the work done by the stress, and the third term of the rate of change of the kinetic energy represents the work done by the thermostat force. The rate of change of the total internal energy is

$$\begin{aligned} \frac{dU}{dt} &= -\sum_{i=1}^{N} \frac{\mathbf{p}_i \mathbf{p}_i}{m_i} : \nabla \mathbf{v} + \frac{1}{2} \sum_{i=1}^{N} \sum_{i=1}^{N} \left(\mathbf{r}_{ij} \mathbf{F}_{ij}\right)^T : \nabla \mathbf{v} - \alpha \sum_{i=1}^{N} \frac{\mathbf{p}_i \cdot \mathbf{p}_i}{m_i} \\ &= -V\mathbf{P}^T : \nabla \mathbf{v} - \alpha \sum_{i=1}^{N} \frac{\mathbf{p}_i \cdot \mathbf{p}_i}{m_i}, \end{aligned} \quad (3.37)$$

where the microscopic expression for the pressure tensor has been used (see Equation (4.47)). If we identify the thermal kinetic energy with the definition of the kinetic temperature (which is assumed to be kept constant by the thermostat), we have

$$K = \sum_{i=1}^{N} \frac{\mathbf{p}_i \cdot \mathbf{p}_i}{2m_i} = \frac{3N}{2} k_B T, \tag{3.38}$$

giving the final expression as

$$\begin{aligned} \frac{dU}{dt} &= -V\mathbf{P}^T : \nabla \mathbf{v} - 3N\alpha k_B T \\ &= -V\mathbf{P}^T : \nabla \mathbf{v} + k_B T \Lambda, \end{aligned} \tag{3.39}$$

using Equation (3.31). We now have sufficient information to continue with the derivation of the nonlinear response. The Liouville equation in Lagrangian form is

$$\frac{df(\mathbf{\Gamma}(t), t)}{dt} = -f(\mathbf{\Gamma}(t), t)\Lambda(\mathbf{\Gamma}). \tag{3.40}$$

Note that the argument of the phase space probability density here is the phase vector at time t, because the time derivative is the streaming derivative which includes changes in the probability density due to explicit time dependence as well as changes due to the motion of the phase vector through phase space. The solution of this equation is easily found to be

$$\ln\left(\frac{f(\mathbf{\Gamma}(t), t)}{f(\mathbf{\Gamma}(0), 0)}\right) = -\int_0^t \Lambda(\mathbf{\Gamma}(s), s)\, ds \tag{3.41}$$

or

$$f(\mathbf{\Gamma}(t), t) = f(\mathbf{\Gamma}(0), 0)\exp\left(-\int_0^t \Lambda(\mathbf{\Gamma}(s), s)\, ds\right), \tag{3.42}$$

where the initial probability density is assumed to be the equilibrium isokinetic distribution function

$$f(\mathbf{\Gamma}(0), 0) = \frac{1}{Z}\exp(-\beta U(\mathbf{\Gamma}))\,\delta[K(\mathbf{\Gamma}) - K_0]. \tag{3.43}$$

By integrating the internal energy balance equation, we can find an expression for the integral of the phase density compression function

$$\beta U(t) = \beta U(0) - \beta \int_0^t V\mathbf{P}^T(s) : \nabla \mathbf{v}(s)\, ds + \int_0^t \Lambda(s)\, ds, \tag{3.44}$$

where $\beta = \frac{1}{k_B T}$. This gives the phase space probability density as

$$f(\mathbf{\Gamma}(t), t) = f(\mathbf{\Gamma}(0), 0)\exp\left(\beta U(0) - \beta U(t) - \beta \int_0^t V\mathbf{P}^T(s) : \nabla \mathbf{v}(s)\, ds\right). \tag{3.45}$$

This can be simplified because we can combine the initial probability density with the internal energy terms to give

$$\begin{aligned} f(\mathbf{\Gamma}(0), 0) \exp(\beta U(\mathbf{\Gamma}(0)) - \beta U(\mathbf{\Gamma}(t))) &= \frac{1}{Z} \exp(-\beta U(\mathbf{\Gamma}(0))) \\ &\quad \times \exp(\beta U(\mathbf{\Gamma}(0)) - \beta U(\mathbf{\Gamma}(t))) \\ &= \frac{1}{Z} \exp(-\beta U(\mathbf{\Gamma}(t))) \\ &= f_0(\mathbf{\Gamma}(t)), \end{aligned} \tag{3.46}$$

where Z is the partition function. This is the equilibrium probability density at the phase point occupied by the nonequilibrium trajectory at time t. Then the phase space probability density at time t is

$$f(\mathbf{\Gamma}(t), t) = f_0(\mathbf{\Gamma}(t)) \exp\left(-\beta V \int_0^t \mathbf{P}^T(s) : \nabla \mathbf{v}(s)\, ds\right).$$

To obtain the phase space probability density at a fixed point in phase space we perform a phase space coordinate transformation and a change of integration variable, giving the final result [44]:

$$f(\mathbf{\Gamma}(0), t) = f_0(\mathbf{\Gamma}(0)) \exp\left(-\beta V \int_0^t \mathbf{P}^T(-s) : \nabla \mathbf{v}(t-s)\, ds\right). \tag{3.47}$$

3.4.2 Nonlinear Response for more General Equations of Motion

We will now generalise the results of the previous section so that they apply to systems with equations of motion having the more general form

$$\dot{\mathbf{r}}_i = \frac{\mathbf{p}_i}{m_i} + \mathbf{C}_i F^e \tag{3.48a}$$

$$\dot{\mathbf{p}}_i = \mathbf{F}_i + \mathbf{D}_i F^e - \alpha \mathbf{p}_i, \tag{3.48b}$$

where F^e is the magnitude of the external field and $\mathbf{C}$ and $\mathbf{D}$ account for the coupling of the external field to the particle dynamics. This general form includes all of the commonly used homogeneous NEMD algorithms. To arrive at the generalised equation for the nonlinear response, we only require a few important results. The formal solution of the Liouville equation is the same as was found previously, i.e. Equation (3.42), but we need a generalised expression for the relationship between the phase space probability density compression and the internal energy. To obtain the usual linear response relationship we will require

$$\frac{dU}{dt} = -JF^e + k_B T \Lambda, \tag{3.49}$$

where J is usually called the dissipative flux. Therefore we should first find the internal energy derivative that results from the generalised equations of motion.

The time derivative of the kinetic energy of the system with a thermostat applied is given by

$$\begin{aligned} \frac{d}{dt} \sum_{i=1}^{N} \frac{\mathbf{p}_i^2}{2m_i} &= \sum_{i=1}^{N} \frac{\mathbf{p}_i \cdot \dot{\mathbf{p}}_i}{m_i} \\ &= \sum_{i=1}^{N} \frac{\mathbf{p}_i}{m_i} \cdot \left(\mathbf{F}_i^{\phi} + \mathbf{D}_i F^e - \alpha \mathbf{p}_i\right). \end{aligned} \tag{3.50}$$

The thermostat term can be expressed in terms of the kinetic temperature as

$$-\alpha\sum_{i=1}^{N}\frac{\mathbf{p}_i\cdot\mathbf{p}_i}{m_i} = -2\alpha K_0 = -3N\alpha k_B T, \tag{3.51}$$

where K_0 is the initial value of the thermal kinetic energy, which is constrained to be constant by the isokinetic thermostat. Note that $3N$ should be replaced by $3N-4$ if we want to account for the degrees of freedom lost due to momentum conservation and the constraint on the thermal kinetic energy, but for large systems, this correction is negligible. The time derivative of the intermolecular potential energy is

$$\begin{aligned}\frac{1}{2}\frac{d}{dt}\sum_{i=1}^{N}\sum_{j\neq i}^{N}\phi_{ij}\left(r_{ij}\right) &= \frac{1}{2}\sum_{i=1}^{N}\sum_{j\neq i}^{N}\left(\frac{\partial\phi_{ij}}{\partial\mathbf{r}_i}\cdot\dot{\mathbf{r}}_i+\frac{\partial\phi_{ij}}{\partial\mathbf{r}_j}\cdot\dot{\mathbf{r}}_j\right)\\ &= -\frac{1}{2}\sum_{i=1}^{N}\sum_{j\neq i}^{N}\left(\mathbf{F}_{ij}^{\phi}\cdot\dot{\mathbf{r}}_i+\mathbf{F}_{ji}^{\phi}\cdot\dot{\mathbf{r}}_j\right)\\ &= -\sum_{i=1}^{N}\mathbf{F}_i^{\phi}\cdot\frac{\mathbf{p}_i}{m_i}-\sum_{i=1}^{N}\mathbf{F}_i^{\phi}\cdot\mathbf{C}_iF^e.\end{aligned} \tag{3.52}$$

This gives the total rate of change of the internal energy as

$$\frac{dU}{dt} = \left(\sum_i\frac{\mathbf{p}_i}{m_i}\cdot\mathbf{D}_i-\sum_i\mathbf{F}_i^{\phi}\cdot\mathbf{C}_i\right)F^e-3N\alpha k_BT. \tag{3.53}$$

We also need to calculate the rate of compression of the phase space probability density, which may, in general, have two contributions: one due to the external field, and one due to the thermostat

$$\begin{aligned}\Lambda &= \sum_{i=1}^{N}\frac{\partial}{\partial\boldsymbol{\Gamma}_i}\cdot\dot{\boldsymbol{\Gamma}}_i\\ &= \left(\sum_{i=1}^{N}\frac{\partial}{\partial\mathbf{r}_i}\cdot\mathbf{C}_i+\sum_{i=1}^{N}\frac{\partial}{\partial\mathbf{p}_i}\cdot\mathbf{D}_i\right)F^e-3N\alpha\\ &= \Lambda_e+\Lambda_t.\end{aligned} \tag{3.54}$$

Using this with the expression for the internal energy derivative, we find that the phase space probability density compression factor can then be written as

$$\begin{aligned}k_BT\Lambda &= \frac{dU}{dt}-\left(\sum_i\frac{\mathbf{p}_i}{m_i}\cdot\mathbf{D}_i-\sum_i\mathbf{F}_i^{\phi}\cdot\mathbf{C}_i\right)F^e+k_BT\Lambda_e\\ &= \frac{dU}{dt}+JF^e,\end{aligned} \tag{3.55}$$

where

$$JF^e = -\left(\sum_i\frac{\mathbf{p}_i}{m_i}\cdot\mathbf{D}_i-\sum_i\mathbf{F}_i^{\phi}\cdot\mathbf{C}_i\right)F^e+k_BT\left(\sum_i\frac{\partial}{\partial\mathbf{r}_i}\cdot\mathbf{C}_i+\sum_i\frac{\partial}{\partial\mathbf{p}_i}\cdot\mathbf{D}_i\right) \tag{3.56}$$

and J is the dissipative flux. This general expression simplifies considerably in most practical cases. For almost all of the homogeneous NEMD algorithms in common use, the phase space compression factor due to the external field is zero. This condition is called the adiabatic incompressibility of phase space or AIΓ, since the compressibility of the phase space probability density in the absence of a thermostat is then zero.

Following the same steps as in the previous section, we obtain the Eulerian form of the solution to the Liouville equation, which gives the phase space probability density at a fixed point in phase space as a function of time for the generalised equations of motion

$$f(\mathbf{\Gamma}(0), t) = f_0(\mathbf{\Gamma}(0)) \exp\left(-\beta \int_0^t J(-s) F^e(t-s)\, ds\right). \tag{3.57}$$

3.4.3 The Transient Time Correlation Function Approach

There are several established theoretical approaches towards nonlinear response formulations in statistical mechanics, and these are treated in considerable detail in the book by Evans and Morriss [2]. These treatments include the original contribution by Kubo [14] and the significant advances made by Kawasaki and colleagues [45–47]. The main limitation in all these treatments was that they were difficult to apply, either to experimentally verifiable results, or to computer simulation, in which the role of thermostats or numerical limitations need to be explicitly considered. The Kawasaki formulation is indeed a powerful theoretical result, but it suffers from the fact that because the distribution function involves the exponential of an integral that is extensive, averaging of phase variables becomes a computationally daunting task. Renormalisation of the distribution function can help solve some of these problems [2], but from a practical computing perspective, the Kawasaki approach is still formidable. Because of this, we will not consider this or other approaches further. The book by Evans and Morriss [2] treats this in considerable detail and interested readers should consult it.

An alternative approach, derived independently by several researchers (Visscher [48], Dufty and Lindenfeld [49], Cohen [50], Morriss and Evans [51, 52]), has come to be known as the transient-time correlation function (TTCF) procedure. It was formalised and applied to thermostatted steady-state systems by Morriss and Evans [23, 51, 52]. Its usefulness in computer simulation is that it provides simple expressions based on time-correlation functions of appropriate phase variables and dissipative fluxes that do not involve the actual distribution function explicitly. Hence there is no exponential term to average over, which makes the formal expressions tractable for NEMD simulation.

The value of nonlinear response theory is twofold: (1) it provides a theoretical basis for the equations of motion used in NEMD simulation (see for example the response-theory proof of the SLLOD equations of motion for shear flow [21]), and (2) it provides a practical means for computing steady-state averages of phase variables at relatively weak field strength. A common criticism of NEMD is that in order to obtain good signal to noise ratios one needs to use field strengths that are physically unreasonable. While the common approach to tackle this is to extrapolate results to the weak field limit, this in itself can be problematic if the weak field limit exhibits a behavior that follows a different trend to what one might have with moderate to high field strengths. As we will see in Chapters 6 and 8 this is likely to be the case in determining whether or not simple fluids behave analytically in the weak-field limit, or the approach to Newtonian behaviour in systems of macromolecules, such as polymer melts. In these cases one may need to simulate at extremely weak field strengths – fields that are actually closer to experimentally realizable values – in which case direct averaging of NEMD data becomes futile due to the excessively poor signal to noise ratio. The application of nonlinear response theory to such situations provides a practical solution to this computational limitation. In particular, application of TTCF algorithms provides an efficient means of probing the weak-field regime whilst still ensuring that all nonlinear phenomena can be captured (unlike linear response theory). The TTCF expressions are in fact very similar to the corresponding linear response expressions; the major difference is that averaging occurs in the nonequilibrium ensemble in the case of TTCF, rather than the equilibrium ensemble as in linear response theory.

In what follows we will derive the general TTCF relations for both time independent and time-dependent flows. In Chapter 6 we will apply the general relations to specific cases of shear and elongational flows and present some results.

3.4.3.1 Time-independent Flow

In Section 3.4.2 it was shown that the distribution function for the Gaussian isothermal (isokinetic) ensemble subject to an external time-dependent field is given by Equation (3.57). It was previously demonstrated that this expression could be further generalised for tensorial time-independent external fields $\mathbf{F}^e$, where the dissipation rate is now given as $-\mathbf{J} : \mathbf{F}^e$ [53]. In this case, Equation (3.57) can be expressed as

$$f(\mathbf{\Gamma}(0), t) = f_0(\mathbf{\Gamma}(0)) \exp\left[-\beta \sum_{i,j} F_{ij}^e \int_0^t ds J_{ji}(-s)\right], \tag{3.58}$$

where the sum over i, j is the sum over Cartesian dimensions x, y, z.

Having determined the form of the governing distribution function we can now take the ensemble average of any phase variable $B(t)$. Thus,

$$\begin{aligned} \langle B(t)\rangle &= \int d\mathbf{\Gamma} B(0) f(\mathbf{\Gamma}(0), t) \\ &= \int d\mathbf{\Gamma} B(0) f_0(\mathbf{\Gamma}(0)) \exp\left[-\beta \sum_{i,j} F_{ij}^e \int_0^t ds J_{ji}(-s)\right]. \end{aligned} \tag{3.59}$$

Taking the time derivative of this expression gives us

$$\begin{aligned}\frac{\partial \langle B(t)\rangle}{\partial t} &= \frac{\partial}{\partial t}\int d\mathbf{\Gamma} B(0) f_0(\mathbf{\Gamma}(0)) \exp\left[-\beta \sum_{i,j} F_{ij}^e \int_0^t ds J_{ji}(-s)\right] \\ &= \int d\mathbf{\Gamma} B(0) f_0(\mathbf{\Gamma}(0)) \left[-\beta \sum_{i,j} F_{ij}^e J_{ji}(-t)\right] \\ &\equiv -\beta \left\langle B(0) \sum_{i,j} F_{ij}^e J_{ji}(-t)\right\rangle \\ &= -\beta \sum_{i,j} F_{ij}^e \left\langle B(0) J_{ji}(-t)\right\rangle. \end{aligned} \tag{3.60}$$

We can make use of the identity $\langle X(t) Y\rangle = \langle XY(-t)\rangle$ to give

$$\frac{\partial \langle B(t)\rangle}{\partial t} = -\beta \sum_{i,j} F_{ij}^e \left\langle B(t) J_{ji}(0)\right\rangle, \tag{3.61}$$

which we integrate to obtain the general form for the TTCF expression for any arbitrary phase variable $B(t)$ evolving under the general equations of motion given by Equation (3.48) and a Gaussian thermostat:

$$\langle B(t)\rangle = \langle B(0)\rangle - \beta \sum_{i,j} F_{ij}^e \int_0^t ds \left\langle B(s) J_{ji}(0)\right\rangle. \tag{3.62}$$

We will apply the above expression to the specific homogeneous flows of planar shear and several elongational flows in Chapter 6. The important point to realise in the derivation above is that the explicit dependence of the phase variable on the distribution function was removed by taking the time derivative of the ensemble average of $B(t)$. Doing this allowed us to eliminate the integral in time (for a time-independent external field). Integrating Equation (3.61) over time in principle still leaves us with an integration over $\mathbf{\Gamma}$. However, as with all MD simulations, as long as the system is ergodic then the ensemble average can be equated to a time average. Thus Equation (3.62) can be interpreted as a time correlation function of the phase variable of interest, $B(t)$ with the dissipative flux at $t = 0$. It is termed a "transient-time correlation function" because the correlation function is computed from the time origin of the nonequilibrium system. Thus, it will take some time before $B(t)$ reaches the steady-state, and this time will depend upon the specific details of the type of system being studied (e.g. atomic or molecular fluid, type of potential interactions involved, etc). We will discuss the practical application of this expression further in Chapter 6.

3.4.3.2 Time-dependent Flow

A general time-dependent response theory is somewhat involved and difficult to implement practically (see reference [2] for a detailed discussion of time-dependent response theory). As our main purpose is to develop, explain and implement theories that are of *practical* use for simulators, we will limit our discussion of nonlinear response theory to one useful application: response of a system to a time-periodic field. In particular, Petravic and Evans [54–58] developed a practical TTCF theory for such systems which can be straightforwardly implemented in NEMD simulation algorithms. They also developed a time-dependent version of the so-called "Kawasaki" form of nonlinear response theory and demonstrated their equivalence [59]. In what follows we sketch a derivation of the time-dependent TTCF algorithm. Readers who wish to see a more mathematical treatment should refer to the original papers by Petravic and Evans [54–59]. As we will see, the mathematical treatment is largely similar to the derivation detailed in the time-independent case, described in the previous section, due to the periodicity of the field.

At the outset we point out that the TTCF method we describe here is limited to time-periodic flows only. It is also limited to systems in which only one frequency of the external field is allowable. Practical applications in situations involving very strong fields, and hence higher harmonics in phase variables, cannot be treated by the following approach. Having said this, the current methodology is still the most practical application of nonlinear response theory for time-dependent flows, and has been successfully applied to oscillatory colour conduction [55], oscillatory shear [57] and elongational [60] flows, and also to demonstrate the time periodicity inherent in Lees-Edwards periodic boundary conditions [56] (see Chapter 5, Section 5.1.5, for a discussion of periodic boundary conditions).

Consider a system evolving under equilibrium dynamics, whose phase space is denoted by $\mathbf{\Gamma}_e$. At various times along this trajectory, new "daughter" trajectories are spawned. The daughters, however, are under the influence of a time-periodic external field. Furthermore, each daughter is separated by its adjacent "sister" by some phase angle $\Delta\varphi$, as shown in Figure 3.1.

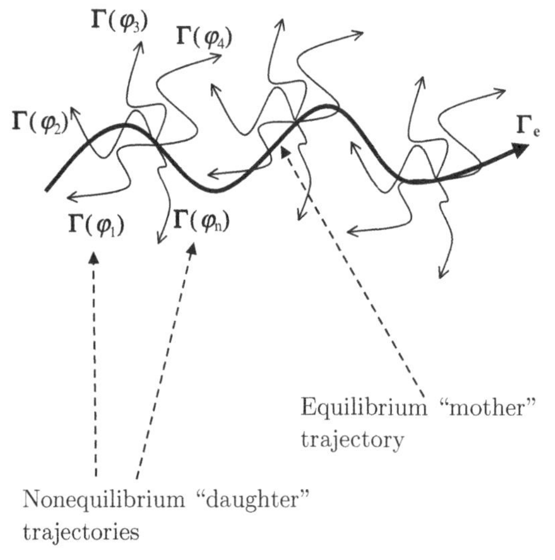

Figure 3.1 Schematic depiction of equilibrium mother and nonequilibrium daughter trajectories in extended phase space. Daughter trajectories are separated from each other by phase angle $\Delta\varphi$.

Now consider the evolution of each of the daughter trajectories. As each daughter is different to each other only to the extent that they originate at a different phase angle, the overall form of each trajectory is, apart from statistical variation (i.e. different atomic coordinates and momenta for individual particles), identical in terms of the extended phase-space of the system. This is important because it gives us a very important realisation, one that allows us to consider time-independent TTCF to be a specific case of the more general time-dependent situation. It is also what ensures that the mathematical derivation and form of the final TTCF expression are similar to that involved in the derivation of Equation (3.62).

Consider now that we are interested in some phase variable $B(t)$ on each of the daughters. For the purposes of illustration, assume we are interested in the value of $B(t)$ at its maximum amplitude, where $\varphi(t) = \varphi_p$ as shown in Figure 3.2. For each different value of initial phase angle φ_i, the evolution of B as a function of time is shown by the curves labelled (i)–(iv). If we plot the value of $\langle B[\mathbf{\Gamma}(t; \varphi(t) = \varphi_p)]\rangle$ against time we end up with curve (v). This last curve depicts the phase variable evaluated at the specific phase angle φ_p averaged over *all* daughter trajectories. When viewed from this "extended phase-space" perspective, in which the phase angle as well as the positions and momenta are part of the phase-space trajectory vector, $\langle B[\mathbf{\Gamma}(t; \varphi(t) = \varphi_p)]\rangle$ evolves *as if it were under the influence of a time-independent field*. This is only possible because the field is periodic in time with a specific frequency ω. It is straightforward now to write down the TTCF expression for the time evolution of $\langle B[\mathbf{\Gamma}(t; \varphi(t) = \varphi_p)]\rangle$, since

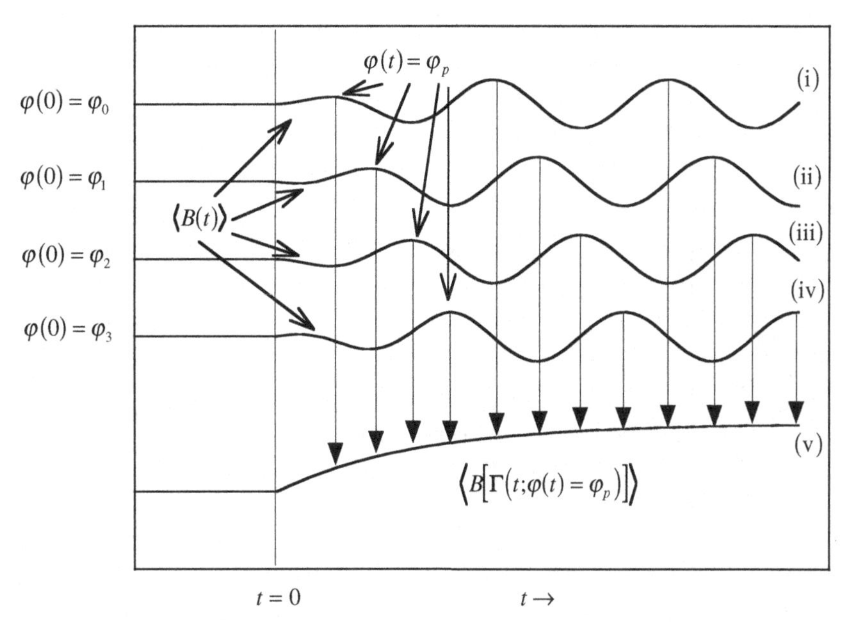

Figure 3.2 Schematic diagram showing the time evolution of phase variable B in extended phase space.

the derivation was already accomplished in the previous section:

$$\begin{aligned}\langle B\left[\mathbf{\Gamma}\left(t;\ \varphi(t)=\varphi_p\right)\right]\rangle &= \langle B\left[\mathbf{\Gamma}\left(0;\ \varphi(0)=\varphi_p\right)\right]\rangle \\ &\quad - \beta \sum_{i,j} \int_0^t ds F_{ij}^e\left(\varphi_p - \omega s\right) \\ &\quad \times \langle B\left[\mathbf{\Gamma}\left(s;\ \varphi(s)=\varphi_p\right)\right] J_{ji}\left[\mathbf{\Gamma}\left(0;\ \varphi(0)=\varphi_p-\omega s\right)\right]\rangle .\end{aligned} \tag{3.63}$$

The derivation is "straightforward" in the sense that the derivation of Equation (3.63) analogously follows the derivation of Equation (3.62), because the evolution of B evaluated at some fixed phase angle is essentially time-independent due to the periodicity of the field. The derivation can be made mathematically rigorous by explicitly including the time-evolution of the phase angle in the equations of motion [54, 55], expressing the distribution function in terms of frequency, and then following steps similar to those detailed above for time-independent fields to obtain Equation (3.63). Note also that now the field term is inside the time integral, unlike the field term in the time-independent version, Equation (3.62).

With an appreciation of the notion of extended phase space, we now consider the interpretation of the symbols under the integral sign, and indeed the interpretation of the integration itself. We closely follow the illustrative example presented

$\varphi \longrightarrow$ $\downarrow t$	0	$\omega\delta t$	$2\omega\delta t$	$3\omega\delta t$	$4\omega\delta t$	$5\omega\delta t$
0	$F^e(0)$ $B(0,0)$ $J(0)$	$F^e(\omega\delta t)$ $B(0,\omega\delta t)$ $J(\omega\delta t)$	$F^e(2\omega\delta t)$ $B(0,2\omega\delta t)$ $J(2\omega\delta t)$	$F^e(3\omega\delta t)$ $B(0,3\omega\delta t)$ $J(3\omega\delta t)$	$F^e(4\omega\delta t)$ $B(0,4\omega\delta t)$ $J(4\omega\delta t)$	$F^e(5\omega\delta t)$ $B(0,5\omega\delta t)$ $J(5\omega\delta t)$
δt	$F^e(5\omega\delta t)$ $B(\delta t,0)$ $J(5\omega\delta t)$	$F^e(0)$ $B(\delta t,\omega\delta t)$ $J(0)$	$F^e(\omega\delta t)$ $B(\delta t,2\omega\delta t)$ $J(\omega\delta t)$	$F^e(2\omega\delta t)$ $B(\delta t,3\omega\delta t)$ $J(2\omega\delta t)$	$F^e(3\omega\delta t)$ $B(\delta t,4\omega\delta t)$ $J(3\omega\delta t)$	$F^e(4\omega\delta t)$ $B(\delta t,5\omega\delta t)$ $J(4\omega\delta t)$
$2\delta t$	$F^e(4\omega\delta t)$ $B(2\delta t,0)$ $J(4\omega\delta t)$	$F^e(5\omega\delta t)$ $B(2\delta t,\omega\delta t)$ $J(5\omega\delta t)$	$F^e(0)$ $B(2\delta t,2\omega\delta t)$ $J(0)$	$F^e(\omega\delta t)$ $B(2\delta t,3\omega\delta t)$ $J(\omega\delta t)$	$F^e(2\omega\delta t)$ $B(2\delta t,4\omega\delta t)$ $J(2\omega\delta t)$	$F^e(3\omega\delta t)$ $B(2\delta t,5\omega\delta t)$ $J(3\omega\delta t)$
$3\delta t$	$F^e(3\omega\delta t)$ $B(3\delta t,0)$ $J(3\omega\delta t)$	$F^e(4\omega\delta t)$ $B(3\delta t,\omega\delta t)$ $J(4\omega\delta t)$	$F^e(5\omega\delta t)$ $B(3\delta t,2\omega\delta t)$ $J(5\omega\delta t)$	$F^e(0)$ $B(3\delta t,3\omega\delta t)$ $J(0)$	$F^e(\omega\delta t)$ $B(3\delta t,4\omega\delta t)$ $J(\omega\delta t)$	$F^e(2\omega\delta t)$ $B(3\delta t,5\omega\delta t)$ $J(2\omega\delta t)$
$4\delta t$	$F^e(2\omega\delta t)$ $B(4\delta t,0)$ $J(2\omega\delta t)$	$F^e(3\omega\delta t)$ $B(4\delta t,\omega\delta t)$ $J(3\omega\delta t)$	$F^e(4\omega\delta t)$ $B(4\delta t,2\omega\delta t)$ $J(4\omega\delta t)$	$F^e(5\omega\delta t)$ $B(4\delta t,3\omega\delta t)$ $J(5\omega\delta t)$	$F^e(0)$ $B(4\delta t,4\omega\delta t)$ $J(0)$	$F^e(\omega\delta t)$ $B(4\delta t,5\omega\delta t)$ $J(\omega\delta t)$
$5\delta t$	$F^e(\omega\delta t)$ $B(5\delta t,0)$ $J(\omega\delta t)$	$F^e(2\omega\delta t)$ $B(5\delta t,\omega\delta t)$ $J(2\omega\delta t)$	$F^e(3\omega\delta t)$ $B(5\delta t,2\omega\delta t)$ $J(3\omega\delta t)$	$F^e(4\omega\delta t)$ $B(5\delta t,3\omega\delta t)$ $J(4\omega\delta t)$	$F^e(5\omega\delta t)$ $B(5\delta t,4\omega\delta t)$ $J(5\omega\delta t)$	$F^e(0)$ $B(5\delta t,5\omega\delta t)$ $J(0)$
$6\delta t$	$F^e(0)$ $B(6\delta t,0)$ $J(0)$	$F^e(\omega\delta t)$ $B(6\delta t,\omega\delta t)$ $J(\omega\delta t)$	$F^e(2\omega\delta t)$ $B(6\delta t,2\omega\delta t)$ $J(2\omega\delta t)$	$F^e(3\omega\delta t)$ $B(6\delta t,3\omega\delta t)$ $J(3\omega\delta t)$	$F^e(4\omega\delta t)$ $B(6\delta t,4\omega\delta t)$ $J(4\omega\delta t)$	$F^e(5\omega\delta t)$ $B(6\delta t,5\omega\delta t)$ $J(5\omega\delta t)$

Figure 3.3 Schematic representation of the TTCF integral in Equation (3.63).

in reference [60] which is more easily explained diagrammatically with the aid of Figure 3.3. Consider a system which is driven by a time-periodic field $F^e(t, \varphi_0)$, such that $F^e(t, \varphi_0) = F^e(t + \tau, \varphi_0)$, where τ is the period and φ_0 is an initial phase angle. Consider a simulation time of $\tau = 6\delta t$, as shown in Figure 3.3, where δt is the integration time-step. The vertical and horizontal axes of the cells respectively portray time and phase angle evolutions of B. In this example, we consider the evolution of a daughter trajectory starting at $t = 0$ with an initial phase angle $\varphi_0 = 2\omega\delta t$. As will become apparent, the TTCF algorithm requires the evolution of a number of different nonequilibrium trajectories, each starting at $t = 0$ with different initial phase angles, φ_0, such that they span the entire range $\varphi \in [0, \Phi_\tau]$, where $\Phi_\tau = \omega\tau$. In this example, the range is divided into only six discrete angles in Figure 3.3, each separated by $\omega\delta t$. Each daughter trajectory in extended phase space is represented by a diagonal arrow in a matrix of cells, commencing at $(t, \varphi) = (0, \varphi_0)$ and evolving down and to the right. Thus, the arrow commences at the point $(t, \varphi) = (0, 2\omega\delta t)$. Each contiguous diagonal cell represents a forward advance in time of δt in an NEMD simulation, and in phase angle of $\omega\delta t$, over the previous one. Furthermore, each contiguous diagonal cell represents an array containing $F^e(0, \varphi(0) = \varphi_0) \times B[\mathbf{\Gamma}(t;\ \varphi(t))] \times J[\mathbf{\Gamma}(0;\ \varphi(0) = \varphi_0)]$. Importantly, the integration in Equation (3.63) is *not* along the diagonal (i.e. not along the single trajectory indicated by the arrow), but rather down the columns of Figure 3.3, from $s = 0$ up to any desired value of the time, $s = t$. The integration occurs only for those values of B evaluated at a *constant value of the phase angle* $\varphi(t) = \varphi_p$. So to calculate the value of $\langle B[\mathbf{\Gamma}(t = 3\delta t;\ \varphi(t) = 5\omega\delta t)]\rangle$ one integrates over each of the shaded array elements in the last column of Figure 3.3 and then average over the total number of daughter time origin points. Doing this for each value of t produces TTCF values of $\langle B[\mathbf{\Gamma}(t;\ \varphi(t) = \varphi_p)]\rangle$ which correspond to the direct values generated for curve (v) of Figure 3.2.

It is important to appreciate that each value of $\langle B[\mathbf{\Gamma}(t;\ \varphi(t) = \varphi_p)]\rangle$ in any one column of Figure 3.3 originates from a distinct and independent trajectory in the full ensemble of extended phase space. Equation (3.63) represents an integration along contiguous trajectories originating from different initial values of the phase angle $\varphi_0 = \varphi_p - \omega s$. The field and dissipative flux are evaluated at $t = 0$ and $\varphi = \varphi_0$ for each trajectory, hence their representation as $F^e(\varphi_p - \omega s)$ and $J[\mathbf{\Gamma}(0;\ \varphi(0) = \varphi_p - \omega s)]$ in Equation (3.63). Analogous to time-independent TTCF, the first term on the right-hand side of Equation (3.63) is the ensemble average of the equilibrium value of B at the time origin. The cells themselves are periodic, with period τ. The simulation only needs to generate trajectories spanning the extended phase space $(\mathbf{\Gamma}, \varphi)$, where $\varphi \in [0, \Phi_\tau]$. As seen in Figure 3.3, $\Phi_\tau = 6\omega\delta t$, so the diagonal arrow corresponding to a single trajectory repeats itself from the left of the array when $\mathrm{mod}(\varphi(t), 6\omega\delta t) = 0$. The complete evolution of B as a function of time is obtained by computing (integrating) Equation (3.63) for each value of φ_p depicted by the shaded columns in Figure 3.3. In our example we obtain the values of $\langle B[\mathbf{\Gamma}(t;\ \varphi(t) = \varphi_p)]\rangle$ for all times up to $t = 5\delta t$. In principle one can (numerically) integrate to $t \to \infty$.

In this illustrative example we limited ourselves to calculating $\langle B[\mathbf{\Gamma}(t;\ \varphi(t))]\rangle$ with initial phase angle $\varphi_0 = 2\omega\delta t$. However, our simulation contains all the information to compute $\langle B[\mathbf{\Gamma}(t;\ \varphi(t))]\rangle$ for any system commencing at $t = 0$ and with any φ_0. This approach also generates a vast amount of data from which many physical properties of interest can be extracted. We will apply the time-dependent TTCF method described here to the specific example of oscillatory elongational flow in Chapter 6 and from this extract the normal stresses. From these normal stresses one can compute the frequency and strain-rate dependent elongational viscosity. Extrapolation to zero frequency would then give us the time-independent elongational viscosity. Exactly the same procedure could be performed for oscillatory shear flow to extract the steady-state zero-frequency shear viscosity.

3.4.4 Linear Response as a Limiting Case of the Nonlinear Response

The linear response of a system to the external field applied in a nonequilibrium molecular dynamics simulation can be easily obtained from the full nonlinear response. To make the discussion concrete, we will again discuss the response of a system subjected to the SLLOD equations of motion (this time for the special case of shear flow) before discussing the more general case. We begin with the TTCF form of the nonlinear response to a time-independent field, given by Equation (3.62). For the particular case of shear flow generated by the SLLOD equations of motion, the external field that generates shear flow in the x direction with the velocity gradient in the y direction is $\partial v_x(y)/\partial y = \dot{\gamma}$ and the dissipative flux defined by Equation (3.56) is VP_{yx}. Then the TTCF expression for the average of a phase variable B from nonlinear response theory is

$$\begin{aligned}\langle B(t)\rangle &= \langle B(0)\rangle - \beta\dot{\gamma}V\int_0^t \langle B(s)P_{yx}(0)\rangle ds \\ &= \langle B(0)\rangle - \beta\dot{\gamma}V\int_0^t \int B(s)P_{yx}(0)f(\mathbf{\Gamma},0)d\mathbf{\Gamma}ds. \end{aligned} \tag{3.64}$$

The second line makes it clear that the initial states are sampled from the equilibrium distribution, and the phase variable is propagated with the full, field-dependent nonequilibrium propagator, i.e. $B(s) = e^{iL_p s}B(0)$. To obtain the linear response from this, we would like to expand the propagator in powers of the field. First we write the p-Liouville operator as a sum of its field-independent and field-dependent parts

$$iL_p = \sum_{i=1}^{N}\left(\dot{\mathbf{r}}_i\cdot\frac{\partial}{\partial\mathbf{r}_i} + \dot{\mathbf{p}}_i\cdot\frac{\partial}{\partial\mathbf{p}_i}\right) = iL_{pe} + iL_{pne}, \tag{3.65}$$

where explicit expressions for iL_{pe} and iL_{pne} are found by inserting the equations of motion and separating the terms that depend on $\dot{\gamma}$ from those that do not. Using the Dyson decomposition of a propagator in which the Liouville operator can be expressed as a sum of two Liouville operators [2], the propagator operating on the phase variable then becomes

$$e^{iL_p s} = e^{iL_{pe}s} + \int_0^s e^{iL_p(s-s_1)}iL_{pne}e^{iL_{pe}s_1}ds_1. \tag{3.66}$$

By only keeping the first term, we obtain the average value of B in the small field or linear response limit,

$$\langle B(t)\rangle = \langle B(0)\rangle - \beta\dot{\gamma}V\int_0^t \langle B(s)P_{yx}(0)\rangle ds, \tag{3.67}$$

where the ensemble average is done with the equilibrium distribution function and the phase variable is propagated with the equilibrium propagator. Therefore, the right-hand side is an equilibrium correlation function, and the field does not even need to be applied to compute the linear response. If the phase variable B is taken as the shear pressure P_{yx}, which has an average value of zero at equilibrium for an isotropic fluid, this becomes

$$P_{yx} = -\frac{\dot{\gamma}V}{k_BT}\int_0^\infty \langle P_{yx}(s)P_{yx}(0)\rangle ds. \tag{3.68}$$

Using Newton's law of viscosity, which defines the viscosity, we find

$$\eta = -\frac{P_{yx}}{\dot{\gamma}} = \frac{V}{k_BT}\int_0^\infty \langle P_{yx}(s)P_{yx}(0)\rangle ds. \tag{3.69}$$

This is an example of a Green–Kubo relation for a linear transport coefficient.

For more general equations of motion that generate a dissipative flux J (defined as an extensive quantity in Equation (3.56)) by the application of an external field F^e, the linear response for a phase variable B is

$$\langle B(t)\rangle = \langle B(0)\rangle - \frac{F^e}{k_BT}\int_0^t \langle B(s)J(0)\rangle ds, \tag{3.70}$$

which is crucial in establishing the relationships between the fluxes, external fields and linear transport coefficients when nonequilibrium molecular dynamics algorithms are being developed.

3.5 Green–Kubo Methods for Linear Transport

The linear response formulae for transport coefficients can all be obtained as a limiting behavior of the more general nonlinear response formulae. The most useful of these formulae are the Green–Kubo expressions for transport, which include those for diffusion, shear and bulk viscosities, and thermal conductivity. These can all be obtained as simple limits as the driving thermodynamic force tends to zero. As we are more concerned with nonequilibrium transport in this book, we do not proceed here to derive the various Green–Kubo formulae, but rather just list them for convenience at the end of this section. Derivations of these formulae can be found in the books by Evans and Morriss [2], Hansen and McDonald [61] and Heyes [62], for example, as well as the numerous references contained therein. Furthermore, the Green–Kubo expressions can be simply obtained from the generalised hydrodynamics approach. For example, in Chapter 11 we derive the wavevector and frequency dependent viscosity in terms of the autocorrelation function of the shear stress tensor (Equation (11.23)). It is trivial to show that in the limit of zero wavevector and frequency, one obtains the Green–Kubo expression for the equilibrium (zero-shear) shear viscosity, namely

$$\eta_0 = \frac{V}{10k_BT}\int_0^\infty \left\langle \mathbf{P}^{ts}(t) : \mathbf{P}^{ts}(0)\right\rangle dt, \tag{3.71}$$

where $\mathbf{P}^{ts}$ is the traceless symmetric part of the pressure tensor. In the above expression we have used the full tensorial nature of the shear stress autocorrelation function, as this gives far greater statistical accuracy compared to averaging over just one component of the full stress tensor. To see how this comes about, we proceed as follows (see for example, reference [63]).

The shear viscosity is most generally defined as a fourth-rank tensor, using the Green–Kubo relation

$$\boldsymbol{\eta}_0 = \frac{V}{k_BT}\int_0^\infty \left\langle \mathbf{P}^{ts}(t)\,\mathbf{P}^{ts}(0)\right\rangle dt, \tag{3.72}$$

where $\boldsymbol{\eta}_0$ is the fourth-rank shear viscosity tensor. For an isotropic material, we assume that the transport coefficient tensors are isotropic. The viscosity tensor is a fourth-rank polar tensor, and the most general form of an isotropic fourth-rank polar tensor is

$$\eta_{\alpha\beta\gamma\delta} = c_1\delta_{\alpha\beta}\delta_{\gamma\delta} + c_2\delta_{\alpha\gamma}\delta_{\beta\delta} + c_3\delta_{\alpha\delta}\delta_{\beta\gamma}, \tag{3.73}$$

where c_1, c_2 and c_3 are the three scalar invariants of the isotropic viscosity tensor. Due to the symmetry properties of the Green–Kubo integral, these coefficients are not independent. The Green–Kubo relation in Equation (3.72) is an outer product of the two symmetric traceless pressure tensors. Therefore, the fourth-rank viscosity tensor must be symmetric with respect to the first two indices and again with respect to the second two indices. This means $\eta_{\alpha\beta\gamma\delta} = \eta_{\beta\alpha\gamma\delta} = \eta_{\alpha\beta\delta\gamma}$, giving

$$\begin{aligned}\eta_{\alpha\beta\gamma\delta} &= c_1\delta_{\alpha\beta}\delta_{\gamma\delta} + c_2\delta_{\alpha\gamma}\delta_{\beta\delta} + c_3\delta_{\alpha\delta}\delta_{\beta\gamma}\\ &= c_1\delta_{\alpha\beta}\delta_{\gamma\delta} + c_2\delta_{\beta\gamma}\delta_{\alpha\delta} + c_3\delta_{\beta\delta}\delta_{\alpha\gamma}\\ &= c_1\delta_{\alpha\beta}\delta_{\gamma\delta} + c_2\delta_{\alpha\delta}\delta_{\beta\gamma} + c_3\delta_{\alpha\gamma}\delta_{\beta\delta},\end{aligned} \tag{3.74}$$

which results in $c_2 = c_3$. The fourth-rank viscosity tensor must also be traceless with respect to the first two and second two indices. This gives

$$\begin{aligned}\sum_\alpha \eta_{\alpha\alpha\gamma\delta} &= \sum_\alpha \left(c_1\delta_{\alpha\alpha}\delta_{\gamma\delta} + 2c_2\delta_{\alpha\gamma}\delta_{\alpha\delta}\right)\\ &= 3c_1\delta_{\gamma\delta} + 2c_2\delta_{\gamma\delta}\\ &= 0\end{aligned} \tag{3.75}$$

or $c_2 = -\frac{3}{2}c_1$, which gives the viscosity tensor as

$$\eta_{\alpha\beta\gamma\delta} = c_1\delta_{\alpha\beta}\delta_{\gamma\delta} - \frac{3}{2}c_1\left(\delta_{\alpha\gamma}\delta_{\beta\delta} + \delta_{\alpha\delta}\delta_{\beta\gamma}\right). \tag{3.76}$$

This can be written in terms of the usual scalar shear viscosity by recognising that the usual Green–Kubo relation in terms of only the yx component of the pressure tensor corresponds to $\eta = \eta_{2121} = -\frac{3}{2}c_1$, which gives

$$\eta_{\alpha\beta\gamma\delta} = -\frac{2}{3}\eta\delta_{\alpha\beta}\delta_{\gamma\delta} + \eta\left(\delta_{\alpha\gamma}\delta_{\beta\delta} + \delta_{\alpha\delta}\delta_{\beta\gamma}\right). \tag{3.77}$$

As was mentioned earlier, a fourth-rank isotropic polar tensor can generally be expressed as a function of the three scalar invariants. Three general scalar invariants of a fourth-rank tensor are easily found by representing the tensor as an outer product of four vectors, $\boldsymbol{\eta} = \mathbf{abcd}$, and then forming scalar combinations of the constituent vectors. This gives the scalar invariants as

$$\begin{aligned} I_1 &= (\mathbf{a}\cdot\mathbf{b})(\mathbf{c}\cdot\mathbf{d}) \\ I_2 &= (\mathbf{a}\cdot\mathbf{c})(\mathbf{b}\cdot\mathbf{d}) \\ I_3 &= (\mathbf{a}\cdot\mathbf{d})(\mathbf{b}\cdot\mathbf{c}). \end{aligned} \tag{3.78}$$

For the shear viscosity tensor, the first of these is zero because the trace with respect to the first and last two indices is zero. Also, the second and the third are equal due to the additional symmetry. Using the Green–Kubo formula, the third scalar invariant can be written as

$$\begin{aligned} I_3 &= \sum_\delta\sum_\gamma\sum_\beta\sum_\alpha \eta_{\alpha\beta\gamma\delta}\delta_{\alpha\delta}\delta_{\beta\gamma} \\ &= \eta_{1111} + \eta_{1221} + \eta_{1331} + \eta_{2112} + \eta_{2222} + \eta_{2332} + \eta_{3113} + \eta_{3223} + \eta_{3333} \\ &= \frac{V}{k_B T}\int_0^\infty \left\langle \mathbf{P}^{ts}(t) : \mathbf{P}^{ts}(0)\right\rangle dt. \end{aligned} \tag{3.79}$$

Using the general form of the viscosity tensor in terms of the scalar viscosity coefficient gives

$$\begin{aligned} I_3 &= \eta_{1111} + \eta_{1221} + \eta_{1331} + \eta_{2112} + \eta_{2222} + \eta_{2332} + \eta_{3113} + \eta_{3223} + \eta_{3333} \\ &= \frac{4}{3}\eta + \eta + \eta + \eta + \eta + \frac{4}{3}\eta + \eta + \eta + \frac{4}{3}\eta \\ &= 10\eta \\ &= \frac{V}{k_B T}\int_0^\infty \left\langle \mathbf{P}^{ts}(t) : \mathbf{P}^{ts}(0)\right\rangle dt \end{aligned} \tag{3.80}$$

or

$$\eta_0 = \frac{V}{10 k_B T}\int_0^\infty \left\langle \mathbf{P}^{ts}(t) : \mathbf{P}^{ts}(0)\right\rangle dt. \tag{3.81}$$

Due to the symmetry of the traceless symmetric pressure tensor, it only has five independent components. These are $P_{xx}^{ts}, P_{yy}^{ts}, P_{xy}^{ts}, P_{xz}^{ts}, P_{yz}^{ts}$. For efficiency of computation, we can express η_{3333} in terms of these components of the pressure tensor:

$$P_{zz}^{ts} = -\left(P_{xx}^{ts} + P_{yy}^{ts}\right) \tag{3.82}$$

$$\eta_{3333} = \eta_{1111} + \eta_{1122} + \eta_{2211} + \eta_{2222}. \tag{3.83}$$

Therefore, we can efficiently compute viscosity from

$$\eta_0 = \frac{1}{10}\left(\eta_{1111} + \eta_{1221} + \eta_{1331} + \eta_{2112} + \eta_{2222} + \eta_{2332} + \eta_{3113} + \eta_{3223} + \eta_{3333}\right)$$
$$= \frac{1}{5}\left(\eta_{1111} + \eta_{2222} + \frac{1}{2}\eta_{1122} + \frac{1}{2}\eta_{2211} + \eta_{1221} + \eta_{1331} + \eta_{2332}\right). \quad (3.84)$$

In the case of the thermal conductivity, the Green–Kubo relation that is obtained when no symmetry is assumed takes the form:

$$\boldsymbol{\lambda} = \frac{V}{k_B T^2} \int_0^\infty \left\langle \mathbf{J}_q(t)\, \mathbf{J}_q(0) \right\rangle dt, \quad (3.85)$$

where $\boldsymbol{\lambda}$ is the second-rank thermal conductivity tensor. If the material is isotropic, then the material property tensor should also be isotropic. The only possible form of an isotropic second-rank polar tensor is $\boldsymbol{\lambda} = \lambda \mathbf{1}$, where λ is the scalar invariant of $\boldsymbol{\lambda}$. In this case, there is only one scalar invariant, and it is given by $\lambda = \frac{1}{3}\mathrm{Tr}(\boldsymbol{\lambda})$. Therefore the Green–Kubo relation for the thermal conductivity tensor of an isotropic material is

$$\boldsymbol{\lambda} = \left(\frac{V}{3k_B T^2} \int_0^\infty \left\langle \mathbf{J}_q(t) \cdot \mathbf{J}_q(0) \right\rangle dt\right) \mathbf{1} = \lambda \mathbf{1} \quad (3.86)$$

so that the scalar thermal conductivity for an isotropic material is

$$\lambda = \frac{V}{3k_B T^2} \int_0^\infty \left\langle \mathbf{J}_q(t) \cdot \mathbf{J}_q(0) \right\rangle dt. \quad (3.87)$$

The two other most useful and commonly used Green–Kubo expressions are those for the bulk viscosity (η_v) and self-diffusion (D) coefficients, which we here simply write down as:

$$\eta_v = \frac{1}{V k_B T} \int_0^\infty \left\langle [p(t)V(t) - \langle pV \rangle][p(0)V(0) - \langle pV \rangle] \right\rangle dt \quad (3.88)$$

$$D = \frac{1}{3} \int_0^\infty \left\langle \mathbf{v}_i(t) \cdot \mathbf{v}_i(0) \right\rangle dt. \quad (3.89)$$

3.6 Fluctuation Theorems

3.6.1 The Fluctuation Theorem

In 1993 a significant discovery was published that, for the first time, provided a mathematical foundation for the microscopic origin of the second law of thermodynamics. This discovery, made by Evans, Cohen and Morriss [64], demonstrated, *on the basis of observations made on NEMD simulation data*, that the ratio of probabilities for forward to reverse trajectories for an ensemble of microscopic systems is exponential in time. This is one of the few examples in science in which a quantitative natural law has been discovered purely on the basis of computer simulation. It is analogous, for example, to the discovery made by Edward Lorenz, on the basis of his simple computational model of coupled differential equations, that chaos is fundamental to global climate patterns [65]. It demonstrates the power of computational science, and in this particular case, the power of NEMD simulation. In their paper what is now known as the Fluctuation Theorem was essentially proposed as

$$\frac{P\left[\bar{\Omega}(t) = A\right]}{P\left[\bar{\Omega}(t) = -A\right]} = e^{At}, \tag{3.90}$$

where $P[...]$ denotes the probability of the quantity in brackets, $\bar{\Omega}(t)$ is the time averaged dissipation at time t, and t is defined to be positive.[1] Note that in this original paper the fluctuation theorem was not given in this general form and was postulated by arguments founded in dynamical systems theory and specifically computed for a system of WCA[2] atoms under planar Couette flow. A formal proof was derived by Evans and Searles for transient systems originating at equilibrium and evolving to a nonequilibrium steady-state (the so-called "transient fluctuation theorem") [67]. In a slightly different context, a proof based on dynamical systems theory was produced by Gallavotti and Cohen [68] for systems in the steady-state.

The significance of this expression is clear if one takes the limit as either $t \to \infty$ or $\Omega \to \infty$. In the case of the former, the implication is that as time becomes macroscopic in scale, the probability of observing a system in which entropy *decreases* becomes insignificant. This is equivalent to the observation that in nature time runs forward, or that the second law of thermodynamics is valid. While there is always a nonzero probability that a trajectory could run in reverse (e.g. water *could* in principle run up a waterfall, rather than down it), the probability for such an event occurring on macroscopic (observable) timescales is so small that it can effectively be regarded as impossible. However, for *small* timescales, such as what happens in microscopic systems in which fluctuations regularly occur between nano to milli seconds, entropy reversing trajectories *are* observable, and indeed this was reported in a noted experiment on colloidal particles [69], which confirmed the validity of the fluctuation theorem. In a similar way, Equation (3.90) also says that as the system becomes macroscopic in size, the *extensive* total dissipation A becomes very large, leading to a rapidly diverging exponential. This says that for macroscopically observable systems, it is highly unlikely that one can observe violations in the second law of thermodynamics. Thus, Equation (3.90) quantifies the second law of thermodynamics. It is of note that Maxwell intuitively knew the second law was only valid in the thermodynamic limit of large time and system size. He in fact wrote: "*Hence the Second Law of thermodynamics is continually being violated and that to a considerable extent in any sufficiently small group of molecules belonging to any real body. As the number of molecules in the group is increased, the deviations from the mean of the whole become smaller and less frequent; and when the number is increased till the group includes a sensible portion of the body, the probability of a*

[1] For a one component isothermal system the dissipation is proportional to the entropy production.

[2] A shifted-truncated Lennard-Jones potential named after the authors who first proposed it, Weeks-Chandler-Andersen [66].

measurable variation from the mean occurring in a finite number of years becomes so small that it may be regarded as practically an impossibility. This calculation belongs of course to molecular theory and not to pure thermodynamics, but it shows that we have reason for believing the truth of the second law to be of the nature of a strong probability, which, though it falls short of certainty by less than any assignable quantity, is not an absolute certainty [70]." It was a precocious insight that was only quantified over a hundred years later, and only made possible because of the advent of computer simulation as an invaluable tool in modern science.

Since the discovery of the fluctuation theorem, there have been many studies done on this, and other fluctuation relations, including those for quantum systems. It is beyond the intention of this book to discuss most of these any further, but we refer interested readers to some excellent reviews [71, 72], as well as the second edition of the book by Evans and Morriss [2] and the recently published book by Evans, Searles and Williams [73].

3.6.2 The Dissipation Function and Dissipation Theorem

Apart from its intrinsic significance in connecting statistical mechanics to thermodynamics, the fluctuation theorem has also proven to be a useful theoretical tool. One can, for instance, derive the (linear response) Green–Kubo relations from it [74]. It was also shown that a proof could also be extended to the nonlinear regime and applied to derive the so-called Kawasaki and TTCF forms of response theory. In order to do this Evans and colleagues first defined what they termed the "dissipation function" as [75]

$$\bar{\Omega}_t \equiv \frac{1}{t}\left\{\ln\left[\frac{f(\mathbf{\Gamma}(0),0)}{f(\mathbf{\Gamma}(t),0)}\right] - \int_0^t \Lambda(\mathbf{\Gamma}(s))\,ds\right\} = \int_0^t \Omega(\mathbf{\Gamma}(s))ds, \tag{3.91}$$

where Λ is the phase-space compression factor and f the probability distribution function, as previously defined. Noting that $\bar{\Omega}_t$ is just the argument of the probability functions in Equation (3.90) above (i.e. the time averaged total dissipation), they were able to deduce the second law inequality, namely $\langle\bar{\Omega}_t\rangle \geq 0$, and further, that the Kawasaki and TTCF expressions (e.g. TTCF given by Equation (3.62)) could be simply derived. In terms of the dissipation function, one has for some arbitrary phase variable $B(t)$

$$\langle B(t)\rangle = \left\langle B(0)\exp\left[-\int_0^{-t}\Omega(\Gamma(s))\,ds\right]\right\rangle \tag{3.92}$$

and

$$\langle B(t)\rangle = \langle B(0)\rangle + \int_0^t ds\,\langle\Omega(0)B(s)\rangle \tag{3.93}$$

respectively, where here $\Omega(0) = -\beta J(0)F^e$. The derivation of these two expressions is known as the "dissipation theorem", even though the expressions themselves were previously known and derived by alternative means, as we have previously seen in the case of nonlinear response theory (see also reference [2]).

3.6.3 The Jarzynski and Crooks Equalities

In classical thermodynamics it is well known that for a quasi-static process in which a system evolves from thermodynamic state A to state B *infinitely slowly*, then the Helmholtz free-energy difference between these two states is given by

$$W_\infty = \Delta F = F_B - F_A. \tag{3.94}$$

This of course is *not* the case for systems that evolve over finite times. For such systems dissipation occurs and so one has the inequality $W \geq \Delta F$.

In 1997 Jarzynski derived an *exact* equality between the work done on a system and its free energy difference [76, 77]. While the equality given in Equation (3.94) is for a quasi-static path between two equilibrium states, A and B, his new expression is independent of the path and is given by

$$\langle \exp(-\beta W) \rangle = \exp(-\beta\, \Delta F) \tag{3.95}$$

where $\langle ... \rangle$ indicates an average over all possible nonequilibrium ensembles. Thus, an exact relationship was derived between an equilibrium thermodynamic free energy difference, and a nonequilibrium process in which work is done in going from state A to B at a finite rate. This expression has proven to be very useful and has enabled efficient computations of free energy differences that otherwise would be cumbersome to calculate.

The link between Jarzynski's equality and the Fluctuation Theorem was shown by Crooks shortly after [78], even though his original formulation was derived without prior knowledge of it [79]. His formulation is based upon a realisation that the entropy production (rather than the production rate) can be expressed over some finite period of time as

$$\Sigma = \beta (W - \Delta F) \tag{3.96}$$

and that the probability of forward (F) to reverse (R) entropy producing states can be expressed as

$$\frac{P_F(+\Sigma)}{P_R(-\Sigma)} = \mathrm{e}^{+\Sigma}. \tag{3.97}$$

Noting that

$$\begin{aligned} \left\langle \mathrm{e}^{-\Sigma} \right\rangle &= \int P_F(+\Sigma)\, e^{-\Sigma} d\Sigma \\ &= \int P_R(-\Sigma)\, e^{+\Sigma} e^{-\Sigma} d\Sigma \\ &= 1, \end{aligned} \tag{3.98}$$

substitution of Equation (3.96) into the left-hand side of Equation (3.98) results in the Jarzynski equality, Equation (3.95). In this derivation one takes note that the free energy difference is a state function and thus can be taken outside of the ensemble average. Equation (3.97) with entropy production expressed as Equation (3.96) is known as the Crooks Fluctuation Theorem. These, and other, fluctuation theorems have been experimentally verified, see for example the review by Bustamante *et al.* and references therein [72].

4 Temperature and Thermodynamic Fluxes

In a nonequilibrium molecular dynamics simulation, we generally compute the response of a system to some sort of perturbation that prevents it from returning to its equilibrium state. It is often the case that the perturbation can be characterised as a thermodynamic force, while the response takes the form of a flux. This may be a momentum flux, a heat flux, a mass flux or the flux of some other quantity, depending on the type of perturbation applied. To evaluate these fluxes, we require microscopic expressions for the fluxes in terms of the positions and momenta that we compute in the simulations. We may also need to monitor the thermodynamic state of the system by computing the temperature, so a microscopic expression for it in terms of the positions and momenta is also required. In this chapter, we discuss derivations of these microscopic expressions, and the practical methods for computing them.

4.1 Temperature

Temperature is one of the most fundamental thermal properties. The concept of temperature is well understood from both the thermodynamic and statistical-mechanical points of view in the context of thermodynamic equilibrium, but the definition, uniqueness and even the existence of temperature for systems that are far from equilibrium are controversial [80]. It has even been proposed that systems that are far from equilibrium are best described by formalisms that deliberately avoid the introduction of nonequilibrium temperature [27]. However, we can safely apply equilibrium concepts whenever the local equilibrium hypothesis is valid. For our present discussion, we will assume that this is the case.

The definition of temperature for a system with a fixed number of particles in classical equilibrium thermodynamics is

$$\frac{1}{T} = \left(\frac{\partial S}{\partial U}\right)_V, \tag{4.1}$$

where V, S and U are the volume, entropy and internal energy. For a system that is out of equilibrium, but obeying local thermodynamic equilibrium, this also defines the temperature at any point in the fluid, provided that we use the local values of thermodynamic variables to define the state.

This definition can be used to obtain a general microscopic expression for the temperature, by following a procedure similar to that outlined by Jepps, Ayton and Evans [81] or Rickayzen and Powles [82], for example. We begin by recalling the microcanonical expression for the entropy, $S = k_B \ln \Omega$ with $\Omega = \int \delta\,(H(\mathbf{\Gamma}) - U)\,d\mathbf{\Gamma}$ where $H(\mathbf{\Gamma})$ is the phase variable for the internal energy. Note that the definition of Ω should strictly include a constant that reduces the right-hand side to a dimensionless number, but this constant does not affect the final results, so it is omitted. This expression for the entropy can be substituted into the thermodynamic definition of temperature giving,

$$\frac{1}{T} = \frac{k_B}{\Omega}\left(\frac{\partial \Omega}{\partial U}\right)_V. \tag{4.2}$$

The derivative can be taken when we use the microscopic expression for the entropy

$$\begin{aligned}\frac{\partial \Omega}{\partial U} &= \frac{\partial}{\partial U}\int \delta\,(H(\mathbf{\Gamma}) - U)\,d\mathbf{\Gamma} \\ &= -\int \frac{\partial}{\partial H}\delta\,(H(\mathbf{\Gamma}) - U)\,d\mathbf{\Gamma}.\end{aligned} \tag{4.3}$$

This can be simplified by writing

$$\frac{\partial}{\partial \mathbf{\Gamma}}\delta\,(H(\mathbf{\Gamma}) - U) = \frac{\partial H}{\partial \mathbf{\Gamma}}\frac{\partial}{\partial H}\delta\,(H(\mathbf{\Gamma}) - U). \tag{4.4}$$

If we now form the scalar product of both sides of this equation with a vector field $\mathbf{B}(\mathbf{\Gamma})$, we obtain a scalar multiplied by the derivative of the delta function on the right-hand side,

$$\mathbf{B}(\mathbf{\Gamma})\cdot\frac{\partial}{\partial \mathbf{\Gamma}}\delta\,(H(\mathbf{\Gamma}) - U) = \mathbf{B}(\mathbf{\Gamma})\cdot\frac{\partial H}{\partial \mathbf{\Gamma}}\frac{\partial}{\partial H}\delta\,(H(\mathbf{\Gamma}) - U). \tag{4.5}$$

This expression can now be simplified and the desired derivative is

$$\frac{\partial}{\partial H}\delta\,(H(\mathbf{\Gamma}) - U) = \frac{\mathbf{B}(\mathbf{\Gamma})}{\mathbf{B}(\mathbf{\Gamma})\cdot\frac{\partial H}{\partial \mathbf{\Gamma}}}\frac{\partial}{\partial \mathbf{\Gamma}}\delta\,(H(\mathbf{\Gamma}) - U). \tag{4.6}$$

Finally, if we use this result in the expression for the derivative of the entropy with respect to internal energy and then integrate by parts, noting that the boundary term is zero, we find

$$\begin{aligned}\frac{1}{k_B T} &= \frac{1}{\Omega}\int \frac{\partial}{\partial \mathbf{\Gamma}}\cdot\left[\frac{\mathbf{B}(\mathbf{\Gamma})}{\mathbf{B}(\mathbf{\Gamma})\cdot\frac{\partial H}{\partial \mathbf{\Gamma}}}\right]\delta\,(H(\mathbf{\Gamma}) - U)\,d\mathbf{\Gamma} \\ &= \left\langle \frac{\partial}{\partial \mathbf{\Gamma}}\cdot\left[\frac{\mathbf{B}(\mathbf{\Gamma})}{\mathbf{B}(\mathbf{\Gamma})\cdot\frac{\partial H}{\partial \mathbf{\Gamma}}}\right]\right\rangle \\ &= \left\langle \frac{\frac{\partial}{\partial \mathbf{\Gamma}}\cdot\mathbf{B}(\mathbf{\Gamma})}{\mathbf{B}(\mathbf{\Gamma})\cdot\frac{\partial H}{\partial \mathbf{\Gamma}}}\right\rangle - \left\langle \frac{\mathbf{B}(\mathbf{\Gamma})\cdot\frac{\partial \mathbf{B}(\mathbf{\Gamma})}{\partial \mathbf{\Gamma}}\cdot\frac{\partial H}{\partial \mathbf{\Gamma}} + \mathbf{B}(\mathbf{\Gamma})\cdot\frac{\partial^2 H}{\partial \mathbf{\Gamma}\partial \mathbf{\Gamma}}\cdot\mathbf{B}(\mathbf{\Gamma})}{\left(\mathbf{B}(\mathbf{\Gamma})\cdot\frac{\partial H}{\partial \mathbf{\Gamma}}\right)^2}\right\rangle,\end{aligned} \tag{4.7}$$

where the second term of the last line is of lower order in N than the first, and is therefore negligible. The vector field $\mathbf{B}$ essentially determines the direction in phase space in which we choose to vary the energy, i.e. the direction in which we move to change the energy when taking the derivative, Equation (4.1). The restrictions on $\mathbf{B}$ have been discussed in detail by Jepps, Ayton and Evans [81] and Rickayzen and Powles [82].

Rugh [83] showed that a natural choice for $\mathbf{B}$ is $\mathbf{B}(\boldsymbol{\Gamma}) = \frac{\partial H}{\partial \boldsymbol{\Gamma}}$. With this choice, the direction of $\mathbf{B}$ is normal to the energy surface, leading to the name "normal temperature" [81]. The specific cases that are of greatest relevance here are those that lead to the kinetic and configurational temperatures and their directional components. In the first case, we may choose $\mathbf{B}(\boldsymbol{\Gamma}) = \frac{\partial K}{\partial \boldsymbol{\Gamma}} = \left(0, \ldots, 0, \frac{\mathbf{p}_1}{m_i} \ldots, \frac{\mathbf{p}_N}{m_N}\right)$, where K is the kinetic energy. Neglecting the second term on the right-hand side of Equation (4.7), we find

$$\frac{1}{k_B T_K} = \left\langle \frac{\sum \frac{3}{m_i}}{\sum \frac{\mathbf{p}_i^2}{m_i^2}} \right\rangle. \tag{4.8}$$

Specialising to the case where the masses of all particles are equal, we obtain the expression for the kinetic temperature in the microcanonical ensemble

$$\frac{1}{k_B T_K} = 3N \left\langle \frac{1}{\sum \frac{\mathbf{p}_i^2}{m}} \right\rangle, \tag{4.9}$$

which, for sufficiently large N reduces to the usual expression for the kinetic temperature,

$$k_B T_K = \frac{1}{3N} \left\langle \sum \frac{\mathbf{p}_i^2}{m} \right\rangle, \tag{4.10}$$

which we will show later, would be obtained directly in the canonical ensemble. The directional components of the kinetic temperature are obtained by considering $\mathbf{B}(\boldsymbol{\Gamma}) = \frac{\partial K_x}{\partial \boldsymbol{\Gamma}} = \left(0, \ldots, 0, \frac{p_{1x}}{m_i}, 0, 0, \ldots, \frac{p_{Nx}}{m_N}, 0, 0\right)$ which leads to

$$\frac{1}{k_B T_K} = \left\langle \frac{\sum \frac{1}{m_i}}{\sum \frac{p_{ix}^2}{m_i^2}} \right\rangle. \tag{4.11}$$

Again, for equal masses and large N this result simplifies to the expression that would be directly obtained in the canonical ensemble

$$k_B T_K = \frac{1}{N} \left\langle \sum \frac{p_{ix}^2}{m} \right\rangle \tag{4.12}$$

with similar expressions for the y and z components.

The configurational temperature is obtained by considering

$$\mathbf{B}(\boldsymbol{\Gamma}) = \frac{\partial \Phi}{\partial \boldsymbol{\Gamma}} = -\left(\mathbf{F}_1, \ldots, \mathbf{F}_N, 0 \ldots, 0\right), \tag{4.13}$$

where $\mathbf{F}_i = -\frac{\partial \Phi}{\partial \mathbf{r}_i}$, which results in

$$\frac{1}{k_B T_C} = -\left\langle \frac{\sum_i \frac{\partial}{\partial \mathbf{r}_i} \cdot \mathbf{F}_i}{\sum_i \mathbf{F}_i^2} + \frac{\sum_j \sum_i \mathbf{F}_i \cdot \frac{\partial \mathbf{F}_j}{\partial \mathbf{r}_i} \cdot \mathbf{F}_j}{\left(\sum_i \mathbf{F}_i^2\right)^2} \right\rangle. \tag{4.14}$$

Again, the second term is negligible in comparison with the first for large systems, and the configurational temperature can be written as

$$k_B T_C = -\left\langle \frac{\sum_i \mathbf{F}_i^2}{\sum_i \frac{\partial}{\partial \mathbf{r}_i} \cdot \mathbf{F}_i} \right\rangle. \tag{4.15}$$

The configurational temperature can also be resolved into three directional components, as was suggested by Baranyai [84]. In this case, we consider $\mathbf{B}(\mathbf{\Gamma}) = -(F_{1x}, 0, 0, \ldots, F_{Nx}, 0 \ldots, 0)$, with the result

$$k_B T_{Cx} = -\left\langle \frac{\sum_i F_{ix}^2}{\sum_i \frac{\partial F_{ix}}{\partial x_i}} \right\rangle. \tag{4.16}$$

These derivations can also be done in the canonical ensemble. Consider the following integral,

$$\int \frac{\partial}{\partial \mathbf{\Gamma}} \cdot \left(\mathbf{B}(\mathbf{\Gamma}) e^{-\beta H}\right) d\dot{\mathbf{\Gamma}} = \int \frac{\partial}{\partial \mathbf{\Gamma}} \cdot \mathbf{B}(\mathbf{\Gamma}) e^{-\beta H} d\mathbf{\Gamma} - \beta \int \mathbf{B}(\mathbf{\Gamma}) \cdot \frac{\partial H}{\partial \mathbf{\Gamma}} e^{-\beta H} d\mathbf{\Gamma}. \tag{4.17}$$

This is equal to zero by Gauss's theorem and the fact that the probability density goes to zero at the limits of integration, so we can express the temperature as

$$\frac{1}{k_B T} = \frac{\left\langle \frac{\partial}{\partial \mathbf{\Gamma}} \cdot \mathbf{B}(\mathbf{\Gamma}) \right\rangle}{\left\langle \mathbf{B}(\mathbf{\Gamma}) \cdot \frac{\partial H}{\partial \mathbf{\Gamma}} \right\rangle}. \tag{4.18}$$

Comparing this expression with Equation (4.7), we see that they only differ by terms that are of order $1/N$, in that the canonical ensemble expression is a ratio of two ensemble averages whereas the microcanonical expression is the ensemble average of a ratio. Jepps *et al.* [81] have compared them and found that the numerical properties of the canonical (fractional) expression for the configurational temperature,

$$k_B T_{Cf} = -\frac{\left\langle \sum_i \mathbf{F}_i^2 \right\rangle}{\left\langle \sum_i \frac{\partial}{\partial \mathbf{r}_i} \cdot \mathbf{F}_i \right\rangle}, \tag{4.19}$$

are generally superior.

4.2 Pressure Tensor and Heat Flux Vector

In the theory of nonequilibrium thermodynamics, the thermodynamic fluxes and forces are identified by analysing the bilinear expression for the entropy production. For simple viscous flow, the relevant term in the entropy production, given by Equation (2.47) is

$$\sigma_{visc} = -\frac{1}{T}\left(\mathbf{P} - p_0\mathbf{1}\right) : \nabla\mathbf{v}, \tag{4.20}$$

where $\mathbf{P}$ is the pressure tensor, p_0 is the equilibrium isotropic pressure [4] and $(\mathbf{P} - p_0\mathbf{1})$ is the nonequilibrium part of the pressure tensor. Nonequilibrium molecular dynamics simulations of viscous flow are usually (but not always [2]) performed in an ensemble in which the velocity gradient tensor (the thermodynamic force) is the independent variable and the nonequilibrium pressure tensor (the thermodynamic flux) is the dependent variable. The velocity gradient tensor occurs as a parameter in the equations of motion and the pressure tensor must be computed from the positions and momenta of the particles in the system. Similarly, heat flow simulations are usually done by applying a heat field and computing the response in the form of the heat flux. Therefore, we need expressions for the pressure tensor and heat flux vector that can be calculated from the microscopic variables. Although this task may appear simple at first, there are many varieties of inter- and intra-molecular forces, different representations of the microscopic densities, and other issues that arise in connection with periodic boundary conditions, that all seem to require special treatment in the derivation of the microscopic expressions for the fluxes. In this section we will attempt to give a unified and complete, but still relatively simple, discussion of these issues summarizing information that is scattered throughout the literature.

Our discussion of the microscopic expressions for the pressure tensor and heat flux vector begins with definitions of the microscopic densities of mass, momentum and total energy. We choose to work with the total energy here rather than the internal energy because it is easier to derive the heat flux vector using the balance equation for the total energy rather than the internal energy. For simplicity, we will only consider a single component fluid. Initially, we will assume that the mass of each atom or interaction site (for coarse-grained models that do not include explicit atomic detail) is localised at a point corresponding to the position of that atom or interaction site. This leads to a definition of the fluxes in the so-called atomic representation. Later, we will also discuss the molecular representation of the fluxes.

The mass, momentum and total energy densities in the atomic representation are given respectively by

$$\rho\left(\mathbf{r},t\right) = \sum_{i=1}^{N} m_i \delta\left(\mathbf{r} - \mathbf{r}_i\right) \tag{4.21}$$

$$\mathbf{J}\left(\mathbf{r},t\right) = \rho\mathbf{v}\left(\mathbf{r},t\right) = \sum_{i=1}^{N} m_i \mathbf{v}_i \delta\left(\mathbf{r} - \mathbf{r}_i\right) \tag{4.22}$$

$$\rho e\left(\mathbf{r},t\right) = \sum_{i=1}^{N} e_i \delta\left(\mathbf{r} - \mathbf{r}_i\right), \tag{4.23}$$

where m_i, $\mathbf{v}_i$ and e_i are the mass, velocity and total energy of particle i and N is the number of atoms in the system. The total energy of an atom consists of its kinetic energy, its intermolecular potential energy and its potential energy due to interactions with conservative external fields,

$$e_i = \frac{1}{2} m_i v_i^2 + \frac{1}{2} \sum_j \phi_{ij} + \psi_i. \tag{4.24}$$

The definitions of the mass, momentum and energy densities also serve as microscopic definitions of the streaming velocity $\mathbf{v}$ and the local specific total energy e. The mass density ρ, momentum density $\mathbf{J}$ and the total energy density ρe defined in these equations are local, instantaneous quantities that depend on the initial microscopic state of the system $\boldsymbol{\Gamma}(0)$ in addition to the explicitly shown arguments of position and time. When these densities are averaged over an (equilibrium) ensemble of initial states, the dependence on the initial microscopic state is removed and we obtain the usual hydrodynamic (averaged) densities of nonequilibrium thermodynamics [2, 85]. Both the microscopic and hydrodynamic (averaged) densities of mass, momentum and internal energy obey the same set of balance equations, given by

$$\frac{\partial \rho}{\partial t} = -\nabla \cdot \mathbf{J} = -\nabla \cdot (\rho \mathbf{v}) \tag{4.25}$$

$$\frac{\partial (\rho \mathbf{v})}{\partial t} = -\nabla \cdot [\mathbf{P} + \rho \mathbf{v}\mathbf{v}] + \rho \mathbf{F}^e \tag{4.26}$$

$$\frac{\partial (\rho e)}{\partial t} = -\nabla \cdot \left[\mathbf{J}_q + \rho e \mathbf{v} + \mathbf{P} \cdot \mathbf{v}\right], \tag{4.27}$$

where $\rho \mathbf{F}^e$ is the local body force density due to external forces, which is expressed in terms of microscopic variables for a single component fluid as

$$\rho \mathbf{F}^e(\mathbf{r}, t) = \sum_{i=1}^{N} \mathbf{F}_i^e \delta(\mathbf{r} - \mathbf{r}_i) \tag{4.28}$$

and $\mathbf{F}_i^e$ is the external force applied to particle i in the microscopic equations of motion. We assume that only conservative external forces are present, so the total energy of the system is conserved.

The pressure tensor and heat flux vector represent the diffusive fluxes of momentum and internal energy. By substituting the microscopic densities into the balance equations, we can obtain microscopic expressions for these fluxes. We will do this explicitly for each of these quantities in the following sections.

4.2.1 Pressure Tensor

The derivation is simplest in k-space, so we first Fourier transform the momentum density balance Equation (4.26), giving

$$\frac{\partial}{\partial t} [\widetilde{\rho \mathbf{v}} (\mathbf{k}, t)] = i\mathbf{k} \cdot \tilde{\mathbf{P}} (\mathbf{k}, t) + i\mathbf{k} \cdot [\widetilde{\rho \mathbf{v}\mathbf{v}} (\mathbf{k}, t)] + \widetilde{\rho \mathbf{F}^e} (\mathbf{k}, t), \tag{4.29}$$

where we define the Fourier transform of any function $f(x)$ as

$$F\{f(x)\} = \tilde{f}(k) = \int_{-\infty}^{\infty} e^{ikx} f(x)\, dx. \tag{4.30}$$

We now take the Fourier transform of the microscopic expression for the momentum density and evaluate the partial derivative with respect to time:

$$\frac{\partial}{\partial t}\left[\widetilde{\rho\mathbf{v}}\left(\mathbf{k},t\right)\right] = \frac{\partial}{\partial t}\sum_{i=1}^{N} m_i \mathbf{v}_i e^{i\mathbf{k}\cdot\mathbf{r}_i}$$
$$= \sum_{i=1}^{N} m_i \left(\dot{\mathbf{v}}_i + i\mathbf{k}\cdot\mathbf{v}_i\mathbf{v}_i\right) e^{i\mathbf{k}\cdot\mathbf{r}_i}. \tag{4.31}$$

The equations of motion in their most general Newtonian form are

$$\mathbf{F}_i = m_i\dot{\mathbf{v}}_i, \tag{4.32}$$

where $\mathbf{F}_i$ is the total force on atom i, including intermolecular forces, constraint forces and any synthetic (i.e. not naturally-occurring) forces added to the equations of motion to maintain a constant temperature or generate nonequilibrium perturbations. For the present discussion, we will restrict our attention to forces that can be expressed in the form:

$$\mathbf{F}_i = \mathbf{F}_i^{\phi} + \mathbf{F}_i^{C} + \mathbf{F}_i^{e} \tag{4.33}$$

where the first term represents forces due to atomic potentials (which may include Lennard-Jones, bond stretching and bending, dihedral or others), the second term represents internal constraint forces (e.g. bond length and bond angle constraints) and the third term represents external forces, which may be applied via the equations of motion. The first term in the second line of Equation (4.31) can be written as

$$\sum_{i=1}^{N} \mathbf{F}_i e^{i\mathbf{k}\cdot\mathbf{r}_i} = \sum_{i=1}^{N}\left(\mathbf{F}_i^{\phi} + \mathbf{F}_i^{C} + \mathbf{F}_i^{e}\right) e^{i\mathbf{k}\cdot\mathbf{r}_i}. \tag{4.34}$$

The term involving the external force is equal to the corresponding body force term on the right-hand side of the momentum balance equation, and does not contribute to the expression for the pressure tensor. The other two terms will be considered in turn. The term due to interatomic potential forces (e.g. Lennard-Jones and dihedral potentials) will become

$$\sum_{i=1}^{N} \mathbf{F}_i^{\phi} e^{i\mathbf{k}\cdot\mathbf{r}_i} = \sum_{i=1}^{N} \mathbf{F}_i^{\phi}\left(1 + i\mathbf{k}\cdot\mathbf{r}_i + \frac{1}{2}\left(i\mathbf{k}\cdot\mathbf{r}\right)^2 + \cdots\right)$$
$$= i\mathbf{k}\cdot\sum_{i=1}^{N}\mathbf{r}_i\left(1 + \frac{1}{2}i\mathbf{k}\cdot\mathbf{r}_i + \cdots\right)\mathbf{F}_i^{\phi}$$
$$= i\mathbf{k}\cdot\sum_{i=1}^{N}\mathbf{r}_i\mathbf{F}_i^{\phi} + O\left(k^2\right), \tag{4.35}$$

where we have used the fact that the sum of any type of internal force over the whole system is zero. Similarly, we can write

$$\sum_{i=1}^{N} \mathbf{F}_i^C e^{i\mathbf{k}\cdot\mathbf{r}_i} = i\mathbf{k} \cdot \sum_{i=1}^{N} \mathbf{r}_i \mathbf{F}_i^C + O\left(k^2\right) \tag{4.36}$$

for the intramolecular constraint forces (e.g. molecular bond length and bond angle constraint forces).

Now we consider the convective term in the momentum balance equation. It is convenient to introduce the peculiar velocity here, by the definition

$$\mathbf{c}_i = \mathbf{v}_i - \mathbf{v}\left(\mathbf{r}_i\right), \tag{4.37}$$

where $\mathbf{v}_i$ is the velocity of particle i relative to the laboratory frame, $\mathbf{v}\left(\mathbf{r}_i\right)$ is the streaming velocity at the position of atom i and we will assume that the equations of motion ensure that the sum of the peculiar momenta remains equal to zero for all time,

$$\sum_{i=1}^{N} m_i \mathbf{c}_i = 0. \tag{4.38}$$

For flowing molecular fluids, the correct determination of the atomic peculiar velocity is a nontrivial exercise, because internal motions such as rotation may possess a systematic or streaming component. We will discuss this issue in more detail later.

The definition of the peculiar velocity allows us to write

$$\sum_{i=1}^{N} m_i \mathbf{v}_i \mathbf{v}_i e^{i\mathbf{k}\cdot\mathbf{r}_i} = \sum_{i=1}^{N} m_i \left[\mathbf{c}_i + \mathbf{v}\left(\mathbf{r}_i\right)\right]\left[\mathbf{c}_i + \mathbf{v}\left(\mathbf{r}_i\right)\right] e^{i\mathbf{k}\cdot\mathbf{r}_i}. \tag{4.39}$$

If we assume that the streaming velocity is given as a function of position and time, substitute the microscopic expression for the momentum density into the convective term of the momentum balance equation and use the properties of the delta function when taking the Fourier transform, we can show that

$$\begin{aligned}
\widetilde{\rho\mathbf{v}\mathbf{v}}\left(\mathbf{k}, t\right) &= F\left\{\left[\rho\mathbf{v}\right]\mathbf{v}\right\} \\
&= F\left\{\left[\sum_{i=1}^{N} m_i \mathbf{v}_i \delta\left(\mathbf{r} - \mathbf{r}_i\right)\right]\mathbf{v}\left(\mathbf{r}, t\right)\right\} \\
&= \sum_{i=1}^{N} m_i \mathbf{v}_i \mathbf{v}\left(\mathbf{r}_i, t\right) e^{i\mathbf{k}\cdot\mathbf{r}_i} \\
&= \sum_{i=1}^{N} m_i \left[\mathbf{c}_i + \mathbf{v}\left(\mathbf{r}_i, t\right)\right]\mathbf{v}\left(\mathbf{r}_i, t\right) e^{i\mathbf{k}\cdot\mathbf{r}_i} \\
&= \sum_{i=1}^{N} m_i \left[\mathbf{c}_i \mathbf{v}\left(\mathbf{r}_i, t\right) + \mathbf{v}\left(\mathbf{r}_i, t\right)\mathbf{v}\left(\mathbf{r}_i, t\right)\right] e^{i\mathbf{k}\cdot\mathbf{r}_i}.
\end{aligned} \tag{4.40}$$

Similarly,

$$\widetilde{\rho \mathbf{v}\mathbf{v}}(\mathbf{k},t) = F\{\mathbf{v}[\rho\mathbf{v}]\} = \sum_{i=1}^{N} m_i [\mathbf{v}(\mathbf{r}_i,t)\mathbf{c}_i + \mathbf{v}(\mathbf{r}_i,t)\mathbf{v}(\mathbf{r}_i,t)] e^{i\mathbf{k}\cdot\mathbf{r}_i} \tag{4.41}$$

and

$$\widetilde{\rho \mathbf{v}\mathbf{v}}(\mathbf{k},t) = F\{\rho[\mathbf{v}\mathbf{v}]\} = \sum_{i=1}^{N} m_i \mathbf{v}(\mathbf{r}_i,t)\mathbf{v}(\mathbf{r}_i,t) e^{i\mathbf{k}\cdot\mathbf{r}_i}. \tag{4.42}$$

Combining these results and expanding the exponential to lowest order in wavevector, we obtain

$$\begin{aligned} \sum_{i=1}^{N} m_i \mathbf{v}_i \mathbf{v}_i e^{i\mathbf{k}\cdot\mathbf{r}_i} &= i\mathbf{k}\cdot\sum_{i=1}^{N} m_i \mathbf{v}_i \mathbf{v}_i + O\left(k^2\right) \\ &= i\mathbf{k}\cdot\sum_{i=1}^{N} m_i [\mathbf{c}_i\mathbf{c}_i + \mathbf{v}(\mathbf{r}_i)\mathbf{v}(\mathbf{r}_i)] + O\left(k^2\right). \end{aligned} \tag{4.43}$$

Substituting these results into the momentum balance equation and collecting terms to first order in wavevector, we obtain

$$\begin{aligned} i\mathbf{k}\cdot\tilde{\mathbf{P}}(\mathbf{k},t) &= \frac{\partial}{\partial t}[\widetilde{\rho\mathbf{v}}(\mathbf{k},t)] - i\mathbf{k}\cdot[\widetilde{\rho\mathbf{v}\mathbf{v}}(\mathbf{k},t)] - \widetilde{\rho\mathbf{F}^e}(\mathbf{k},t) \\ &= i\mathbf{k}\cdot\left(\sum_{i=1}^{N} m_i\mathbf{c}_i\mathbf{c}_i + \sum_{i=1}^{N}\mathbf{r}_i\mathbf{F}_i^{\phi} + \sum_{i=1}^{N}\mathbf{r}_i\mathbf{F}_i^{C}\right) + O\left(k^2\right). \end{aligned} \tag{4.44}$$

In real space, this equation relates the divergence of the local, instantaneous pressure tensor to the divergence of a microscopic quantity. It also provides us with an expression, to within an arbitrary divergenceless quantity, for the zero wavevector pressure tensor itself,

$$\tilde{\mathbf{P}}(\mathbf{k}=0,t) = V\mathbf{P}(t) = \sum_{i=1}^{N} m_i\mathbf{c}_i\mathbf{c}_i + \sum_{i=1}^{N}\mathbf{r}_i\mathbf{F}_i^{\phi} + \sum_{i=1}^{N}\mathbf{r}_i\mathbf{F}_i^{C}, \tag{4.45}$$

where V is the system volume. The arbitrary divergenceless quantity is usually chosen to be equal to zero, as we have done here, giving a pressure tensor that is consistent with thermodynamics. Different choices of the divergenceless quantity that may be added to the pressure tensor are analogous to the choice of gauge in electrodynamics [8]. For a simple atomic fluid with two-body central forces between atoms and no internal constraint forces, the forces can be written as

$$\mathbf{F}_i^{\phi} = \sum_j \mathbf{F}_{ij}^{\phi} = -\sum_j \frac{\partial\phi\left(r_{ij}\right)}{\partial\mathbf{r}_i} = \sum_j \frac{\partial\phi\left(r_{ij}\right)}{\partial r_{ij}}\frac{\mathbf{r}_{ij}}{r_{ij}}, \tag{4.46}$$

where $\mathbf{r}_{ij}$ is defined as $\mathbf{r}_{ij} \equiv \mathbf{r}_j - \mathbf{r}_i$. Since $\mathbf{F}_{ij} = -\mathbf{F}_{ji}$, we then have the usual expression for the volume-averaged instantaneous pressure tensor in a simple atomic fluid:

$$V\mathbf{P}(t) = \sum_{i=1}^{N} m_i \mathbf{c}_i \mathbf{c}_i - \frac{1}{2} \sum_{i=1}^{N} \sum_{j \neq i}^{N} \mathbf{r}_{ij} \mathbf{F}_{ij}^{\phi}. \tag{4.47}$$

If the derivation is carried out without making the small wavevector approximation, the expression for the wavevector dependent pressure tensor for a simple atomic fluid with only two-body central forces becomes

$$\tilde{\mathbf{P}}(\mathbf{k}, t) = \sum_{i=1}^{N} m_i \mathbf{c}_i \mathbf{c}_i e^{i\mathbf{k}\cdot\mathbf{r}_i} - \frac{1}{2} \sum_{i}^{N} \sum_{j \neq i}^{N} \mathbf{r}_{ij} \mathbf{F}_{ij}^{\phi} \left(\frac{e^{i\mathbf{k}\cdot\mathbf{r}_{ij}} - 1}{i\mathbf{k}\cdot\mathbf{r}_{ij}} \right) e^{i\mathbf{k}\cdot\mathbf{r}_i}. \tag{4.48}$$

This expression must be used when nonzero wavevector components of the pressure tensor are evaluated, for example when we want to compute the response to a sinusoidal transverse force.

The full expression for the pressure tensor of an atomic fluid interacting via central two-body forces in reciprocal space is given by Equation (4.48). This expression is exact and valid for all k. To obtain the pressure tensor in real space one would take the inverse Fourier transform of this expression, and in principle it is this quantity we would compute if we were not only interested in the zero-wavevector pressure tensor. Keeping the explicit time-dependence for the moment, to emphasise that the following refer to *instantaneous* quantities, the inverse transform of Equation (4.48) is thus

$$\mathbf{P}(\mathbf{r}, t) = \sum_{i} m_i \mathbf{c}_i(t) \mathbf{c}_i(t) \delta(\mathbf{r} - \mathbf{r}_i) - \frac{1}{2} \sum_{i} \sum_{j \neq i} \mathbf{r}_{ij}(t) \mathbf{F}_{ij}^{\phi}(t) O_{ij} \delta(\mathbf{r} - \mathbf{r}_i) \Big|_{\mathbf{r}_i(t)=\mathbf{r}}, \tag{4.49}$$

where $O_{ij}(t)$ is a differential operator given by

$$\begin{aligned} O_{ij}(t) &= 1 - \frac{1}{2!} \mathbf{r}_{ij}(t) \cdot \frac{\partial}{\partial \mathbf{r}} + \cdots + \frac{1}{n!} \left(-\mathbf{r}_{ij}(t) \cdot \frac{\partial}{\partial \mathbf{r}} \right)^{n-1} + \cdots \\ &= \sum_{n=1}^{\infty} \frac{1}{n!} \left(-\mathbf{r}_{ij}(t) \cdot \frac{\partial}{\partial \mathbf{r}} \right)^{n-1}. \end{aligned} \tag{4.50}$$

While Equation (4.49) is exact and can in principle be used to compute the pressure tensor for an inhomogeneous fluid arbitrarily far from equilibrium, in practice it is difficult to compute because of the O_{ij} operator, which results from an expansion of a difference in delta function terms, $\delta(\mathbf{r} - \mathbf{r}_j) - \delta(\mathbf{r} - \mathbf{r}_i)$. For a homogeneous fluid we have identically $O_{ij} = 1$, which gives us the standard Irving-Kirkwood pressure tensor. However, for an inhomogeneous fluid this is not the case (the so-called IK1 approximation) and this can lead to significant errors in pressure computations [86].

To derive an alternative expression for the configurational part of the local, instantaneous pressure tensor, we will need to express the difference between two delta functions as a divergence. This can be done by the following method. By the fundamental theorem of calculus, we can write

$$\delta\left(\mathbf{r}-\mathbf{r}_j\right)-\delta\left(\mathbf{r}-\mathbf{r}_i\right)=\int_0^1 \frac{\partial}{\partial\lambda}\delta\left(\mathbf{r}-\left[\mathbf{r}_i+\lambda\mathbf{r}_{ij}\right]\right)d\lambda \tag{4.51}$$

but

$$\frac{\partial}{\partial\lambda}=\frac{\partial\left(\mathbf{r}_i+\lambda\mathbf{r}_{ij}\right)}{\partial\lambda}\cdot\frac{\partial}{\partial\left(\mathbf{r}_i+\lambda\mathbf{r}_{ij}\right)} \tag{4.52}$$

so we can write

$$\begin{aligned}\delta\left(\mathbf{r}-\mathbf{r}_j\right)-\delta\left(\mathbf{r}-\mathbf{r}_i\right)&=\int_0^1 \frac{\partial}{\partial\lambda}\delta\left(\mathbf{r}-\left[\mathbf{r}_i+\lambda\mathbf{r}_{ij}\right]\right)d\lambda\\&=\mathbf{r}_{ij}\cdot\int_0^1\frac{\partial}{\partial\left(\mathbf{r}_i+\lambda\mathbf{r}_{ij}\right)}\delta\left(\mathbf{r}-\left[\mathbf{r}_i+\lambda\mathbf{r}_{ij}\right]\right)d\lambda\\&=-\frac{\partial}{\partial\mathbf{r}}\cdot\mathbf{r}_{ij}\int_0^1\delta\left(\mathbf{r}-\left[\mathbf{r}_i+\lambda\mathbf{r}_{ij}\right]\right)d\lambda.\end{aligned} \tag{4.53}$$

If the delta function is Fourier transformed and the integral over λ is carried out, we obtain the $\left(\frac{e^{i\mathbf{k}\cdot\mathbf{r}_{ij}}-1}{i\mathbf{k}\cdot\mathbf{r}_{ij}}\right)e^{i\mathbf{k}\cdot\mathbf{r}_i}$ factor in Equation (4.48).

This leads to an alternative form for the instantaneous local pressure tensor, in which the differential O_{ij} operator is recast in an integral form [2, 87]

$$\mathbf{P}(\mathbf{r},t)=\sum_i m_i\mathbf{c}_i(t)\,\mathbf{c}_i(t)\,\delta\left(\mathbf{r}-\mathbf{r}_i\right)-\frac{1}{2}\sum_i\sum_{j\neq i}\mathbf{r}_{ij}(t)\,\mathbf{F}_{ij}^{\phi}(t)\int_0^1 d\lambda\delta\left(\mathbf{r}-\mathbf{r}_i-\lambda\mathbf{r}_{ij}\right). \tag{4.54}$$

For a homogeneous fluid the O_{ij} and integral operators are both identically unity and the standard Irving-Kirkwood form for the pressure tensor is obtained [85], i.e. Equations (4.47) or (4.88) in the case of periodic boundary conditions.

4.2.2 Heat Flux Vector

The heat flux vector can be derived in a very similar way to the pressure tensor. We begin with the Fourier transform of the balance equation for the total energy,

$$\frac{\partial}{\partial t}\left[\widetilde{\rho e}(\mathbf{k},t)\right]=i\mathbf{k}\cdot\tilde{\mathbf{J}}_q(\mathbf{k},t)+i\mathbf{k}\cdot\left[\widetilde{\rho e\mathbf{v}}(\mathbf{k},t)\right]+i\mathbf{k}\cdot\left[\widetilde{\mathbf{P}\cdot\mathbf{v}}(\mathbf{k},t)\right]. \tag{4.55}$$

The time derivative of the transformed total energy density is

$$\begin{aligned}\frac{\partial}{\partial t}\left[\widetilde{\rho e}(\mathbf{k},t)\right]&=\frac{\partial}{\partial t}\sum_i e_i e^{i\mathbf{k}\cdot\mathbf{r}_i}\\&=\sum_i\dot{e}_i e^{i\mathbf{k}\cdot\mathbf{r}_i}+i\mathbf{k}\cdot\sum_i e_i\mathbf{v}_i e^{i\mathbf{k}\cdot\mathbf{r}_i}.\end{aligned} \tag{4.56}$$

The first term is

$$\begin{aligned}\sum_i \dot{e}_i e^{i\mathbf{k}\cdot\mathbf{r}_i} &= \sum_i m_i \mathbf{v}_i \cdot \dot{\mathbf{v}}_i e^{i\mathbf{k}\cdot\mathbf{r}_i} - \frac{1}{2}\sum_i \sum_{j\neq i} \left(\mathbf{F}_{ij}^{\phi} \cdot \mathbf{v}_i + \mathbf{F}_{ji}^{\phi} \cdot \mathbf{v}_j\right) e^{i\mathbf{k}\cdot\mathbf{r}_i} \\ &= \sum_i \mathbf{v}_i \cdot \mathbf{F}_i e^{i\mathbf{k}\cdot\mathbf{r}_i} - \frac{1}{2}\sum_i \sum_{j\neq i} \left(\mathbf{F}_{ij}^{\phi} \cdot \mathbf{v}_i + \mathbf{F}_{ji}^{\phi} \cdot \mathbf{v}_j\right) e^{i\mathbf{k}\cdot\mathbf{r}_i}. \end{aligned} \tag{4.57}$$

We will assume that the total force on an atom only consists of intermolecular pair forces. Intramolecular constraint forces can easily be included, as we have seen for the pressure tensor. We will also exclude external forces because, as we saw in the derivation of the expression for the pressure tensor, they do not contribute to the final expression for the flux. This results in

$$\begin{aligned}\sum_i \dot{e}_i e^{i\mathbf{k}\cdot\mathbf{r}_i} &= \frac{1}{2}\sum_i \sum_j \left(\mathbf{F}_{ij}^{\phi} \cdot \mathbf{v}_i - \mathbf{F}_{ji}^{\phi} \cdot \mathbf{v}_j\right) e^{i\mathbf{k}\cdot\mathbf{r}_i} \\ &= \frac{1}{2}\sum_i \sum_{j\neq i} \mathbf{F}_{ij}^{\phi} \cdot \mathbf{v}_i \left(e^{i\mathbf{k}\cdot\mathbf{r}_i} - e^{i\mathbf{k}\cdot\mathbf{r}_j}\right). \end{aligned} \tag{4.58}$$

Since we are mainly interested here in the average heat flux vector for a homogeneous system, we can take the small wavevector limit, so we expand each of the exponentials. The sum of the zero wavevector terms is $\sum_i \dot{e}_i = 0$. This is consistent with energy conservation, i.e. in the absence of nonconservative external forces, the total energy of the system is conserved. Collecting terms of first order in the wavevector, we find

$$\sum_i \dot{e}_i e^{i\mathbf{k}\cdot\mathbf{r}_i} = i\mathbf{k} \cdot \left(\sum_i \mathbf{r}_i \mathbf{F}_i^{C} \cdot \mathbf{v}_i - \frac{1}{2}\sum_i \sum_{j\neq i} \mathbf{r}_{ij} \mathbf{F}_{ij}^{\phi} \cdot \mathbf{v}_i\right) + O\left(k^2\right). \tag{4.59}$$

The Fourier transform of the convective energy transport term is

$$i\mathbf{k} \cdot [\widetilde{\rho e \mathbf{v}}(\mathbf{k}, t)] = i\mathbf{k} \cdot F\left\{\sum_i e_i \delta(\mathbf{r} - \mathbf{r}_i) \mathbf{v}(\mathbf{r}, t)\right\} = i\mathbf{k} \cdot \sum_i e_i \mathbf{v}(\mathbf{r}_i, t) e^{i\mathbf{k}\cdot\mathbf{r}_i}. \tag{4.60}$$

The term accounting for work done by the pressure tensor is

$$\begin{aligned} i\mathbf{k} \cdot \widetilde{\mathsf{P} \cdot \mathbf{v}}(\mathbf{k}, t) = i\mathbf{k} \cdot F &\left\{\left[\sum_i m_i \mathbf{c}_i \mathbf{c}_i \delta(\mathbf{r} - \mathbf{r}_i)\right.\right. \\ &\left.\left. - \frac{1}{2}\sum_i \sum_{j\neq i} \mathbf{r}_{ij} \mathbf{F}_{ij}^{\phi} O_{ij} \delta(\mathbf{r} - \mathbf{r}_i)\right] \cdot \mathbf{v}(\mathbf{r})\right\}, \end{aligned} \tag{4.61}$$

where the O_{ij} differential operator was defined previously. This can be simplified further. The term involving the kinetic part of the pressure tensor is

$$i\mathbf{k} \cdot \widetilde{\mathsf{P}^{K} \cdot \mathbf{v}}(\mathbf{k}, t) = i\mathbf{k} \cdot \sum_i m_i \mathbf{c}_i \mathbf{c}_i \cdot \mathbf{v}(\mathbf{r}_i, t) + O\left(k^2\right). \tag{4.62}$$

The term involving the configurational part of the pressure tensor is

$$i\mathbf{k}\cdot\widetilde{\mathbf{P}^{\phi}\cdot\mathbf{v}}(\mathbf{k},t) = -\frac{1}{2}i\mathbf{k}\cdot\sum_{i}\sum_{j\neq i}\mathbf{r}_{ij}\mathbf{F}_{ij}^{\phi}\cdot F\left\{\mathbf{v}(\mathbf{r})O_{ij}\delta(\mathbf{r}-\mathbf{r}_i)\right\}. \tag{4.63}$$

By substituting the explicit form of the O_{ij} operator into this equation and carrying out the Fourier transform, we find

$$\int_{-\infty}^{\infty} e^{i\mathbf{k}\cdot\mathbf{r}}\mathbf{v}(\mathbf{r})O_{ij}\delta(\mathbf{r}-\mathbf{r}_i)\,d\mathbf{r} = \sum_{n=0}\frac{(-1)^n}{(n+1)!}\int_{-\infty}^{\infty} e^{i\mathbf{k}\cdot\mathbf{r}}\mathbf{v}(\mathbf{r})\left(\mathbf{r}_{ij}\cdot\frac{\partial}{\partial\mathbf{r}}\right)^n\delta(\mathbf{r}-\mathbf{r}_i)\,d\mathbf{r}. \tag{4.64}$$

Now if we use the identity

$$\int_{-\infty}^{\infty} f(x)\frac{\partial^n}{\partial x^n}\delta(x-a)\,dx = (-1)^n f^n(a) \tag{4.65}$$

we find

$$\int_{-\infty}^{\infty} e^{i\mathbf{k}\cdot\mathbf{r}}\mathbf{v}(\mathbf{r})O_{ij}\delta(\mathbf{r}-\mathbf{r}_i)\,d\mathbf{r} = \sum_{n=0}\frac{1}{(n+1)!}\left[\left(\mathbf{r}_{ij}\cdot\frac{\partial}{\partial\mathbf{r}}\right)^n\left(\mathbf{v}(\mathbf{r})e^{i\mathbf{k}\cdot\mathbf{r}}\right)\right]_{\mathbf{r}=\mathbf{r}_i} \tag{4.66}$$

and similarly,

$$\begin{aligned}\int_{-\infty}^{\infty} O_{ij}\delta(\mathbf{r}-\mathbf{r}_i)e^{i\mathbf{k}\cdot\mathbf{r}}d\mathbf{r} &= \sum_{n=0}\frac{1}{(n+1)!}\left[\left(\mathbf{r}_{ij}\cdot\frac{\partial}{\partial\mathbf{r}}\right)^n e^{i\mathbf{k}\cdot\mathbf{r}}\right]_{\mathbf{r}=\mathbf{r}_i}\\ &\equiv g(i\mathbf{k}\cdot\mathbf{r}_i)e^{i\mathbf{k}\cdot\mathbf{r}_i}.\end{aligned} \tag{4.67}$$

The term due to the work done by the configurational part of the pressure tensor is then

$$\begin{aligned}i\mathbf{k}\cdot\widetilde{\mathbf{P}^{\phi}\cdot\mathbf{v}}(\mathbf{k},t) &= -\frac{1}{2}i\mathbf{k}\cdot\sum_{i}\sum_{j\neq i}\mathbf{r}_{ij}\mathbf{F}_{ij}^{\phi}\cdot F\left\{\mathbf{v}(\mathbf{r})O_{ij}\delta(\mathbf{r}-\mathbf{r}_i)\right\}\\ &= -\frac{1}{2}i\mathbf{k}\cdot\sum_{i}\sum_{j\neq i}\mathbf{r}_{ij}\mathbf{F}_{ij}^{\phi}\cdot\sum_{n=0}\frac{1}{(n+1)!}\left[\left(\mathbf{r}_{ij}\cdot\frac{\partial}{\partial\mathbf{r}}\right)^n\left(\mathbf{v}(\mathbf{r})e^{i\mathbf{k}\cdot\mathbf{r}}\right)\right]_{\mathbf{r}=\mathbf{r}_i}.\end{aligned} \tag{4.68}$$

To lowest order in the wavevector, this becomes

$$\begin{aligned}&i\mathbf{k}\cdot\widetilde{\mathbf{P}^{\phi}\cdot\mathbf{v}}(\mathbf{k},t)\\ &= -\frac{1}{2}i\mathbf{k}\cdot\sum_{i}\sum_{j\neq i}\mathbf{r}_{ij}\mathbf{F}_{ij}^{\phi}\cdot\sum_{n=0}\frac{1}{(n+1)!}\left[\left(\mathbf{r}_{ij}\cdot\frac{\partial}{\partial\mathbf{r}}\right)^n\mathbf{v}(\mathbf{r})\right]_{\mathbf{r}=\mathbf{r}_i} + O\left(k^2\right)\end{aligned} \tag{4.69}$$

and in the case of a uniform velocity field, this reduces to

$$i\mathbf{k}\cdot\widetilde{\mathbf{P}^{\phi}\cdot\mathbf{v}}(\mathbf{k},t) = -\frac{1}{2}i\mathbf{k}\cdot\sum_{i}\sum_{j\neq i}\mathbf{r}_{ij}\mathbf{F}_{ij}^{\phi}\cdot\mathbf{v}(\mathbf{r}_i) + O\left(k^2\right). \tag{4.70}$$

The final expression for the Fourier transformed heat flux vector is obtained by substituting all of these results into the energy balance equation, and then equating terms inside the dot product with $i\mathbf{k}$. For the kinetic component, we find

$$V\tilde{\mathbf{J}}_q^K(t) = \sum_i e_i \mathbf{v}_i - \sum_i e_i \mathbf{v}(\mathbf{r}_i, t) - \sum_i m_i \mathbf{c}_i \mathbf{c}_i \cdot \mathbf{v}(\mathbf{r}_i, t) = \sum_i u_i \mathbf{c}_i, \tag{4.71}$$

where we have used the result

$$\frac{1}{2}\sum_i m_i v_i^2 \mathbf{c}_i - \sum_i m_i \mathbf{c}_i \mathbf{c}_i \cdot \mathbf{v}(\mathbf{r}_i, t) = \frac{1}{2}\sum_i m_i c_i^2 \mathbf{c}_i \tag{4.72}$$

to express the diffusive flux of total kinetic energy in terms of the diffusive flux of thermal kinetic energy and the internal energy per particle u_i is given by $u_i = \frac{1}{2}m_i \mathbf{c}_i^2 + \frac{1}{2}\sum_j \phi_{ij}$.

For the configurational component, we find

$$\begin{aligned} V\tilde{\mathbf{J}}_q^\phi(t) &= -\frac{1}{2}\sum_i \sum_{j\neq i} \mathbf{r}_{ij}\mathbf{F}_{ij}^\phi \cdot \mathbf{v}_i + \frac{1}{2}\sum_i \sum_{j\neq i} \mathbf{r}_{ij}\mathbf{F}_{ij}^\phi \cdot \mathbf{v}(\mathbf{r}_i, t) \\ &= -\frac{1}{2}\sum_i \sum_{j\neq i} \mathbf{r}_{ij}\mathbf{F}_{ij}^\phi \cdot \mathbf{c}_i \end{aligned} \tag{4.73}$$

so the final expression for the heat flux vector of a homogeneous, simple fluid with a uniform or zero velocity field is

$$V\tilde{\mathbf{J}}_q(t) = \sum_i u_i \mathbf{c}_i - \frac{1}{2}\sum_i \sum_{j\neq i} \mathbf{r}_{ij}\mathbf{F}_{ij}^\phi \cdot \mathbf{c}_i. \tag{4.74}$$

In the more general case of nonzero wavevector and arbitrary velocity field, we have

$$\begin{aligned} \tilde{\mathbf{J}}_q(\mathbf{k}, t) &= \sum_i u_i \mathbf{c}_i e^{i\mathbf{k}\cdot\mathbf{r}_i} \\ &- \frac{1}{2}\sum_i \sum_{j\neq i} \mathbf{r}_{ij}\mathbf{F}_{ij}^\phi \cdot \sum_{n=0} \frac{1}{(n+1)!}\left[\left(\mathbf{r}_{ij} \cdot \frac{\partial}{\partial \mathbf{r}}\right)^n \left(e^{i\mathbf{k}\cdot\mathbf{r}}\left(\mathbf{v}_i - \mathbf{v}(\mathbf{r}, t)\right)\right)\right]_{\mathbf{r}=\mathbf{r}_i}. \end{aligned} \tag{4.75}$$

We could add to this any quantity for which the dot product with $i\mathbf{k}$ is equal to zero (or in real space, any divergence free quantity) and it would still satisfy the energy balance equation. To this extent, the heat flux vector, the pressure tensor and other diffusive fluxes are incompletely specified.

Using the integral form of $O_{ij}\delta(\mathbf{r} - \mathbf{r}_i)$, the real space heat flux vector can be expressed as [2, 87]

$$\begin{aligned} \mathbf{J}_q(\mathbf{r}, t) = \sum_i u_i \mathbf{c}_i \delta(\mathbf{r} - \mathbf{r}_i) &- \frac{1}{2}\sum_i \sum_{j\neq i} \mathbf{r}_{ij}\mathbf{F}_{ij}^\phi \cdot \left[\mathbf{c}_i + \mathbf{v}(\mathbf{r}_i) - \mathbf{v}(\mathbf{r})\right] \\ &\times \int_0^1 d\lambda \delta\left(\mathbf{r} - \mathbf{r}_i - \lambda \mathbf{r}_{ij}\right). \end{aligned} \tag{4.76}$$

4.2.3 Effect of Periodic Boundary Conditions

Let us now consider what effect periodic boundary conditions will have on the calculation of the pressure tensor. Similar considerations apply to the heat flux vector. In the previous discussion, it has been assumed that the number of particles in the system and the volume are finite. Under these circumstances, it does not matter whether the potential part of the spatially averaged pressure tensor is computed using the expression $\mathbf{P}^{\phi}(t) = \frac{1}{V}\sum_{i=1}^{N}\mathbf{r}_i\mathbf{F}_i^{\phi}$ or $\mathbf{P}^{\phi}(t) = -\frac{1}{2V}\sum_{i=1}^{N}\sum_{j\neq i}^{N}\mathbf{r}_{ij}\mathbf{F}_{ij}^{\phi}$. However, under periodic boundary conditions, the total system volume is infinite. If we restrict the sum to particles within one periodic box, the volume is then finite, but the first expression will give incorrect results and the second is only correct if we use a special definition of the separation of a pair of interacting particles, i.e. the minimum image separation.

In computer simulations of homogeneous systems, we usually consider an infinite periodic system of particles constructed by replicating a set of particles in the primary simulation box over an infinite lattice. Once the initial positions and momenta have been allocated, each particle that is not in the primary box can be regarded as either a copy of a particle in the primary box, or as an independent particle with the same initial conditions and equations of motion. In principle, each primary particle should maintain its periodic relationship with every image of itself indefinitely.

This is easily seen if we consider two different ways of solving the equations of motion. For simplicity, we will discuss a two-dimensional example, as illustrated in Figure 4.1. Imagine that we start by positioning the particles on a lattice with four identical square unit cells in the primary simulation box. We make the momentum of each image particle identical to that of the particle in the first cell, so that the simulation box contains four identical periodic copies of each particle, one in each cell. One way of solving the equations of motion would be to compute the new positions and momenta by explicitly solving the equations of motion for all particles in a primary simulation box consisting of all four unit cells and applying periodic boundary conditions to particles outside this box. The second is to compute only the positions and momenta of the particles in one of the unit cells (treating this unit cell as the primary simulation box), and apply periodic boundary conditions to particles outside this smaller box. If the solution of the equations of motion were perfect, both methods would give the same particle trajectories. In practice, at long times, the trajectories of the two systems would eventually diverge, due to numerical and discretisation error and Lyapunov instability. Thus, we see that periodic boundary conditions can be regarded as a way of imposing exact periodicity on the solutions of the equations of motion for image particles.

To derive the form of the pressure tensor in an infinite periodic system, we begin with an expression for the momentum density. The momentum density at an arbitrary point $\mathbf{r}$ in a system of N particles replicated with a spatial period L in all directions to infinity is given by

$$\mathbf{J}(\mathbf{r},t) = \rho\mathbf{v}(\mathbf{r},t) = \sum_{\nu=-\infty}^{\infty}\sum_{i=1}^{N} m_i\mathbf{v}_{i\nu}\delta(\mathbf{r}-\mathbf{r}_{i\nu}), \tag{4.77}$$

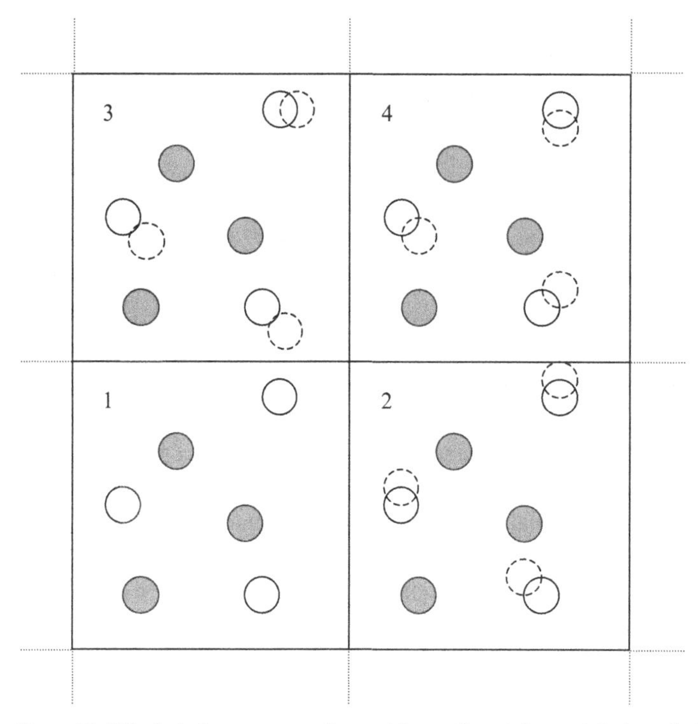

Figure 4.1 Filled circles represent the positions of particles and their periodic images at some time t_0. Unfilled circles represent the positions of the particles at some later time t. If box 1 is treated as the primary simulation box, and we solve the equations of motion numerically to obtain the positions of the particles at time t using periodic boundary conditions to determine the positions of the image particles in boxes 2, 3 and 4, then the positions of the image particles at time t are given by the unfilled circles with full lines. If instead of this, a larger box consisting of boxes 1–4 is regarded as the primary simulation box and the equations of motion for all of the particles are solved numerically, then numerical error will result in a departure of particle trajectories from their exact periodic relationship indicated schematically by the circles with dashed lines. Only an exact analytical solution of the equations of motion would maintain the periodic relationships.

where the sum over the integer vector $\boldsymbol{\nu}$ is an abbreviated notation representing a triple sum over all images of the primary simulation box in each Cartesian direction. If we Fourier transform the momentum density, substitute into the momentum balance equation and proceed with the derivation as before (assuming that only simple pair forces are present), we find

$$\tilde{\mathbf{P}}(\mathbf{k},t) = \sum_{\boldsymbol{\nu}=-\infty}^{\infty}\sum_{i=1}^{N} m_i \mathbf{c}_i \mathbf{c}_i e^{i\mathbf{k}\cdot\mathbf{r}_{i\nu}} - \frac{1}{2}\sum_{\boldsymbol{\nu}=-\infty}^{\infty}\sum_{i=1}^{N}\sum_{\boldsymbol{\mu}=-\infty}^{\infty}\sum_{j=1}^{N} \mathbf{r}_{i\nu j\mu}\mathbf{F}_{i\nu j\mu}\left(\frac{e^{i\mathbf{k}\cdot\mathbf{r}_{i\nu j\mu}}-1}{i\mathbf{k}\cdot\mathbf{r}_{i\nu j\mu}}\right) e^{i\mathbf{k}\cdot\mathbf{r}_{i\nu}}, \tag{4.78}$$

where $\mathbf{F}_{i\nu}$ is the total force on image $\boldsymbol{\nu}$ of particle i and $\mathbf{F}_{i\nu j\mu}$ is the force on $i\boldsymbol{\nu}$ due to $j\boldsymbol{\mu}$ and periodicity of the peculiar momenta has been assumed.

The potential part of the pressure tensor contains a quadruple summation over all pairs of periodic images of every pair of particles, but if the force is short-ranged, we can use the well-known minimum image convention to simplify the expression to a double summation over the number of particles N in one periodic box. This becomes possible when the force is zero for particles of separation greater than or equal to $L/2$, because no more than one image of a particle j can then interact with a given image of particle i. In other words, for each value of $\boldsymbol{\nu}$ the force between the relevant images of particles i and j can only be nonzero for one value of $\boldsymbol{\mu}$, such that $\mathbf{F}_{i\nu j\mu} = \mathbf{F}_{ij}$ for $\boldsymbol{\mu} - \boldsymbol{\nu} = \mathbf{m}_{ij}$ and $\mathbf{F}_{i\nu j\mu} = 0$ otherwise. The separation of the interacting images of particles i and j is called the minimum image distance, and it is defined as

$$\mathbf{r}_{i\nu j\mu} = \mathbf{r}_j + \boldsymbol{\nu}L - \mathbf{r}_i - \boldsymbol{\mu}L = \mathbf{r}_{ij} + \mathbf{m}_{ij}L = \mathbf{d}_{ij}. \tag{4.79}$$

Thus, the summation over $\boldsymbol{\mu}$ reduces to a single term for each value of $\boldsymbol{\nu}$ and the expression for the pressure tensor becomes

$$\tilde{\mathsf{P}}(\mathbf{k}, t) = \sum_{\nu=-\infty}^{\infty} \left[\sum_{i=1}^{N} m_i \mathbf{c}_i \mathbf{c}_i e^{i\mathbf{k}\cdot\mathbf{r}_{i\nu}} - \frac{1}{2} \sum_{i=1}^{N} \sum_{j=1}^{N} \mathbf{d}_{ij} \mathbf{F}_{ij} g(i\mathbf{k} \cdot \mathbf{d}_{ij}) e^{i\mathbf{k}\cdot\mathbf{r}_{i\nu}} \right], \tag{4.80}$$

where

$$g(i\mathbf{k} \cdot \mathbf{d}_{ij}) = \frac{e^{i\mathbf{k}\cdot\mathbf{d}_{ij}} - 1}{i\mathbf{k} \cdot \mathbf{d}_{ij}}. \tag{4.81}$$

If we now recall that $\mathbf{r}_{i\nu} = \mathbf{r}_i + \boldsymbol{\nu}L$ and use the identity

$$\sum_{\nu=-\infty}^{\infty} e^{i\mathbf{k}\cdot\boldsymbol{\nu}L} = \frac{2\pi}{L^3} \sum_{\mathbf{n}=-\infty}^{\infty} \delta\left(\mathbf{k} - \mathbf{k_n}\right), \tag{4.82}$$

where $\mathbf{k}_n = 2\pi\mathbf{n}/L$, and the components of the vector $\mathbf{n}$ are integers, we see that the periodicity of the particle coordinates results in a discrete set of wavevectors. Thus, the usual expression for the wavevector dependent pressure tensor must be replaced by

$$\tilde{\mathsf{P}}(\mathbf{k}, t) = \frac{2\pi}{L^3} \sum_{\mathbf{n}=-\infty}^{\infty} \left[\sum_{i=1}^{N} m_i \mathbf{c}_i \mathbf{c}_i e^{i\mathbf{k}\cdot\mathbf{r}_i} - \frac{1}{2} \sum_{i=1}^{N} \sum_{j=1}^{N} \mathbf{d}_{ij} \mathbf{F}_{ij} g(i\mathbf{k} \cdot \mathbf{d}_{ij}) e^{i\mathbf{k}\cdot\mathbf{r}_i} \right] \delta(\mathbf{k} - \mathbf{k_n}). \tag{4.83}$$

A periodic function with a Fourier series representation given by

$$A(\mathbf{r}, t) = \sum_{\nu=-\infty}^{\infty} \sum_{i=1}^{N} A_i \delta(\mathbf{r} - \mathbf{r}_{i\nu}) = \sum_{\mathbf{n}=-\infty}^{\infty} A_{\mathbf{n}} e^{-i\mathbf{k_n}\cdot\mathbf{r}} \tag{4.84}$$

can be Fourier transformed, giving

$$\tilde{A}(\mathbf{k}, t) = F \left\{ \sum_{\mathbf{n}=-\infty}^{\infty} A_{\mathbf{n}} e^{-i\mathbf{k_n}\cdot\mathbf{r}} \right\} = 2\pi \sum_{\mathbf{n}=-\infty}^{\infty} A_{\mathbf{n}} \delta(\mathbf{k} - \mathbf{k_n}), \tag{4.85}$$

where the Fourier coefficients are given by

$$A_{\mathbf{n}} = \frac{1}{L^3} \sum_i A_i e^{i\mathbf{k_n}\cdot\mathbf{r}_i}. \tag{4.86}$$

Comparing Equations (4.83) and (4.85), we see that the nth Fourier coefficient of the pressure tensor is

$$\mathbf{P_n}(t) = \frac{1}{L^3}\left[\sum_{i=1}^{N} m_i \mathbf{c}_i \mathbf{c}_i e^{i\mathbf{k_n}\cdot\mathbf{r}_i} - \frac{1}{2}\sum_{i=1}^{N}\sum_{j=1}^{N} \mathbf{d}_{ij}\mathbf{F}_{ij} g(i\mathbf{k_n}\cdot\mathbf{d}_{ij}) e^{i\mathbf{k_n}\cdot\mathbf{r}_i}\right]. \tag{4.87}$$

The pressure that we actually compute in molecular dynamics simulations of homogeneous systems with periodic boundary conditions and the minimum image convention is then clearly seen to be the zeroth Fourier coefficient of a Fourier series expansion

$$\mathbf{P}(t) = \mathbf{P}_0(t) = \frac{1}{L^3}\left[\sum_{i=1}^{N} m_i \mathbf{c}_i \mathbf{c}_i - \frac{1}{2}\sum_{i=1}^{N}\sum_{j=1}^{N} \mathbf{d}_{ij}\mathbf{F}_{ij}\right]. \tag{4.88}$$

4.3 Method of Planes Techniques for Inhomogeneous Fluids

Equations (4.54) and (4.76) are exact and can in principle be used for any inhomogeneous fluid. They are the most general useful forms when the flow geometry is not simple. However, a significant simplification can be made for planar surfaces in which inhomogeneity is only apparent in one direction, such as fluids confined between parallel walls in which the flow is tangential to the walls. Common examples of such flow include planar Couette and Poiseuille flows. Not only can one obtain simple yet exact expressions for the pressure tensor, but in fact for all relevant thermodynamic fluxes. This treatment is commonly known as the "method of planes" formalism (MoP) [86, 88, 89], which we now proceed to derive.

4.3.1 Pressure Tensor

Consider a three-dimensional system of interacting particles under flow such that the flow is in the x-direction and any inhomogeneity is only in the y-direction. We first consider only standard pair-potentials, but will reformulate the derivation for the inclusion of 3-body forces in Section 4.3.5. We start first with the momentum balance equation, Equation (4.26), and the microscopic definition of the momentum density, Equation (4.22), which we again write here as:

$$\frac{\partial(\rho\mathbf{v})}{\partial t} = -\nabla\cdot\mathbf{P} - \nabla\cdot(\rho\mathbf{v}\mathbf{v}) + \rho\mathbf{F}^e \tag{4.89}$$

and

$$\mathbf{J}(\mathbf{r},t) = \rho\mathbf{v}(\mathbf{r},t) = \sum_{i=1}^{N} m_i \mathbf{v}_i \delta(\mathbf{r}-\mathbf{r}_i). \tag{4.90}$$

Since the system is homogeneous in both the x and z directions, momentum transport will only be a function of y. As such, we can average over the $x-z$ dimensions. As with the previous derivation of the pressure tensor for homogeneous fluids, it is simpler to work in reciprocal space. Thus, the partial Fourier transform of the microscopic momentum density in the y-direction, averaged over the $x-z$ system dimensions, gives us [86]

$$\tilde{J}_\alpha\left(k_y,t\right)=\frac{1}{A}\sum_i m_i v_{\alpha i} e^{ik_y y_i}, \tag{4.91}$$

where A is the area in the $x-z$ plane and α is any of x, y or z. Note again that the velocities $v_{\alpha i}$ refer to laboratory – not peculiar – velocities. If we now take the partial Fourier transform of the momentum balance equation we obtain

$$\frac{\partial \tilde{J}_\alpha\left(k_y,t\right)}{\partial t}=ik_y\left[\tilde{P}_{y\alpha}\left(k_y,t\right)+F\left\{\rho\left(y\right)v_y\left(y\right)v_\alpha\left(y\right)\right\}\right]+F\left\{\rho\left(y\right)F_\alpha^e\left(y\right)\right\}, \tag{4.92}$$

where v_α and v_y refer to components of the streaming velocity of the fluid and $F\{\cdots\}$ again denotes the Fourier transform. Taking the time-derivative of Equation (4.91) and substituting this into Equation (4.92) leads to the identification of the potential and kinetic contributions to the pressure tensor, $P^\phi_{y\alpha}$ and $P^K_{y\alpha}$, respectively:

$$P^\phi_{y\alpha}\left(k_y,t\right)=\frac{1}{A}\sum_i \frac{F^\phi_{\alpha i}}{ik_y}e^{ik_y y_i} \tag{4.93}$$

and

$$P^K_{y\alpha}\left(k_y,t\right)=\frac{1}{A}\sum_i \frac{m_i v_{\alpha i}}{ik_y}\frac{d}{dt}e^{ik_y y_i}-\widetilde{\rho v_\alpha v_y}. \tag{4.94}$$

Note that the contribution from the external force vanishes, since the time-derivative of the laboratory velocity of particle i on the left-hand side of Equation (4.92) (by substitution of Equation (4.91)) results in forces due to the potential interactions and the external force. This external force term exactly cancels the external force term appearing in the right-hand side of that equation.

To obtain computationally useful expressions for the configurational and kinetic contributions to the pressure tensor, we now take the inverse Fourier transform of these expressions. In the case of the configurational component, we make use of the relation

$$\frac{1}{\pi}\int_{-\infty}^{\infty}\frac{e^{ik_y y}}{ik_y}dk_y=\mathrm{sgn}\left(y\right) \tag{4.95}$$

to give the final expression in real space

$$P^\phi_{y\alpha}\left(y,t\right)=\frac{1}{2A}\sum_i F^\phi_{\alpha i}\mathrm{sgn}\left(y_i-y\right). \tag{4.96}$$

This can be expressed in terms of the forces between pairs of particles, giving

$$P^{\phi}_{y\alpha}(y,t) = \frac{1}{4A}\sum_{ij} F^{\phi}_{\alpha ij}\left[\operatorname{sgn}(y_i - y) - \operatorname{sgn}(y_j - y)\right]$$
$$= \frac{1}{2A}\sum_{ij} F^{\phi}_{\alpha ij}\left[\Theta(y_i - y)\Theta(y - y_j) - \Theta(y_j - y)\Theta(y - y_i)\right]. \quad (4.97)$$

In effect, Equation (4.97) states that the configurational part of the pressure on any particular plane located at y in the fluid is given by the sum over all pair interactions ij, for which particles i and j are located on opposite sides of the plane. Any pairs of particles on the same side of the plane make zero contribution to the pressure on that plane.

This computation is depicted in Algorithm 4.1, in which the first expression in Equation (4.97) is used. There are various different techniques one could use to form the sum over particles i and j, depending on the type of neighbour list used, the kind of processors available (CPU/GPU), and the like, and this is best implemented by individual users, as is the use of either the first or second of the two equivalent expressions in Equation (4.97). However, it is likely to be more computationally efficient to first compute the interaction forces of all pairs ij *before* determining the locations y of the various planes lying between particles i and j. In practice, one would at the start of the program determine the number and location of planes (e.g. by dividing the channel width by the desired separation of the planes (the spatial resolution of the planes)). If one is using a neighbour list, then in the force routine the book-keeping of all pairs of particles ij and their forces are already computed and stored. It is then a relatively straightforward matter to determine which plane or planes any vector $\mathbf{r}_{ij}$ intersects and then pick out the relevant value(s) of $F^{\phi}_{\alpha ij}$ from the array of stored forces. Also notice that Equation (4.97) is an *instantaneous* expression for the $y\alpha$ component of the pressure acting on the plane located at y. The total steady-state pressure component is given by time-averaging over the length of the entire simulation.

The kinetic contribution to the pressure tensor is likewise obtained through inverse Fourier transforming Equation (4.94), which leads to

$$P^{K}_{y\alpha}(y,t) = \frac{1}{2A}\sum_{i} m_i v_{\alpha i}\frac{d}{dt}\operatorname{sgn}(y_i - y) - \rho v_y v_\alpha. \quad (4.98)$$

This expression tells us that kinetic contributions only occur if particles cross planes located at $y_i = y$. If no particle crosses a plane then the time-derivative in Equation (4.98) is zero. Application of the chain rule to the sgn function gives

$$P^{K}_{y\alpha}(y,t) = \frac{1}{2A}\sum_{i} m_i v_{yi} v_{\alpha i}\frac{d}{dy}\operatorname{sgn}(y_i - y) - \rho v_y v_\alpha$$
$$= \frac{1}{A}\sum_{i} m_i v_{yi} v_{\alpha i}\delta(y_i - y) - \rho v_y v_\alpha. \quad (4.99)$$

For a single-component system with equal masses, this expression can be written in terms of the peculiar velocities as

$$P^K_{y\alpha}(y,t) = \frac{1}{A}\sum_i m_i c_{yi} c_{\alpha i} \delta(y_i - y), \tag{4.100}$$

which is just the Irving-Kirkwood kinetic contribution given by Equation (4.49), except that now the kinetic contributions are limited to intersections of particles with planes at $y_i = y$, rather than at a specific point $\mathbf{r}$. As it is currently expressed, this equation is not useful for computer simulation because of the delta function. However, if we denote the times at which particle i crosses a plane located at y as $\{t_{i,m};\ i \in [1,N];\ m \in [1,\infty];\ i,m \in Z\}$, and use the sign of the y-component of the peculiar velocity to indicate whether a particle is crossing a plane from left to right, or vice-versa, then Equation (4.100) can be rewritten as

$$\begin{aligned} P^K_{y\alpha}(y,t) &= \frac{1}{2A}\sum_i m_i c_{\alpha i}\frac{d}{dt}\mathrm{sgn}\left[y_i(t) - y\right] \\ &= \frac{1}{A}\sum_i\sum_m m_i c_{\alpha i}(t_{i,m})\,\delta(t_{i,m} - t)\mathrm{sgn}\left[c_{yi}(t_{i,m})\right]. \end{aligned} \tag{4.101}$$

Taking the time-average over a total averaging time τ gives

$$\left\langle P^K_{y\alpha}(y)\right\rangle = \lim_{\tau\to\infty}\frac{1}{A\tau}\sum_i\sum_m m_i c_{\alpha i}(t_{i,m})\mathrm{sgn}\left[c_{yi}(t_{i,m})\right]. \tag{4.102}$$

This expression states that the kinetic contribution to the $y\alpha$ component of the pressure tensor at a plane located at y is equal to the sum of the peculiar momenta of all particles that intersect the plane, time averaged over the total length of the simulation and divided by the area of the plane to convert an impulsive force into a pressure. As time is discrete in the simulation (with time step Δt) one should use an interpolation method (e.g. Newton-Raphson) to predict the time at which any particle i intersects the plane. From the predicted time one can compute the momentum of the particle at that time by calling the integrator routine. Pseudocode for this algorithm in the case of a single plane is also presented in Algorithm 4.1.

Equations (4.97) and (4.102) are easily programmable and useful for finding the $y\alpha$ elements of the pressure tensor for an inhomogeneous fluid with planar geometry, such as fluids at interfaces or in narrow pores. They are valid both at equilibrium and nonequilibrium. In the case of the latter, they are applicable when the driving field is tangential to the surface of the planes and can readily be derived for cylindrical rather than planar geometry [90], which could be useful for the case of flow in carbon nanotubes, for example. They have also been derived and used for systems of spherical geometry [91]. We will demonstrate practical applications of these expressions in Chapter 9. Note also that the method of planes derivation leads naturally to the concept of pressure being a "force across a unit area" without the need for any heuristic assumptions [86].

Heyes *et al.* [92] have since shown how the method of planes can be derived from the limiting case of the so-called volume averaging method [93, 94], in which the pressureis

Algorithm 4.1 Pseudocode for the method of planes pressure tensor

$P^{\phi}_{y\alpha}(y,t) \leftarrow 0$;
$P^{K}_{y\alpha}(y,t) \leftarrow 0$;
$\left\langle P^{\phi}_{y\alpha}(y)\right\rangle \leftarrow 0$;
$\left\langle P^{K}_{y\alpha}(y)\right\rangle \leftarrow 0$;
$\left\langle P_{y\alpha}(y)\right\rangle \leftarrow 0$;
for $i, j = 1, N$ **do** {configurational components}
  if $(\{y_i < y < y_j\} || \{y_j < y < y_i\})$ **then**
    $s_{ij}(y) = \mathrm{sgn}(y_i - y) - \mathrm{sgn}(y_j - y)$;
    $P^{\phi}_{y\alpha}(y,t) \leftarrow F^{\phi}_{\alpha ij} s_{ij}(y)$;
  end if
end for
for $i = 1, N$ **do** {kinetic components}
  call interpolation routine;
  $P^{K}_{y\alpha}(y,t) \leftarrow m_i c_{\alpha i} \mathrm{sgn}(c_{yi})$; {sum over all planes y crossed by particles i}
end for
time average at end of simulation
$\left\langle P^{\phi}_{y\alpha}(y)\right\rangle \leftarrow \sum_{t \le \tau} P^{\phi}_{y\alpha}(y,t)/4AN_{av}$; {$N_{av}$ = number of averaging time steps}
$\left\langle P^{K}_{y\alpha}(y)\right\rangle \leftarrow \sum_{t_{i,m} \le \tau} P^{K}_{y\alpha}(y,t)/A\tau$;
$\left\langle P_{y\alpha}(y)\right\rangle \leftarrow \left\langle P^{\phi}_{y\alpha}(y)\right\rangle + \left\langle P^{K}_{y\alpha}(y)\right\rangle$;

computed in bins of finite width suitably scaled to include only that fraction of $\mathbf{r}_{ij}$ that lies within a bin of volume $\Omega < V$, where V is the overall system volume. By taking the limit as a bin of width $\Delta y \to 0$, the method of planes is recovered. The advantage of the volume averaging method is that it allows for the computation of any component of the pressure tensor, not just $y\alpha$ components.

4.3.2 Heat Flux Vector

The expression for the heat flux vector for homogeneous fluids was previously derived in Section 4.2.2 and expressed by Equation (4.75). The method of planes procedure to compute the heat flux vector at planes normal to the direction of inhomogeneity is analogous to the derivation of the pressure tensor at planes [88]. We again start with the energy balance equation and microscopic definition of the total energy density, given by Equations (4.27) and (4.23) respectively but will omit the potential energy due to conservative external forces as it has no effect on the final result. We write the energy balance as

$$\frac{\partial\left[\rho e(\mathbf{r},t)\right]}{\partial t} = -\nabla\cdot\left[\mathbf{J}_q(\mathbf{r},t) + \rho(\mathbf{r},t)\, e(\mathbf{r},t)\,\mathbf{v}(\mathbf{r},t) + \mathsf{P}(\mathbf{r},t)\cdot\mathbf{v}(\mathbf{r},t)\right]. \quad (4.103)$$

Taking the Fourier transform of this expression gives

$$\frac{\partial}{\partial t}\widetilde{\rho e}(\mathbf{k},t) = i\mathbf{k}\cdot\left[\widetilde{\mathbf{J}}_q(\mathbf{k},t) + F\{\rho e\mathbf{v}\} + F\{\mathbf{P}\cdot\mathbf{v}\}\right], \tag{4.104}$$

where the total energy density is given as Equation (4.23) and the total energy of particle i is defined by Equation (4.24). Note we again assume that only pairwise interactions are present (3-body terms are considered in Section 4.3.5). The time-derivative of the Fourier transform of the energy density (i.e. the left-hand side of Equation (4.104)) is

$$\frac{\partial}{\partial t}\widetilde{\rho e}(\mathbf{k},t) = \frac{\partial}{\partial t}\sum_i e_i e^{i\mathbf{k}\cdot\mathbf{r}_i} = i\mathbf{k}\cdot\sum_i \mathbf{v}_i e_i e^{i\mathbf{k}\cdot\mathbf{r}_i} + \frac{1}{2}\sum_{ij}\mathbf{v}_i\cdot\mathbf{F}_{ij}^{\phi}\left(e^{i\mathbf{k}\cdot\mathbf{r}_i} - e^{i\mathbf{k}\cdot\mathbf{r}_j}\right). \tag{4.105}$$

The second term on the right-hand side of Equation (4.104) is

$$F\{\rho e\mathbf{v}(\mathbf{k},t)\} = \sum_i e_i\mathbf{v}(\mathbf{r}_i,t)\,e^{i\mathbf{k}\cdot\mathbf{r}_i}, \tag{4.106}$$

where we have used Equation (4.23) and the sifting property of the delta function. This term is the convective component of the energy transport. If we now substitute Equations (4.105) and (4.106) into Equation (4.104) and take the zero-wavevector limit in the x and z directions (i.e. assume planar symmetry and average over the x-z plane), we obtain

$$\begin{aligned} ik_y J_{qy}(k_y,t) &= \frac{ik_y}{A}\sum_i \left(v_{yi} - v_y(y_i)\right)e_i e^{ik_y y_i} \\ &\quad + \frac{1}{2A}\sum_{ij}\mathbf{v}_i\cdot\mathbf{F}_{ij}^{\phi}\left(e^{ik_y y_i} - e^{ik_y y_j}\right) - ik_y\left[F\{\mathbf{P}\cdot\mathbf{v}\}\right]_y. \end{aligned} \tag{4.107}$$

Dividing by ik_y and taking the inverse Fourier transform gives

$$\begin{aligned} J_{qy}(y,t) &= \frac{1}{A}\sum_i \left(v_{yi} - v_y(y_i)\right)e_i\delta(y-y_i) \\ &\quad - \frac{1}{4A}\sum_{ij}\mathbf{v}_i\cdot\mathbf{F}_{ij}^{\phi}\left[\mathrm{sgn}(y-y_i) - \mathrm{sgn}(y-y_j)\right] - [\mathbf{P}\cdot\mathbf{v}]_y. \end{aligned} \tag{4.108}$$

In order to obtain the required expression for the heat flux vector, we need to substitute the $y\alpha$ element of the pressure tensor into Equation (4.108). Substituting Equations (4.97) and (4.100) into the above expression, and splitting the contributions into kinetic and potential terms respectively, gives

$$J_{qy}(y,t) = J_{qy}^{K}(y,t) + J_{qy}^{\phi}(y,t), \tag{4.109}$$

where

$$J_{qy}^{K}(y,t) = \frac{1}{A}\sum_i \left(v_{yi} - v_y(y_i)\right)u_i\delta(y-y_i) \tag{4.110}$$

and

$$J^{\phi}_{qy}(y,t) = -\frac{1}{4A}\sum_{ij}\left[\mathbf{v}_i - \mathbf{v}(y)\right]\cdot\mathbf{F}^{\phi}_{ij}\left[\operatorname{sgn}(y-y_i) - \operatorname{sgn}\left(y-y_j\right)\right]$$
$$= \frac{1}{4A}\sum_{ij}\left[\mathbf{v}_i - \mathbf{v}(y)\right]\cdot\mathbf{F}^{\phi}_{ij}\left[\operatorname{sgn}(y_i-y) - \operatorname{sgn}\left(y_j-y\right)\right]. \quad (4.111)$$

In this expression u_i is the total internal energy of particle i, defined such that the streaming kinetic energy is now subtracted out of the energy density, i.e.

$$u_i = \frac{1}{2}m\mathbf{c}_i^2 + \frac{1}{2}\sum_j \phi_{ij}. \quad (4.112)$$

In deriving these expressions, we have also used the relation

$$\mathbf{v}^2(\mathbf{r},t)\sum_i m_i(\mathbf{v}_i - \mathbf{v})\,\delta(\mathbf{r}-\mathbf{r}_i) = \mathbf{v}^2(\mathbf{r},t)\left[\rho\mathbf{v}(\mathbf{r},t) - \rho\mathbf{v}(\mathbf{r},t)\right] = 0. \quad (4.113)$$

By using a similar procedure for the kinetic contribution of the pressure tensor, it is straightforward to show that the time averaged kinetic term for the heat flux vector can be written as

$$\left\langle J^{K}_{qy}(y)\right\rangle = \lim_{\tau\to\infty}\frac{1}{A\tau}\sum_i\sum_m u_i \operatorname{sgn}\left[c_{yi}\left(t_{i,m}\right)\right]. \quad (4.114)$$

The configurational and kinetic contributions to the heat flux vector are depicted in pseudocode in Algorithm 4.2 for a single plane. Note here that it is assumed the streaming velocity is computed at the position y. This can be done in various ways, such as the method-of-planes (see Section 4.3.3 below) or by assuming a functional form for the streaming velocity (e.g. quadratic for Poiseuille flow) and finding the coefficients of the velocity profile by least-squares fitting (as done in references [86, 88]), or even at the post-processing stage as long as all required information is stored. Note also that some variables are already defined in Algorithm 4.1.

It is important to understand the interpretation of these expressions for the kinetic and potential contributions to the heat flux vector. Equation (4.110) states that when a particle crosses a plane at $y_i = y$ an instantaneous kinetic energy flux will be transferred through it. Similarly, Equation (4.111) states that at any time t if there are two particles i and j, such that a straight line drawn through their centres intersects the plane located at y, then a flow of energy can take place across the plane.

Furthermore, it is crucial to understand that the term $\mathbf{v}(y)$ is *not* to be confused with the streaming velocity of particle i, $\mathbf{v}(y_i)$. It is rather the streaming velocity of the fluid evaluated at the plane located at y. In reference [88] all velocities $\mathbf{v}_i - \mathbf{v}(y)$ were termed "plane peculiar velocities", to distinguish them from the usual peculiar velocities defined as $\mathbf{c}_i \equiv \mathbf{v}_i - \mathbf{v}(y_i)$. The plane peculiar velocities are velocities of particles relative to the streaming velocity evaluated at a plane. For the kinetic contribution of the heat flux vector we find $\mathbf{v}(y) = \mathbf{v}(y_i)$ due to the delta function term in Equation (4.110), but this is not the case for the potential contribution. In fact, this

Algorithm 4.2 Pseudocode for method of planes heat flux vector

$J_{qy}^{\phi}(y,t) \leftarrow 0;$
$J_{qy}^{K}(y,t) \leftarrow 0;$
$\left\langle J_{qy}^{\phi}(y)\right\rangle \leftarrow 0;$
$\left\langle J_{qy}^{K}(y)\right\rangle \leftarrow 0;$
$\left\langle J_{qy}(y)\right\rangle \leftarrow 0;$
for $i, j = 1, N$ **do** {configurational components}
  if $(\{y_i < y < y_j\} || \{y_j < y < y_i\})$ **then**
    $J_{qy}^{\phi}(y,t) \leftarrow [\mathbf{v}_i - \mathbf{v}(y)] \cdot \mathbf{F}_{ij}^{\phi} s_{ij}(y);$
  end if
end for
for $i = 1, N$ **do** {kinetic components}
  $J_{qy}^{K}(y,t) \leftarrow u_i \mathrm{sgn}\left[c_{yi}(t_{i,m})\right];$ {sum over all planes y crossed by particles i}
end for
time average at end of simulation
$\left\langle J_{qy}^{\phi}(y)\right\rangle \leftarrow \sum_{t \le \tau} J_{qy}^{\phi}(y,t)/4AN_{av};$
$\left\langle J_{qy}^{K}(y)\right\rangle \leftarrow \sum_{t_{i,m} \le \tau} J_{qy}^{K}(y,t)/A\tau;$
$\left\langle J_{qy}(y)\right\rangle \leftarrow \left\langle J_{qy}^{\phi}(y)\right\rangle + \left\langle J_{qy}^{K}(y)\right\rangle$

would be a very important point in the evaluation of heat fluxes in highly confined fluids where curvature in the streaming velocity profile is significant over the range of the intermolecular potential energy interaction. In such situations the plane peculiar velocities could be significantly different to the peculiar velocities, and so use of the latter in Equation (4.111) would lead to wrong values for the heat flux vector at y, where y is close to the walls (and the streaming velocity varies substantially over small distances). Use of peculiar velocities, rather than plane peculiar velocities when the curvature in $\mathbf{v}$ is small, would probably not generate significant error when results are time-averaged. Nevertheless, as a matter of principle, the correct plane peculiar velocities should *always* be used in these expressions. Examples of the use of these expressions will be given in Chapter 9.

4.3.3 Generalised Form: Density, Velocity and Temperature Profiles

One can use the techniques developed above to compute *any* of the densities for an inhomogeneous system. Some quantities, apart from the pressure tensor and heat flux vector described above, which are useful for nonequilibrium conditions are the hydrodynamic densities of mass, momentum and energy. From these quantities one is able to extract the density, streaming velocity and the temperature of a nonequilbrium fluid as a function of the dimension of inhomogeneity (in our case, the y dimension) [89]. Each of these quantities is essential knowledge for an NEMD simulation. In particular, the streaming velocity is often needed to be computed on the fly to determine the temperature or elements of the pressure tensor, whereas the temperature or stress profiles themselves can give important information as to whether the system has reached steady-state, for example.

In the case of the mass density, defined by Equation (4.21), we again average over the x-z plane and time interval τ to obtain the relevant expression for the y-dependent density:

$$\begin{aligned}\rho(y) &= \frac{1}{\tau A}\int_0^{\tau} ds \int_{-L_x/2}^{L_x/2} dx \int_{-L_z/2}^{L_z/2} dz\, \rho(\mathbf{r}, s) \\ &= \frac{1}{\tau A}\int_0^{\tau} ds \sum_i m_i \int_{-L_x/2}^{L_x/2} \delta\left[x - x_i(s)\right] dx \int_{-L_z/2}^{L_z/2} \delta\left[z - z_i(s)\right] dz \times \delta\left[y - y_i(s)\right] \\ &= \frac{1}{\tau A}\int_0^{\tau} ds \sum_i m_i \delta\left[y - y_i(s)\right] \end{aligned} \tag{4.115}$$

where we assume here that the simulation box is centred at the origin with lengths L_x and L_z in the x and z dimensions, respectively. We can now write the remaining delta function in y as a sum of delta functions in time,

$$\delta\left[y - y_i(t)\right] = \sum_{\alpha} \frac{\delta\left(t - t_{i\alpha}\right)}{\left|\dot{y}_i\left(t_{i\alpha}\right)\right|} \tag{4.116}$$

where $t_{i\alpha}$ are times when particle i intersects the plane located at $y_i = y$ and $\dot{y}_i$ is of course just the y component of the velocity of particle i. Substituting Equation (4.116) into Equation (4.115) and evaluating the integral over time gives the time-averaged value of the fluid density at a plane located at y:

$$\rho(y) = \frac{1}{\tau A}\sum_i \sum_{\alpha} \frac{m_i}{\left|\dot{y}_i\left(t_{i\alpha}\right)\right|}. \tag{4.117}$$

Similarly, from the microscopic definition of the momentum density given in Equation (4.22), it is straightforward to show that the momentum density at a plane (assuming flow is in the x-direction) is

$$J_x(y) = \frac{1}{\tau A}\sum_i \sum_{\alpha} \frac{m_i v_{xi}}{\left|\dot{y}_i\left(t_{i\alpha}\right)\right|}. \tag{4.118}$$

The streaming velocity is then computed as $J_x(y)/\rho(y)$. Finally, the kinetic temperature, defined as $T(\mathbf{r}, t) = 2K(\mathbf{r}, t)/[3k_B n(\mathbf{r}, t)]$, where $n(\mathbf{r}, t)$ is the number density, can be evaluated from the *peculiar* kinetic energy density, $K(\mathbf{r}, t)$. At a plane located at y this peculiar kinetic energy density can be readily shown to be

$$K(y) = \frac{1}{\tau A}\sum_i \sum_{\alpha} \frac{m_i\left[v_{xi} - v_x(y_i)\right]^2}{2\left|\dot{y}_i\left(t_{i\alpha}\right)\right|}. \tag{4.119}$$

Note that the number density is, for a single-component system, just the mass density given by Equation (4.117) divided by the particle mass.

Examples of these quantities for a confined fluid undergoing planar Poiseuille flow will be given in Chapter 9.

4.3.4 Diffusion

The method of planes can also be used to compute the diffusion coefficient. In particular, it can be useful to obtain the position-dependent diffusion coefficient for confined fluids in the direction perpendicular to the surface of the wall, $D_\perp$. In the case of planar walls, Hartkamp *et al.* [95] demonstrated how the position-dependent self-diffusion coefficient can be computed. By labeling half the atoms for a simple one-component atomic fluid "red" and the other half "blue", they obtained an expression for the diffusive mass flux of a particular species crossing planes as

$$J_{D_\perp}(y) = \lim_{\tau\to\infty}\frac{1}{A\tau}\sum_{i}^{N_c}\sum_{m} m_i \operatorname{sgn}\left[c_{yi}\left(t_{i,m}\right)\right], \tag{4.120}$$

where all terms have the same meaning as defined previously, except that here N_c refers to the number of labelled (coloured) species (e.g. "blue" particles, etc.). The density of labelled species (the so-called "colour" density, which we define here as ρ_c) can be obtained as in Equation (4.117). Thus the diffusion coefficient in the direction perpendicular to the planar walls is obtained as

$$D_\perp(y) = -\frac{J_{D_\perp}(y)}{\partial\rho_c(y)/\partial y}. \tag{4.121}$$

The method can of course be generalised for cylindrical geometry to be useful for diffusion in nanotubes. It could also presumably be further generalised to compute mutual diffusion coefficients in confined geometries.

An important limitation must be stressed here. This method computes only a *local* diffusion coefficient. As such it is not truly representative of diffusive transport for very highly confined fluids, such as nanochannels separated by only a few atomic diameters in width. For such systems, transport becomes nonlocal in nature and generalised hydrodynamics approaches must be used to compute the nonlocal transport properties. This is discussed further in Chapter 11. Nevertheless the method could be useful for transport in channels of width greater than around five atomic diameters [96].

4.3.5 Inclusion of Many-body Interaction Terms

The derivation of method of planes expressions for the pressure tensor and heat flux vector can be extended to include n-body interactions, rather than just pair potential interactions. It has been shown that in the case of noble-gas fluids, 3-body terms can be significant for the pressure, but less so for viscosity [97–99]. Similarly, the heat flux vector was shown not to have a large contribution from 3-body terms for a single-component fluid of argon [100]. This may not be the case for more realistic molecular fluids, particularly liquid metals where n-body terms become very important. In such situations, MoP expressions could be very useful to compute the pressure, shear stress (hence viscosity) and heat flux in highly confined fluids, though to our knowledge this has not yet been explored. In what follows we derive the expressions for the pressure tensor and heat flux vector under the influence of 3-body forces, as presented in reference [100]. We will also extend the pressure tensor derivation to arbitrary n-body terms. While it is certainly possible to extend the heat flux vector derivation to arbitrary order, for the sake of brevity we limit ourselves to just 3-body terms.

4.3.5.1 Pressure Tensor

Consider the arrangement of a triplet of interacting particles labeled by 1, 2 and 3, as depicted in Figure 4.2. A plane located at $y = y_0$ comes between particles 1 and 2 and 1 and 3. Therefore, contributions to the pressure at the plane will come from all these particles along the vectors $\mathbf{r}_{12}$ and $\mathbf{r}_{13}$. As particles 2 and 3 are on the same side of the plane, a line drawn through their centres does not intersect it and so no contribution along the vector $\mathbf{r}_{23}$ is included. We define the 2- and 3-body force contributions to the total interatomic force as $\mathbf{F}_i^{(2)}$ and $\mathbf{F}_i^{(3)}$, respectively. We also define $\mathbf{F}_{ij}^{(3)}$ to be the total 3-body force on particle i due to particle j. If the 2- and 3-body potentials are denoted as $\phi_{ij}^{(2)} = \phi^{(2)}(\mathbf{r}_i, \mathbf{r}_j)$ and $\phi_{ijk}^{(3)} = \phi^{(3)}(\mathbf{r}_i, \mathbf{r}_j, \mathbf{r}_k)$ respectively, then

$$\mathbf{F}_i^{(2)} \equiv \sum_j \mathbf{F}_{ij}^{(2)} = -\sum_j \left(\frac{\partial \phi_{ij}^{(2)}}{\partial \mathbf{r}_i} \right) \tag{4.122}$$

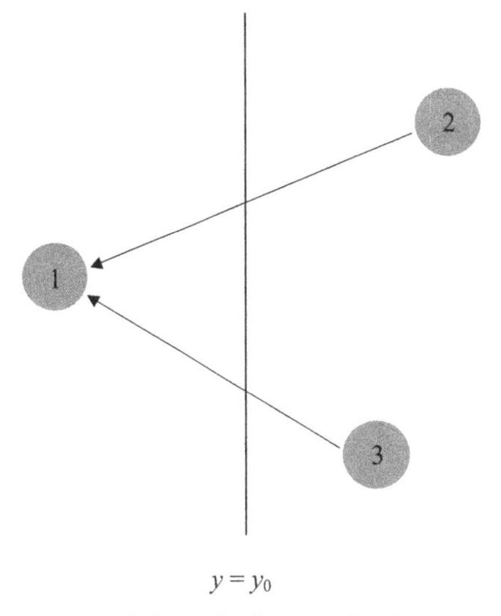

Figure 4.2 Schematic diagram showing potential interaction terms between a triplet of particles across a plane located at $y = y_0$. Reprinted with permission from reference [100]. Copyright 2004 by the American Physical Society.

and

$$\mathbf{F}_i^{(3)} \equiv \sum_{jk} \left(\mathbf{F}_{ij}^{(3)} + \mathbf{F}_{ik}^{(3)} \right) = - \left[\sum_{jk} \left\{ \left(\frac{\partial \phi_{ijk}^{(3)}}{\partial \mathbf{r}_{ij}} \right) + \left(\frac{\partial \phi_{ijk}^{(3)}}{\partial \mathbf{r}_{ik}} \right) \right\} \right] \quad (\mathbf{r}_{ij} \equiv \mathbf{r}_i - \mathbf{r}_j) \tag{4.123}$$

where

$$\mathbf{F}_{ij}^{(3)} \equiv - \frac{\partial \phi_{ijk}^{(3)}}{\partial \mathbf{r}_{ij}}. \tag{4.124}$$

In Equation (4.96) we showed that the instantaneous configurational part of the pressure tensor is given as

$$P_{y\alpha}^{\phi}(y) = \frac{1}{2A} \sum_i F_{\alpha i}^{\phi} \operatorname{sgn}(y_i - y).$$

This derivation was based only on the microscopic definition of the momentum density and the momentum balance equations and is independent of the specific type of interaction potential. We can thus split this expression into contributions from 2- and 3-body interaction forces:

$$\begin{aligned} P_{y\alpha}^{\phi}(y) &= \frac{1}{2A} \sum_i F_{\alpha i}^{\phi} \operatorname{sgn}(y_i - y) \\ &= \frac{1}{2A} \sum_i (F_{\alpha i}^{(2)} + F_{\alpha i}^{(3)}) \operatorname{sgn}(y_i - y) \\ &= P_{y\alpha}^{(2)}(y) + P_{y\alpha}^{(3)}(y). \end{aligned} \tag{4.125}$$

The 3-body contribution can be expressed as

$$\begin{aligned} P_{y\alpha}^{(3)}(y) &= \frac{1}{2A} \sum_i F_{\alpha i}^{(3)} \operatorname{sgn}(y_i - y) \\ &= \frac{1}{6A} \left[\sum_i F_{\alpha i}^{(3)} \operatorname{sgn}(y_i - y) + \sum_j F_{\alpha j}^{(3)} \operatorname{sgn}(y_j - y) + \sum_k F_{\alpha k}^{(3)} \operatorname{sgn}(y_k - y) \right]. \end{aligned} \tag{4.126}$$

Substituting Equation (4.123) into Equation (4.126) yields the expression for the 3-body contribution to the potential component of the pressure tensor:

$$\begin{aligned} P_{y\alpha}^{(3)}(y) = \frac{1}{6A} \Bigg\{ & \sum_{ij} F_{\alpha ij}^{(3)} [\operatorname{sgn}(y_i - y) - \operatorname{sgn}(y_j - y)] \\ & + \sum_{ik} F_{\alpha ik}^{(3)} [\operatorname{sgn}(y_i - y) - \operatorname{sgn}(y_k - y)] \\ & + \sum_{jk} F_{\alpha jk}^{(3)} [\operatorname{sgn}(y_j - y) - \operatorname{sgn}(y_k - y)] \Bigg\}. \end{aligned} \tag{4.127}$$

This can in turn be expressed in terms of Heaviside step functions as

$$P^{(3)}_{y\alpha}(y) = \frac{1}{3A}\left\{\sum_{ij} F^{(3)}_{\alpha ij}[\Theta(y_i - y)\Theta(y - y_j) - \Theta(y_j - y)\Theta(y - y_i)]\right.$$
$$+\sum_{ik} F^{(3)}_{\alpha ik}[\Theta(y_i - y)\Theta(y - y_k) - \Theta(y_k - y)\Theta(y - y_i)]$$
$$\left.+\sum_{jk} F^{(3)}_{\alpha jk}[\Theta(y_j - y)\Theta(y - y_k) - \Theta(y_k - y)\Theta(y - y_j)]\right\}. \quad (4.128)$$

The 2-body and kinetic contributions to the pressure tensor are just given by Equations (4.97) and (4.102), respectively, and the sum of all these terms gives the total pressure tensor.

Calculations showing the relative contributions of the 2- and 3-body terms to the configurational part of the isotropic pressure and shear stress are given in reference [100], for a fluid interacting via a Barker-Fisher-Watts 2-body potential [101] plus 3-body Axilrod-Teller potential [102] for a fluid of argon atoms confined by atomistic walls. The yy component of the pressure tensor (i.e. the component normal to the surface of the walls) shows a significant dependence on 3-body terms, whereas the shear stress (hence viscosity) shows very weak dependence. This is in agreement with previous studies for homogeneous fluids showing similar behaviour [99]. However, as already pointed out, the dependence on 3- and higher-body terms is likely to become far more important for heavier atomic nuclei, such as liquid metals, though to our knowledge this has not yet been demonstrated for highly confined fluids.

One can generalise the above derivation to the case of n-body forces, leading to the following expression:

$$P^{\phi(n)}_{y\alpha}(y) = \frac{1}{2nA}\left\{\sum_{ij} F^{\phi(n)}_{\alpha ij}\left[\operatorname{sgn}(y_i - y) - \operatorname{sgn}(y_j - y)\right]\right.$$
$$+\sum_{ik} F^{\phi(n)}_{\alpha ik}\left[\operatorname{sgn}(y_i - y) - \operatorname{sgn}(y_k - y)\right]$$
$$+\sum_{il} F^{\phi(n)}_{\alpha il}\left[\operatorname{sgn}(y_i - y) - \operatorname{sgn}(y_l - y)\right] + \cdots$$
$$+\sum_{jk} F^{\phi(n)}_{\alpha jk}\left[\operatorname{sgn}(y_j - y) - \operatorname{sgn}(y_k - y)\right]$$
$$+\sum_{jl} F^{\phi(n)}_{\alpha jl}\left[\operatorname{sgn}(y_j - y) - \operatorname{sgn}(y_l - y)\right]$$
$$\left.+\sum_{jm} F^{\phi(n)}_{\alpha jm}\left[\operatorname{sgn}(y_j - y) - \operatorname{sgn}(y_m - y)\right] + \cdots + \cdots\right\} \quad (4.129)$$

where we note that there are a total of $\frac{1}{2}n(n-1)$ terms in the sum.

4.3.5.2 Heat Flux Vector

We start again with the same fundamental balance equation transformed in Fourier space, namely the energy balance equation given by Equation (4.104) and the total energy for particle i, now given as

$$e_i = \frac{\mathbf{p}_i^2}{2m_i} + \frac{1}{2}\sum_j \phi_{ij}^{(2)} + \frac{1}{3}\sum_{jk} \phi_{ijk}^{(3)}, \tag{4.130}$$

where again the $\mathbf{p}_i$ are defined as laboratory momenta. From Section 4.3.2 we note that $\widetilde{\rho e \mathbf{v}}(\mathbf{k}, t) = \sum_i e_i \mathbf{v}(\mathbf{r}_i, t)\, e^{i\mathbf{k}\cdot\mathbf{r}_i}$ and $\widetilde{\rho e}(\mathbf{k}, t) = \sum_i e_i e^{i\mathbf{k}\cdot\mathbf{r}_i}$, so the time-derivative of the total energy density can be shown to be [100]

$$\begin{aligned}\frac{\partial \widetilde{\rho e}(\mathbf{k}, t)}{\partial t} &= i\mathbf{k} \cdot \sum_i \mathbf{v}_i e_i e^{i\mathbf{k}\cdot\mathbf{r}_i} + \frac{1}{2}\sum_{ij} \mathbf{v}_i \cdot \mathbf{F}_{ij}^{(2)} \left(e^{i\mathbf{k}\cdot\mathbf{r}_i} - e^{i\mathbf{k}\cdot\mathbf{r}_j}\right) + \sum_i \mathbf{v}_i \cdot \mathbf{F}_i^{(3)} e^{i\mathbf{k}\cdot\mathbf{r}_i} \\ &\quad + \frac{1}{3}\sum_{ijk}\left(\dot{\mathbf{r}}_i \cdot \frac{\partial \phi_{ijk}^{(3)}}{\partial \mathbf{r}_i} + \dot{\mathbf{r}}_j \cdot \frac{\partial \phi_{ijk}^{(3)}}{\partial \mathbf{r}_j} + \dot{\mathbf{r}}_k \cdot \frac{\partial \phi_{ijk}^{(3)}}{\partial \mathbf{r}_k}\right) e^{i\mathbf{k}\cdot\mathbf{r}_i}.\end{aligned} \tag{4.131}$$

We can simplify this expression by symmetrising both terms involving 3-body forces. The first term can be expressed as:

$$\begin{aligned}\sum_i \mathbf{v}_i \cdot \mathbf{F}_i^{(3)} e^{i\mathbf{k}\cdot\mathbf{r}_i} &= \frac{1}{3}\left[\sum_i \mathbf{v}_i \cdot \mathbf{F}_i^{(3)} e^{i\mathbf{k}\cdot\mathbf{r}_i} + \sum_j \mathbf{v}_j \cdot \mathbf{F}_j^{(3)} e^{i\mathbf{k}\cdot\mathbf{r}_j} + \sum_k \mathbf{v}_k \cdot \mathbf{F}_k^{(3)} e^{i\mathbf{k}\cdot\mathbf{r}_k}\right] \\ &= \frac{1}{3}\left[\sum_{ijk} \mathbf{v}_i \cdot \left(\mathbf{F}_{ij}^{(3)} + \mathbf{F}_{ik}^{(3)}\right) e^{i\mathbf{k}\cdot\mathbf{r}_i} + \sum_{ijk} \mathbf{v}_j \cdot \left(\mathbf{F}_{ji}^{(3)} + \mathbf{F}_{jk}^{(3)}\right) e^{i\mathbf{k}\cdot\mathbf{r}_j}\right. \\ &\quad \left. + \sum_{ijk} \mathbf{v}_k \cdot \left(\mathbf{F}_{ki}^{(3)} + \mathbf{F}_{kj}^{(3)}\right) e^{i\mathbf{k}\cdot\mathbf{r}_k}\right]\end{aligned} \tag{4.132}$$

whereas the second term can be expanded as

$$\begin{aligned}&\frac{1}{3}\sum_{ijk}\left(\dot{\mathbf{r}}_i \cdot \frac{\partial \phi_{ijk}^{(3)}}{\partial \mathbf{r}_i} + \dot{\mathbf{r}}_j \cdot \frac{\partial \phi_{ijk}^{(3)}}{\partial \mathbf{r}_j} + \dot{\mathbf{r}}_k \cdot \frac{\partial \phi_{ijk}^{(3)}}{\partial \mathbf{r}_k}\right) e^{i\mathbf{k}\cdot\mathbf{r}_i} \\ &= \frac{1}{3}\sum_{ijk}\left[\mathbf{v}_i \cdot \left(\frac{\partial \phi_{ijk}^{(3)}}{\partial \mathbf{r}_{ij}} + \frac{\partial \phi_{ijk}^{(3)}}{\partial \mathbf{r}_{ik}}\right) + \mathbf{v}_j \cdot \left(\frac{\partial \phi_{ijk}^{(3)}}{\partial \mathbf{r}_{ji}} + \frac{\partial \phi_{ijk}^{(3)}}{\partial \mathbf{r}_{jk}}\right)\right. \\ &\quad \left. + \mathbf{v}_k \cdot \left(\frac{\partial \phi_{ijk}^{(3)}}{\partial \mathbf{r}_{ki}} + \frac{\partial \phi_{ijk}^{(3)}}{\partial \mathbf{r}_{kj}}\right)\right] e^{i\mathbf{k}\cdot\mathbf{r}_i} \\ &= -\frac{1}{3}\sum_{ijk}\left(\mathbf{v}_i \cdot \mathbf{F}_{ij}^{(3)} + \mathbf{v}_i \cdot \mathbf{F}_{ik}^{(3)} - \mathbf{v}_j \cdot \mathbf{F}_{ij}^{(3)} + \mathbf{v}_j \cdot \mathbf{F}_{jk}^{(3)} - \mathbf{v}_k \cdot \mathbf{F}_{ik}^{(3)} - \mathbf{v}_k \cdot \mathbf{F}_{jk}^{(3)}\right) e^{i\mathbf{k}\cdot\mathbf{r}_i}.\end{aligned} \tag{4.133}$$

Defining Σ to be the sum of Equations (4.132) and (4.133) and permuting the triplet indices such that all velocities are in terms of the index i, we obtain

$$\Sigma = \frac{1}{3}\sum_{ik} \mathbf{v}_i \cdot \mathbf{F}_i^{(3)} \left(e^{i\mathbf{k}\cdot\mathbf{r}_i} - e^{i\mathbf{k}\cdot\mathbf{r}_k}\right) + \frac{1}{3}\sum_{ij} \mathbf{v}_i \cdot \mathbf{F}_i^{(3)} \left(e^{i\mathbf{k}\cdot\mathbf{r}_i} - e^{i\mathbf{k}\cdot\mathbf{r}_j}\right). \quad (4.134)$$

Substituting Equation (4.134) back into Equation (4.131) gives

$$\frac{\partial \widetilde{\rho e}(\mathbf{k},t)}{\partial t} = i\mathbf{k}\cdot\sum_i \mathbf{v}_i e_i e^{i\mathbf{k}\cdot\mathbf{r}_i} + \frac{1}{2}\sum_{ij} \mathbf{v}_i \cdot \mathbf{F}_{ij}^{(2)} \left(e^{i\mathbf{k}\cdot\mathbf{r}_i} - e^{i\mathbf{k}\cdot\mathbf{r}_j}\right)$$
$$+ \frac{1}{3}\sum_{ik} \mathbf{v}_i \cdot \mathbf{F}_i^{(3)} \left(e^{i\mathbf{k}\cdot\mathbf{r}_i} - e^{i\mathbf{k}\cdot\mathbf{r}_k}\right) + \frac{1}{3}\sum_{ij} \mathbf{v}_i \cdot \mathbf{F}_i^{(3)} \left(e^{i\mathbf{k}\cdot\mathbf{r}_i} - e^{i\mathbf{k}\cdot\mathbf{r}_j}\right). \quad (4.135)$$

We now substitute this expression back into the Fourier transformed energy balance equation (4.104), integrate over the x-z plane and inverse transform to give the kinetic and potential contributions to the total heat flux vector

$$J_{qy}(y,t) = J_{qy}^{K}(y,t) + J_{qy}^{\phi}(y,t). \quad (4.136)$$

The kinetic contribution is found to be

$$J_{qy}^{K}(y,t) = \frac{1}{A}\sum_i \left(v_{yi} - v(y)\right) u_i \delta(y - y_i), \quad (4.137)$$

where u_i is now the total internal energy, defined as

$$u_i = \frac{1}{2} m_i \mathbf{c}_i^2 + \frac{1}{2}\sum_j \phi_{ij}^{(2)} + \frac{1}{3}\sum_{jk} \phi_{ijk}^{(3)}. \quad (4.138)$$

This can be time-averaged, resulting in the same expression as given in Equation (4.114). The potential contribution is found to be

$$J_{qy}^{\phi}(y,t) = -\frac{1}{4A}\sum_{ij} (\mathbf{v}_i - \mathbf{v}(y)) \cdot \mathbf{F}_{ij}^{(2)} \left[\mathrm{sgn}(y - y_i) - \mathrm{sgn}(y - y_j)\right]$$
$$- \frac{1}{6A}\sum_{ij} (\mathbf{v}_i - \mathbf{v}(y)) \cdot \mathbf{F}_{ij}^{(3)} \left[\mathrm{sgn}(y - y_i) - \mathrm{sgn}(y - y_j)\right]$$
$$- \frac{1}{6A}\sum_{ik} (\mathbf{v}_i - \mathbf{v}(y)) \cdot \mathbf{F}_{ik}^{(3)} \left[\mathrm{sgn}(y - y_i) - \mathrm{sgn}(y - y_k)\right]$$
$$- \frac{1}{6A}\sum_{ijk} \mathbf{v}_i \cdot \mathbf{F}_{ij}^{(3)} \left[\mathrm{sgn}(y - y_i) - \mathrm{sgn}(y - y_k)\right]$$
$$- \frac{1}{6A}\sum_{ijk} \mathbf{v}_i \cdot \mathbf{F}_{ik}^{(3)} \left[\mathrm{sgn}(y - y_i) - \mathrm{sgn}(y - y_j)\right]$$
$$+ \frac{1}{6A}\sum_{jk} \mathbf{v}(y) \cdot \mathbf{F}_{jk}^{(3)} \left[\mathrm{sgn}(y - y_j) - \mathrm{sgn}(y - y_k)\right]. \quad (4.139)$$

Note that in order to obtain this expression we also have to substitute the 2- and 3-body terms for the pressure tensor given by Equations (4.97) and (4.127) into the inverse-transformed balance equation (isolated to extract the heat flux vector), as we did for the 2-body derivation of the heat flux vector in Section 4.3.2. Finally, Equation (4.139) can be rewritten in a more concise form as [100]

$$\begin{aligned}
J_{qy}^{\phi}(y,t) = &-\frac{1}{2A}\sum_{i}(\mathbf{v}_i - \mathbf{v}(y))\cdot\mathbf{F}_i^{(2)}\mathrm{sgn}(y-y_i) \\
&+\frac{1}{4A}\sum_{ij}\mathbf{v}_i\cdot\mathbf{F}_{ij}^{(2)}\left[\mathrm{sgn}(y-y_i)+\mathrm{sgn}(y-y_j)\right] \\
&-\frac{1}{2A}\sum_{i}(\mathbf{v}_i - \mathbf{v}(y))\cdot\mathbf{F}_i^{(3)}\mathrm{sgn}(y-y_i) \\
&+\frac{1}{6A}\sum_{ijk}\mathbf{v}_i\cdot\mathbf{F}_i^{(3)}\left[\mathrm{sgn}(y-y_i)+\mathrm{sgn}(y-y_j)+\mathrm{sgn}(y-y_k)\right], \qquad (4.140)
\end{aligned}$$

though it is a matter of programming preference which form to use (i.e. if one prefers only to compute forces between particles, or total forces acting on particles). Most likely, the former is the more convenient form to code up. Note again the meaning of the quantity $\mathbf{v}_i - \mathbf{v}(y)$, which implies the velocities are "plane-peculiar", as described in Section 4.3.2. Finally, we point out that the original derivation presented in reference [100] contained a typographical error in the presentation of Equation (4.140), in which the second term on the right-hand side was missing.

The above expressions for the heat flux vector have been validated against NEMD simulations for a confined fluid of argon undergoing planar Poiseuille flow [100]. 3-body terms for argon were shown not to be significant for the heat flux vector of a low-density fluid, but the effect on high-density, highly confined fluids has not yet been studied, nor have simulations been performed on heavier atomic nuclei inhomogeneous fluids, where n-body terms should dominate. However, we note that the original Irving-Kirkwood heat flux derivation has recently been generalised to n-body forces, suitable for application in homogeneous liquids, including molecular fluids [103].

The algorithms to code the additional 3-body terms for the pressure tensor and heat flux vector involve straightforward extensions to those presented in Algorithms 4.1 and 4.2, where now sums are over pairs and triplets rather than just pairs, and all forces and momenta are already computed in the numerical integration routine.

4.4 Volume Averaged Form of the Local Pressure Tensor

The integral representation of the configurational part of the pressure tensor given in Equation (4.54) cannot be directly evaluated at a point, but any amount of spatial averaging results in integration of the delta function, giving a numerical result. When this averaging over a small volume is carried out, we obtain the volume-averaged representation of the local pressure tensor, which is very useful as an alternative to the method of planes or for situations where planar symmetry is absent [92–94, 104, 105].

Let us consider the configurational part of the pressure tensor,

$$\mathbf{P}^{\phi}(\mathbf{r},t) = -\frac{1}{2}\sum_{i}^{N}\sum_{j\neq i}^{N}\mathbf{r}_{ij}\mathbf{F}_{ij}\int_{0}^{1}\delta\left(\mathbf{r}-\mathbf{r}_{i}-\lambda\mathbf{r}_{ij}\right)d\lambda. \tag{4.141}$$

The volume integral of the position dependent part of this is

$$l_{ij} = \int_{\delta V}\int_{0}^{1}\delta\left(\mathbf{r}-\mathbf{r}_{i}-\lambda\mathbf{r}_{ij}\right)d\lambda d\mathbf{r}, \tag{4.142}$$

which has the interpretation that it is equal to the fraction of the distance separating particles i and j that is enclosed by the volume δV [94]. The volume average of the kinetic part of the pressure tensor is more easily evaluated since it is essentially a density. Integrating over the delta function in the kinetic part gives a value of 1 for each particle within the volume δV and zero for each particle outside that volume. The volume average of the left-hand side of the expression for the local pressure is $\int_{\delta V}\mathbf{P}(\mathbf{r},t)dV = \delta V\mathbf{P}_{V}(\mathbf{r},t)$. Equating this to the microscopic expressions, the full expression for the volume averaged pressure tensor is then

$$\mathbf{P}_{V}(\mathbf{r},t) = \frac{1}{\delta V}\left[\sum_{i}^{N}m_{i}\mathbf{c}_{i}\mathbf{c}_{i}\theta_{i} - \frac{1}{2}\sum_{i}^{N}\sum_{j\neq i}^{N}\mathbf{r}_{ij}\mathbf{F}_{ij}l_{ij}\right], \tag{4.143}$$

where θ_i is equal to 1 when particle i is inside δV and zero otherwise.

4.5 Inclusion of Electrostatic Forces

For systems involving charged, dipolar or multipolar molecules, the pressure tensor and heat flux vector must be modified to account for the long-ranged Coulombic interactions. The most standard procedure to deal with electrostatics is to use the Ewald summation method [106], though there are other methods such as those proposed by Lekner [107, 108], Wolf [109, 110] and reaction field methods [111, 112]. The Ewald technique applied to liquids has been well described in a number of articles and books (see for example [5, 62]) and we refer readers to these for specific details. An algorithm is also presented in the book by Sadus [8], which we will not repeat here. In this section we instead provide details of how electrostatics are incorporated into the computation of the pressure tensor and heat flux vector, via the Ewald and shifted force (Wolf) methods.

4.5.1 Interaction Energy

The Ewald method involves superimposing a smooth screening charge distribution onto the Coulomb point-charge distribution of the atomic charges in the system to produce an effective short-ranged electrostatic interaction. An opposing charge distribution that exactly compensates for the screening charge distribution is also added, maintaining charge neutrality. The screening distribution is typically represented by a sum of Gaussians (though Heyes [113] has shown that any number of functional forms can be used), giving rise to the appearance of the error function and complementary error function in the final expressions for the electrostatic energy and force. This results in two separate terms for the interaction energy, one a real space component involving interatomic pairs that decays rapidly as a function of separation distance within the primary simulation box, and the other a rapidly converging series in reciprocal space representing the long-range interactions over multiple periodic images of the primary box. One also corrects for atomic self-interactions, intramolecular bonded interactions and the permittivity of the background continuum [62]. If we assume that the background is a good conductor, this latter term is zero and the resulting expression for the total electrostatic interaction energy is

$$\Phi^e = \Phi^r + \Phi^k + \Phi^s + \Phi^b, \tag{4.144}$$

where Φ^r, Φ^k, Φ^s and Φ^b are the real space, Fourier space, self-interaction and intramolecular bonded interaction terms, respectively, given as

$$\Phi^r = \sum_{i=1}^{N}\sum_{\alpha=1}^{N_s}\sum_{j>i}^{N}\sum_{\beta=1}^{N_s} \frac{q_{i\alpha} q_{j\beta}}{d_{i\alpha j\beta}} \operatorname{erfc}\left(\kappa d_{i\alpha j\beta}\right) \tag{4.145}$$

$$\Phi^k = \frac{2\pi}{V} \sum_{\mathbf{k_n}\neq 0}^{\infty} k_{\mathbf{n}}^{-2} \exp\left(-k_{\mathbf{n}}^2/4\kappa^2\right) \left|\sum_{i=1}^{N}\sum_{\alpha=1}^{N_s} q_{i\alpha} \exp\left(i\mathbf{k_n}\cdot\mathbf{r}_{i\alpha}\right)\right|^2 \tag{4.146}$$

$$\Phi^s = -\frac{\kappa}{\sqrt{\pi}} \sum_{i=1}^{N}\sum_{\alpha=1}^{N_s} q_{i\alpha}^2 \tag{4.147}$$

$$\Phi^b = -\frac{1}{2} \sum_{i=1}^{N}\sum_{\alpha=1}^{N_s}\sum_{\beta\neq\alpha}^{N_s} \frac{q_{i\alpha} q_{i\beta}}{r_{i\alpha i\beta}} \operatorname{erf}\left(\kappa r_{i\alpha i\beta}\right). \tag{4.148}$$

Here the sums extend over all N_s sites α, β of N molecules i, j. $q_{i\alpha}$ is the charge on site $i\alpha$, $d_{i\alpha j\beta}$ is the minimum image separation of the pair $i\alpha j\beta$ that is within the cutoff distance r_c (see Equation (4.79)) and κ is the half-width of the Gaussian. If κ is chosen such that the minimum image convention can be applied to the real space term (typically of the order $\kappa L \sim 5$), only one pair of images will be within the cutoff distance and the sum over all periodic boxes reduces to the sum over pairs for which one of the atoms is within the primary simulation box. $\mathbf{k_n} = 2\pi\left(n_1\mathbf{M}_1 + n_2\mathbf{M}_2 + n_3\mathbf{M}_3\right)$ is the wave vector for the periodic system and $\mathbf{M}_1$, $\mathbf{M}_2$, and $\mathbf{M}_3$, are the reciprocal lattice vectors which, in general, may be non-orthogonal [114].

The form of Φ^k is not particularly convenient for computation, especially as we will need it to calculate the heat flux vector in a following subsection. A more convenient form is found by expressing the complex modulus as

$$\begin{aligned}\left|\sum_{i=1}^{N}\sum_{\alpha=1}^{N_s} q_{i\alpha}\exp\left(i\mathbf{k}\cdot\mathbf{r}_{i\alpha}\right)\right|^2 &= \sum_{i=1}^{N}\sum_{\alpha=1}^{N_s} q_{i\alpha}\exp\left(-i\mathbf{k}\cdot\mathbf{r}_{i\alpha}\right)\sum_{i=1}^{N}\sum_{\beta=1}^{N_s} q_{i\beta}\exp\left(i\mathbf{k}\cdot\mathbf{r}_{i\beta}\right)\\ &= \sum_{i=1}^{N}\sum_{\alpha=1}^{N_s} q_{i\alpha}\mathrm{Re}\left[\exp\left(-i\mathbf{k}\cdot\mathbf{r}_{i\alpha}\right)\sum_{i=1}^{N}\sum_{\beta=1}^{N_s} q_{i\beta}\exp\left(i\mathbf{k}\cdot\mathbf{r}_{i\beta}\right)\right],\end{aligned} \tag{4.149}$$

where we use the fact that

$$\mathrm{Im}\left[\sum_{i=1}^{N}\sum_{\alpha=1}^{N_s} q_{i\alpha}\exp\left(-i\mathbf{k}\cdot\mathbf{r}_{i\alpha}\right)\sum_{i=1}^{N}\sum_{\beta=1}^{N_s} q_{i\beta}\exp\left(i\mathbf{k}\cdot\mathbf{r}_{i\beta}\right)\right] = 0. \tag{4.150}$$

A further simplification can be made by using the symmetry of the Fourier term with respect to inversion of $\mathbf{k}$. Since this term does not change when $\mathbf{k}$ is replaced by $-\mathbf{k}$ only half of the terms need to be computed, and the result is multiplied by two to get the final result, given as

$$\begin{aligned}\Phi^k &= \frac{4\pi}{V}\sum_{\substack{\mathbf{k_n}\neq 0\\ n_x>0}}^{\infty} k_{\mathbf{n}}^{-2}\exp\left(-k_{\mathbf{n}}^2/4\kappa^2\right)\\ &\times\sum_{i=1}^{N}\sum_{\alpha=1}^{N_s} q_{i\alpha}\mathrm{Re}\left[\exp\left(-i\mathbf{k_n}\cdot\mathbf{r}_{i\alpha}\right)\sum_{i=1}^{N}\sum_{\beta=1}^{N_s} q_{i\beta}\exp\left(i\mathbf{k_n}\cdot\mathbf{r}_{i\beta}\right)\right],\end{aligned} \tag{4.151}$$

where the sum over wavevectors now only goes over those nonzero wavevectors with positive x-components. This is efficiently computed by following Wheeler *et al.* [114]. Defining

$$C_{\mathbf{n}} = \frac{4\pi}{V}\frac{\exp\left(-k_{\mathbf{n}}^2/4\kappa^2\right)}{k_{\mathbf{n}}^2} \tag{4.152}$$

and

$$S\left(\mathbf{k_n}\right) = \sum_{i=1}^{N}\sum_{\alpha=1}^{N_s} q_{i\alpha}\exp\left(i\mathbf{k_n}\cdot\mathbf{r}_{i\alpha}\right) \tag{4.153}$$

leads to a computationally efficient expression for the Fourier series contribution to the internal energy

$$\Phi^k = \sum_{i=1}^{N}\sum_{\alpha=1}^{N_s}\phi_{i\alpha} = \sum_{i=1}^{N}\sum_{\alpha=1}^{N_s}\sum_{\substack{\mathbf{k_n}\neq 0\\ n_x>0}}^{\infty} C_{\mathbf{n}}q_{i\alpha}\mathrm{Re}\left[\exp\left(-i\mathbf{k_n}\cdot\mathbf{r}_{i\alpha}\right)S\left(\mathbf{k_n}\right)\right]. \tag{4.154}$$

The final, compact expression for this term is obtained by defining

$$\phi_n = \sum_{i=1}^{N}\sum_{\alpha=1}^{N_s} C_{\mathbf{n}}q_{i\alpha}\mathrm{Re}\left[\exp\left(-i\mathbf{k_n}\cdot\mathbf{r}_{i\alpha}\right)S\left(\mathbf{k_n}\right)\right], \tag{4.155}$$

such that

$$\Phi^k = \sum_{\substack{\mathbf{k_n} \neq 0 \\ n_x > 0}}^{\infty} \phi_n. \tag{4.156}$$

It is important to again emphasise the following points. The self-energy term accounts for the fact that the Fourier space energy includes the interaction energy between the Gaussian charge distribution and itself. These interactions should be excluded, because any given charge is not allowed to interact with itself. Such interactions are already excluded from the real space term, but they also need to be subtracted from the Fourier series term. The second correction term (the bonded pair term) accounts for the fact that the Fourier space energy also includes the energies of interaction between bonded pairs of atoms within a molecule and atoms interacting through bending and dihedral potential energy functions. These interactions are also normally excluded. The real space term already excludes them through a molecular interaction mask, but this does not apply to the Fourier space term, hence this correction must be added.

4.5.2 Interaction Forces

The total electrostatic force on site α of molecule i in the Ewald summation method for a molecular fluid can be expressed as the sum of three terms

$$\mathbf{F}^e_{i\alpha} = \mathbf{F}^r_{i\alpha} + \mathbf{F}^k_{i\alpha} + \mathbf{F}^b_{i\alpha}, \tag{4.157}$$

which represent the real and Fourier space contributions and a correction to the Fourier series contribution to account for charges within a molecule that are constrained or interacting through a specific bond force. The real space contribution to the force α of molecule i due to interactions with other charged sites is truncated so that it satisfies the minimum image convention and is given by

$$\mathbf{F}^r_{i\alpha} = -\sum_{j=1}^{N}\sum_{\beta=1}^{N_s} \frac{q_{i\alpha} q_{j\beta} \mathbf{d}_{i\alpha j\beta}}{d^2_{i\alpha j\beta}} \left(\frac{\operatorname{erfc}\left(\kappa d_{i\alpha j\beta}\right)}{d_{i\alpha j\beta}} + \frac{2\kappa}{\sqrt{\pi}} \exp\left(-\kappa^2 d^2_{i\alpha j\beta}\right)\right), \tag{4.158}$$

whereas the Fourier series force term is

$$\begin{aligned}
\mathbf{F}^k_{i\alpha} &= -\frac{4\pi}{V} \sum_{\mathbf{k_n} \neq 0}^{\infty} \mathbf{k_n} k_{\mathbf{n}}^{-2} \exp\left(-k_{\mathbf{n}}^2 / 4\kappa^2\right) \\
&\quad \times q_{i\alpha} \operatorname{Im}\left[\exp\left(-i\mathbf{k_n} \cdot \mathbf{r}_{i\alpha}\right) \sum_{j=1}^{N}\sum_{\beta=1}^{N_s} q_{j\beta} \exp\left(i\mathbf{k_n} \cdot \mathbf{r}_{j\beta}\right)\right] \\
&= -\frac{8\pi}{V} \sum_{\substack{\mathbf{k_n} \neq 0 \\ n_x > 0}}^{\infty} \mathbf{k_n} k_{\mathbf{n}}^{-2} \exp\left(-k_{\mathbf{n}}^2 / 4\kappa^2\right) \\
&\quad \times q_{i\alpha} \operatorname{Im}\left[\exp\left(-i\mathbf{k_n} \cdot \mathbf{r}_{i\alpha}\right) \sum_{j=1}^{N}\sum_{\beta=1}^{N_s} q_{j\beta} \exp\left(i\mathbf{k_n} \cdot \mathbf{r}_{j\beta}\right)\right].
\end{aligned} \tag{4.159}$$

This can be expressed in a more efficiently computable form as

$$\mathbf{F}_{i\alpha}^{k} = -2 \sum_{\substack{\mathbf{k_n} \neq 0 \\ n_x > 0}}^{\infty} \mathbf{k_n} C_\mathbf{n} q_{i\alpha} \mathrm{Im}\left[\exp\left(-i\mathbf{k_n} \cdot \mathbf{r}_{i\alpha}\right) S\left(\mathbf{k_n}\right)\right], \tag{4.160}$$

which includes interactions between all charges.

The correction that must be added to the Fourier series term to allow for bonded atoms that do not interact electrostatically within a molecule is [114]

$$\mathbf{F}_{i\alpha}^{b} = \sum_{\beta \in \{bonded\}}^{N_s} \frac{q_{i\alpha} q_{i\beta} \mathbf{d}_{i\alpha i\beta}}{d_{i\alpha i\beta}^2} \left(\frac{\mathrm{erf}\left(\kappa d_{i\alpha i\beta}\right)}{d_{i\alpha i\beta}} - \frac{2\kappa}{\sqrt{\pi}} \exp\left(-\kappa^2 d_{i\alpha i\beta}^2\right) \right), \tag{4.161}$$

where the summation extends over atoms excluded from interacting electrostatically with atom α of molecule i. There is no force due to the self-interaction term because the self-interaction potential energy does not depend on the positions.

4.5.3 Pressure Tensor

Now that the electrostatic force terms have been derived, we can include their effects into the pressure tensor. In what follows we derive the relevant terms for the pressure tensor for both atomic and molecular fluids, as well as both the atomic and molecular representations for the molecular fluid.

4.5.3.1 Atomic Fluid

As discussed in Section 4.2.1, the pressure tensor can be divided into two parts, one that corresponds to transfer of momentum by thermal motion of particles (the kinetic part) and one that corresponds to the transfer of momentum through the direct action of intermolecular forces (the potential or interaction part). As derived in that section, for an atomic fluid the kinetic part is given by

$$V\mathbf{P}^{K} = \sum_{i=1}^{N} m_i \mathbf{c}_i \mathbf{c}_i, \tag{4.162}$$

whereas the potential part of the pressure tensor can be further divided into two parts, one that satisfies the minimum image convention (commonly called the real space part) and another that includes the effects of long-ranged forces. For a force that satisfies the minimum image convention, such as the short-ranged part of the Ewald decomposition of the electrostatic force, the potential part of the pressure tensor can be expressed as

$$V\mathbf{P}^{r} = -\frac{1}{2} \sum_{i=1}^{N} \sum_{j \neq i}^{N} \mathbf{d}_{ij} \mathbf{F}_{ij}^{r}, \tag{4.163}$$

with $\mathbf{d}_{ij}$ previously defined as the minimum image separation between atoms i and j. The long-ranged part must be handled in a similar way to the long-ranged part of the potential energy, i.e. expressed as a Fourier series:

$$V\mathbf{P}^k = \sum_{\substack{\mathbf{k_n}\neq 0\\ n_x>0}}^{\infty} \sum_{i=1}^{N} C_\mathbf{n} \mathbf{B_n} q_i \mathrm{Re}\left[\exp\left(-i\mathbf{k_n}\cdot\mathbf{r}_i\right) S\left(\mathbf{k_n}\right)\right], \tag{4.164}$$

where

$$\mathbf{B_n} \equiv \mathbf{1} - 2\mathbf{k_n}\mathbf{k_n}\left(\frac{1}{k_\mathbf{n}^2} + \frac{1}{4\kappa^2}\right) \tag{4.165}$$

and $S(\mathbf{k_n})$ is as defined in Equation (4.153), but this time only for an atomic fluid (i.e. there is no second summation over intramolecular sites). This can be alternatively expressed as

$$V\mathbf{P}^k = \sum_{\substack{\mathbf{k_n}\neq 0\\ n_x>0}}^{\infty} \phi_\mathbf{n} \mathbf{B_n} \tag{4.166}$$

with

$$\phi_\mathbf{n} = \sum_{i=1}^{N} C_\mathbf{n} q_i \mathrm{Re}\left[\exp\left(-i\mathbf{k_n}\cdot\mathbf{r}_i\right) S\left(\mathbf{k_n}\right)\right], \tag{4.167}$$

due to its similarity to the computation of the internal energy.

4.5.3.2 Molecular Fluid: Atomic Representation

The kinetic part, the real space part and Fourier series part of the pressure tensor in the atomic representation for a molecular fluid can be calculated in exactly the same way as previously for an atomic fluid. However, for a molecular fluid, the electrostatic interactions between bonded atoms are usually eliminated and replaced by either bond constraints or specific bond forces. In the real space part of the pressure tensor, these terms are removed by applying an interaction mask to the pairs of bonded atoms. The Fourier space part as written above would include these contributions, so they must be excluded by subtracting the following term from the Fourier space pressure tensor

$$V\mathbf{P}^{(A),b} = \sum_{i=1}^{N} \sum_{\alpha=1}^{N_s} \sum_{\beta\in\{bonded\}}^{N_s} \mathbf{r}_{i\alpha i\beta} \mathbf{F}_{i\alpha i\beta}^{k,b} \tag{4.168}$$

where the correction to the electrostatic force due to the exclusion of bonded pairs is defined in Equation (4.161).

4.5.3.3 Molecular Fluid: Molecular Representation

In the molecular representation, we write the kinetic part of the molecular pressure tensor as

$$V\mathbf{P}^{(M),K} = \sum_{i=1}^{N} m_i \mathbf{c}_i \mathbf{c}_i, \tag{4.169}$$

where the mass m_i is the mass of molecule i and now $\mathbf{c}_i$ is the thermal (peculiar) centre of mass velocity of molecule i.

The electrostatic contribution is given as a sum of a short-ranged (real space) component ($\mathbf{P}^{(M),r}$), the Fourier series part of the atomic representation given previously ($\mathbf{P}^{(A),k}$), and the correction to the Fourier series part from the atomic-to-molecular representation ($\mathbf{P}^{(M-A),k}$) [114, 115]:

$$V\mathbf{P}^{(M),e} = V\mathbf{P}^{(M),r} + V\mathbf{P}^{(A),k} + V\mathbf{P}^{(M-A),k}, \tag{4.170}$$

where

$$V\mathbf{P}^{(M),r} = -\frac{1}{2}\sum_{i=1}^{N}\sum_{\alpha=1}^{N_s}\sum_{j\neq i}^{N}\sum_{\beta=1}^{N_s}\mathbf{d}_{ij}\mathbf{F}^{r}_{i\alpha j\beta} \tag{4.171}$$

and

$$\begin{aligned} V\mathbf{P}^{(M-A),k} &= \frac{1}{2}\sum_{i=1}^{N}\sum_{\alpha=1}^{N_s}\sum_{j=1}^{N}\sum_{\beta=1}^{N_s}\mathbf{r}_{i\alpha j\beta}\mathbf{F}^{k}_{i\alpha j\beta} - \frac{1}{2}\sum_{i=1}^{N}\sum_{\alpha=1}^{N_s}\sum_{j=1}^{N}\sum_{\beta=1}^{N_s}\mathbf{r}_{ij}\mathbf{F}^{k}_{i\alpha j\beta} \\ &= -\sum_{i=1}^{N}\sum_{\alpha=1}^{N_s}\left(\mathbf{r}_{i\alpha} - \mathbf{r}_i\right)\mathbf{F}^{k}_{i\alpha}, \end{aligned} \tag{4.172}$$

where the force is the full, uncorrected Fourier series contribution to the electrostatic force. This is essentially the expression given by Alejandre *et al.* [115], which corrects minor errors in those given by Heyes [116] and Nosé and Klein [117]. However, note the order of the force and position vectors in the dyadic product in the last term is different from that given by Alejandre *et al.* This could be important when the antisymmetric part of the molecular pressure tensor is of interest.

4.5.4 Heat Flux Vector

As discussed in Section 4.2.2, the heat flux vector consists of two terms: a kinetic term that transports the internal energy by atomic motions, and a potential part that transports internal energy through the interaction forces. The derivations for the inclusion of electrostatic forces has been performed by Galamba *et al.* [118] and Petravic [119], and the discussion that follows is based on these references.

In the case of an atomic fluid, the electrostatic contribution to the kinetic part of the heat flux vector is given as

$$\mathbf{J}_q^{e,K} = \frac{1}{V}\sum_{i=1}^{N}\phi_i^e\mathbf{c}_i, \tag{4.173}$$

where the electrostatic contribution to the internal energy of particle i is given by the sum of the real, Fourier and self-interaction potential energy terms

$$\phi_i^e = \phi_i^r + \phi_i^k + \phi_i^s \tag{4.174}$$

and where the individual components are

$$\phi_i^r = \sum_{j \neq i}^{N} \frac{q_i q_j}{d_{ij}} \operatorname{erfc}\left(\kappa d_{ij}\right) \tag{4.175}$$

$$\phi_i^k = \frac{2\pi}{V} \sum_{\mathbf{k_n} \neq 0}^{\infty} k_{\mathbf{n}}^{-2} \exp\left(-k_{\mathbf{n}}^2/4\kappa^2\right) |q_i \exp\left(i\mathbf{k_n} \cdot \mathbf{r}_i\right)|^2 \tag{4.176}$$

$$\phi_i^s = -\frac{\kappa}{\sqrt{\pi}} q_i^2. \tag{4.177}$$

It is important to realise that the above only expresses the *electrostatic* potential energy contributions to the kinetic part of the heat flux vector. To obtain the kinetic contribution to the total heat flux vector, one would have to add in the thermal kinetic energy contribution and any nonelectrostatic contributions (such as van der Waals interactions) to Equation (4.174).

In a similar way to the force and pressure tensor calculations, the potential part of the heat flux vector can also be divided into two terms,

$$\mathbf{J}_q^{e,\phi} = \mathbf{J}_q^{e,r} + \mathbf{J}_q^{e,k} \tag{4.178}$$

representing the real space and Fourier series contributions. The self-interaction energy does not contribute to this part of the heat flux vector, because it does not result in a force. The two contributions to the potential part of the heat flux vector will now be considered. The force in the real space part satisfies the minimum image convention and can be written as

$$V\mathbf{J}_q^{e,r} = -\frac{1}{2} \sum_{i=1}^{N} \sum_{j \neq i}^{N} \mathbf{d}_{ij} \mathbf{F}_{ij}^r \cdot \mathbf{c}_j, \tag{4.179}$$

while the Fourier space contribution can be written as

$$V\mathbf{J}^{e,k} = \sum_{\substack{\mathbf{k_n} \neq 0 \\ n_x > 0}}^{\infty} C_{\mathbf{n}} \mathbf{B_n} \cdot \sum_{i=1}^{N} \sum_{j=1}^{N} q_i q_j \mathbf{c}_j \exp\left(i\mathbf{k_n} \cdot \mathbf{r}_{ij}\right). \tag{4.180}$$

Equation (4.180) is however not efficient for computation. Petravic [119] has shown how to transform the double sum into a more computationally convenient form. This is more convenient when the pressure tensor is also being calculated, since some quantities are the same for both calculations. Since the heat flux vector must be real, the result is

$$V\mathbf{J}^{e,k} = \sum_{\substack{\mathbf{k_n} \neq 0 \\ n_x > 0}}^{\infty} C_{\mathbf{n}} \mathbf{B_n} \cdot \mathbf{A_n}, \tag{4.181}$$

where

$$\mathbf{A_n} = \mathrm{Re}\left[\sum_{i=1}^{N}\sum_{j=1}^{N} q_i q_j \mathbf{c}_j \exp\left(i\mathbf{k_n}\cdot\mathbf{r}_{ij}\right)\right]$$
$$= \mathrm{Re}\left[S\left(\mathbf{k_n}\right)\right]\mathrm{Re}\left[\mathbf{S}_v\left(\mathbf{k_n}\right)\right] + \mathrm{Im}\left[S\left(\mathbf{k_n}\right)\right]\mathrm{Im}\left[\mathbf{S}_v\left(\mathbf{k_n}\right)\right] \tag{4.182}$$

and

$$S\left(\mathbf{k_n}\right) \equiv \sum_{i=1}^{N} q_i \exp\left(i\mathbf{k_n}\cdot\mathbf{r}_i\right) = \sum_{i=1}^{N} q_i \cos\left(\mathbf{k_n}\cdot\mathbf{r}_i\right) + i\sum_{i=1}^{N} q_i \sin\left(\mathbf{k_n}\cdot\mathbf{r}_i\right)$$
$$\mathbf{S}_v\left(\mathbf{k_n}\right) \equiv \sum_{i=1}^{N} q_i\mathbf{c}_i \exp\left(i\mathbf{k_n}\cdot\mathbf{r}_i\right) = \sum_{i=1}^{N} q_i\mathbf{c}_i \cos\left(\mathbf{k_n}\cdot\mathbf{r}_i\right) + i\sum_{i=1}^{N} q_i\mathbf{c}_i \sin\left(\mathbf{k_n}\cdot\mathbf{r}_i\right). \tag{4.183}$$

4.5.4.1 Molecular Fluid: Atomic Representation

In the case of the atomic representation of a molecular fluid, as with the case of the pressure tensor we should exclude electrostatic interactions between bonded atoms. Thus the kinetic part of the heat flux vector is expressible as

$$V\mathbf{J}_q^{K,e} = \sum_{i=1}^{N}\sum_{\alpha=1}^{N_s} \phi_{i\alpha}^{e}\mathbf{c}_{i\alpha} \tag{4.184}$$

where the potential energy per particle $\phi_{i\alpha}^{e}$ is corrected to account for this. The calculation of the real space contribution to the potential part of the heat flux vector must also exclude the contributions from bonded pairs within a molecule. This can be expressed as a sum of two contributions, intermolecular and intramolecular (excluding bonded pairs)

$$V\mathbf{J}_q^{e,r} = -\frac{1}{2}\sum_{i=1}^{N}\sum_{\alpha=1}^{N_s}\sum_{j\neq i}^{N}\sum_{\beta=1}^{N_s} \mathbf{d}_{i\alpha j\beta}\mathbf{F}_{i\alpha j\beta}^{r}\cdot\mathbf{c}_{j\beta} - \frac{1}{2}\sum_{i=1}^{N}\sum_{\alpha=1}^{N_s}\sum_{j\neq i}^{N}\sum_{\beta|\alpha\beta\notin\{bonded\}}^{N_s} \mathbf{d}_{i\alpha i\beta}\mathbf{F}_{i\alpha i\beta}^{r}\cdot\mathbf{c}_{j\beta}. \tag{4.185}$$

The Fourier series part of the heat flux vector for a molecular fluid in the atomic representation is given by

$$V\mathbf{J}_q^{e,k} = \sum_{\substack{\mathbf{k_n}\neq 0\\ n_x>0}}^{\infty} C_\mathbf{n}\left[\mathbf{1} - 2\mathbf{k_n}\mathbf{k_n}\left(\frac{1}{k_\mathbf{n}^2} + \frac{1}{4\kappa^2}\right)\right]\cdot\sum_{i=1}^{N}\sum_{\alpha=1}^{N_s}\sum_{j=1}^{N}\sum_{\beta=1}^{N_s} q_{i\alpha}q_{j\beta}\mathbf{c}_{j\beta}\exp\left(i\mathbf{k_n}\cdot\mathbf{r}_{i\alpha j\beta}\right). \tag{4.186}$$

This is the same as the expression for an atomic fluid, but with the atomic and molecular indices explicit, and it can be calculated in the same way. However, as it includes all interacting pairs this must be corrected by adding the following term to the heat flux

$$V\mathbf{J}^{k,b} = \sum_{i=1}^{N}\sum_{\alpha=1}^{N_s}\sum_{j\neq i}^{N}\sum_{\beta|\alpha,\beta\in\{bonded\}}^{N_s} \mathbf{r}_{i\alpha i\beta}\mathbf{F}_{i\alpha i\beta}^{k,b}\cdot\mathbf{c}_{j\beta}. \tag{4.187}$$

4.5.4.2 Molecular Fluid: Molecular Representation

In the molecular representation, the kinetic part of the heat flux vector is

$$V\mathbf{J}_q^{K,e} = \sum_{i=1}^{N}\sum_{\alpha=1}^{N_s} \phi_{i\alpha}^e \mathbf{c}_i = \sum_{i=1}^{N} \phi_i^e \mathbf{c}_i, \tag{4.188}$$

which is easily computed by summing the internal energy over each molecule before multiplying by the centre of mass molecular velocity and doing the summation over molecules. The potential part of the molecular representation is different from the atomic representation in two ways. The real space contribution to the potential part of the molecular heat flux vector must be calculated as

$$V\mathbf{J}_q^{e,r} = -\frac{1}{2}\sum_{i=1}^{N}\sum_{\alpha=1}^{N_s}\sum_{j\neq i}^{N}\sum_{\beta=1}^{N_s} \mathbf{d}_{ij}\mathbf{F}_{i\alpha j\beta}^r \cdot \mathbf{c}_{j\beta}. \tag{4.189}$$

For the Fourier series part, the contribution to the atomic heat flux vector is calculated first, and then a correction factor is added to convert it into the molecular form. The correction is calculated similarly to the one for the pressure tensor. The motivation can be seen as follows. For a force that satisfies the minimum image convention, we can write

$$\begin{aligned}\sum_{i=1}^{N}\sum_{\alpha=1}^{N_s}\sum_{j\neq i}^{N}\sum_{\beta=1}^{N_s} \mathbf{d}_{ij}\mathbf{F}_{i\alpha j\beta}^r \cdot \mathbf{c}_{j\beta} &= \sum_{i=1}^{N}\sum_{\alpha=1}^{N}\sum_{j\neq i}^{N}\sum_{\beta=1}^{N_s} \mathbf{d}_{i\alpha j\beta}\mathbf{F}_{i\alpha j\beta}^r \cdot \mathbf{c}_{j\beta} \\ &- \sum_{i=1}^{N}\sum_{\alpha=1}^{N_s}\sum_{j\neq i}^{N}\sum_{\beta=1}^{N_s} \left(\mathbf{d}_{i\alpha j\beta} - \mathbf{d}_{ij}\right)\mathbf{F}_{i\alpha j\beta}^r \cdot \mathbf{c}_{j\beta}.\end{aligned} \tag{4.190}$$

With a little manipulation the right-hand side of Equation (4.190) can be rewritten as

$$\begin{aligned}&\sum_{i=1}^{N}\sum_{\alpha=1}^{N_s}\sum_{j\neq i}^{N}\sum_{\beta=1}^{N_s} \mathbf{d}_{i\alpha j\beta}\mathbf{F}_{i\alpha j\beta}^r \cdot \mathbf{c}_{j\beta} \\ &+ \sum_{i=1}^{N}\sum_{\alpha=1}^{N_s} \left(\mathbf{d}_{i\alpha} - \mathbf{d}_i\right)\left(\mathbf{F}_{i\alpha}^r \cdot \mathbf{c}_{i\alpha} + \sum_{j\neq i}^{N}\sum_{\beta=1}^{N_s} \mathbf{F}_{i\alpha j\beta}^r \cdot \mathbf{c}_{j\beta}\right).\end{aligned} \tag{4.191}$$

If we now assume that this correction can be applied to the Fourier series part of the atomic heat flux vector, we get the final expression for the electrostatic contribution to the heat flux vector for a molecular fluid in the molecular representation

$$V\mathbf{J}_q^{(M)e} = V\left[\mathbf{J}_q^{(M)e,K} + \mathbf{J}_q^{(M)e,r} + \mathbf{J}_q^{(A)e,k} + \mathbf{J}_q^{(A-M)e,k}\right], \tag{4.192}$$

where

$$V\mathbf{J}_q^{(M)e,K} = \sum_{i=1}^{N}\sum_{\alpha=1}^{N_s}\phi_{i\alpha}^{e}\mathbf{c}_i = \sum_{i=1}^{N}\phi_i^e\mathbf{c}_i \tag{4.193}$$

$$V\mathbf{J}_q^{(M)e,r} = -\frac{1}{2}\sum_{i=1}^{N}\sum_{\alpha=1}^{N_s}\sum_{j\neq i}^{N}\sum_{\beta=1}^{N_s}\mathbf{d}_{ij}\mathbf{F}_{i\alpha j\beta}^{r}\cdot\mathbf{c}_{j\beta} \tag{4.194}$$

$$V\mathbf{J}_q^{(A)e,k} = \sum_{\substack{\mathbf{k_n}\neq 0\\ n_x>0}}^{\infty} C_{\mathbf{n}}\mathbf{B_n}\cdot\mathbf{A_n} \tag{4.195}$$

$$V\mathbf{J}_q^{(A-M)e,k} = -\frac{1}{2}\sum_{i=1}^{N}\sum_{\alpha=1}^{N_s}(\mathbf{d}_{i\alpha}-\mathbf{d}_i)\left(\mathbf{F}_{i\alpha}^{r}\cdot\mathbf{c}_{i\alpha}+\sum_{j\neq i}^{N}\sum_{\beta=1}^{N_s}\mathbf{F}_{i\alpha j\beta}^{r}\cdot\mathbf{c}_{j\beta}\right). \tag{4.196}$$

4.5.5 Shifted Force Method

The implementations of the Ewald method to computing energy, forces, pressure tensor and heat flux vector described above are exact for systems that can be represented infinitely periodically in space. Other methods do exist to handle systems that are not infinitely periodic in all three spatial dimensions, such as may be the case for confined fluids (see for example [120–122]). The Ewald method is however computationally expensive. Wolf and colleagues [109, 110] suggested an alternative method that avoids the use of complicated procedures to handle long-range forces. The basis of the Wolf implementation was the realisation that computational power has increased exponentially since the Ewald algorithms were first developed. Thus, researchers were no longer restricted to small systems replicated infinitely in space and that radial cut-off methods could now be utilised and could reach convergence for much larger systems than previously realisable. This, coupled with the observation that that the electrostatic interaction is effectively short-ranged in soft matter systems, allowed for the use of shifted radial cut-off methods, analogous to the Weeks-Chandler-Andersen (WCA) implementation for Lennard-Jones fluids [66]. The significant advantage in the Wolf implementation is that one no longer needs to use complicated mathematical "tricks" to compute the forces, pressure tensor and heat flux vector. Instead, they are computed in the usual manner, as one would with standard short-ranged potentials.

The disadvantage is that the method is not exact and thus there is some residual error that should be considered. For inhomogeneous fluids, where the charge distribution may not be isotropic, the error could be significant. Caution should be exercised, because truncation methods can lead to inaccuracies in the calculation of certain properties, as shown for example by Muscatello and Bresme [123].

We now outline the implementation of the Wolf method developed by Fennel and Gezelter [124]. Readers should consult the original papers for details. Once again the electrostatic energy is split into a sum of two parts. The first part is the damped interaction, identical to the Ewald method

$$\Phi^{e,r} = \sum_{i=1}^{N}\sum_{\alpha=1}^{N_s}\sum_{j>i}^{N}\sum_{\beta=1}^{N_s} \frac{q_{i\alpha}q_{j\beta}}{d_{i\alpha j\beta}} \operatorname{erfc}\left(\kappa d_{i\alpha j\beta}\right). \tag{4.197}$$

The second part consists of two terms that incorporate both the shifting and truncating of the potential to zero at the cut-off distance d_c:

$$\Phi^{e,c} = \sum_{i=1}^{N}\sum_{\alpha=1}^{N_s}\sum_{j>i}^{N}\sum_{\beta=1}^{N_s} q_{i\alpha}q_{j\beta} \times \left[-\frac{\operatorname{erfc}(\kappa d_c)}{d_c} + \left(\frac{\operatorname{erfc}(\kappa d_c)}{d_c^2} + \frac{2\kappa}{\sqrt{\pi}}\frac{\exp\left(-\kappa^2 d_c^2\right)}{d_c}\right)\left(d_{i\alpha j\beta} - d_c\right)\right], \tag{4.198}$$

where here we have the condition that $d_{i\alpha j\beta} \leq d_c$. The force on atomic site $i\alpha$ due to all other sites $j\beta$ within this interaction radius is thus

$$\begin{aligned}\mathbf{F}_{i\alpha} &= -\sum_{j>i}^{N}\sum_{\beta=1}^{N_s} \hat{\mathbf{d}}_{i\alpha j\beta}\frac{\partial \phi_{i\alpha j\beta}}{\partial d_{i\alpha j\beta}} \\ &= -\sum_{j>i}^{N}\sum_{\beta=1}^{N_s} \hat{\mathbf{d}}_{i\alpha j\beta} q_{i\alpha}q_{j\beta} \times \left[\frac{\operatorname{erfc}\left(\kappa d_{i\alpha j\beta}\right)}{d_{i\alpha j\beta}^2} + \frac{2\kappa}{\sqrt{\pi}}\frac{\exp\left(-\kappa^2 d_{i\alpha j\beta}^2\right)}{d_{i\alpha j\beta}}\right. \\ &\quad \left. -\left(\frac{\operatorname{erfc}(\kappa d_c)}{d_c^2} + \frac{2\kappa}{\sqrt{\pi}}\frac{\exp\left(-\kappa^2 d_c^2\right)}{d_c}\right)\right],\end{aligned} \tag{4.199}$$

where $\hat{\mathbf{d}}_{i\alpha j\beta}$ is the unit vector directed along the line connecting atomic site $i\alpha$ to site $j\beta$. The force is zero when $d_{i\alpha j\beta} > d_c$. Furthermore, the exclusion of intramolecular interactions can be achieved with an interaction mask, as for example in the case of short-ranged Lennard-Jones interactions.

Since the electrostatic force is reduced to a relatively short-ranged pair force, there is no need for complicated corrections to the pressure tensor and heat flux vector. The usual expressions for these quantities in the atomic and molecular representations for potentials that obey the minimum image convention can now be used.

5 Homogeneous Flows for Atomic Fluids: Theory

In this chapter, we introduce homogeneous nonequilibrium molecular dynamics simulation techniques by discussing the theoretical background to the SLLOD equations of motion. When these equations of motion are used in conjunction with compatible periodic boundary conditions and a homogeneous thermostat, they provide a very robust, reliable and well-understood method for studying fluids subjected to homogeneous flows. Here, we introduce the SLLOD equations of motion for the simple case of atomic fluids. This provides the groundwork for our discussion of methods for simulating homogeneous flows of molecular fluids in Chapter 8.

5.1 The SLLOD Equations of Motion

5.1.1 Background

To conduct microscopic simulations of flows driven by boundaries, mimicking real physical systems (e.g. Couette or elongational flows) we must explicitly include the walls. This inevitably induces density inhomogeneities into the fluid. If one is interested in nano-confined flow, then this is an appropriate simulation strategy since spatial inhomogeneity needs to be explicitly included in the simulation. However, if one is concerned with computing bulk properties such as mass, momentum and heat transport coefficients that we do not want to be distorted by surface effects, then the explicit use of boundaries is inappropriate.

An alternative to using atomistic wall boundaries is to generate flow through a suitable implementation of periodic boundary conditions. The first and most popular method of inducing flow through the periodic boundary conditions employs the so called Lees-Edwards boundary conditions [15] to generate planar shear flow. In such a scheme, a simulation box is replicated in all directions by periodic images. Atoms interact through their interatomic forces, but no other external forces are explicitly imposed on them via the equations of motion. Unlike the case for standard equilibrium MD, the periodic image boxes are translated with respect to the simulation box by an amount $\pm L_y \dot{\gamma} t$, where $\pm L_y$ is the length of the box in the y-direction, $\dot{\gamma}$ is the strain rate and t is the simulation time (see Figure 5.1). This translation of atoms as they move from top to bottom (or vice versa) is what induces a linear streaming velocity profile for low Reynolds number flows. However, this approach has two serious shortcomings. First,

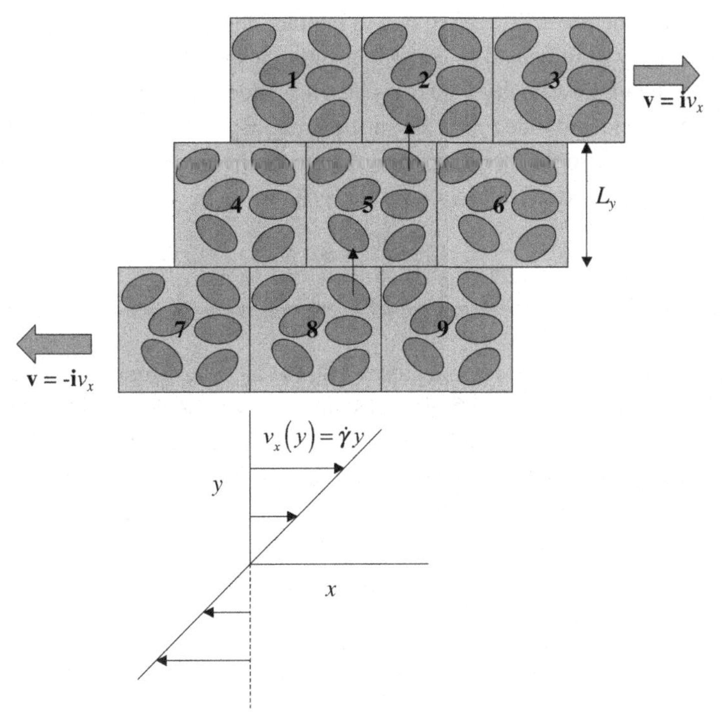

Figure 5.1 Two dimensional representation of a simulation box (box number 5) surrounded by its periodic images (all other numbered boxes) undergoing planar Couette (shear) flow. Flow is in the x-direction and is linear in y with streaming velocity $\mathbf{v} = \mathbf{i}\dot{\gamma}y$. Image boxes above and below box 5 are displaced by $\pm L_y \dot{\gamma}\, t$. A molecule that moves out of the top of box 5 into box 2 is displaced by an amount $-L_y \dot{\gamma}\, \Delta t$ (where Δt is the integration timestep) in the x-direction and L_y in the y-direction. It will re-emerge from box 8 into box 5. A similar but opposite translation occurs if a molecule moves out of the bottom of the simulation box. Molecules that move out of the left or right boundaries are translated by $\pm L_x$ respectively.

there is no connection with response theory, so link to the statistical mechanics of transport (e.g. the Green–Kubo expressions for shear viscosity) is broken. Secondly, it takes time for the effects of translation of atoms between boundaries to communicate throughout the fluid. This is approximately the time taken for the transverse momentum to diffuse across the simulation box. Therefore a linear streaming velocity profile will not be imposed immediately, but will evolve only after a sufficiently long time. This also makes it difficult to study time-dependent transport properties with boundary-driven methods.

It was because of these problems that the first of the homogeneous NEMD algorithms was invented, namely the DOLLS Hamiltonian method, proposed by Hoover *et al.* [20]. In this algorithm the role of a boundary that drives the flow is replaced with a fictitious external field. The field is itself engineered to ensure that the required streaming velocity profile is indefinitely sustained. The DOLLS Hamiltonian is

$$H_{DOLLS}\left(\mathbf{\Gamma}^N, t\right) = \phi\left(\mathbf{r}^N\right) + \sum_{i=1}^{N} \frac{\mathbf{p}_i^2}{2m_i} + \sum_{i=1}^{N} \mathbf{r}_i \cdot \nabla\mathbf{v} \cdot \mathbf{p}_i \Theta(t), \tag{5.1}$$

where $\phi(\mathbf{r}^N)$ is the potential energy due to interactions between all N atoms, $\mathbf{r}_i$ and $\mathbf{p}_i$ are the laboratory position and peculiar momentum respectively of atom i, $\nabla\mathbf{v}$ is the gradient of the streaming velocity $\mathbf{v}$, $\Theta(t)$ is the Heaviside step function, and it is assumed that flow commences at time $t = 0$. The peculiar momentum is defined to be the thermal momentum relative to the streaming momentum of the fluid. This Hamiltonian generates the DOLLS equations of motion for the system:

$$\dot{\mathbf{r}}_i = \frac{\mathbf{p}_i}{m_i} + \mathbf{r}_i \cdot \nabla\mathbf{v}$$
$$\dot{\mathbf{p}}_i = \mathbf{F}_i^{\phi} - \nabla\mathbf{v} \cdot \mathbf{p}_i, \tag{5.2}$$

where $\mathbf{F}_i^{\phi}$ is the interatomic force on atom i due to all other atoms. Evans and Morriss [125] pointed out that while DOLLS was suitable for simulating flows in the linear response limit, it was unsuitable for generating physically realistic shear flow at higher rates of strain. Errors in computed properties grow quadratically in strain rate and are observable in the normal stress differences [21]. Hoover *et al.* [126] later demonstrated that compared to boundary driven flow, the DOLLS algorithm displays the wrong sign in the normal stress differences.

Evans and Morriss [21] and Ladd [127] (in the case of molecular fluids) proposed that the correct set of equations are given by

$$\dot{\mathbf{r}}_i = \frac{\mathbf{p}_i}{m_i} + \mathbf{r}_i \cdot \nabla\mathbf{v}$$
$$\dot{\mathbf{p}}_i = \mathbf{F}_i^{\phi} - \mathbf{p}_i \cdot \nabla\mathbf{v}, \tag{5.3}$$

where now the only difference between Equations (5.2) and (5.3) is that the last term in the DOLLS force equation is transposed, i.e. $\nabla\mathbf{v} \cdot \mathbf{p}_i \rightarrow \mathbf{p}_i \cdot \nabla\mathbf{v}$. Evans and Morriss [2] performed a number of definitive theoretical and simulation studies to prove that these equations of motion (which they termed "SLLOD" to indicate the transpose of DOLLS) did give the correct nonlinear response for shear flow.[1] It was also clear that the SLLOD equations of motion could not be generated from a system Hamiltonian [128]. We will derive the SLLOD equations formally and discuss their relationship to Hamiltonian mechanics in the next section.

It is important to appreciate that the boundaries of the simulation box do not generate the flow induced by the SLLOD equations of motion. The equations of motion and momentum conservation are in themselves sufficient to generate the correct velocity gradient, as we will see shortly. However, to prevent the periodic boundary conditions (pbcs) from interfering with particle trajectories, the SLLOD equations of motion *must* be used in conjunction with suitably compatible pbcs. For planar shear flow (or planar Couette flow (PCF)) a suitable pbc scheme is the one proposed by Lees and Edwards [15]. Use of the SLLOD equations in conjunction with Lees-Edwards pbcs guarantees the generation of the desired velocity gradient for low Reynolds number flows.

[1] As an historical curiosity, the term "dolls" was used by Hoover to denote the Hamiltonian as it contained q's and p's, sounding similar to a "cupie" doll. There is no mythical Dr Dolls to have originated these well known equations of motion!

Similarly, for planar elongational flows the so-called Kraynik-Reinelt boundary conditions [129], suitably implemented for NEMD simulations [130–133], can also be used. This is true for any generalised homogeneous, divergenceless flow as long as suitably compatible boundary conditions are used, such as for example planar mixed (shear and elongation) flow [134]. The compatibility between equations of motion and boundary conditions is essential to ensure that the boundaries themselves do not perturb the system and hence invalidate the use of response theory.

The SLLOD algorithm succeeds in transforming a boundary driven flow into one that is driven solely by a synthetic external field. It has been demonstrated [2] that in fact SLLOD generates the correct nonlinear response for adiabatic (i.e. unthermostatted) shear flow and is equivalent to a system with an initial local equilibrium distribution function, with a superposed linear streaming velocity profile propagated in time with Newton's equations of motion [2, 45, 46]. Apart from the rigour of the SLLOD equations of motion, they are also highly useful because they are compatible with nonlinear response theory. This has the practical benefit of allowing the extraction of transport properties under field strengths much weaker than can normally be used for direct averaging of NEMD fluxes via use of the transient-time correlation function (TTCF) formalism [2, 23, 51–60] (see Sections 3.4.3 and 6.5). We note here that boundary-driven algorithms (such as wall-driven Couette flow) can make use of response theory as long as the correct formulation of the dissipation is utilised, as demonstrated in references [135, 136].

In the following section we present a simple, direct derivation of the SLLOD equations of motion for generalised homogeneous flows. This derivation [1, 128] was partly motivated by a series of papers in the literature that questioned the validity of the SLLOD equations for extensional flows [137–140]. Our derivation serves as a proof of the validity of the SLLOD algorithm for generalised homogeneous flows, such as planar shear and elongational flows, and mixed flows that can be simulated by a combination of flow geometries and suitable boundary conditions.

5.1.2 Derivation of the SLLOD Equations of Motion from First Principles

In this section we will show that for arbitrary homogeneous flows, the SLLOD equations of motion are identical to Newton's equations of motion for a fluid in the presence of an external force. Let $\mathbf{G}(\mathbf{r}, t)$ be an external body force density (i.e. the force per unit volume) applied to a fluid of infinite extent at the laboratory position $\mathbf{r}$ and time t. The local, instantaneous, equation of motion can be expressed as [2, 4]

$$\frac{\partial \mathbf{J}(\mathbf{r}, t)}{\partial t} = -\nabla \cdot \mathbf{P}(\mathbf{r}, t) - \nabla \cdot [\rho(\mathbf{r}, t)\mathbf{v}(\mathbf{r}, t)\mathbf{v}(\mathbf{r}, t)] + \mathbf{G}(\mathbf{r}, t) \tag{5.4}$$

where $\mathbf{J}(\mathbf{r}, t)$ is the local instantaneous momentum density as defined in Equation (4.22). $\mathbf{G}(\mathbf{r}, t)$ is given by

$$\mathbf{G}(\mathbf{r}, t) = \sum_i \mathbf{F}_i^e \delta(\mathbf{r} - \mathbf{r}_i), \tag{5.5}$$

where $\mathbf{F}_i^e$ is the external force on particle i, and the index i ranges over all particles in the fluid. The wavevector dependent form of Equation (5.4) is obtained by Fourier transformation, giving

$$\frac{\partial \tilde{\mathbf{J}}(\mathbf{k},t)}{\partial t} = i\mathbf{k}\cdot\tilde{\mathbf{P}}(\mathbf{k},t) + i\mathbf{k}\cdot\left[\widetilde{\rho\mathbf{v}\mathbf{v}}(\mathbf{k},t)\right] + \tilde{\mathbf{G}}(\mathbf{k},t). \tag{5.6}$$

This equation tells us that zero-wavevector contributions to the momentum density can *only* come from the external field.

The equation of motion for the zero-wavevector component of the momentum density is

$$\frac{d}{dt}\sum_i m_i\mathbf{v}_i = \sum_i \mathbf{F}_i^e, \tag{5.7}$$

where it is important to appreciate that $\mathbf{v}_i$ is the laboratory velocity of particle i, i.e. its total velocity comprising of (microscopic) thermal and (macroscopic) streaming motion. We now define the peculiar, or thermal, velocity $\mathbf{c}_i$ for particle i, with a suitable change of variable:

$$\begin{aligned}\mathbf{v}_i &= \mathbf{c}_i + \mathbf{v}(\mathbf{r}_i)\,\Theta(t)\\ &= \mathbf{c}_i + \mathbf{r}_i\cdot\nabla\mathbf{v}\,\Theta(t)\end{aligned} \tag{5.8}$$

in which the second line is valid for a homogeneous velocity gradient (see also Chapter 4). Furthermore, the total thermal momentum and its derivative are both zero, i.e.

$$\sum_i m_i\mathbf{c}_i = 0, \quad \frac{d}{dt}\sum_i m_i\mathbf{c}_i = 0 \tag{5.9}$$

and we assume that they are conserved by the equations of motion (conservation of the thermal component of momentum).

In this derivation we have also assumed that the velocity gradient is applied as a step function at time $t = 0$, where the Heaviside step function is defined as

$$\Theta(t) = \begin{cases} 0 & ;\ t < 0 \\ 1 & ;\ t \geq 0 \end{cases}. \tag{5.10}$$

The SLLOD equations of motion will emerge elegantly when the zero-wavevector momentum obeys Equation (5.7) and when the peculiar (thermal) velocity satisfies Equations (5.8) and (5.9).

Taking the time derivative of the zero wavevector momentum, we find

$$\begin{aligned}\frac{d}{dt}\sum_i m_i\mathbf{v}_i &= \frac{d}{dt}\sum_i m_i\left[\mathbf{c}_i + \mathbf{r}_i\cdot\nabla\mathbf{v}\,\Theta(t)\right]\\ &= \frac{d}{dt}\sum_i m_i\mathbf{r}_i\cdot\nabla\mathbf{v}\,\Theta(t),\end{aligned} \tag{5.11}$$

where we have implicitly used the conservation of thermal momentum to obtain the second line. Taking the derivative of the product inside the sum, substituting Equation (5.8) for the velocity and again using conservation of thermal momentum, leads to

$$\begin{aligned}\frac{d}{dt}\sum_i m_i\mathbf{v}_i &= \sum_i \left[m_i\mathbf{v}_i \cdot \nabla\mathbf{v}\,\Theta(t) + m_i\mathbf{r}_i \cdot \frac{d}{dt}(\nabla\mathbf{v}\,\Theta(t))\right] \\ &= \sum_i [m_i\mathbf{r}_i \cdot \nabla\mathbf{v} \cdot \nabla\mathbf{v}\,\Theta(t) + m_i\mathbf{r}_i \cdot \nabla\mathbf{v}\,\delta(t)]. \end{aligned} \tag{5.12}$$

The last line of Equation (5.12) is obtained because the total peculiar momentum is conserved and the product of the step function with itself is just the step function. In an actual computer simulation, finite precision numerics and discretisation error in the solution of the ordinary differential equations mean that conservation of thermal momentum may not be strictly true, though Equation (5.12) is formally correct. This will be discussed further in Section 5.1.4. Comparison of the right-hand sides of Equations (5.7) and (5.12) indicates that the sum of the external forces over all particles is given by the right-hand side of Equation (5.12). If we demand that the equations of motion are spatially homogeneous such that they have the same functional form for every particle in the system, then the external force acting on each particle required to generate the correct zero-wavevector momentum density, subject to the conservation of thermal momentum, is given by

$$\mathbf{F}_i^e = m_i\mathbf{r}_i \cdot \nabla\mathbf{v}\,\delta(t) + m_i\mathbf{r}_i \cdot \nabla\mathbf{v} \cdot \nabla\mathbf{v}\,\Theta(t). \tag{5.13}$$

We can now identify that the total external force per particle consists of two components: an impulse force at time $t = 0$ at the moment the field is applied, and an additional term that is zero before the application of the field and a constant multiplied by the particle's laboratory position afterwards. This expression for the external force is the *only* one that gives the same equations of motion for each particle and simultaneously satisfies Equations (5.7)–(5.9). It also excludes terms of the form $m_i\mathbf{c}_i \cdot \nabla\mathbf{v}\,\Theta(t)$ with zero sum which cannot contribute to the zero-wavevector component of the momentum density or the total external force.

Expressed in Newtonian form, the equations of motion are

$$\mathbf{F}_i = m_i\frac{d\mathbf{v}_i}{dt}, \tag{5.14}$$

where the total force on particle i is

$$\mathbf{F}_i = \mathbf{F}_i^{\phi} + \mathbf{F}_i^e \tag{5.15}$$

and the external force is given by Equation (5.13).

However, the SLLOD equations of motion are not usually expressed in this form. Instead, a change of variable is performed, in which we define the peculiar or thermal velocity by

$$\dot{\mathbf{r}}_i = \mathbf{c}_i + \mathbf{r}_i \cdot \nabla\mathbf{v}\,\Theta(t). \tag{5.16}$$

Substituting Equation (5.16) into Newton's second law (Equation (5.14)), and then inserting the forces given by Equation (5.15) into the left-hand side of Equation (5.14), gives us a first-order differential equation for the peculiar velocity

$$m_i \dot{\mathbf{c}}_i = \mathbf{F}_i^{\phi} - m_i \mathbf{c}_i \cdot \nabla \mathbf{v} \, \Theta(t) . \tag{5.17}$$

For times $t \geq 0$ we thus have

$$\begin{aligned} \dot{\mathbf{r}}_i &= \mathbf{c}_i + \mathbf{r}_i \cdot \nabla \mathbf{v} \\ m_i \dot{\mathbf{c}}_i &= \mathbf{F}_i^{\phi} - m_i \mathbf{c}_i \cdot \nabla \mathbf{v}. \end{aligned} \tag{5.18}$$

Equations (5.18) are just the first-order version of the SLLOD equations of motion, identical to Equation (5.3) when we define the peculiar momentum as $\mathbf{p}_i = m_i \mathbf{c}_i$. These are normally expressed without the explicit inclusion of the step function, which we have deliberately included in Equations (5.17) for pedagogical purposes. It is important to appreciate that we have solved for the *laboratory* position and the *thermal* velocity, both of which are calculated *relative* to the laboratory reference frame. In a moving reference frame, the streaming velocity would be zero. This is clearly *not* the case in the SLLOD equations of motion, which explicitly include a nonzero streaming velocity. We will again discuss the SLLOD equations of motion in a moving reference frame in Section 5.3, after we introduce the concept of profile biased and unbiased thermostats. This will again emphasise that the SLLOD equations are no more than an expression of Newton's equations of motion with the inclusion of an external force.

We will now consider two specific types of homogeneous flow, namely planar shear and planar elongational flow. For planar shear with flow in the x-direction and gradient in the y-direction, we have

$$\nabla \mathbf{v} = \begin{pmatrix} 0 & 0 & 0 \\ \dot{\gamma} & 0 & 0 \\ 0 & 0 & 0 \end{pmatrix}, \tag{5.19}$$

where $\dot{\gamma} = \partial v_x / \partial y$ is the magnitude of the velocity gradient. In this case $\nabla \mathbf{v} \cdot \nabla \mathbf{v} = 0$ and so the external force described by Equation (5.13) consists of the impulse term only. The flow is generated by an impulse at $t = 0$ and, because the system is infinite in extent (the use of periodic boundary conditions ensures this), the impulse results in a persistent nondecaying zero-wavevector momentum current for all times thereafter. There is no need for a constant driving external field at $t > 0$. This must be the case because there is no zero-wavevector acceleration of the collective system of particles in planar shear flow after the impulse, as expressed by Equation (5.12). Equation (5.6) also shows that we can ignore the effect of stresses because they occur at first order, not zeroth order, in wavevector. The SLLOD equations for this flow geometry reduce to

$$\begin{aligned} \dot{\mathbf{r}}_i &= \mathbf{c}_i + \mathbf{i} \dot{\gamma} y_i \\ m_i \dot{\mathbf{c}}_i &= \mathbf{F}_i^{\phi} - \mathbf{i} \dot{\gamma} m_i c_{yi}, \end{aligned} \tag{5.20}$$

where $\mathbf{i}$ is the unit vector in the x-direction.

For planar elongational flow (PEF), with expansion in the x-direction, contraction in the y-direction, and no field in the z-direction, the velocity gradient tensor is

$$\nabla \mathbf{v} = \begin{pmatrix} \dot{\epsilon} & 0 & 0 \\ 0 & -\dot{\epsilon} & 0 \\ 0 & 0 & 0 \end{pmatrix}, \tag{5.21}$$

where now

$$\nabla \mathbf{v} \cdot \nabla \mathbf{v} = \begin{pmatrix} \dot{\epsilon}^2 & 0 & 0 \\ 0 & \dot{\epsilon}^2 & 0 \\ 0 & 0 & 0 \end{pmatrix} \tag{5.22}$$

and $\dot{\epsilon}$ is the elongational strain rate. For PEF the external force given by Equation (5.13) also has the initial impulse, but in addition contains a term proportional to the laboratory position of the molecule. For the individual components we have

$$\begin{aligned} F_{i,x}^{e} &= m_i \dot{\epsilon} x_i \delta(t) + m_i \dot{\epsilon}^2 x_i \Theta(t) \\ F_{i,y}^{e} &= -m_i \dot{\epsilon} y_i \delta(t) + m_i \dot{\epsilon}^2 y_i \Theta(t) \\ F_{i,z}^{e} &= 0. \end{aligned} \tag{5.23}$$

For planar elongation an impulse force alone is not sufficient in itself to sustain an indefinite flow. An additional external force proportional to $\mathbf{r}_i$ must apply at all times to each particle. This external force induces a zero-wavevector component of the fluid's acceleration that manifests itself as the hyperbolic streaming velocity profile observed in PEF, in which both the magnitude and direction of the velocity of a small element of fluid continually change with time. This is distinct from planar shear flow, in which they remain constant. Equations (5.14) and (5.17) are equivalent ways of expressing Newton's second law and are completely general for homogeneous flows.

For PEF, the SLLOD equations given by Equation (5.18) reduce to

$$\begin{aligned} \dot{\mathbf{r}}_i &= \mathbf{c}_i + \dot{\epsilon}\,(\mathbf{i}x_i - \mathbf{j}y_i) \\ m_i \dot{\mathbf{c}}_i &= \mathbf{F}_i^{\phi} - m_i \dot{\epsilon}\left(\mathbf{i}c_{xi} - \mathbf{j}c_{yi}\right), \end{aligned} \tag{5.24}$$

where $\mathbf{i}$ and $\mathbf{j}$ are unit vectors in the x and y directions, respectively.

The SLLOD algorithm with appropriate periodic boundary conditions is completely compatible with a homogeneous stress tensor. We can see this by first writing down the momentum continuity equation, which is valid for any flow arbitrarily far from equilibrium

$$\rho(\mathbf{r},t)\frac{d\mathbf{v}(\mathbf{r},t)}{dt} = -\nabla \cdot \mathbf{P}(\mathbf{r},t) + \rho(\mathbf{r},t)\,\mathbf{F}^{e}. \tag{5.25}$$

For PEF we have the streaming velocity components

$$\mathbf{v}(\mathbf{r},t) = \dot{\epsilon}\,(x, -y, 0) \tag{5.26}$$

and the acceleration of the fluid at any point $\mathbf{r}$ and time t is

$$\begin{aligned}\frac{d\mathbf{v}(\mathbf{r},t)}{dt} &= \dot{\epsilon}\,(\dot{x},\ -\dot{y},\ 0) \\ &= \dot{\epsilon}\left(v_x,\ -v_y,\ 0\right) \\ &= \dot{\epsilon}^2\,(x,\ y,\ 0) \\ &\neq 0.\end{aligned} \tag{5.27}$$

From Equation (5.25) it is clear that if there is no external force acting on the fluid then $\nabla \cdot \mathbf{P} \neq 0$, which contradicts the assumption of homogeneity. Therefore homogeneous elongational flow can *only* be simulated with the inclusion of a nonzero external force. It is also important to appreciate that elongational flow implies an acceleration in the velocity streamlines. The implications of this in devising the SLLOD algorithm for elongational flow will be discussed in Sections 5.3.3 and 5.4.

In the case of shear flow, we have

$$v_x\,(y) = \dot{\gamma} y \tag{5.28}$$

and so

$$\mathbf{v}\,(\mathbf{r},t) = (\dot{\gamma} y,\ 0,\ 0)\,. \tag{5.29}$$

Therefore,

$$\begin{aligned}\frac{d\mathbf{v}(\mathbf{r},t)}{dt} &= (\dot{\gamma}\dot{y},\ 0,\ 0) \\ &= \left(\dot{\gamma}\, v_y,\ 0,\ 0\right) \\ &= 0,\end{aligned} \tag{5.30}$$

since $v_y = 0$. Unlike elongational flow, there is no acceleration for shear flow *anywhere* in the fluid. The implication here is that planar shear can be sustained in the zero-wavevector limit by an initial impulse at $t = 0$, where $\rho\,(\mathbf{r},t)\,\mathbf{F}^e = 0$ for $t > 0$. This expresses Newton's first law of motion at times $t > 0$.

One could of course drive such systems with Lees-Edwards (shear flow) or Kraynik-Reinelt (planar elongation) boundary conditions and solve Newton's equations without an external driving field. This would generate either steady shear or elongational flows after the decay of initial transients. However, the absence of appropriate terms in the equations of motion means that the use of either Lees-Edwards or Kraynik-Reinelt boundary conditions alone leads to perturbations of the particle trajectories. Since these perturbations do not appear in the equations of motion the link with response theory is lost. This link is important because response theory is the *only* rigorous statistical mechanical means of determining the validity of the equations of motion. Response theory applied to SLLOD dynamics has been proven to yield excellent comparisons with direct NEMD averaging for shear and elongation stresses [2, 21, 23, 51–60, 141] (see Section 6.5).

5.1.3 Relationship to Hamiltonian Mechanics

We previously noted that the SLLOD equations of motion can not be derived from a Hamiltonian. For a Hamiltonian to exist, the external force that appears on the right-hand side of Equation (5.13) should be expressible as a gradient of a potential energy function, or equivalently have zero curl. In the case of the impulsive force, it must satisfy the condition

$$\begin{aligned}\nabla_i \times \mathbf{r}_i \cdot \nabla \mathbf{v} &= \mathbf{i}\left(\frac{\partial v_z}{\partial y} - \frac{\partial v_y}{\partial z}\right) - \mathbf{j}\left(\frac{\partial v_z}{\partial x} - \frac{\partial v_x}{\partial z}\right) + \mathbf{k}\left(\frac{\partial v_y}{\partial x} - \frac{\partial v_x}{\partial y}\right) \\ &= 0. \end{aligned} \tag{5.31}$$

This is satisfied only if the velocity gradient tensor is symmetric. Thus, for planar shear it is not satisfied, whereas for planar elongational flow it is. For PEF this implies an impulse potential energy contribution of

$$\begin{aligned} V &= -\int \mathbf{F} \cdot d\mathbf{r}_i \\ &= -\delta(t)\, m_i \int \mathbf{r}_i \cdot \nabla \mathbf{v} \cdot d\mathbf{r}_i \\ &= -\frac{m_i \dot{\epsilon}}{2}\left(x_i^2 - y_i^2\right)\delta(t). \end{aligned} \tag{5.32}$$

The nonimpulsive external force term must satisfy

$$\begin{aligned}\nabla_i \times \mathbf{r}_i \cdot \nabla \mathbf{v} \cdot \nabla \mathbf{v} &= \mathbf{i}\left(\frac{\partial \mathbf{v}}{\partial y} \cdot \nabla v_z - \frac{\partial \mathbf{v}}{\partial z} \cdot \nabla v_y\right) - \mathbf{j}\left(\frac{\partial \mathbf{v}}{\partial x} \cdot \nabla v_z - \frac{\partial \mathbf{v}}{\partial z} \cdot \nabla v_x\right) \\ &\quad + \mathbf{k}\left(\frac{\partial \mathbf{v}}{\partial x} \cdot \nabla v_y - \frac{\partial \mathbf{v}}{\partial y} \cdot \nabla v_x\right) \\ &= 0, \end{aligned} \tag{5.33}$$

only if the velocity gradient tensor is symmetric. This is true for all types of elongational flow, but is irrelevant for shear flow which has no nonimpulsive external force. Therefore there is no potential energy corresponding to this part of the external force for shear flow, while for planar elongational flow it is

$$\begin{aligned} V &= -\int \mathbf{F} \cdot d\mathbf{r}_i \\ &= -\Theta(t)\, m_i \int \mathbf{r}_i \cdot \nabla \mathbf{v} \cdot \nabla \mathbf{v} \cdot d\mathbf{r}_i \\ &= -\frac{m_i \dot{\epsilon}^2}{2}\left(x_i^2 - y_i^2\right)\Theta(t). \end{aligned} \tag{5.34}$$

This simple analysis shows that a Hamiltonian description for SLLOD dynamics does not exist for planar shear flow, but does exist for planar elongational flow. In fact it exists for all types of elongational flow, such as uniaxial or biaxial stretching.

The general form of the Hamiltonian for flows with a symmetric velocity gradient tensor is

$$H\left(\Gamma',t\right)=\phi\left(\mathbf{r}^{N}\right)+K\left(\mathbf{p}'^{N}\right)+V\left(\mathbf{r}^{N},t\right), \tag{5.35}$$

where ϕ is the intermolecular potential energy, K is the kinetic energy and V is the potential energy due to the external field given by

$$V\left(\mathbf{r}^{N},t\right)=-\tfrac{1}{2}\sum_{i}m_i\left(\mathbf{r}_i\cdot\nabla\mathbf{v}\right)^2\Theta\left(t\right)-\tfrac{1}{2}\sum_{i}m_i\mathbf{r}_i\cdot\nabla\mathbf{v}\cdot\mathbf{r}_i\delta\left(t\right). \tag{5.36}$$

The conjugate momentum $\mathbf{p}'$ is related to the velocity by the usual classical mechanics definition, $\mathbf{p}'_i=\frac{\partial L}{\partial\dot{\mathbf{r}}_i}=m_i\dot{\mathbf{r}}$, consistent with the system Lagrangian. Hamilton's equations of motion applied to Equation (5.35) result in equations of motion equivalent to Equation (5.15), except that the velocity gradient in the impulse term of the external force is just the symmetric part of the velocity gradient. This is consistent with the previous discussion in which we found that a Hamiltonian only exists for flows with a symmetric velocity gradient tensor. Furthermore, if we perform the canonical transformation

$$\begin{aligned}\mathbf{q}_i&=\mathbf{r}_i\\ \mathbf{p}_i&=\mathbf{p}'-m_i\mathbf{r}_i\cdot\nabla\mathbf{v}\Theta\left(t\right)\end{aligned} \tag{5.37}$$

with a generating function for the canonical transformation [42] given by

$$F\left(\Gamma,t\right)=\sum_{i}\mathbf{r}_i\cdot\mathbf{p}_i+\frac{1}{2}\sum_{i}m_i\left(\mathbf{r}_i\cdot\nabla\mathbf{v}\right)^2\Theta\left(t\right), \tag{5.38}$$

we obtain the Hamiltonian in terms of the new variables as

$$\begin{aligned}H_{DOLLS}\left(\Gamma,t\right)&=H+\frac{\partial F}{\partial t}\\ &=\phi\left(\mathbf{q}^{N}\right)+\sum\frac{\mathbf{p}_i^2}{2m_i}+\sum\mathbf{q}_i\cdot\nabla\mathbf{v}\cdot\mathbf{p}_i\Theta\left(t\right),\end{aligned} \tag{5.39}$$

which is just the DOLLS tensor Hamiltonian [20]. This results in the DOLLS equations of motion. This then proves that SLLOD and DOLLS are only equivalent for symmetric velocity gradients, such as is the case for elongation flow. The key – and somewhat subtle – point here is that the explicit time dependence of the Hamiltonian must be taken into account in order to obtain correct results. We will discuss what can go wrong when such subtleties are overlooked or misunderstood in Section 5.4, once the basis of our complete algorithmic derivation is established.

5.1.4 Conservation of Momentum and Dissipation of Energy

Summing over all molecules in Equation (5.17) demonstrates the conservation of total thermal momentum for SLLOD dynamics. This is an exact result and is true for all forms of homogeneous flow. However, finite precision arithmetic can lead to instability problems for elongation flows [132]. The equation of motion for the total thermal momentum given by the SLLOD equations of motion is $\dot{\mathbf{P}}=-\mathbf{P}\cdot\nabla\mathbf{v}$ where

$\mathbf{P} = \sum_i m_i \mathbf{c}_i$ is the total peculiar momentum of the system. If we allow the initial value of the total thermal momentum to be nonzero – contradicting the exact equations of motion, but allowing for a small numerical error – the equation of motion for the thermal momentum for a system under PEF can be solved for each Cartesian component to give [132]

$$\begin{aligned} P_x(t) &= P_x(0) \exp(-\dot{\epsilon} t) \\ P_y(t) &= P_y(0) \exp(\dot{\epsilon} t) \\ P_z(t) &= P_z(0) . \end{aligned} \tag{5.40}$$

If the initial thermal momentum is identically zero, the evolution of the system will always result in zero total thermal momentum at all times. This is an idealisation, and in practice $P_{x,y,z}(0)$ is never identically zero in a computer simulation due to finite numerical precision. Because any nonzero error in total initial momentum in the y-direction grows exponentially in time, this causes serious problems for the simulation if it remains uncorrected. Fortunately, this numerical error can be easily corrected by any one of several numerical procedures, such as periodically subtracting the mean y-component of the peculiar momentum from each particle, by adding proportional feedback to correct the momentum, or by adding a constraint to prevent it from growing [132]. It has also been suggested [142, 143] that fluctuations could be an inherently chaotic source of instability for elongational flows in general. Indeed, simulation studies of the microscopic chaos for particles undergoing planar elongational flow indicates an inherently higher degree of chaoticity compared to those undergoing planar shear flow [144]. No such problem will occur in the x-direction because the total momentum is forced to converge exponentially to zero.

As an example of numerical divergence of the total momentum for planar elongation, consider the case where $P_y(0) = 10^{-12}$ and $\dot{\epsilon} = 0.5$. To reach a value of $P_y(t) = 0.1$ only requires around 50 000 timesteps, where we use an integration timestep of $\Delta t = 10^{-3}$ (all units are in reduced form). This exponential growth was clearly reported and explained in reference [12]. Also shown in that work is $P_x(t)$, which is forced to converge to zero at all times. It is instructive also to observe that this numerical problem is intimately related to dynamical systems theory and chaos, and can in fact be directly related to the Arnold cat map which, as we will discuss further in Section 5.2.1, is equivalent to the Kraynik-Reinelt scheme of boundary conditions [129], but gives greater physical insight into the nature of the flow [142, 143].

Fortunately the solution to this numerical problem is quite simple, as we have already noted. For example, for a single component atomic system, after the calculation of the momenta at each timestep one can subtract the average y-component of momentum from the y-momentum of each particle

$$p_{yi} \rightarrow p_{yi} - P_y(t)/N. \tag{5.41}$$

This perturbation is very small and does not affect the particle dynamics in any meaningful way [144] but does guarantee that momentum is conserved at all times. Alternatively one could apply either a constraint or proportional feedback [132].

In the case of planar shear flow we have

$$\begin{aligned}\dot{P}_x(t) &= -\dot{\gamma}P_y(t)\\ \dot{P}_y(t) &= 0\\ \dot{P}_z(t) &= 0.\end{aligned} \tag{5.42}$$

The second and third equations above imply that $P_y(t) = P_y(0) = 0$ and $P_z(t) = P_z(0) = 0$, because the initial peculiar momenta are set to zero as demanded by SLLOD. This then implies that $P_x(t) = P_x(0) = 0$. This is the exact result for SLLOD and the result that would be achieved if computers had infinite numerical precision. Finite numerical precision of course means that none of $P_{x,y,z}(0)$ are identically zero. There is always some round off error, which means the values of $P_{x,y,z}(0)$ are very small (depending on the floating point precision of the processing chip used) but nonzero. Therefore, assuming small but nonzero $P_{x,y,z}(0)$ we have from the first line of Equation (5.42)

$$P_x(t) = P_x(0) - \dot{\gamma}P_y(0)t. \tag{5.43}$$

In this we see that the numerical error in the total momentum in the x-direction grows linearly in time. Because $P_y(0)$ is very small, and because the growth in the sum of momenta is linear in time, in practice no NEMD simulation of planar shear has ever been significantly affected by lack of momentum conservation. One would need to run a calculation for trillions of timesteps before any problems could be detected. As an example, consider an NEMD simulation with $P_x(0)$ and $P_y(0)$ of the order of 10^{-12} (typically). If $\dot{\gamma} = 0.5$ and the integration timestep is $\Delta t = 10^{-3}$ then it will take two hundred trillion (2×10^{14}) timesteps for $P_x(t)$ to reach a value of 0.1 where stability problems will be catastrophic. There has never been an NEMD simulation of this duration to date. Even when such long simulations are routine this problem of momentum round-off can always be accounted for by suitable rescaling of total momenta, as discussed above for elongational flow.

It is reasonable from the macroscopic nature of the flow (i.e., from the streamlines of the flow themselves) that numerical errors in elongational flow *should* evolve exponentially in time, whereas errors for shear flow evolve linearly. As shear involves linear streamlines there is no nonzero Lyapunov exponent associated with the mapping scheme for periodic boundary conditions (i.e. the macroscopic flow is not chaotic, even though the microscopic flow is [143, 144]). That errors in total momenta grow linearly is one consequence of the nonchaotic nature of the macroscopic flow field. Elongation, on the other hand, has hyperbolic streamlines and nonzero Lyapunov exponents associated with the Kraynik-Reinelt/cat-map mapping scheme, taking on values of $\pm\lambda$ [142–144]. The macroscopic flow is therefore chaotic and must involve the exponential divergence of neighbouring values of the total system momentum. An exponential growth in numerical error is actually compatible with the requirements of a hyperbolic, accelerating, inherently chaotic streaming velocity profile.

It is important to appreciate that this lack of conservation is not due to the SLLOD dynamics. It is solely a feature of finite precision numerics and is unavoidable. In fact, the analysis performed above is itself a simplification of the true picture. In practice one simulates NEMD systems not with adiabatic equations (as the currently written SLLOD equations are), but coupled with a thermostat to extract dissipated heat. Therefore another term $-\alpha \mathbf{p}_i$, where α is the thermostat multiplier, is added on the right hand-side of the second equation in Equation (5.18) in the case of a kinetic (rather than configurational) thermostat, which makes such a simple analysis impossible (thermostats are considered in greater detail in Section 5.3). Furthermore, again due to finite precision numerics, $\sum_i F_{i\beta}^{\phi}$ is not identically zero, where β represents any of the Cartesian coordinates. However, this tends to be a very small effect and can be fairly safely neglected in the above analysis [132].

We conclude this discussion by stressing that the aim of a synthetic NEMD algorithm is to *replace* the boundary conditions that are responsible for generating macroscopic fluxes with synthetic forces in the equations of motion. The goal is to generate the appropriate dissipative flux such that the link with linear response theory becomes explicit and tractable. This is the great virtue of SLLOD: the direct link between classical mechanics that defines the microscopic dynamics of the system and statistical mechanics and irreversible thermodynamics. It is therefore essential that the rate of change of internal energy should only consist of a bilinear expression involving the appropriate flux and synthetic force.

Finally, we observe that the SLLOD equations of motion do generate the correct rate of energy dissipation. This is obtained by taking the time derivative of the total internal energy $U = \sum_i \frac{1}{2} m_i \mathbf{c}_i^2 + \frac{1}{2} \sum_{ij} \phi_{ij}(r_{ij})$ and is found to be [1, 2, 143]

$$\dot{U}^{SLLOD} = -V\mathbf{P}^T:\nabla\mathbf{u} = \begin{cases} -VP_{xy}\dot{\gamma} & planar\,shear \\ -V\dot{\epsilon}\left(P_{xx} - P_{yy}\right) & planar\,elongation \end{cases}. \tag{5.44}$$

This is of course precisely the energy dissipation obtained from classical irreversible thermodynamics (see for example Equation (2.37) and the derivation thereof in Section 2.3.3).

5.1.5 Compatible Periodic Boundary Conditions

We now consider the implementation of suitable periodic boundary conditions that are compatible with the SLLOD equations of motion for generalised homogeneous flows. There is one governing principle which dictates how the boundaries must evolve in time, namely they must be compatible with the imposed streaming velocity profile. To see this clearly, we first write down the strain rate tensor as

$$\nabla \mathbf{v} = \begin{pmatrix} \frac{\partial v_x}{\partial x} & \frac{\partial v_y}{\partial x} & \frac{\partial v_z}{\partial x} \\ \frac{\partial v_x}{\partial y} & \frac{\partial v_y}{\partial y} & \frac{\partial v_z}{\partial y} \\ \frac{\partial v_x}{\partial z} & \frac{\partial v_y}{\partial z} & \frac{\partial v_z}{\partial z} \end{pmatrix}. \tag{5.45}$$

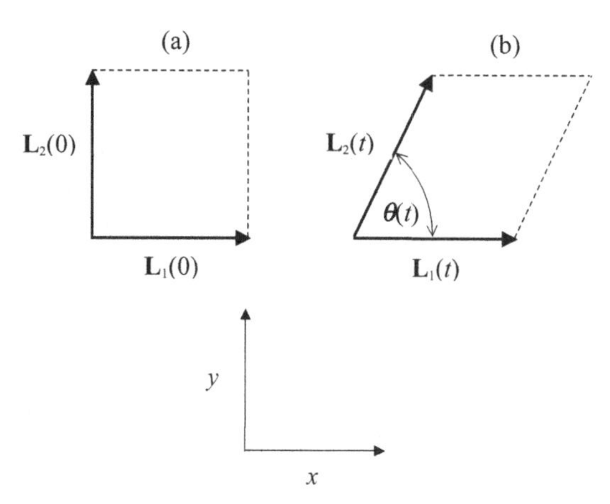

Figure 5.2 Two-dimensional representation of primitive cell lattice vectors, $\mathbf{L}_1$ and $\mathbf{L}_2$, (a) $t = 0$; (b) $t > 0$, under planar shear. As $t \to \infty$, $\theta(t) \to 0$ and $|\mathbf{L}_2| \to \infty$, whereas $\mathbf{L}_1$ remains unchanged.

The SLLOD equation of motion for particle velocities with explicit time dependence included for clarity in the following discussion (Equation (5.16)) is

$$\dot{\mathbf{r}}_i = \frac{\mathbf{p}_i}{m_i} + \mathbf{r}_i \cdot \nabla \mathbf{v}\, \Theta(t) . \tag{5.46}$$

This equation states that the total *laboratory* velocity $\dot{\mathbf{r}}_i$ (*not* the peculiar velocity, which is defined by the first term on the right-hand side of this equation, i.e. $\mathbf{p}_i/m_i$) is a superposition of the streaming velocity at particle *i*'s position ($\mathbf{r}_i \cdot \nabla \mathbf{v}$) and the thermal (peculiar) velocity ($\mathbf{p}_i/m_i$) *relative* to the streaming velocity. The evolution of the boundaries (or indeed of any point not subject to thermal motion) is governed by the *same* equation of motion, except that there is no thermal velocity for the boundaries, i.e.

$$\dot{\mathbf{L}}_k(t) = \mathbf{L}_k(t) \cdot \nabla \mathbf{v}\, \Theta(t) , \tag{5.47}$$

where $\mathbf{L}_k(t) = (L_{kx}(t), L_{ky}(t), L_{kz}(t))$ are the initially orthogonal set of primitive lattice vectors (cell or box vectors) that define the axes of the boundaries for $k = 1, 2, 3$, and we have explicitly included the time dependence for clarity (see Figure 5.2). Thus the spatial evolution of the boundaries is solved by integrating Equation (5.47). The introduction of equations of motion that allowed a flow to be induced by an explicit external force, rather than by direct deformation of the simulation box, has major benefits. One of these is that the nonequilibrium response of the system can be analysed mathematically using linear and nonlinear response theory [2]. It follows from this that if we want our mathematical analysis to be correct, no other external forces than those explicitly imposed on the system should be included. It is therefore essential that the periodic boundary conditions should not impose a force on the system and the way to achieve this is to allow the simulation box to evolve according to the kinematics of the induced flow (Equation (5.47)).

The acceleration of the box vertex must therefore be given by

$$\ddot{\mathbf{L}} = \mathbf{L} \cdot \nabla \mathbf{v} \cdot \nabla \mathbf{v} \, \Theta(t) + \mathbf{L} \cdot \nabla \mathbf{v} \delta(t) . \tag{5.48}$$

This is identical to the acceleration of a noninteracting particle with no thermal momentum given by the SLLOD equations of motion, as is easily shown by setting the peculiar velocity to zero and then taking the time derivative of Equation (5.46). Imagine a noninteracting particle with an initial velocity of zero. At time $t = 0$, the external field is applied. With properly formulated equations of motion, this particle's motion at subsequent times should be solely determined by the external force that generates the flow. In this case, the equation of motion given by the SLLOD equations would be

$$\begin{aligned} m_i \frac{d^2 \mathbf{r}_i}{dt^2} &= m_i \mathbf{r}_i . \nabla \mathbf{v} \, \delta(t) + m_i \mathbf{r}_i \cdot \nabla \mathbf{v} \cdot \nabla \mathbf{v} \, \Theta(t) \\ &= \frac{d}{dt} \left(m_i \mathbf{r}_i . \nabla \mathbf{v} \, \Theta(t) \right) . \end{aligned} \tag{5.49}$$

Integrating this gives the expected particle velocity, $\mathbf{v}(t) = \mathbf{r}_i . \nabla \mathbf{v} \; \Theta(t)$. In Section 5.4 we will give a brief discussion on a number of problems that occur with incorrectly formulated equations of motion that lead to particle trajectories incompatible with the evolving boundary conditions and energy dissipation inconsistent with what is known from hydrodynamics. We now consider the explicit cases of shear and elongational flows and the evolution of periodic boundary conditions.

5.1.5.1 Shear Flow

For the shear flow depicted in Figure 5.1, the strain rate tensor (Equation (5.45)) is given by Equation (5.19). Equation (5.47) therefore reduces to

$$\begin{aligned} \dot{L}_{kx}(t) &= L_{ky}(t) \dot{\gamma} \\ \dot{L}_{ky}(t) &= 0 \\ \dot{L}_{kz}(t) &= 0. \end{aligned} \tag{5.50}$$

The second and third lines in Equation (5.50) imply that the box vectors in the y and z directions do not change in time. For a simulation of length t_s (starting from $t = 0$), the boundaries of the simulation box in the x-direction grow linearly in time as

$$L_{kx}(t_s) = L_{ky} \dot{\gamma} \, t_s, \tag{5.51}$$

where $L_{ky} \equiv L_{ky}(0)$. This evolution of L_{kx} is depicted in Figure 5.2, in which the two initially orthogonal vectors are depicted as $\mathbf{L}_1$ and $\mathbf{L}_2$. An initially square simulation box (drawn in two-dimensions but equally valid arguments apply for a three-dimensional cube) is deformed in the x-direction (Figure 5.2(b)). If atoms leave the left/right faces of the box, they are returned into the box through the right/left faces. If they move through the top/bottom faces, they are returned to the bottom/top faces, but displaced in the x-direction by an amount $\mp L_{2y} \dot{\gamma} \, \Delta t$ where Δt is the simulation timestep. If $\mathbf{L}_1$ is parallel to the x-axis at $t = 0$, it is clear from Equation (5.50) that $\mathbf{L}_1$ is the same for all times afterwards. Since $\mathbf{L}_2$ is orthogonal to $\mathbf{L}_1$ at $t = 0$ (hence parallel to the y-axis at $t = 0$), Equation (5.50) tells us that it will evolve for $t > 0$. Denoting $\theta(t)$ as the

Algorithm 5.1 Pseudocode for Lees-Edwards sliding brick pbcs for shear

$\delta x = 0$; {set strain to zero at start of simulation}
for t = start $\rightarrow$ end **do** {loop over simulation time}
  $\delta x \leftarrow \delta x + \dot{\gamma} * \Delta t$; {compute strain at time t}
  $\delta x \leftarrow \delta x - (\text{int})\delta x$; {reset to zero after one box length traversed}
  call integrator;
  while executing integrator **do**
    for $i = 1 \rightarrow N$ **do** {loop over particles}
      $x_i \leftarrow x_i - \delta x * \text{nint}\left(y_i/L_y\right) * L_y$ {top/bottom crossings, see Figure 5.3}
      $x_i \leftarrow x_i - \text{nint}\left(x_i/L_x\right) * L_x$; {left/right crossings, see Figure 5.3}
      $y_i \leftarrow y_i - \text{nint}\left(y_i/L_y\right) * L_y$;
      $z_i \leftarrow z_i - \text{nint}\left(z_i/L_z\right) * L_z$;
    end for
  end while
end for

angle between $\mathbf{L}_1$ and $\mathbf{L}_2$, then simple geometry gives $\theta(t_s) = \tan^{-1}(1/\dot{\gamma}\, t_s)$. Clearly the situation becomes untenable as $t \rightarrow \infty$ because $\theta(t) \rightarrow 0$ and $|\mathbf{L}_2| \rightarrow \infty$.

Computationally this is very inefficient as one deals with increasingly larger numbers in the implementation of the equations of motion, in which the $\mathbf{r}_i$ can now become very large in magnitude, even though the $\mathbf{p}_i$ do not. Fortunately there are two simple geometrical operations that save the simulator from this problem. These two operations can be implemented in two different but equivalent algorithms for the application of periodic boundary conditions. The first of these is to let the simulation box deform until it reaches some angle, $\theta_c(t)$, at which time the box is transformed back into the original square (cubic) shape. In this way transformations occur relatively infrequently and box side lengths do not grow too large. An implementation of this algorithm has been performed for $\theta_c(t) = \pi/4$, and a slight modification allows the box to be transformed back to an initial angle of $-\pi/4$, thus halving again the number of transformations [145].

The second, equivalent, method is to just use the Lees-Edwards "sliding brick" implementation of pbcs. As already discussed, image boxes above and below the simulation box are displaced relative to it by $\mp L_{2y}\dot{\gamma}\,\Delta t$ respectively. Pseudocode for this scheme is presented in Algorithm 5.1 and the equivalence of both schemes is depicted in Figure 5.3. Note that the only difference between these pbcs and those for an equilibrium system is the first line in the for-loop, which accounts for crossings of particles through the top/bottom boundaries and requires a shift in the x-coordinate due to the shear flow field. The algorithm is shown for particle boundary crossings, but exactly the same procedure is used for minimum image distance computations, except that now these distances might be computed within the force and neighbour list routines instead of within the numerical integrator.

Both Lagrangian-rhomboid and sliding brick schemes are similarly efficient computationally, though the former is more useful for implementation in parallel computing

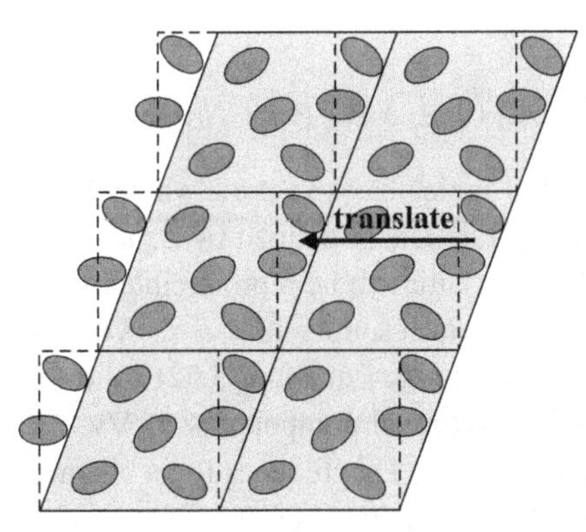

Figure 5.3 Equivalence of Lagrangian-rhomboid (LR) and sliding-brick (SB) periodic boundary conditions depicted in 2D. SB boundaries are constructed from LR boundaries by trimming the triangular regions shown and translating by L_x to create squares (cubes in 3D). Stacking LR cells or SB cells on top of each other results in two entirely equivalent systems because the relative distances between all molecules remain the same. This results in the equivalence of all thermodynamic properties.

environments [145, 146]. It is of note that for both schemes the boundary conditions are explicitly time-dependent or nonautonomous in nature [56]. For sufficiently large systems, this time dependence is not observable and poses no problem in the computation of time-independent properties of thermodynamic systems far from equilibrium.

5.1.5.2 Elongational Flow

The problem with performing NEMD simulations of a fluid undergoing elongational flow is that, if one uses conventional "square" periodic boundary conditions, the simulation can only continue for a finite time. At a time given by $t_{\max} = \frac{1}{\dot{\epsilon}} \ln\left(\frac{2r_c}{L_y}\right)$ [147] the box length in the contracting direction reaches its minimum allowable value of twice the cut-off radius. This finite simulation time meant that steady-state simulations of important molecularly structured fluids, such as polymer melts, in which elongational flow is central to processing rheology, could never be attained due to the large molecular relaxation times. Various inventive schemes were devised to try to extend this maximum time [147, 148] but the issue was not satisfactorily addressed until an analogous scheme for the implementation of nonautonomous pbcs for elongational flow was devised. This scheme is based upon the Kraynik-Reinelt (KR) boundary conditions. Kraynik and Reinelt [129] demonstrated that several types of lattice structures are reproducible under planar extensional flow. One such structure is the cubic lattice. Todd and Daivis [130] and, independently, Baranyai and Cummings [133] showed that this reproducibility can be successfully utilised as periodic boundary conditions for NEMD simulations of indefinite planar elongational flow. In a later paper Todd and Daivis [131] developed an efficient scheme of pbcs for planar elongational flow analogous to both the sliding brick and deforming box schemes used for shear.

In their derivation of a suitable set of spatio-temporally reproducible lattices, Kraynik and Reinelt [129] first construct an arbitrary lattice consisting of points $\mathbf{L}_i(t)$ governed by the evolution equation

$$\mathbf{L}_i(t) = \mathbf{L}_i(0) \cdot \mathbf{\Lambda} = N_{i1}\mathbf{L}_1(0) + N_{i2}\mathbf{L}_2(0) + N_{i3}\mathbf{L}_3(0). \tag{5.52}$$

$\mathbf{L}_i(0)$ are the linearly independent initial lattice basis vectors, where $\mathbf{\Lambda} = \exp(\nabla\mathbf{v}t)$ is the evolution matrix and $\nabla\mathbf{v}$ is the strain rate tensor defined by Equation (5.45), which for PEF is given by Equation (5.21). For a lattice to be reproducible (periodic in space and time), the constants $N_{ij}(t)$ must be integers at some time $t = \tau_p$. Kraynik and Reinelt demonstrated which sets of integers $N_{ij}(\tau_p)$ make Equation (5.52) valid for a variety of extensional flows described by different diagonal components of $\nabla\mathbf{v}$. These integers define the mapping, hereafter termed KR pbcs. This reduces to an eigenvalue problem with corresponding eigenvectors. For PEF the eigenvalues are

$$\lambda_1 = \frac{k + \sqrt{k^2 - 4}}{2}; \quad \lambda_2 = \lambda_1^{-1} = \frac{k - \sqrt{k^2 - 4}}{2}, \tag{5.53}$$

where k is an integer with allowable values $k \geq 3$. Not all values of k are allowed but there are an infinite number of them.

The directions of the two orthogonal eigenvectors, $\mathbf{e}_1$ and $\mathbf{e}_2$, define the directions of the expanding and compressing fields with respect to the orientation of the basis vectors of the square lattice. If the x-axis is chosen as the direction of expansion at a rate $\dot{\epsilon}$, and the y-axis is the direction of contraction at a rate $-\dot{\epsilon}$, then the angle between $\mathbf{e}_1$ (parallel to the x-axis) and the lattice basis vector $\mathbf{L}_1(0)$ is the orientation angle θ (see Figure 5.4). There are an infinite number of these "magic" angles which the initial lattice (or, equivalently, simulation cell) can be aligned with respect to the flow fields that guarantee the simulation cell is infinitely periodic in both space and time. It is this spatio-temporal periodicity, coupled with the SLLOD equations of motion, that allows us to perform NEMD simulations of PEF indefinitely. If τ_p is the lattice strain period, then the actual mapping of the extended simulation cell back into its original cell shape occurs when the value of the strain, $\epsilon = \dot{\epsilon}\tau_p$, equals the so-called Hencky strain

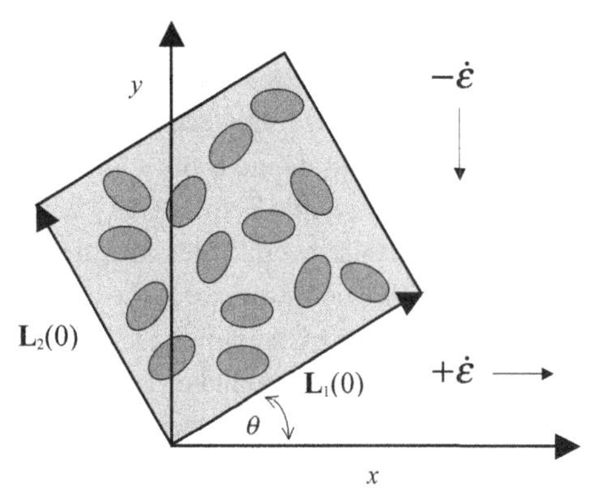

Figure 5.4 Initial orientation of a simulation box with KR boundary conditions [129–131]. The x and y axes define the directions of expansion and contraction respectively, with an elongational strain rate $\dot{\epsilon}$. The simulation box is oriented at an angle θ with respect to the x-axis.

[129], ϵ_p. This will be made clearer in the discussion that follows. Unfortunately this spatio-temporal mapping only works for PEF. There are no reproducible (periodic) lattices for other types of extensional flow such as uniaxial or biaxial stretching [129], though reproducible lattices for mixed PEF and shear can be found [134]. A 2D depiction of a simulation box for implementation with KR pbcs is shown in Figure 5.4. While the primitive lattice vectors $\mathbf{L}_1$ and $\mathbf{L}_2$ are again orthogonal, they are now no longer parallel to the x and y axes, which themselves are defined as the directions of expansion and contraction, respectively. For $t > 0$ Equation (5.47) therefore becomes

$$\begin{aligned}\dot{L}_{kx}(t) &= L_{kx}(t)\,\dot{\epsilon} \\ \dot{L}_{ky}(t) &= -L_{ky}(t)\,\dot{\epsilon} \\ \dot{L}_{kz}(t) &= 0,\end{aligned} \tag{5.54}$$

which, when solved, leads to the exponentially growing/contracting evolution of lattice vectors for planar elongation

$$\begin{aligned}L_{kx}(t) &= L_{kx}(0)\exp(\dot{\epsilon}\,t) \\ L_{ky}(t) &= L_{ky}(0)\exp(-\dot{\epsilon}\,t) \\ L_{kz}(t) &= L_{kz}(0)\,.\end{aligned} \tag{5.55}$$

The simulation box deforms (stretching in x and contracting in y) until such time that the total strain equals the periodic strain amplitude, ϵ_p, and the box is remapped back into its original shape. This is demonstrated in Figure 5.5. The re-mapping preserves all relative distances and ensures there are no discontinuities in computed physical properties. An alternative derivation to that of Kraynik and Reinelt was performed by Hunt and Todd [142, 143], in which the equivalence of the KR scheme and the famous Arnold cat map of dynamical systems theory was demonstrated. The cat map derivation is simpler and more intuitive, and we discuss this and the general relationship between dynamical maps and pbcs in Section 5.2.1 below. Evans and Morriss also provide an alternative derivation in the second edition of their book [2].

The implementation of pbcs in the reference frame shown in Figure 5.5 is not efficient. Efficiency can be improved significantly by performing a rotation of the simulation box about the origin and performing all necessary pbc operations (computation of minimum image distances and translation of particles as they cross boundaries) in the rotated frame [131]. It is important to appreciate that *no particle dynamics takes place in this rotated frame*. Only pbc operations that are dynamics-independent take place in the rotated frame. Once pbc operations are performed, the box is rotated back to its original orientation and the simulation (i.e. particle dynamics) continues for the next time-step, and so on indefinitely. This scheme is illustrated in Figure 5.6. Let $\mathbf{r}_i$ be the position vector of atom i at time t in the simulation box and $\mathbf{r}'_i$ be its rotated (transformed) position vector. $\mathbf{r}_i$ and $\mathbf{r}'_i$ are related to each other via a transformation matrix $\mathbf{M}(t)$ such that

$$\mathbf{r}'_i = \mathbf{M}(t)\,\mathbf{r}_i, \tag{5.56}$$

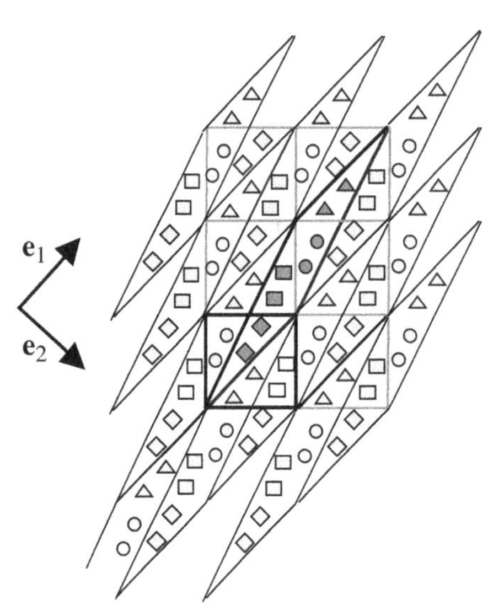

Figure 5.5 Schematic representation of the KR simulation box and periodic boundary conditions. At $t = 0$ the simulation box is square (bottom left square box). An elongation field is applied for $t \geq 0$, with direction and magnitude given by eigenvectors $\mathbf{e}_1$ (direction of expansion, x-axis) and $\mathbf{e}_2$ (direction of compression, y-axis). As time evolves the square box deforms into a parallelogram. At a time corresponding to a strain given by $\epsilon = \epsilon_p$ (the Hencky strain, or equivalently in the cat map scheme, the positive Lyapunov exponent of the mapping operation) the cell is stretched to its maximum extent, depicted by the thick black boundaries. At this time the cell is mapped back onto the original square, and all molecules are likewise mapped back onto the square. The mapping preserves all relative spaces between molecules (each molecule is represented by different shapes). In this way the KR simulation box and its replicated periodic images (depicted by thin grey boundaries and semi-transparent molecules) is *identical* to the system of the mapped square cell and its replicated periodic images (depicted by thick black boundaries and *exactly* the same semi-transparent molecules). The replicated parallelograms and squares and their associated molecules extend infinitely in space.

where the transformation matrix is

$$\mathbf{M}(t) = \begin{pmatrix} \cos\theta(t) & \sin\theta(t) & 0 \\ -\sin\theta(t) & \cos\theta(t) & 0 \\ 0 & 0 & 1 \end{pmatrix} \tag{5.57}$$

and $\theta(t)$ is the angle between $\mathbf{L}_1$ and the x-axis. After minimum image distances and pbcs are applied in the rotated frame the periodically imaged particles $P(\mathbf{r}_i)$ are inverse-rotated back into the original simulation frame

$$P(\mathbf{r}_i) = \mathbf{M}^{-1}(t) P'\left(\mathbf{r}_i'\right), \tag{5.58}$$

where

$$\mathbf{M}^{-1}(t) = \begin{pmatrix} \cos\theta(t) & -\sin\theta(t) & 0 \\ \sin\theta(t) & \cos\theta(t) & 0 \\ 0 & 0 & 1 \end{pmatrix} \tag{5.59}$$

and $P'(\mathbf{r}_i')$ is the periodically imaged particle i in the $\mathbf{r}'$ frame. Once again dynamics are applied, where now the time has advanced to $t + \Delta t$, and the procedure is repeated. It is

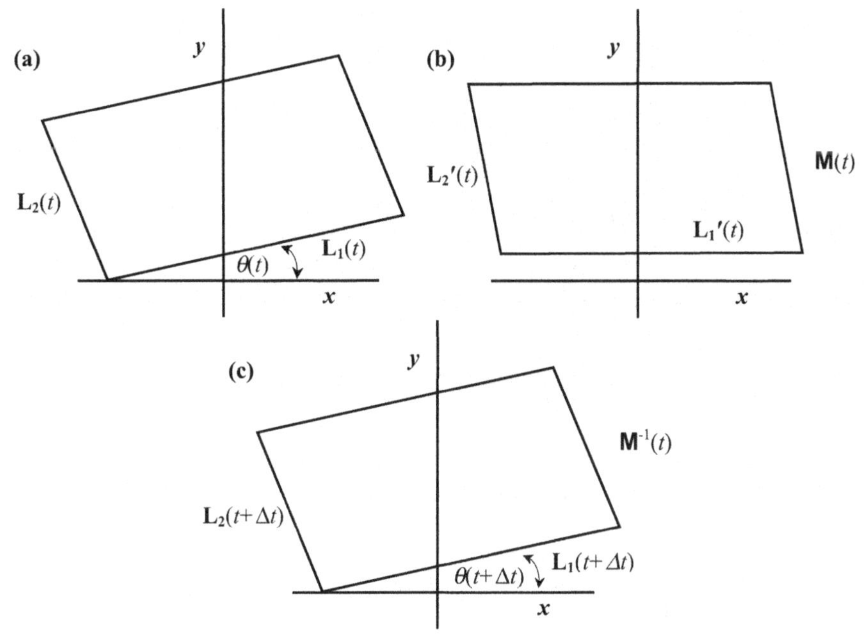

Figure 5.6 Schematic diagram for application of pbcs and minimum image distance computations for PEF. In (a) the simulation has been running for time t. The vectors $\mathbf{L}_1$ and $\mathbf{L}_2$ define the boundaries of the cell. Expansion occurs in the x-direction, contraction in the y-direction, and the cell is oriented at an angle $\theta(t)$ with respect to $\mathbf{L}_1$. (b) The cell is now rotated by $\theta(t)$ with respect to the x-axis. pbcs and minimum image distances are computed in this reference frame but no particle dynamics takes place. This is accomplished via the transformation operation $\mathbf{M}(t)$ given by Equations (5.56) and (5.57). (c) Once pbcs and minimum image distances have been computed, the cell is rotated by $-\theta(t)$ back to the configuration in (a) and dynamics proceeds for the next incremental timestep Δt. The inverse rotation is accomplished by the operation $\mathbf{M}^{-1}(t)$, given by Equation (5.58). Steps (b) and (c) are then repeated on this new configuration, and so on until the end of the simulation. Note that at times corresponding to $t = n\tau_p$, where n is an integer, the fully extended KR cell is mapped back into the original square cell, as described in Figure 5.5.

simple to show that θ evolves as

$$\begin{aligned}\theta(t) &= \tan^{-1}\left(L_{1y}(t)/L_{1x}(t)\right) \\ &= \tan^{-1}\left(L_{1y}(0)/L_{1x}(0)\right)\exp(-2\dot{\epsilon}t).\end{aligned} \tag{5.60}$$

When the strain is equivalent to ϵ_p the entire simulation box is mapped back into the original shape, i.e. $\mathbf{L}_k(nt_p) \to \mathbf{L}_k(0)$ where t_p is the time that corresponds to a strain of ϵ_p (the Hencky strain) and n is a positive integer. ϵ_p is itself computed from the logarithm of the positive Lyapunov exponent [143] which is easily determined by solving the eigenvalue equation for the mapping operation [142]. It is this mapping operation that allows for indefinite simulation times and is analogous to the mapping of a shear cell back into the original shape for any particular desired value of $\theta(t)$ (see previous section). Note that there are in principle an infinite number of initial orientations of the

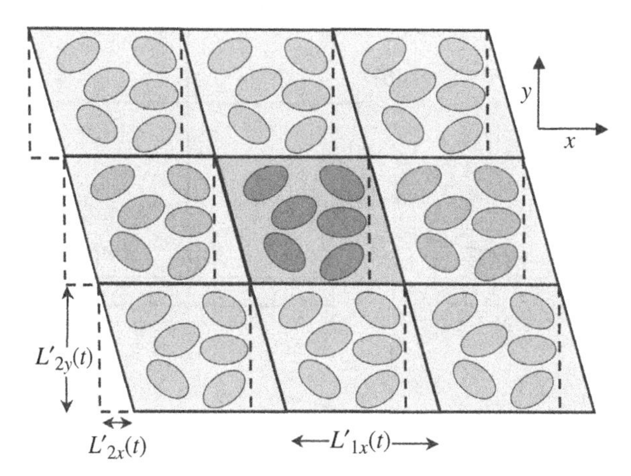

Figure 5.7 Schematic diagram of Lagrangian rhomboid (LR) and deforming brick (DB) periodic boundary conditions. The simulation cell is the central box with molecules, surrounded by an infinite number of periodic image boxes. The LR boundaries are outlined in thick solid lines, whereas the DB boundaries are dashed. The equivalence of the two schemes is seen by cutting the triangular shaped region in the simulation cell, translating it by $-L'_{1x}(t)$ and then replicating the "brick" infinitely in space.

simulation cell (i.e. infinitely many possible $\theta(0)$), but in practice one chooses values of $\theta(0)$ that ensure the minimum extension of the simulation cell at $t = nt_p$ must be $\ll 2r_c$, where r_c is the interaction potential cut-off radius. This ensures that there is no violation of the minimum image convention.

As there are two equivalent schemes for implementation of pbcs in shear (sliding brick and deforming cube) so too are there two equivalent schemes for elongation [131]. These schemes are implemented in the rotated reference frame, and are shown in Figure 5.7. In the deformed box (or Lagrangian rhomboid, LR) scheme, particles that move out of the left/right sides are translated back into the right/left sides by an amount $\pm L'_{1x}(t)$. Particles that move out of the top/bottom are imaged back into the bottom/top but displaced by $\pm L'_{2x}(t)$ in the x-direction and $\mp L'_{2y}(t)$ in the y-direction. Equivalently, the particles can move out of boundaries defined by the dashed lines in Figure 5.7. In this case, known as the deforming brick (DB) scheme, the cell is a rectangular brick that deforms in time. The translation of particles in and out of box sides is treated in exactly the same way, except that it is slightly simpler because the box boundaries are parallel to the y-axis. Pseudocode for the LR implementation is given in Algorithm 5.2.

Note that while Algorithms 5.1 and 5.2 are presented primarily for the application of pbcs for atomic coordinates, they are equally applicable for minimum image vectors $\mathbf{r}_{ij}$ between pairs of particles. See references [125, 131] for further details of minimum image implementations. See also reference [131] for implementation of the elongational flow algorithm for the DB scheme of pbcs. Furthermore, it is assumed that calls to force and neighbour list routines will also be made within the integrator routine, and pbcs for

Algorithm 5.2 Pseudocode for Lagrangian-rhomboid pbcs for elongation

$t_p = 0$; {initialise lattice strain time to zero}
for t = start $\rightarrow$ end **do** {loop over simulation time}
  $t_p \leftarrow t_p + \Delta t$;
  if $t_p = t_p$ **then** {map box back to original shape for integer multiples of strain period}
    $t_p = 0$; {reset lattice strain time to zero}
    $L_{1x} \leftarrow L_{1x}(0)$;
    $L_{1y} \leftarrow L_{1y}(0)$;
    $L_{2x} \leftarrow L_{2x}(0)$;
    $L_{2y} \leftarrow L_{2y}(0)$;
    $\theta \leftarrow \theta(0)$;
    $\mathbf{L}'_{1,2} \leftarrow \mathbf{M}(t)\mathbf{L}_{1,2}$; {rotate box lattice vectors to align with x-axis via Equation (5.56)}
    for $i = 1 \rightarrow N$ **do** {loop over particles}
      $\mathbf{r}'_i \leftarrow \mathbf{M}(t)\mathbf{r}_i$; {rotate particle coordinates to align with rotated box vectors via Equation (5.56)}
      $x'_i \leftarrow x'_i - \text{nint}\left(y'_i/L'_{2y}\right) * L'_{2x}$; {top/bottom translations}
      $y'_i \leftarrow y'_i - \text{nint}\left(y'_i/L'_{2y}\right) * L'_{2y}$; {top/bottom translations}
      $\delta x'_i \leftarrow \left(x'_i - y'_i * L'_{2x}/L'_{2y}\right)$;
      $x'_i \leftarrow x'_i - \text{nint}\left(\delta x'_i/L'_{1x}\right) * L'_{1x}$; {left/right translations}
      $z'_i \leftarrow z'_i - \text{nint}\left(z'_i/L'_z\right) * L'_z$; {all z translations}
      $\mathbf{r}_i \leftarrow \mathbf{M}^{-1}(t)\mathbf{r}'_i$; {rotate particle coordinates back to simulation box coordinate system via Equation (5.58)}
    end for
    $\mathbf{L}_{1,2} \leftarrow \mathbf{M}^{-1}(t)\mathbf{L}'_{1,2}$; {rotate box lattice vectors back to simulation box coordinate system via Equation (5.58)}
  end if
  $L_{1x} \leftarrow L_{1x}(0) * \exp\left(\dot{\epsilon}t_p\right)$; {evolve box lattice vectors}
  $L_{1y} \leftarrow L_{1y}(0) * \exp\left(-\dot{\epsilon}t_p\right)$;
  $L_{2x} \leftarrow L_{2x}(0) * \exp\left(\dot{\epsilon}t_p\right)$;
  $L_{2y} \leftarrow L_{2y}(0) * \exp\left(-\dot{\epsilon}t_p\right)$;
  $\theta \leftarrow \arctan\left(L_{1y}/L_{1x}\right)$; {evolve box angle}
  call integrator;
  while executing integrator **do**
    for $i = 1 \rightarrow N$ **do** {loop over particles}
      apply pbcs as above to all particles and minimum image vectors $\mathbf{r}_{ij}$;
    end for
  end while
end for

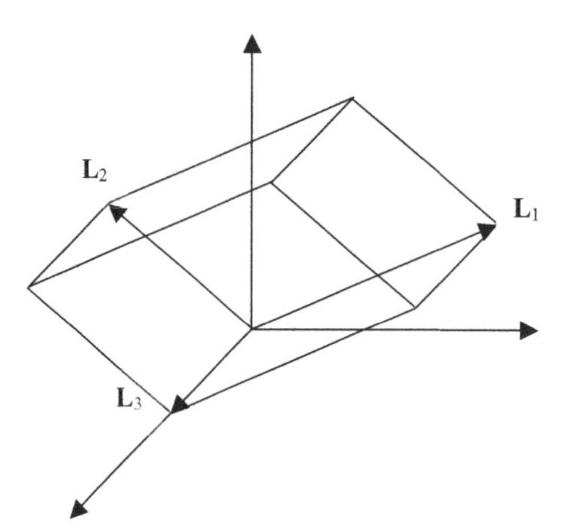

Figure 5.8 Simulation box constructed from nonorthogonal box vectors.

particle coordinates and minimum image distances will similarly be implemented within them by the same algorithms depicted here.

5.1.5.3 Periodic Boundary Conditions for Arbitrary Parallelepiped Boxes

When we apply pbcs to the particle position vectors $\mathbf{r}_i$ or to interparticle separation vectors to determine the minimum image separation of a pair of particles, it may be convenient to use a method that applies generally, regardless of whether the flow is an elongational, shear, or bulk (isotropic) deformational flow. The method described here is a general method suitable for this purpose. The essential idea behind this method is that we express the particle position as a linear combination of the box vectors $\mathbf{L}_1$, $\mathbf{L}_2$ and $\mathbf{L}_3$, which are not necessarily orthogonal (see Figure 5.8)

$$\mathbf{r}_i = a_i\mathbf{L}_1 + b_i\mathbf{L}_2 + c_i\mathbf{L}_3. \tag{5.61}$$

The values of a_i, b_i, and c_i, will be real numbers and their fractional parts are the coefficients of vectors $\mathbf{L}_1$, $\mathbf{L}_2$ and $\mathbf{L}_3$ that locate the image particle in the primary simulation box. The value of a_i is determined by finding the scalar product of both sides of Equation (5.61) with a vector that is perpendicular to both $\mathbf{L}_2$ and $\mathbf{L}_3$, i.e.

$$\mathbf{r}_i \cdot \mathbf{L}_2 \times \mathbf{L}_3 = a_i\mathbf{L}_1 \cdot \mathbf{L}_2 \times \mathbf{L}_3, \tag{5.62}$$

giving

$$a_i = \frac{\mathbf{r}_i \cdot \mathbf{L}_2 \times \mathbf{L}_3}{\mathbf{L}_1 \cdot \mathbf{L}_2 \times \mathbf{L}_3} = \frac{\mathbf{r}_i \cdot \mathbf{L}_2 \times \mathbf{L}_3}{V}, \tag{5.63}$$

where it is assumed that the box vectors are defined in such a way that the volume $V = \mathbf{L}_1 \cdot \mathbf{L}_2 \times \mathbf{L}_3$ is positive. Similarly, b_i and c_i are found by

$$b_i = \frac{\mathbf{r}_i \cdot \mathbf{L}_3 \times \mathbf{L}_1}{\mathbf{L}_2 \cdot \mathbf{L}_3 \times \mathbf{L}_1}$$
$$c_i = \frac{\mathbf{r}_i \cdot \mathbf{L}_1 \times \mathbf{L}_2}{\mathbf{L}_3 \cdot \mathbf{L}_1 \times \mathbf{L}_2}. \tag{5.64}$$

The position vector of the image of particle i that is located in the primary simulation box is then given by

$$\mathbf{r}'_i = \mathbf{r}_i - ([\text{nint}(a_i - 0.5)]\,\mathbf{L}_1 + [\text{nint}(b_i - 0.5)]\,\mathbf{L}_2 + [\text{nint}(c_i - 0.5)]\,\mathbf{L}_3), \quad (5.65)$$

where nint(x) represents the nearest integer to x.

5.2 Dynamical Maps and the Relationship to Periodic Boundary Conditions

As we have seen so far, periodic boundary conditions are essential for molecular dynamics simulations of spatially homogeneous systems. What is not well known is that these boundary conditions are equivalent to spatio-temporal mappings.

A dynamical map is an operation transforming the state of a system at time t to a well-defined state at time $t + 1$. Maps have been particularly useful in the study of dynamical systems theory and chaos. A map can be expressed as

$$\mathbf{x}_{t+1} = \mathbf{M}(\mathbf{x}_t) \quad (5.66)$$

where $\mathbf{x}$ is an n-dimensional vector describing the state of the system and $\mathbf{M}$ is the mapping operation. In equilibrium MD simulations, one can depict particle positions through a mapping operation (see Figure 5.9) expressed mathematically as [142]

$$\mathbf{x}' = \mathbf{M}^E \mathbf{x} \bmod (L), \quad (5.67)$$

where here $\mathbf{M}^E$ is the unit tensor and L is the length of the simulation cell (assumed in this case to be equal in all three spatial dimensions). The actual mapping/translation of particles with coordinates $\mathbf{x}_i = (x_i, y_i, z_i)$ into those with coordinates $\mathbf{x}'_i = (x'_i, y'_i, z'_i)$ is a result of the modulo operation. This mapping operation is clearly equivalent to the scheme of periodic boundary conditions used in typical equilibrium MD simulations within a square simulation box. It says that a particle inside the box stays inside the box. If it moves outside the box in any direction, it returns to the box through a translation of length L in that direction. In the case of planar shear flow (PSF) with flow in x and velocity gradient in y the mapping scheme can be represented by

$$\mathbf{x}' = \mathbf{M}^{PSF} \mathbf{x} \bmod (L), \quad (5.68)$$

where now the PSF map is given by

$$\mathbf{M}^{PSF} = \begin{pmatrix} 1 & \dot{\gamma} t & 0 \\ 0 & 1 & 0 \\ 0 & 0 & 1 \end{pmatrix}. \quad (5.69)$$

In addition to left-right translations of a particle's position if the left-right boundaries are crossed, we must also account for the convective part of the flow as particles cross the top or bottom boundaries. In this case the x-position of the particle is shifted by an

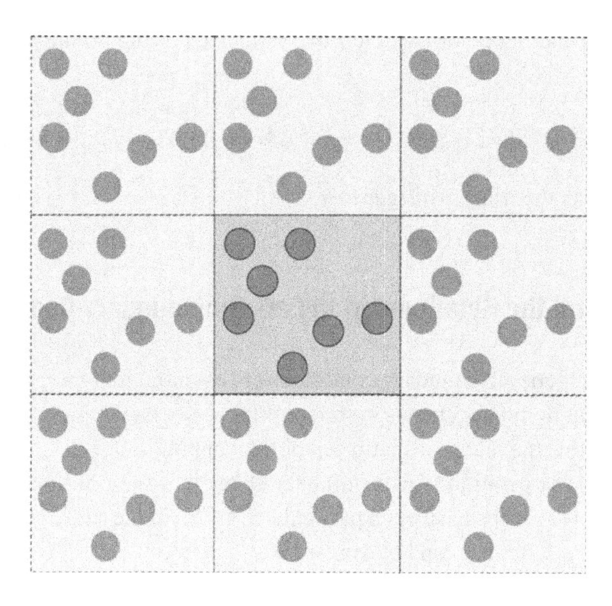

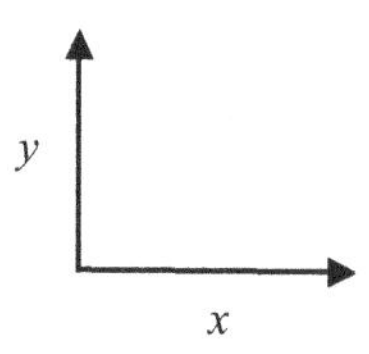

Figure 5.9 Periodic boundary conditions for a two dimensional system of particles in equilibrium. The central square represents the simulation box, while all other boxes are its periodic images.

amount $\pm\dot{\gamma}L\Delta t$, as previously discussed, if the particle moves out of the bottom or top boundary, respectively.

In this respect the maps described by Equations (5.67) and (5.68) represent the convective motion of the fluid. They give us no information about atomic thermal motion which evolve under appropriate equations of motion, such as the SLLOD equations. Rather, they tell us how a macroscopic volume of fluid evolves as a function of time. At equilibrium there is no convective flow, therefore the simulation box remains cubic for all time. For PSF the box deforms in conformity with the macroscopic flow field. It commences as a cube and deforms into a three-dimensional rhomboid. At times $t = \tau_p = \dot{\gamma}^{-1}$ the cell is in fact periodic in space [2], which allows us to map the deformed rhomboid back into the original cubic box as shown in Figure 5.3. This temporal mapping implies that homogeneous simulations of PSF are nonautonomous in nature. However, the effects of periodicity are only observable in very small systems of a few particles [56]. They are insignificant in NEMD simulations of realistic systems. Importantly, however, discontinuities in the flow can *only* be eliminated through an appropriate coupling of the equations of motion to the system's periodic boundary conditions. For PSF the most commonly used combination is the SLLOD equations of motion with Lees-Edwards boundary conditions.

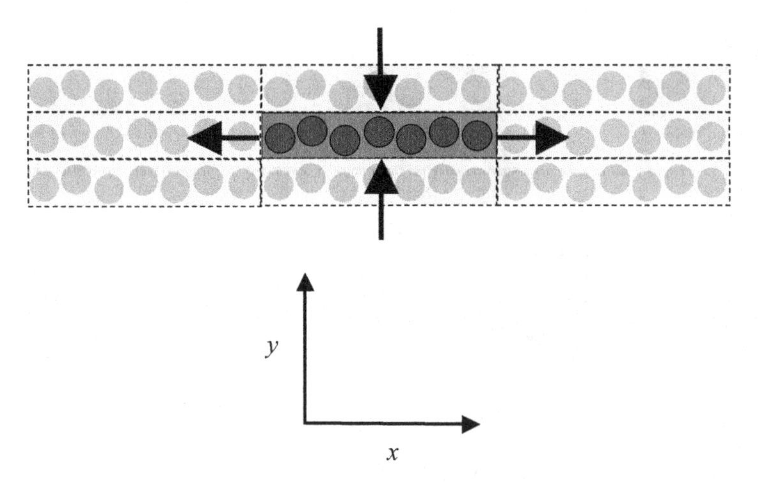

Figure 5.10 Traditional rectangular periodic boundary conditions for PEF in two dimensions. The simulation box is the central dark grey region. Periodic images (lighter grey) surround the cell in all directions. Expansion takes place in x while the box contracts in y.

In the case of planar elongational flow (PEF) the mapping scheme for particle positions is given by

$$\mathbf{x}' = \mathbf{M}^{PEF}\mathbf{x} \bmod (\mathbf{L}), \tag{5.70}$$

where

$$\mathbf{M}^{PEF} = \begin{pmatrix} \exp(\dot{\varepsilon}t) & 0 & 0 \\ 0 & \exp(-\dot{\varepsilon}t) & 0 \\ 0 & 0 & 1 \end{pmatrix} \tag{5.71}$$

and $\mathbf{L} = (L_x, L_y, L_z)^T$, depicted schematically in Figure 5.10. The flow is such that exponential expansion occurs in the x-direction with concomitant exponential contraction in the y-direction, with no flow in the z-direction. For this traditional rectangular geometry, in which the flow fields align with the box axes, the simulation will end when the length in the contracting direction (L_y) reaches its minimum possible value of twice the effective interaction potential radius (r_c). This, as already observed, gives a maximum simulation time of $t_{\max} = \frac{1}{\dot{\epsilon}} \ln\left(\frac{2r_c}{L_y}\right)$ [147], which will be substantially less than typical relaxation times of molecular fluids. As noted earlier, it was this limitation that made NEMD simulations of steady state elongational flow unfeasible for many years, until the introduction of K-R pbcs.

5.2.1 The Arnold Cat Map and Kraynik-Reinelt Boundary Conditions

The Arnold cat map is a famous example of a map of the torus $T^2 = R^2/Z^2$ onto itself, defined by

$$\begin{pmatrix} x' \\ y' \end{pmatrix} = \begin{pmatrix} 1 & 1 \\ 1 & 2 \end{pmatrix} \begin{pmatrix} x \\ y \end{pmatrix} \bmod(1). \tag{5.72}$$

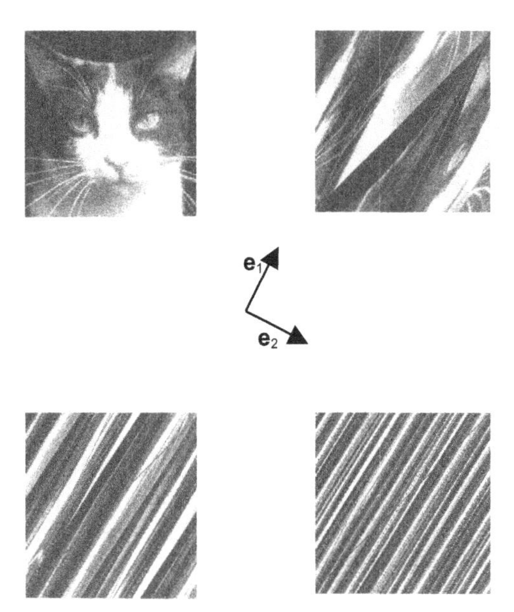

Figure 5.11 The Arnold cat map of Equation (5.72). (a) Unperturbed cat. After (b) one, (c) two and (d) three iterations of the map. This repetitive stretching and folding operation leads to the chaoticity of the map. Figure reproduced by permission from Oxford University Press [149].

There are an infinite number of possible representations of this map and, in this typical depiction, expansion is in the eigendirection $\mathbf{e}_1$ and contraction occurs in the eigendirection $\mathbf{e}_2$ (see Figure 5.11). This map takes a set of points in two dimensions $\mathbf{x} \in R^2$ and stretches it in the direction $\mathbf{e}_1$ while compressing it in the direction $\mathbf{e}_2$. The modulus operation ensures that all points undergoing the cat map deformation are remapped back into the unit cell. This leads to a discrete set of "stretch and fold" operations, rather like stretching and folding dough. It is this stretching and folding operation that makes this map chaotic and a rich example in dynamical systems theory. If we think of this stretching/compressing/folding process as a flow in which periodic boundary conditions operate (via the folding operation), the cat map flow is seen to be opposite to that used by Kraynik and Reinelt and consequent NEMD simulations of planar elongational flow [130, 131, 142].[2] While the flow is in two dimensions, an actual NEMD simulation cell is generally three dimensional in which no flow takes place in the z direction for PEF.

The mapping operation given by Equation (5.72) has two eigenvalues,

$$\lambda_+ = \frac{3+\sqrt{5}}{2} > 1 \quad \text{and} \quad \lambda_+^{-1} \equiv \lambda_- = \frac{3-\sqrt{5}}{2} < 1, \tag{5.73}$$

[2] In the original Kraynik and Reinelt paper [129] stretching occurs in the x-direction, while compression is in y. The same scheme was used by us in the implementation of the scheme for PEF [130, 131]. The traditional Arnold cat map, however, has stretching in $\mathbf{e}_1$ and compression in $\mathbf{e}_2$. Thus stretching is more pronounced in y (vertical), while compression is dominant in x (horizontal), as seen in Figure 5.11.

which are equivalent to a particular pair found by Kraynik and Reinelt. For the cat map, this corresponds to an alignment of the direction of expansion (in this case, parallel to the eigenvector $\mathbf{e}_1$) at an angle of $\theta = 31.7°$ with respect to the y-axis (equivalent to the angle formed between $\mathbf{e}_2$ and the x-axis). Due to the area preserving properties of the map, the product of any conjugate pair of eigenvalues is 1. The eigenvectors of the map described by Equation (5.72) are

$$\mathbf{e}_1 = \begin{pmatrix} s \\ \frac{1+\sqrt{5}}{2}s \end{pmatrix} \quad \text{and} \quad \mathbf{e}_2 = \begin{pmatrix} s \\ \frac{1-\sqrt{5}}{2}s \end{pmatrix}, \tag{5.74}$$

where s is a real number. Typically s is set to unity.

It is important to appreciate that the cat map transformation is equivalent to the mapping that takes place in the Kraynik-Reinelt scheme when the Hencky strain, $\epsilon_p = \dot{\epsilon}\tau_p$, is attained. It is this mapping which makes the Kraynik-Reinelt system reproducible and periodic in both space and time.

The relationship between the cat map and reproducibility of the K-R lattice is made clearer in Figure 5.12. In 5.12(a) the cat map is shown after one iteration, but before the modulo operation (i.e. before folding back into the unit cell). We show different regions of the cell by different symbols (diamonds, squares, circles and triangles) for clarity. In Figure 5.12(b) the modulo operation (folding) is applied, and the deformed cell is remapped into the unit cell. A visual inspection of Figures 5.12(a) and 5.12(b) demonstrates how the different regions of the deformed cell are mapped back into the unit square. The inherent periodicity in space of such a system is further elucidated in Figure 5.12(c), in which the deformed cell of Figure 5.12(a) is again shown, but this time replicated with periodic images of itself. It is now clear that the periodic images of the original unit cell exactly match the sections of the original elongated cell that are mapped back into the unit square by the modulo operation shown in Figure 5.12(b). This visually demonstrates that the modulo operation of the cat map is equivalent to alignment of the elongated cell with infinite periodic images of itself. This is precisely what is done to simulate indefinite PEF by NEMD with K-R boundary conditions. Thus, the cat map and the K-R scheme are identical (compare with Figure 5.5, which depicts the infinitely periodic system described here). This relationship was first pointed out by Hunt and Todd [142] and is another example of how seemingly disparate concepts in physics (in this case, dynamical systems theory and extensional flow) can be analogous and pedagogically instructive.

The example given in Equation (5.72) is a typical but not unique case of this type of map. There are an infinite number of cat maps but each map must satisfy three constraints, namely **M** must be symmetric, it must have integer elements and unit determinant (area preserving). These maps are known collectively as hyperbolic toral automorphisms, which in turn are a set of Anosov diffeomorphisms [150]. The family of lattices found by Kraynik and Reinelt [129] corresponds to a subset of hyperbolic toral automorphisms [142].

It is simpler to compute the eigenvalues, eigenvectors and orientation angles from the cat map formalism [142], rather than the method first used by Kraynik and Reinelt

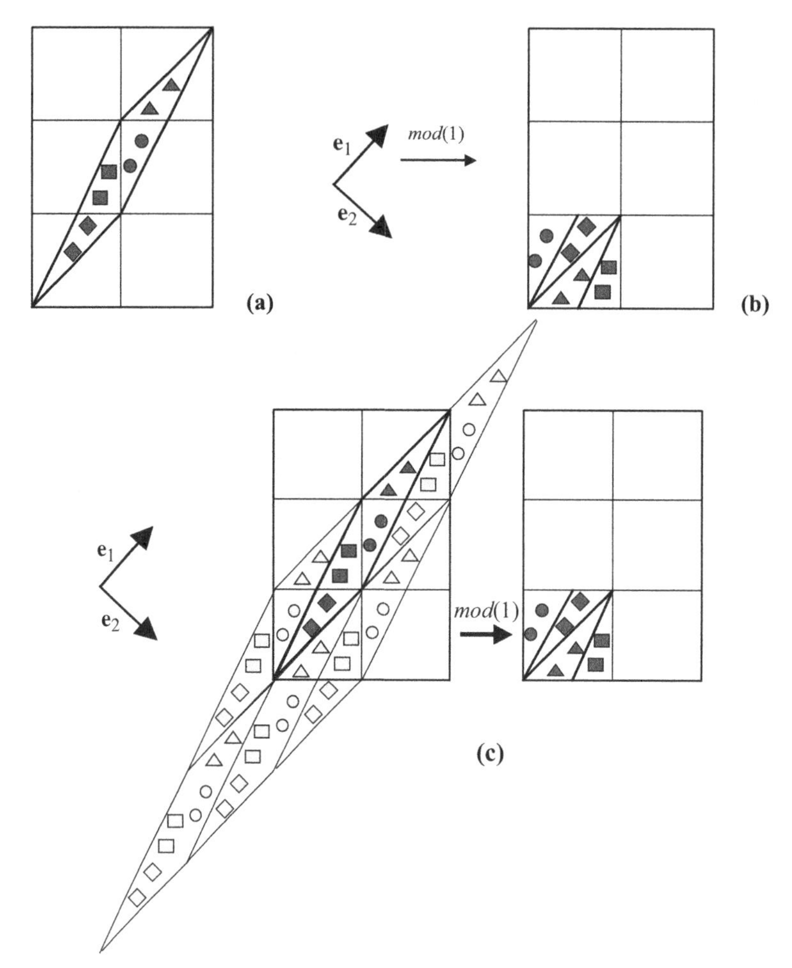

Figure 5.12 (a) The cat map after stretching along $\mathbf{e}_1$ and contraction along $\mathbf{e}_2$ but before the modulo operation. Different regions of the map are drawn with different symbols to aid visualisation of the modulo (i.e. folding) operation. (b) After the modulo operation. (c) Equivalence of the modulo operation with infinite periodic images of the elongated cell in (a).

[129]. To see this, consider again the map given by Equation (5.66) with

$$\mathbf{M} = \begin{pmatrix} m_1 & m_2 \\ m_2 & m_3 \end{pmatrix}, \tag{5.75}$$

being symmetric and $m_1, m_2, m_3 \in Z$. The determinant is

$$|\mathbf{M}| = m_1 m_3 - m_2^2 = 1. \tag{5.76}$$

Solving the eigenvalue equation for the map given by Equation (5.75) gives the conjugate eigenvalue pair

$$\lambda_{\pm} = \frac{(m_1 + m_3) \pm \left[(m_1 + m_3)^2 - 4\right]^{1/2}}{2}. \tag{5.77}$$

The Hencky strain is $\epsilon_p = \ln(\lambda_+)$ [142]. The Hencky strain is thus equivalent to the maximum Lyapunov exponent for the cat map. It is simultaneously the value of strain at which the simulation cell can be mapped back into the original box and the logarithm of the product of the rate of exponential divergence of neighbouring regions of phase space with the mapping iteration time (in this case, unity).

The eigenvalue equation above (5.77) is identical to that found by Kraynik and Reinelt (Equation (5.53)), where we note that $k = \mathrm{Tr}(\mathbf{M})$. The eigenvectors of $\mathbf{M}$ are readily evaluated as [142]

$$\mathbf{e}_1 = \begin{pmatrix} 1 \\ \frac{(\lambda_+ - m_1)}{m_2} \end{pmatrix} \quad \text{and} \quad \mathbf{e}_2 = \begin{pmatrix} 1 \\ \frac{(\lambda_- - m_1)}{m_2} \end{pmatrix}, \tag{5.78}$$

where we have again set $s = 1$ for simplicity and we also note that $m_2 = 0$ does not correspond to an Anosov diffeomorphism [142].

By defining the unit vector in the y-direction as $\hat{\mathbf{y}}$, the cell orientation angle (angle between the expanding direction and the y-axis) is evaluated from

$$\theta = \cos^{-1}\left(\frac{\mathbf{e}_1 \cdot \hat{\mathbf{y}}}{|\mathbf{e}_1|}\right). \tag{5.79}$$

This derivation of the eigenvalues, eigenvectors and orientation angles is simpler and more intuitive than that presented by Kraynik and Reinelt. We note here that Evans and Morriss in the second edition of their book have now generalised this derivation for the infinite number of "cat maps" [2].

As a specific example, consider the Arnold cat map defined by Equation (5.72). It has eigenvalues $\lambda_\pm = \frac{3\pm\sqrt{5}}{2}$ and eigenvectors $\mathbf{e}_1 = \begin{pmatrix} 1 \\ \frac{1+\sqrt{5}}{2} \end{pmatrix}$ and $\mathbf{e}_2 = \begin{pmatrix} 1 \\ \frac{1-\sqrt{5}}{2} \end{pmatrix}$. The Hencky strain/maximum Lyapunov exponent ($\ln \lambda_+$) is 0.9624 and the orientation angle with respect to the y-axis (from Equation (5.79)) is $\theta = 31.7°$. This is equivalent to the K-R mapping used previously by us in NEMD simulations of PEF [130, 131], except that in our geometry expansion occurred in the x-direction, hence θ was defined with respect to the x-axis.

The cat map Lyapunov exponents are only related to the convective part of the flow. They give no information at all about the $2dN$ Lyapunov exponents for a system of N interacting particles in d dimensions. These positive Lyapunov exponents are indicative of microscopic chaos, and are present for equilibrium and shearing systems. However, these two systems have zero Lyapunov exponent for the convective part of their flow. On this basis, we previously postulated that a system of particles undergoing elongational flow will be more chaotic than systems at equilibrium or under shear flow due to their strongly chaotic convective nature [143]. This was later confirmed from NEMD simulations of shear and elongational flow [144, 151, 152] and it is tempting to speculate that this might be the microscopic origin of the rheological instability in polymer solutions and melts subject to elongational flow. Extensional flows are highly unstable and experimental measurements of their physical properties are difficult to perform for this reason. This would be a worthwhile phenomenon to study by the techniques developed here.

Our discussion in the sections above demonstrates the similarities and differences between the boundary conditions for planar elongation and planar shear [131]. In Chapter 6 we will discuss how these periodic boundary conditions can be implemented in efficient NEMD code with some pseudocode, and present a few results for atomic fluids under shear and various forms of elongational flow. We will also discuss the implementation of pbcs for mixed flows of planar shear and elongation.

5.3 Thermostats

Work done by stresses will result in viscous heating. This is true whether the source of the work is an external field (such as is the case for SLLOD) or the boundaries (such as shear flow generated solely by moving boundaries, e.g. Lees-Edwards pbcs without the application of the SLLOD equations). As energy is dissipated in the system its temperature rises. To maintain the system at a thermodynamic steady-state this excess heat must be removed. In a real system with thermostatted boundaries, heat can flow naturally to the boundaries, but in homogeneous flow simulations with periodic boundary conditions in all directions and no walls, this cannot occur. Instead, this is normally accomplished by the use of appropriately formulated synthetic thermostats. Over the years various thermostats have been devised, but for NEMD systems the two most favoured are the Gaussian isokinetic thermostat [153] and the Nosé-Hoover thermostat [154–156]. Recently the so-called configurational thermostat [81, 83, 157–164] has also been shown to be advantageous in certain circumstances. It has been shown that both Gaussian and Nosé-Hoover isokinetic thermostats give identical steady-state averages of material properties [165, 166] and have equivalent nonlinear responses [167]. It has also been demonstrated that material properties computed via artificially thermostatted SLLOD dynamics are identical to those computed in systems that attempt to more faithfully model nature. For example, Liem *et al.* [168] conducted a careful comparison between a homogeneous shearing system under SLLOD thermostatted dynamics and a wall-driven planar shear flow system, in which a liquid was sandwiched between solid walls and the walls allowed to move with equal and opposite constant velocities. By keeping the confinement length large they were able to generate a system that was essentially homogeneous in the central part of the channel. Computations of viscosity and pressure at equivalent thermodynamic state points were in excellent agreement with corresponding values computed via SLLOD dynamics. Padilla and Toxvaerd also performed a similar study [169] in which thermostatted shear flow was investigated for atomic and molecular fluids. They concluded that shear rates should be sufficiently low so that the rate of production of viscous heat is not significantly greater than the characteristic rate of heat transport through the system. Indeed a similar conclusion was also reached by Liem *et al.* [168]. The conclusions reached by these studies demonstrate that as long as one is careful in the implementation of a suitably formulated thermostat and one is working within reasonable rates of energy dissipation, then the synthetic SLLOD algorithm can be trusted to provide accurate values for the linear transport properties of fluids even when they may not be experimentally measurable. It has been shown that properties that depend nonlinearly on the shear rate can, however, depend strongly on the details of the thermostat when the directional components of the thermal kinetic energy are thermostatted [170]. In what follows we briefly review the Gaussian, Nosé-Hoover and configurational thermostats for atomic fluid systems and pay particular attention to their correct implementation. Molecular thermostats will be discussed later in Chapter 8.

5.3.1 Gaussian Thermostat

Thermodynamic constraints are nonholonomic by nature and nonintegrable. Peculiar kinetic energy constraints are one type of nonholonomic constraint that can be used to fix the temperature of a system. In general, for a system under the influence of an external field and constrained by a Gaussian thermostat, the equations of motion can be written as

$$\begin{aligned}\dot{\mathbf{r}}_i &= \frac{\mathbf{p}_i}{m_i} + \mathbf{C}_i \cdot \mathbf{F}^e(t) \\ \dot{\mathbf{p}}_i &= \mathbf{F}_i^{\phi} + \mathbf{D}_i \cdot \mathbf{F}^e(t) - \alpha \mathbf{p}_i. \end{aligned} \tag{5.80}$$

Here $\mathbf{C}_i$ and $\mathbf{D}_i$ are vector phase variables that express the coupling of the field to the system, $\mathbf{F}^e(t)$ is an external generalised tensorial force, α is the thermostat multiplier and the other variables are as previously defined. The equation for the thermostat multiplier α can be straightforwardly derived from Gauss's principle of least constraint [2]. In doing so, it must be assumed that $\mathbf{p}_i$ are defined relative to the streaming velocity of the fluid (i.e. $\mathbf{p}_i \equiv m_i [\mathbf{v}_i - \mathbf{v}(\mathbf{r}_i)]$, where $\mathbf{v}_i$ is the laboratory frame velocity of particle i and $\mathbf{v}(\mathbf{r}_i)$ is the fluid streaming velocity at $\mathbf{r}_i$), that the sum of these peculiar (i.e. thermal) momenta are identically zero (see the discussion in Sections 5.1.2 and 5.1.4 above) and that the total internal energy is given by the sum of the total peculiar kinetic energy and the total interatomic potential energy. The thermostat multiplier is found to be

$$\alpha = \frac{1}{\left(\sum_i \frac{\mathbf{p}_i^2}{m_i}\right)} \left[\sum_i \frac{\mathbf{p}_i}{m_i} \cdot \mathbf{F}_i^{\phi} + \sum_i \frac{\mathbf{p}_i}{m_i} \cdot \mathbf{D}_i \cdot \mathbf{F}^e(t) \right]. \tag{5.81}$$

In the case of the SLLOD equations of motion with no time dependence in the strain rate tensor, Equations (5.80) and (5.81) reduce to

$$\begin{aligned}\dot{\mathbf{r}}_i &= \frac{\mathbf{p}_i}{m_i} + \mathbf{r}_i \cdot \nabla \mathbf{v} \\ \dot{\mathbf{p}}_i &= \mathbf{F}_i^{\phi} - \mathbf{p}_i \cdot \nabla \mathbf{v} - \alpha \mathbf{p}_i \end{aligned} \tag{5.82}$$

and

$$\alpha = \frac{\sum_i \frac{\mathbf{p}_i}{m_i} \cdot \left(\mathbf{F}_i^{\phi} - \mathbf{p}_i . \nabla \mathbf{v}\right)}{\sum_i \frac{\mathbf{p}_i^2}{m_i}}. \tag{5.83}$$

We note here that a time-dependent velocity gradient tensor $\nabla \mathbf{v}(t)$ is completely compatible with SLLOD and has in fact been used to study time-periodic shear [54–59, 171] and elongational flows [60, 147, 172]. This will be formally proven in Chapter 6. Equations (5.80) and (5.81) will keep the temperature fixed to the desired target value instantaneously. A more physically intuitive way of seeing this is to write the total peculiar kinetic energy $\sum_i \frac{\mathbf{p}_i^2}{2m_i}$ which *defines* the kinetic temperature as

$$T_K(t) \equiv \frac{1}{(dN - N_c) k_B} \sum_i \frac{\mathbf{p}_i^2}{m_i}, \tag{5.84}$$

where N_c is the number of constraints on the system (including constraints for conserved quantities), d is the dimensionality of the system and all other terms are as previously defined (see also Equation (3.29)). Taking the time derivative of this expression, substituting $\dot{\mathbf{p}}_i$ in the generalised equations of motion, Equation (5.80), and setting the resulting expression equal to zero (since $\dot{T}_K(t) = 0, \ \forall t$) gives Equation (5.81).

For planar shear flow the Gaussian-thermostatted SLLOD equations of motion are given by

$$\begin{aligned} \dot{\mathbf{r}}_i &= \frac{\mathbf{p}_i}{m_i} + \mathbf{i}\dot{\gamma} y_i \\ \dot{\mathbf{p}}_i &= \mathbf{F}_i^{\phi} - \mathbf{i}\dot{\gamma} p_{yi} - \alpha \mathbf{p}_i, \end{aligned} \tag{5.85}$$

where the thermostat multiplier, α is

$$\alpha = \frac{\sum\limits_i \frac{1}{m_i}\left(\mathbf{F}_i^{\phi} \cdot \mathbf{p}_i - \dot{\gamma} p_{xi} p_{yi}\right)}{\sum\limits_i \frac{\mathbf{p}_i \cdot \mathbf{p}_i}{m_i}}. \tag{5.86}$$

In the case of Gaussian-thermostatted planar elongational flow, the governing SLLOD equations are

$$\begin{aligned} \dot{\mathbf{r}}_i &= \frac{\mathbf{p}_i}{m_i} + \dot{\epsilon}\left(\mathbf{i} x_i - \mathbf{j} y_i\right) \\ \dot{\mathbf{p}}_i &= \mathbf{F}_i^{\phi} - \dot{\epsilon}\left(\mathbf{i} p_{xi} - \mathbf{j} p_{yi}\right) - \alpha \mathbf{p}_i \end{aligned} \tag{5.87}$$

with

$$\alpha = \frac{\sum\limits_i \frac{1}{m_i}\left(\mathbf{F}_i^{\phi} \cdot \mathbf{p}_i - \dot{\epsilon}\left[p_{xi}^2 - p_{yi}^2\right]\right)}{\sum\limits_i \frac{\mathbf{p}_i \cdot \mathbf{p}_i}{m_i}}. \tag{5.88}$$

Gaussian constraints do suffer from one undesirable feature, namely that in the limit as $t \to \infty$ the constrained quantity (in this case the temperature) will drift away from its target value due to the accumulation of numerical error. This is true either at equilibrium or under nonequilibrium conditions. A solution to this problem is straightforward [173] and involves adding a proportional feedback term into the force equation (second equation in Equation (5.80)) via a slightly modified thermostat multiplier, i.e.

$$\alpha \rightarrow \alpha + \alpha' \left[\left(\sum_i \mathbf{p}_i^2 / m_i - (dN - N_c) k_B T \right) \Big/ (dN - N_c) k_B T \right] \tag{5.89}$$

where α' is a weighting term chosen such that it is large enough to correct for numerical drift, but not too large so that the equations of motion do not become stiff. Typical values range between 0.1 and 10. The term in square brackets represents the proportion of deviation between the actual kinetic temperature and the desired target temperature T. If the kinetic temperature and target temperature are identical the second term in Equation (5.89) is zero and no feedback is applied. If they are not the same (as is the case in practice), then the amount of feedback added into the constraint term is directly proportional to the deviation. For practical application of the thermostat, it is advisable to first set $\alpha' = 0$ to make sure the thermostat is working correctly and is not maintaining temperature only because of the feedback term. Once this is tested and confirmed, a nonzero value of α' can be used.

The Gaussian thermostat generates the so-called isokinetic ensemble. At equilibrium this ensemble is characterised by the distribution function

$$f_T(\mathbf{\Gamma}) = \frac{\exp\left[-\beta\phi(\mathbf{\Gamma})\right] \delta\left(K(\mathbf{\Gamma}) - K(\mathbf{\Gamma}_0)\right)}{\int d\mathbf{\Gamma} \exp\left[-\beta\phi(\mathbf{\Gamma})\right] \delta\left(K(\mathbf{\Gamma}) - K(\mathbf{\Gamma}_0)\right)}, \tag{5.90}$$

where K is the kinetic energy and all other terms are as previously defined. The streaming motion of a point in phase space is characterised by a nonzero time derivative df_T/dt and is associated with the compression of phase space.

One can show [174] that there are an infinite number of ways of thermostatting a system, namely by fixing the sum of the momenta raised to an arbitrary power, $\sum_i |p_i|^{\mu+1} = c$, where c is a constant (see also reference [175]). Interestingly, only one unique value of μ, namely $\mu = 1$, minimises the phase space compression. This value of μ corresponds to the traditional Gaussian isokinetic thermostat. It is also the only value of μ for which the conjugate pairing rule [176, 177] remains valid. All other $\mu \neq 1$ thermostats actually perform work on the system and only a $\mu = 1$ thermostat allows for an equilibrium state (see also references [167] and [175] for related work).

5.3.2 Nosé-Hoover Thermostat

In 1984 Nosé developed his now famous integral feedback thermostat and associated equations of motion [154, 155] that preserve an initial canonical distribution for all time and for all system sizes. This approach was made more useful for simulations by Hoover [156], and the resulting thermostat has been known thereafter as the Nosé-Hoover thermostat. The form of the equations of motion is essentially the same as Equation (5.80) except now the thermostat multiplier is not a simple function of positions and momenta (as is the case for Gaussian constraints) but instead is obtained by solving an additional equation of motion. For a system under the influence of an external field, the

Nosé-Hoover equations of motion are given as

$$\begin{aligned}
\dot{\mathbf{r}}_i &= \frac{\mathbf{p}_i}{m_i} + \mathbf{C}_i \cdot \mathbf{F}^e(t) \\
\dot{\mathbf{p}}_i &= \mathbf{F}_i^{\phi} + \mathbf{D}_i \cdot \mathbf{F}^e(t) - \zeta \mathbf{p}_i \\
\dot{\zeta} &= \frac{1}{Q}\left[\sum_i \frac{\mathbf{p}_i^2}{m_i} - N_f k_B T\right].
\end{aligned} \tag{5.91}$$

Here ζ is the Nosé-Hoover thermostat multiplier, and the target temperature T is related to the target kinetic energy K_0 by $T = 2K_0/N_f k_B$, where N_f is the number of degrees of freedom. Q is a parameter associated with an additional degree of freedom coupled to an external heat reservoir and should be chosen to correctly determine the average kinetic energy and its fluctuations [178, 179]. This additional degree of freedom is what essentially scales the particle velocities to the desired kinetic temperature. For the SLLOD equations of motion, $\mathbf{C}_i = \mathbf{r}_i$, $\mathbf{D}_i = -\mathbf{p}_i$ and $\mathbf{F}^e = \nabla \mathbf{v}$.

The field-free Nosé-Hoover equations of motion generate an equilibrium distribution that is canonical in the extended phase space

$$f_c(\mathbf{\Gamma}, \zeta) = \frac{\exp\left[-\beta\left(U + \frac{1}{2}Q\zeta^2\right)\right]}{\int d\mathbf{\Gamma} d\zeta \exp\left[-\beta\left(U + \frac{1}{2}Q\zeta^2\right)\right]}. \tag{5.92}$$

After Nosé's initial work it was shown that the Nosé-Hoover thermostat does not generate ergodicity in the system it is being applied to if the system is small or stiff [156, 180]. Chain thermostats were then developed to address this shortcoming, and we refer the interested reader to the following references that describe this, and other relevant matters [179, 181–183], though we will not pursue them further.

Both the Gaussian and Nosé-Hoover thermostats are kinetic thermostats that are typically implemented to thermostat thermal momenta *relative* to the zero-wavevector momentum current. The most common application of these thermostats for NEMD simulations of homogeneous systems is under SLLOD dynamics, in which the thermostat assumes that the streaming velocity profile defined by the equations of motion is stable. This is true for both shear and elongational flows or combinations of them. For low shear rates and low Reynolds numbers this assumption is justified. For higher flow rates this assumption can no longer be trusted. For a system under planar shear flow at high rates of strain the streaming velocity profile is no longer linear but rather develops an S-shaped profile. The thermostat, however, is designed in such a way that it interprets any deviations away from linearity as excess thermal energy. It therefore tries to correct for this deviation by removing the part of the streaming kinetic energy that it interprets as excess thermal energy. This in turn forces the streaming velocity back to a linear profile, even though it should not by rights be linear. The end result of this is an enhanced ordering effect, in which the liquid structure (even for a simple atomic liquid) displays accentuated alignment with respect to the velocity streamlines. This effect was first observed by Erpenbeck [184] and was initially thought to be a new "string" phase of matter. However, this was soon demonstrated to be caused entirely by the thermostat [185, 186].

Thermostats that assume a fixed, unchanging streaming velocity profile are termed *profile-biased thermostats* (PBT). They are the most widely used thermostats for NEMD simulations. However, in circumstances in which one can *not* assume an unchanging stable velocity profile (e.g. for high strain rates where secondary flows may exist, or inhomogeneous systems) such thermostats are entirely inappropriate. Under such conditions one should use equations of motion that make no assumption on the form of the streaming velocity profile. Thermostats with this characteristic are termed *profile-unbiased thermostats* (PUT). Both PBTs and PUTs can be formulated in either the Gaussian or Nosé-Hoover schemes.

5.3.3 Profile Unbiased Thermostats (PUT)

One can in fact recast the SLLOD equations (or any other type of defining equations of motion) for a PUT. In the case of an atomic fluid these equations become

$$\begin{aligned} \dot{\mathbf{r}}_i &= \mathbf{v}_i \\ m_i\dot{\mathbf{v}}_i &= \mathbf{F}_i - \alpha m_i \left(\mathbf{v}_i - \mathbf{v}\left(\mathbf{r}_i, t\right)\right), \end{aligned} \tag{5.93}$$

where $\mathbf{F}_i$ is the total force on particle i, given as $\mathbf{F}_i = \mathbf{F}_i^{\phi} + \mathbf{F}_i^{e}$. It is important to realise here that the particle velocities, $\mathbf{v_i}$ are the *laboratory* velocities and not peculiar. The streaming velocity must now be computed "on the fly" at each timestep. This can be done by using the definition of the momentum density

$$\mathbf{J}\left(\mathbf{r}, t\right) \equiv \rho\left(\mathbf{r}, t\right)\mathbf{v}\left(\mathbf{r}, t\right) \tag{5.94}$$

and therefore

$$\mathbf{v}\left(\mathbf{r}, t\right) = \frac{\langle \mathbf{J}\left(\mathbf{r}, t\right)\rangle}{\langle \rho\left(\mathbf{r}, t\right)\rangle}. \tag{5.95}$$

The momentum density itself can be defined microscopically as

$$\rho\left(\mathbf{r}, t\right)\mathbf{v}\left(\mathbf{r}, t\right) = \mathbf{v}\left(\mathbf{r}, t\right)\sum_i m_i \delta\left(\mathbf{r} - \mathbf{r}_i\right), \tag{5.96}$$

where the mass density is given as $\rho\left(\mathbf{r}, t\right) = \sum_i m_i \delta\left(\mathbf{r} - \mathbf{r}_i\right)$. Equation (5.96) can be equivalently expressed as

$$\rho\left(\mathbf{r}, t\right)\mathbf{v}\left(\mathbf{r}, t\right) = \sum_i m_i \mathbf{v}_i \delta\left(\mathbf{r} - \mathbf{r}_i\right). \tag{5.97}$$

Substitution of this latter form of the momentum density into Equation (5.95) gives

$$\mathbf{v}\left(\mathbf{r}, t\right) = \frac{\left\langle \sum_i m_i \mathbf{v}_i \delta\left(\mathbf{r} - \mathbf{r}_i\right)\right\rangle}{\left\langle \sum_i m_i \delta\left(\mathbf{r} - \mathbf{r}_i\right)\right\rangle}. \tag{5.98}$$

In practice one integrates the delta functions in Equation (5.98) over bins of finite volume ΔV. Thus, Equation (5.98) is more usefully expressed as

$$\mathbf{v}(\mathbf{r}_{bin}, t) = \frac{\left\langle \sum\limits_{i \in bin} m_i \mathbf{v}_i \right\rangle}{\left\langle \sum\limits_{i \in bin} m_i \right\rangle}. \tag{5.99}$$

$\mathbf{r}_{bin}$ is a coarse-grained "position" centred at the location $\mathbf{r}_{bin,0}$, with volume ΔV. Unless the streaming velocity is allowed to fluctuate in all directions, a partially biased thermostat is obtained and artifacts such as the string phase are observed [187, 188]. Equations (5.94)–(5.99) could all be equivalently cast in terms of number density rather than mass density if all atoms have the same mass. The resulting streaming velocity profile is of course identical using either definition.

Once the streaming velocity is computed at any number of bins its value at any location $\mathbf{r}_i$ can be determined by least-squares fitting. Note now that least squares fitting itself requires the user to supply a functional form for the streaming velocity. This warrants caution and some cross-checking to ensure that the assumed functional form for the streaming velocity does in fact accurately represent it. For example, in the case of high Reynolds number planar shear flow, an S-shaped velocity profile could be fit by an odd polynomial of order n. The fit should then be compared directly with $\mathbf{v}(\mathbf{r}, t)$ computed in bins via Equation (5.99).

It is instructive to write out Equation (5.93) explicitly for planar shear and planar elongational flows. The general form for the SLLOD equations of motion *in the laboratory reference frame* is

$$\begin{aligned} \dot{\mathbf{r}}_i &= \mathbf{v}_i \\ m_i \dot{\mathbf{v}}_i &= \mathbf{F}_i^{\phi} + m_i \mathbf{r}_i \cdot \nabla \mathbf{v} \delta(t) + m_i \mathbf{r}_i \cdot \nabla \mathbf{v}(\mathbf{r}, t) \cdot \nabla \mathbf{v}(\mathbf{r}, t) \Theta(t) \\ &\quad - \alpha m_i (\mathbf{v}_i - \mathbf{v}(\mathbf{r}_i, t)). \end{aligned} \tag{5.100}$$

For planar shear flow the term $m_i \mathbf{r}_i \cdot \nabla \mathbf{v}(\mathbf{r}, t) \cdot \nabla \mathbf{v}(\mathbf{r}, t)$ is zero, so the equations reduce to

$$\begin{aligned} \dot{\mathbf{r}}_i &= \mathbf{v}_i \\ m_i \dot{\mathbf{v}}_i &= \mathbf{F}_i^{\phi} + m_i \mathbf{r}_i \cdot \nabla \mathbf{v} \delta(t) - \alpha m_i (\mathbf{v}_i - \mathbf{v}(\mathbf{r}_i, t)). \end{aligned} \tag{5.101}$$

Consider first the case with no thermostat acting ($\alpha = 0$). If we integrate these equations of motion around some infinitesimal interval at $t = 0$, we have our initial condition at the start of a simulation:

$$\begin{aligned} \int_0^{0+} d\mathbf{r}_i &= \int_0^{0+} \mathbf{v}_i \, dt \\ \int_0^{0+} d\mathbf{v}_i &= \frac{1}{m_i} \int_0^{0+} \mathbf{F}_i^{\phi} dt + \mathbf{r}_i(0) \cdot \nabla \mathbf{v}. \end{aligned} \tag{5.102}$$

The first equation evolves the laboratory position of particle i. The second equation evolves the equilibrium (i.e. peculiar) momentum by integrating the interatomic forces experienced by particle i and then *superimposes* upon this peculiar momentum the streaming velocity at $\mathbf{r}_i(0)$, which is given by the term $\mathbf{r}_i(0) \cdot \nabla\mathbf{v}$. At *all other times* $t > 0$, the equations of motion are just Newton's because the delta function term in Equation (5.101) only acts at $t = 0$ and is zero afterwards. Thus, for $t > 0$ we have

$$\begin{aligned} \dot{\mathbf{r}}_i &= \mathbf{v}_i \\ m_i \dot{\mathbf{v}}_i &= \mathbf{F}_i^{\phi}. \end{aligned} \tag{5.103}$$

This now makes it very clear that an actual NEMD simulation of planar shear flow using the SLLOD algorithm involves only Newton's equations of motion applied to a system at $t = 0$ that has a superimposed linear velocity distribution. The system continues shearing at all $t > 0$ without the application of any external or boundary forces due to the fact that peculiar momentum is conserved, as discussed in Section 5.1.2. The application of a suitably chosen thermostat does not affect this conservation of peculiar momentum, so isothermal flows are possible.

In the low Reynolds number flow regime where profile biased thermostats are valid, for $t > 0$ Equation (5.101) could be validly expressed as

$$\begin{aligned} \dot{\mathbf{r}}_i &= \mathbf{v}_i \\ m_i \dot{\mathbf{v}}_i &= \mathbf{F}_i^{\phi} - \alpha m_i \left(\mathbf{v}_i - \mathbf{i}\dot{\gamma} y_i \right). \end{aligned} \tag{5.104}$$

We now examine the case of unthermostatted elongational flow, Equation (5.100) with $\alpha = 0$, and again consider the initial condition, in which we integrate the equations of motion about $t = 0$:

$$\begin{aligned} \int_0^{0+} d\mathbf{r}_i &= \int_0^{0+} \mathbf{v}_i \, dt \\ \int_0^{0+} d\mathbf{v}_i &= \frac{1}{m_i} \int_0^{0+} \mathbf{F}_i^{\phi} dt + \mathbf{r}_i(0) \cdot \nabla\mathbf{v} + \int_0^{0+} \mathbf{r}_i \cdot \nabla\mathbf{v} \cdot \nabla\mathbf{v} dt. \end{aligned} \tag{5.105}$$

Thus, at $t = 0^+$ the laboratory velocity is a sum of equilibrium peculiar velocities (first term), the superimposed streaming velocity profile (second term), and an additional third term due to the contribution of the external force. At times $t > 0$ the delta function superimposed streaming velocity vanishes, and we are left with

$$\begin{aligned} \dot{\mathbf{r}}_i &= \mathbf{v}_i \\ m_i \dot{\mathbf{v}}_i &= \mathbf{F}_i^{\phi} + m_i \mathbf{r}_i \cdot \nabla\mathbf{v} \cdot \nabla\mathbf{v}, \end{aligned} \tag{5.106}$$

which is none other than Newton's equations of motion for a system acted on by an external force. As already explained in Section 5.1.2 this external force is what drives the nonequilibrium steady-state for elongational flows. Without it, the flow could not be sustained and would decay in time to zero in the absence of elongational pbcs.

For planar elongation for $t > 0$, Equation (5.100) becomes

$$\begin{aligned} \dot{\mathbf{r}}_i &= \mathbf{v}_i \\ m_i \dot{\mathbf{v}}_i &= \mathbf{F}_i^{\phi} + m_i \dot{\epsilon}^2 \left(\mathbf{i} x_i + \mathbf{j} y_i \right) - \alpha m_i \left(\mathbf{v}_i - \mathbf{v}(\mathbf{r}_i, t) \right), \end{aligned} \tag{5.107}$$

which again in the low Reynolds number regime can be accurately expressed as

$$\begin{aligned} \dot{\mathbf{r}}_i &= \mathbf{v}_i \\ m_i \dot{\mathbf{v}}_i &= \mathbf{F}_i^{\phi} + m_i \dot{\epsilon}^2 (\mathbf{i}x_i + \mathbf{j}y_i) - \alpha m_i (\mathbf{v}_i - \dot{\epsilon} [\mathbf{i}x_i - \mathbf{j}y_i]) . \end{aligned} \tag{5.108}$$

An examination of Equations (5.100)–(5.108) is in complete harmony with the discussion in Section 5.1.2, in which it was shown that the adiabatic SLLOD equations of motion are equivalent to Newton's equations of motion with (in the case of elongation) or without (in the case of shear) an external force. The initial impulse term in Equation (5.100) is generated at $t = 0$ in a simulation by superposing the linear streaming velocity profile onto all particle thermal velocities. For all times $t > 0$ the flow velocity is self-sustaining (by Newton's equations of motion alone for shear, or coupled to an additional external force for elongation), can be computed instantaneously as described above (see Equations (5.98) and (5.99)), and fed back into Equation (5.100) as $\mathbf{v}(\mathbf{r}_i, t)$. It is only useful in that it allows for the computation of the thermal momenta (hence kinetic temperature, pressure tensor, heat flux vector etc.) but in itself has no thermodynamic significance.

These observations gleaned from recasting the SLLOD equations in a form suitable for implementation with a PUT (i.e. cast in terms of laboratory momenta rather than the standard peculiar momenta) provide a useful perspective from which to critically examine the SLLOD equations and understand how they work. A convincing simulation study that demonstrates how SLLOD works for a system under planar shear was also performed by Delhommelle *et al.* [189]. This study clearly shows that for planar shear, SLLOD is equivalent to (a) superposing a linear streaming velocity profile onto canonically (or isokinetically) distributed thermal momenta at $t = 0$, and (b) propagating this initial distribution forward in time by Newton's equations alone.

All the kinetic thermostats described above, as well as others not considered in this book, rely upon one crucial property: knowledge of an accurate streaming velocity profile. There are serious consequences if $\mathbf{v}(\mathbf{r}_i, t)$ is incorrectly assumed. The thermostat term in the momentum equation of motion itself can induce an additional perturbation in the system if $\mathbf{v}(\mathbf{r}_i, t)$ is incorrect. Evans and Morriss [185] have shown that in the case of shear flow under SLLOD dynamics coupled with a kinetic thermostat α, the momentum continuity equation has the form

$$\begin{aligned} \frac{\partial \mathbf{J}(\mathbf{r}, t)}{\partial t} =& - \nabla \cdot (\mathbf{P}(\mathbf{r}, t) + \rho(\mathbf{r}, t) \mathbf{v}(\mathbf{r}, t) \mathbf{v}(\mathbf{r}, t)) \\ & - \frac{\alpha}{m} (\mathbf{J}(\mathbf{r}, t) - \rho(\mathbf{r}, t) \mathbf{v}_s(\mathbf{r}, t)) , \end{aligned} \tag{5.109}$$

where $\mathbf{J}(\mathbf{r}, t)$ is the momentum density, $\mathbf{v}_s(\mathbf{r}, t)$ is the *assumed* linear streaming velocity for planar shear flow and the analysis is performed with an atomic fluid of particles each of mass m. It is clear that *only* if the momentum density is identically $\mathbf{J}(\mathbf{r}, t) = \rho(\mathbf{r}, t) \mathbf{v}_s(\mathbf{r}, t) = \mathbf{i} m \dot{\gamma} y \ \forall \mathbf{r}, t$ will the contribution to the stress due to the thermostat be identically zero at all times and all locations. If however $\mathbf{J}(\mathbf{r}, t) \neq \mathbf{i} m \dot{\gamma} y$ anywhere or at any time, then an additional stress will be induced in the system. This stress is purely artificial and undesirable. As previously noted, it has been seen to cause "string" phases in computer simulations of shear flow [184], which are an ordering of particles preferentially in the direction of the velocity streamlines at high rates of strain where a linear velocity profile cannot be assumed. These strings were demonstrated to be artifacts by employing a suitably formulated PUT [185]. Similar problems, as well as some new ones, occur for molecular systems if the thermostatting is not handled correctly. These issues will be discussed later in Chapter 8.

5.3.4 Configurational Thermostat

Other types of thermostats have been devised over the years, but the Gaussian and Nosé-Hoover thermostats (both in their profile-biased and profile-unbiased forms) are the only kinetic thermostats that have found favour in NEMD simulations, primarily because they are rigorous from a statistical mechanical and dynamical point of view. However, recently the configurational thermostat (based on thermostatting the configurational temperature) has been a valuable inclusion. The configurational temperature T_c (discussed earlier in Section 4.1) is determined only from the configurational component of the system's full phase-space, and is given as [157]

$$\frac{1}{k_B T_c} = \left\langle \frac{\sum_i \nabla_i^2 \phi}{\left|\sum_i \nabla_i \phi\right|^2} \right\rangle \tag{5.110}$$

for sufficiently large N (see Section 4.1). Because the system is thermostatted to a configurational quantity involving only particle positions but not momenta, the thermostat term now operates on the equation of motion that governs the evolution of particle positions and, for the SLLOD equations, can be written as [159]

$$\begin{aligned}
\dot{\mathbf{r}}_i &= \frac{\mathbf{p}_i}{m_i} + \mathbf{r}_i \cdot \nabla \mathbf{v} + \frac{s}{T} \frac{\partial T_c}{\partial \mathbf{r}_i} \\
\dot{\mathbf{p}}_i &= \mathbf{F}_i^{\phi} - \mathbf{p}_i \cdot \nabla \mathbf{v} \\
\dot{s} &= -Q \frac{(T_c - T)}{T}.
\end{aligned} \tag{5.111}$$

Here T is the target temperature, s is an additional degree of freedom coupling the heat reservoir to the system and Q is a damping factor.

The above equations of motion are formulated with a Nosé-Hoover feedback scheme rather than a Gaussian formalism due to the relative simplicity of the former implementation and the complexity of the latter for this particular thermostat. The advantage of using a configurational thermostat for an NEMD system is that one does not need to worry about how to compute the streaming velocity profile. As the thermostat acts only on particle positions without reference to the streaming velocity, one can be assured of no artifacts such as string phases, etc. It also simplifies the coding by not requiring binned velocity profiles or least squares fits in the case of PUTs (see above). Detailed studies have addressed many of the advantages of using configurational thermostats for atomic and molecular systems under planar shear flow, including the elimination of string phases [158–161, 186, 189, 190].

A recent development in the formulation of configurational thermostats was made by Braga and Travis [162]. They have derived a new form of the configurational temperature thermostat also based on the Nosé-Hoover equations of motion. Their thermostat is an improvement on the one proposed by Delhommelle and Evans [159] in that it does not contain a term proportional to the spatial gradient of the temperature, which can lead to stiff equations of motion, and also because it generates the canonical ensemble. Their equations of motion for a system at equilibrium are

$$\begin{aligned} \dot{\mathbf{r}}_i &= \frac{\mathbf{p}_i}{m_i} - \zeta \frac{\partial \phi}{\partial \mathbf{r}_i} \\ \dot{\mathbf{p}}_i &= \mathbf{F}_i^{\phi} \\ \dot{\zeta} &= \frac{1}{Q_\zeta}\left[\sum_i \left(\frac{\partial \phi}{\partial \mathbf{r}_i}\right)^2 - k_B T \sum_i \frac{\partial^2 \phi}{\partial \mathbf{r}_i^2}\right]. \end{aligned} \tag{5.112}$$

For planar shear flow these equations can be simply augmented by the usual field terms [163] and can be expressed as

$$\begin{aligned} \dot{\mathbf{r}}_i &= \frac{\mathbf{p}_i}{m_i} + \mathbf{i}\dot{\gamma} y_i - \zeta \frac{\partial \phi}{\partial \mathbf{r}_i} \\ \dot{\mathbf{p}}_i &= \mathbf{F}_i^{\phi} - \mathbf{i}\dot{\gamma} p_{yi} \\ \dot{\zeta} &= \frac{1}{Q_\zeta}\left[\sum_i \left(\frac{\partial \phi}{\partial \mathbf{r}_i}\right)^2 - k_B T \sum_i \frac{\partial^2 \phi}{\partial \mathbf{r}_i^2}\right]. \end{aligned} \tag{5.113}$$

We are unaware of any attempt to use the configurational thermostat for systems under planar elongational flow, but see no reason why it should not be equally successful. The equations of motion for planar elongational flow are

$$\begin{aligned} \dot{\mathbf{r}}_i &= \frac{\mathbf{p}_i}{m_i} + \dot{\epsilon}(\mathbf{i}x_i - \mathbf{j}y_i) - \zeta \frac{\partial \phi}{\partial \mathbf{r}_i} \\ \dot{\mathbf{p}}_i &= \mathbf{F}_i^{\phi} - \dot{\epsilon}(\mathbf{i}p_{xi} - \mathbf{j}p_{yi}) \\ \dot{\zeta} &= \frac{1}{Q_\zeta}\left[\sum_i \left(\frac{\partial \phi}{\partial \mathbf{r}_i}\right)^2 - k_B T \sum_i \frac{\partial^2 \phi}{\partial \mathbf{r}_i^2}\right]. \end{aligned} \tag{5.114}$$

This implementation of the configurational thermostat is shown to be more robust than the original version developed by Delhommelle and Evans and may be useful for systems involving large changes in temperature over small time scales. Braga and Travis also demonstrate that the configurational thermostat is a special case of a general set of Nosé-Hoover equations developed by Kusnezov *et al.* [191]. The method can also be extended to the isobaric-isothermal ensemble [192] and molecular fluids [164].

The issue of exactly what a temperature is for systems far from equilibrium remains unresolved and so all thermostats, either kinetic or configurational, rely upon the standard equilibrium thermodynamic definition of temperature, i.e. $1/T = (\partial S/\partial U)_V$. This definition forms the basis for all of the microscopic expressions for temperature that we introduced in Section 4.1 and have used in this chapter. We will give explicit numerical algorithms for the thermostats discussed above in the following chapter.

5.4 Further Considerations of the SLLOD Equations of Motion

There are several misconceptions about the SLLOD equations of motion. One common one is that the boundary conditions drive the flow. As detailed in the above sections, this is not the case. In fact, it was precisely to eliminate boundary stresses that the SLLOD equations were first devised. Boundary forces have the undesired effect of inducing time dependencies and inhomogeneity into the system. They have the additional disadvantage of not being directly amenable to response theory. Other misconceptions involve the use of so-called molecular versus atomic formulations of the SLLOD equations when simulating systems of molecules under flow, and the use of appropriate thermostats, and these issues will be addressed in Chapter 8.

Perhaps the most serious misconception in the use of SLLOD has been its application in elongational flow. While it is remarkable that a shearing system under SLLOD dynamics is self-sustained for times $t > 0$ without the imposition of an external force, the same is not true for a system under elongational flow, as has been stressed in the sections above. Tuckermann *et al.* [193], and later Edwards and Dressler [140] tried to devise equations of motion that involved no external forces for $t > 0$ for systems under elongation. This led to the so-called g-SLLOD (or "generalised SLLOD") equations of motion. In this case, the equation of motion for the momenta (e.g. second equation in Equations (5.3)) is augmented by the term $-m_i\mathbf{r}_i \cdot \nabla\mathbf{v} \cdot \nabla\mathbf{v}$ for times $t > 0$. It was claimed that this then allowed the system to evolve naturally without the influence of an external force, as is the case with shear flow. This augmentation of the momentum equation of motion was also claimed by Edwards and colleagues [137] to solve the problem of momentum conservation (see Section 5.1.4).

In our discussions above we already demonstrated that elongation *must* be sustained by this additional force in the equations of motion. In fact, if one uses the g-SLLOD equations of motion then the flow is actually driven by the boundaries, which is in complete contradiction to the intentions of using an algorithm like SLLOD in the first place [1, 128], whose primary purpose is to remove the influence of boundary driven flow for the reasons already expressed. It is easy to show that g-SLLOD on its own (i.e. without the influence of the boundaries in sustaining the flow) in fact gives streaming velocity profiles that are incompatible with the flow. If we express the momentum equation of motion in the *laboratory* reference frame (see Equation (5.100)), then for adiabatic g-SLLOD the governing equation of motion for *any generalised homogeneous flow* is

$$m_i\dot{\mathbf{v}}_i = \mathbf{F}_i^{\phi} + m_i\mathbf{r}_i \cdot \nabla\mathbf{v}\delta(t), \tag{5.115}$$

where $\mathbf{v}_i$ is the laboratory velocity of particle i. We can consider what happens to the streaming velocity of the fluid under generalised flow conditions by investigating a system of noninteracting particles, as we did in Section 5.1.5, in which we looked at the implications of compatibility of periodic boundary conditions with the SLLOD equations. If we therefore set $\mathbf{F}_i^{\phi} = 0$ we find

$$\dot{\mathbf{v}}_i \equiv \dot{\mathbf{v}}(\mathbf{r}_i) = \mathbf{r}_i \cdot \nabla \mathbf{v} \delta(t), \tag{5.116}$$

where now $\mathbf{v}(\mathbf{r}_i)$ is just the streaming velocity at $\mathbf{r}_i$ as previously defined. The term $\mathbf{r}_i \cdot \nabla \mathbf{v}$ is also just the streaming velocity at $\mathbf{r}_i$, and considering that the delta function acts only at time $t = 0$ we can express this differential equation as

$$\frac{d\mathbf{v}(\mathbf{r}_i, t)}{dt} = \mathbf{v}(\mathbf{r}_i, t = 0) \delta(t), \tag{5.117}$$

which leads to the trivial solution

$$\mathbf{v}(\mathbf{r}_i, t) = \mathbf{v}(\mathbf{r}_i, t = 0), \quad \forall t. \tag{5.118}$$

Thus, irrespective of the specific flow geometry, the g-SLLOD equations of motion will *always* lead to time-independent streaming velocities, i.e. zero fluid acceleration. As we discussed in Section 5.1.2, this is perfectly correct in the case of shear flow (for which, incidentally, g-SLLOD and SLLOD are equivalent), but is clearly wrong for elongational flow, where the flow results in hyperbolic velocity profiles and acceleration along the streamlines (see Equation (5.27) and the discussion in Sections 5.1.2 and 5.3.3). That accelerations in the streamlines are probably observed in simulations of elongational flow under g-SLLOD is entirely a result of the influence of the periodic boundaries driving the flow. If the boundaries did not sustain the flow, the flow itself would decay to zero. Only in shear flow is an initial impulse sustained indefinitely without the need for an external field (or boundary conditions) to sustain the flow.

It has been claimed [137, 139] that the divergence in the total momentum for elongational flow is symptomatic of an inherent flaw in the SLLOD algorithm and that use of the g-SLLOD equations of motion for elongational flow fixes this problem. In Section 5.1.4 we already discussed that it is actually physically reasonable and expected that the total momentum will suffer from an exponential growth in round-off error in the case of elongational flow, whereas shear flow will still result in round-off error but the growth will be linear and hence much slower. Furthermore, the introduction of the additional term $-m_i \mathbf{r}_i \cdot \nabla \mathbf{v} \cdot \nabla \mathbf{v}$ into the momentum equation for g-SLLOD does not actually solve the so-called problem. It merely transforms an exponential growth of numerical errors into a linear growth of those same errors. If the simulation was run for long enough the same stability problems experienced in elongational flow under SLLOD dynamics would also be experienced for elongational flow under g-SLLOD dynamics (indeed, it would also be experienced for shear, where we note that SLLOD and g-SLLOD are equivalent for this flow geometry). The root of the problem remains unchanged, nor can it ever be changed without computers and algorithms that have infinite numerical precision. A linear growth in numerical error is unphysical for elongational flow and incompatible with the requirements of a hyperbolic, accelerating, inherently chaotic streaming velocity profile.

Finally, we point out that the rate of energy dissipation for the g-SLLOD equations is incompatible with what is known to be the rate of energy dissipation from linear irreversible thermodynamics, given by Equation (2.37), namely:

$$\dot{U} = -V\mathbf{P}^T : \nabla\mathbf{v}. \tag{5.119}$$

This is precisely the rate of energy dissipation obtained via the SLLOD equations as shown in Equation (5.44). The g-SLLOD equations on the other hand give the dissipation rate as

$$\dot{U}^{g-SLLOD} = -V\mathbf{P}^T : \nabla\mathbf{v} - \left(\sum_i \mathbf{p}_i\mathbf{r}_i\right) \cdot \nabla\mathbf{v} : \nabla\mathbf{v}. \tag{5.120}$$

This gives the same dissipation rate as the SLLOD equations of motion for shear, but for planar elongation it is incorrect, since it is known from hydrodynamics that the rate of internal energy generation by viscous processes is given by Equation (5.119) for *all* flows. While the extra term in Equation (5.120) might become negligible in the thermodynamic limit, Equation (5.120) has the disadvantage that the energy balance equation is no longer satisfied exactly and instantaneously, as it is for SLLOD. This means that the work done by the external field will not correspond exactly to the energy removed by the internal energy constraint force in a constant internal energy simulation. Furthermore, because the application of linear response theory requires that the correct dissipation be generated by the equations of motion, the g-SLLOD equations of motion will not give the correct linear response for small systems and the magnitude of the error will not be easy to quantify.

6 Homogeneous Flows for Atomic Fluids: Applications

6.1 Time-independent Flow

A number of alternative ensembles for NEMD computer simulation can be devised, such as isoenergetic, isothermal-isobaric, isobaric-isoenthalpic, constant stress, etc. Of these, the two most useful ensembles for most applications are the isoenergetic and isothermal-isobaric ensembles, and we discuss these, as well as the most commonly used isothermal-isochoric ensemble, in the following sections. Details on other ensembles may be found in the book by Evans and Morriss [2].

6.1.1 (NVU) Ensemble

One can generate a system under flow conditions at constant internal energy U rather than constant temperature by applying an ergostat (rather than a thermostat) to the equations of motion. Running an NEMD simulation with an ergostat is useful in that it allows a check of energy conservation in much the same way as one should always check an equilibrium MD simulation for energy conservation. In order to devise an ergostat one first needs to consider the energy dissipation for a system undergoing SLLOD dynamics. This was shown to be $-V\mathbf{P}^T : \nabla\mathbf{v} - \alpha \sum_i \frac{\mathbf{p}_i^2}{m_i}$ for a thermostatted system (see Equation (3.37)). For an adiabatic system ($\alpha = 0$), this result shows that the total internal energy increases in exactly the way described by the internal energy balance equation (Equation (2.40c)).

The expression for the pressure tensor used to compute the dissipation is not strictly valid for an infinitely periodic system. The correct expression, Equation (4.88), depends on the minimum image separation of interacting images of particles i and j, rather than the absolute position of each particle. If the whole derivation is performed starting from the expression for the internal energy of an infinitely periodic system, and the minimum image convention is used in the same way as in the derivation of Equation (4.88), then the correct result is obtained. In fact, there are several features of simulations in homogeneous periodic systems that need to be carefully taken into account when applying the balance equations. Firstly, the local heat flux vector has a time averaged value of zero everywhere in the system, even when a velocity gradient produces viscous heating. The periodicity of the heat flux vector ensures that the net heat absorbed by conduction through the periodic boundaries of the simulation box is exactly zero at all times, since the flux entering through one face is always exactly matched by the flux leaving through the opposite face. By the divergence theorem, this is equivalent to saying that the integral of the divergence of the heat flux vector over the volume of the periodic simulation box is zero. Similarly, if we consider only simulations of systems that are at equilibrium or in flows with spatially uniform, divergenceless velocity gradient tensors, the densities of mass, momentum and internal energy could be expected to remain uniform (in the absence of unusual flow instabilities) and the integrals over the whole simulation box of the convective terms in the balance equations are also then equal to zero. This means that the change in the total internal energy of the simulation box is entirely due to the thermodynamic work term, as given by the dissipation term above . Fortunately, the net result of these subtle differences between a finite, isolated system and an infinite, periodic system is only a very minor change: the potential part of the expression for the pressure tensor in periodic boundary conditions must contain the minimum image particle separations rather than the absolute particle positions, as previously stated.

For planar shear flow the dissipation is $-P_{yx}\dot{\gamma}V$ whereas for planar elongation it is $-(P_{xx}-P_{yy})\dot{\epsilon}V$. For constant temperature simulations in the steady state, the average rate of viscous heat dissipation is equal to the average rate of energy extraction by the thermostat, i.e.

$$\left\langle -V\mathbf{P}^{T}(t):\nabla\mathbf{v}\right\rangle = \left\langle \alpha\sum_{i}\frac{\mathbf{p}_i^2}{m_i}\right\rangle. \tag{6.1}$$

This is a very useful result because one can compute the shear or elongational viscosities in two ways: firstly via the hydrodynamic constitutive relations

$$\eta_S = -\frac{\langle P_{yx}+P_{xy}\rangle}{2\dot{\gamma}} \tag{6.2}$$

for shear, and

$$\eta_E = -\frac{\langle P_{xx}-P_{yy}\rangle}{4\dot{\epsilon}} \tag{6.3}$$

for planar elongation, and secondly by making use of the equivalence of viscous dissipation and extraction of dissipated heat by the thermostat in the steady-state (Equation (6.1)). Substitution of Equations (6.2) and (6.3) into (6.1) gives the second method of computing the viscosities as

$$\eta_S = \frac{\left\langle \alpha\sum_{i}\frac{\mathbf{p}_i^2}{m_i}\right\rangle}{V\dot{\gamma}^2} \tag{6.4}$$

for shear and

$$\eta_E = \frac{\left\langle \alpha\sum_{i}\frac{\mathbf{p}_i^2}{m_i}\right\rangle}{4V\dot{\epsilon}^2} \tag{6.5}$$

for planar elongation.

Now, instead of keeping the temperature fixed at all times, let the internal energy be fixed instantaneously. In this case the left-hand side of Equation (3.37) is zero, which leads to the result

$$\alpha_U = -\frac{V\dot{\gamma}P_{yx}}{\sum_i \frac{\mathbf{p}_i^2}{m_i}} \tag{6.6}$$

for shear and

$$\alpha_U = -\frac{V\dot{\epsilon}\left(P_{xx} - P_{yy}\right)}{\sum_i \frac{\mathbf{p}_i^2}{m_i}} \tag{6.7}$$

for planar elongation, where now the subscript U indicates the multiplier is an ergostat that constrains the total system internal energy to be a constant of the motion (see also the discussion in Section 8.4). The pseudocode listed in Algorithm 6.1 depicts the scheme to compute the ergostat multipliers for planar shear or elongational flow. These should be computed within the force routine, while the ergostat multiplier itself is applied to each atom during numerical integration of the equations of motion, as described in Algorithm 6.2. It is assumed here that pairs of interacting atoms i and j have already been stored in an array set up in a standard neighbour list routine (see for example the Verlet or cell neighbour lists in Allen and Tildesley [5]). One can of course generalise these multipliers for any other type of homogeneous flow geometry (e.g. uniaxial stretching or mixed shear and elongation etc.). Note that the form of the equations of motion in this example is identical to that expressed for a Gaussian thermostat implementation as in Equation (5.87), except that the multiplier is the ergostat.

6.1.2 (*NVT*) Ensemble

In the case of the canonical or isokinetic ensembles, one applies the SLLOD equations of motion as depicted in Equations (5.85) and (5.87) (for shear, extensional or mixed flows of shear and extension), where now the multiplier is a standard thermostat, for example the Nosé-Hoover thermostat for the classical canonical ensemble or a Gaussian thermostat for the isokinetic ensemble, etc. The algorithms used are thus essentially the same as those depicted in Algorithms 6.1 and 6.2, except that now the thermostat multiplier is computed instead of an ergotstat. This is displayed for both Gaussian and Nosé-Hoover thermostats in Algorithms 6.3 and 6.4 respectively. In the case of the Gaussian thermostat, the multiplier is computed within the force routine, whereas for the Nosé-Hoover thermostat it is computed within the numerical integrator routine, since one needs to solve the differential equation for the multiplier (third line in Equation (5.91)). The strategy for applying pbcs, forces etc. is exactly the same as that depicted in Algorithm 6.1, while that for the SLLOD equations is identical to that of Algorithm 6.2 except that the multipliers are those given in Algorithms 6.3 and 6.4. There is one additional step to correct for numerical drift

Algorithm 6.1 (NVU) pseudocode to compute ergostat multiplier for shear or elongation. This contains pseudocode for the computation of pair interaction forces and pressure tensor.

$\mathbf{F}_i \leftarrow 0$; {initialise force arrays to zero}
$\mathsf{P} \leftarrow 0$; {initialise pressure tensor arrays to zero}
$K \leftarrow 0$; {initialise sum of momenta squared}
for ipair $\leq$ npair **do** {ipair = pair index; npair = total pair number}
  $\mathbf{r}_{ij} \leftarrow \mathbf{r}_j - \mathbf{r}_i$;
  if shear flow **then**
    $\mathbf{r}_{ij} \leftarrow$ pbcsShear $(\mathbf{r}_{ij})$; {apply shear pbcs}
  else
    $\mathbf{r}_{ij} \leftarrow$ pbcsElongation $(\mathbf{r}_{ij})$; {apply elongation pbcs}
  end if
  $\mathbf{F}_{ij} \leftarrow -\nabla\phi\left(\mathbf{r}_{ij}\right)$; {compute pair interaction forces}
  $\mathsf{P} \leftarrow \mathsf{P} - \mathbf{r}_{ij}\mathbf{F}_{ij}$; {configurational component of pressure tensor}
end for
for i $\leq$ N **do** {N = total number of particles}
  for j $\leq$ N-1 **do**
    $\mathbf{F}_i \leftarrow \mathbf{F}_i + \mathbf{F}_{ij}$; {pair summation could be more sophisticated depending on neighbour list used}
  end for
end for
for i $\leq$ N **do**
  $\mathsf{P} \leftarrow \mathsf{P} + \mathbf{p}_i\mathbf{p}_i$; {add kinetic term to configurational pressure tensor}
  $K \leftarrow K + \mathbf{p}_i \cdot \mathbf{p}_i/m_i$; {sum momenta squared divided by particle mass}
end for
$\mathsf{P} \leftarrow \mathsf{P}/V$; {make pressure intensive by dividing by volume}
if shear flow **then**
  $\alpha_U \leftarrow -V\dot{\gamma}P_{yx}/K$; {shear ergostat}
else
  $\alpha_U \leftarrow -V\dot{\epsilon}\left(P_{xx} - P_{yy}\right)/K$; {elongation ergostat}
end if

in the temperature for the Gaussian thermostat (see discussion of numerical drift for Gaussian constraints in Section 5.3.1). The algorithm for the Braga-Travis version of the configurational thermostat is given in Algorithm 6.5. Algorithmically, configurational thermostats are similar to a Nosé-Hoover thermostat, except that now the thermostat multiplier is applied in the equation for the positions, *not* the momenta (see Equations (5.112)–(5.114)) and again one solves for the multiplier within the numerical integrator. Note that for both Nosé-Hoover and configurational thermostats, there is no need to correct for numerical drift since these schemes work on integral feedback mechanisms.

Algorithm 6.2 (NVU) pseudocode for SLLOD equations of motion for shear or elongation, implemented within integration routine

if elongational flow **then**
  pysum $\leftarrow$ 0; {initialise sum of y-component of momenta}
end if
for i $\leq$ N **do**
  if shear flow **then**
    $\mathbf{r}_i \leftarrow$ integrator $(\mathbf{p}_i/m_i + \mathbf{i}\dot{\gamma}y_i)$; {integrate particle positions forward in time}
    $\mathbf{r}_i \leftarrow$ pbcsShear $(\mathbf{r}_i)$; {apply shear pbcs to particle positions}
    $\mathbf{p}_i \leftarrow$ integrator $\left(\mathbf{F}_i - \mathbf{i}\dot{\gamma}p_{yi} - \alpha_U \mathbf{p}_i\right)$; {integrate particle momenta forward in time}
  else {elongational flow}
    $\mathbf{r}_i \leftarrow$ integrator $(\mathbf{p}_i/m_i + \dot{\epsilon}\,(\mathbf{i}x_i - \mathbf{j}y_i))$; {integrate particle positions forward in time}
    $\mathbf{r}_i \leftarrow$ pbcsElongation $(\mathbf{r}_i)$; {apply elongation pbcs to particle positions}
    $\mathbf{p}_i \leftarrow$ integrator $\left(\mathbf{F}_i - \dot{\epsilon}\left(\mathbf{i}p_{xi} - \mathbf{j}p_{yi}\right) - \alpha_U \mathbf{p}_i\right)$; {integrate particle momenta forward in time}
    pysum $\leftarrow$ pysum + p_{yi}; {sum y-component of momenta}
  end if
end for
if elongational flow **then**
  pysum $\leftarrow$ pysum/N;
  for i $\leq$ N **do**
    $p_{yi} \leftarrow p_{yi}$ − pysum; {correct for momentum drift}
  end for
end if

6.1.3 (NpT) Ensemble

Equations (6.4)–(6.7) serve as useful checks that the computer program for an NEMD simulation is behaving correctly. While ergostatically constrained systems are a useful simulation tool for checking the reliability of NEMD code, they are not particularly useful for physical applications of interest to most researchers. Apart from the isothermal-isochoric (NVT) ensemble, the next most physically significant ensemble is the isothermal-isobaric (NpT) ensemble, in which both the temperature and isotropic pressure are constrained. Constant pressure simulations under shear were first performed by Evans and Morriss [125, 194], later extended for molecular systems (decane) [63], and were accomplished by coupling the system to an extended degree of freedom associated with the volume of the simulation box. Thus, the volume, rather than the pressure, is allowed to fluctuate. One can implement constant pressure simulations in either the Gaussian or Nosé-Hoover forms [2]. The barostat is more easily implemented with a Nosé-Hoover feedback scheme which implies that the instantaneous pressure will fluctuate about its mean (target) value. For a thermostatted system, the isothermal-isobaric

Algorithm 6.3 (NVT) pseudocode for Gaussian thermostat multiplier for shear or elongational flow

while executing force routine **do**
  num $\leftarrow$ 0; {initialise numerator for thermostat calculation}
  den $\leftarrow$ 0; {initialise denominator for thermostat calculation}
  for i $\leq$ N **do**
    if shear flow **then**
      num $\leftarrow$ num $+ \mathbf{F}_i \cdot \mathbf{p}_i - \dot{\gamma} p_{xi} p_{yi}$;
      den $\leftarrow$ den $+ \mathbf{p}_i \cdot \mathbf{p}_i / m_i$;
    else {elongational flow}
      num $\leftarrow$ num $+ \mathbf{F}_i \cdot \mathbf{p}_i - \dot{\epsilon}\left(p_{xi}^2 - p_{yi}^2\right)$;
      den $\leftarrow$ den $+ \mathbf{p}_i \cdot \mathbf{p}_i / m_i$;
    end if
  end for
  $\alpha \leftarrow$ num/den;
  $\alpha \leftarrow \alpha + \alpha' [(\text{den} - (3N - N_C) k_B T_0) / ((3N - N_C) k_B T_0)]$; {correct for numerical drift; T_0 is the target temperature}
end while

Algorithm 6.4 (NVT) pseudocode for Nosé-Hoover thermostat multiplier for arbitrary homogeneous flow

while executing integrator routine **do**
  sum $\leftarrow$ 0; {initialise sum variable for thermostat calculation}
  for i $\leq$ N **do**
    sum $\leftarrow$ sum $+ \mathbf{p}_i \cdot \mathbf{p}_i / m_i$;
  end for
  $\mathbf{r}_i \leftarrow$ integrator $(\mathbf{p}_i / m_i + \mathbf{r}_i \cdot \nabla \mathbf{v}_i)$; {integrate particle positions forward in time}
  $\mathbf{r}_i \leftarrow$ pbcs $(\mathbf{r}_i)$; {apply pbcs to particle positions according to type of flow}
  $\mathbf{p}_i \leftarrow$ integrator $(\mathbf{F}_i - \mathbf{p}_i \cdot \nabla \mathbf{v}_i - \zeta \mathbf{p}_i)$; {integrate particle momenta forward in time}
  $\zeta \leftarrow$ integrator $\left(\left[\text{sum} - N_f k_B T_0\right] / Q\right)$; {integrate thermostat multiplier forward in time; T_0 is target temperature, Q is damping term}
end while

SLLOD equations of motion are

$$
\begin{aligned}
\dot{\mathbf{r}}_i &= \frac{\mathbf{p}_i}{m_i} + \mathbf{r}_i \cdot \nabla \mathbf{v} + \dot{\xi} \mathbf{r}_i \\
\dot{\mathbf{p}}_i &= \mathbf{F}_i^{\phi} - \mathbf{p}_i \cdot \nabla \mathbf{v} - \alpha \mathbf{p}_i - \dot{\xi} \mathbf{p}_i \\
\dot{V} &= 3 \dot{\xi} V.
\end{aligned}
\tag{6.8}
$$

Now there are additional constraints imposed on both the $\dot{\mathbf{r}}_i$ and $\dot{\mathbf{p}}_i$ equations to account for the fact that pressure has both configurational and kinetic contributions. The new

Algorithm 6.5 (NVT) pseudocode for the Braga-Travis configurational thermostat multiplier for arbitrary homogeneous flow

while executing integrator routine **do**
  sum1 $\leftarrow 0$;
  sum2 $\leftarrow 0$; {initialise sum variables for thermostat calculation}
  for i $\leq$ N **do**
    sum1 $\leftarrow$ sum1 $+ \mathbf{F}_i^{\phi} \cdot \mathbf{F}_i^{\phi}$;
    sum2 $\leftarrow$ sum2 $+ \nabla_i \cdot \mathbf{F}_i^{\phi}$;
  end for
  $\mathbf{r}_i \leftarrow \text{integrator}\big(\mathbf{p}_i/m_i + \mathbf{r}_i \cdot \nabla \mathbf{v}_i + \zeta \mathbf{F}_i^{\phi}\big)$; {integrate particle positions forward in time}
  $\mathbf{r}_i \leftarrow \text{pbcs}\,(\mathbf{r}_i)$; {apply pbcs to particle positions according to type of flow}
  $\mathbf{p}_i \leftarrow \text{integrator}\,(\mathbf{F}_i - \mathbf{p}_i \cdot \nabla \mathbf{v}_i)$; {integrate particle momenta forward in time}
  $\zeta \leftarrow \text{integrator}\,([\text{sum1} + k_B T_0 \text{sum2}]\,/Q)$; {integrate thermostat multiplier forward in time; T_0 is target temperature, Q is damping term}
end while

barostat multiplier, $\dot{\xi}$, is the rate of dilation that depends on the difference between the instantaneous and target pressures. This dilation rate is obtained by solving the additional differential equation

$$\ddot{\xi} = \frac{(p - p_0)\,V}{QNk_BT}, \tag{6.9}$$

where p is the instantaneous pressure given by $p = 1/3\text{Tr}\,(\mathbf{P})$, p_0 is the desired pressure and Q is a damping factor for which the optimal value will depend on the type of system studied. Thus there are a total of $6N + 2$ differential equations to solve: $3N$ for the atomic positions, $3N$ for the peculiar momenta, one for the system volume and one for the dilation rate.

Care must be taken in coupling the system to a barostat because the pressure can oscillate about the mean value with a period dependent on the value of Q. The choice of the value of Q should take into account two requirements. First, in order to ensure that the thermodynamic properties computed from the phase space trajectories are independent of these oscillations, one must ensure that the simulations are sufficiently long, considerably longer than the wavelength of the oscillations. This can be accomplished by a combination of sufficiently long simulation times and sufficiently small values of Q. Next, Q should be chosen such that the period of oscillation is large compared to the characteristic decay time of dynamical correlation functions (e.g. the velocity autocorrelation function). This ensures that oscillations are decoupled from the particle dynamics. In practice, trial and error is required to find the right balance that satisfies all the above requirements. For systems in which it is important to allow the shape of the simulation box to vary (e.g. solids) an implementation of (NpT) dynamics has been published by Melchionna *et al.* [195].

6.1.3.1 Planar Shear

For shear flow the equations of motion given by Equation (6.8) reduce to

$$\begin{aligned}\dot{\mathbf{r}}_i &= \frac{\mathbf{p}_i}{m_i} + \mathbf{i}\dot{\gamma}y_i + \dot{\xi}\mathbf{r}_i \\ \dot{\mathbf{p}}_i &= \mathbf{F}_i^{\phi} - \mathbf{i}\dot{\gamma}p_{yi} - \alpha\mathbf{p}_i - \dot{\xi}\mathbf{p}_i \\ \dot{V} &= 3\dot{\xi}V.\end{aligned} \tag{6.10}$$

A Gaussian form for the thermostat multiplier, obtained by taking the time derivative of the instantaneous kinetic energy and setting this to zero, is thus

$$\alpha = \frac{\sum_i \frac{1}{m_i}\left(\mathbf{F}_i^{\phi}\cdot\mathbf{p}_i - \dot{\gamma}p_{xi}p_{yi}\right)}{\sum_i \frac{\mathbf{p}_i\cdot\mathbf{p}_i}{m_i}} - \dot{\xi}. \tag{6.11}$$

If V_T is the updated volume that ensures that the average pressure is maintained at its target value of p_0, governed by the solution of Equation (6.9) (V_T is thus the solution of the third line in Equation (6.10)), then the rescaling factor for each lattice vector is

$$\lambda(t) = \left[\frac{V_T}{V(t)}\right]^{1/3}, \tag{6.12}$$

where $V(t)$ is the *un-rescaled* simulation box volume ($V(t) \neq V_T$). One now just rescales the sheared simulation box dimensions such that $L_i \to \lambda L_i$, where $i = x, y$ or z.

In Figure 6.1 we display shear viscosities for a WCA fluid shearing in either the (NVT) or (NpT) ensembles. (NpT) simulations were thermostatted with a Gaussian isokinetic thermostat, whereas the (NVT) simulations need to be thermostatted with

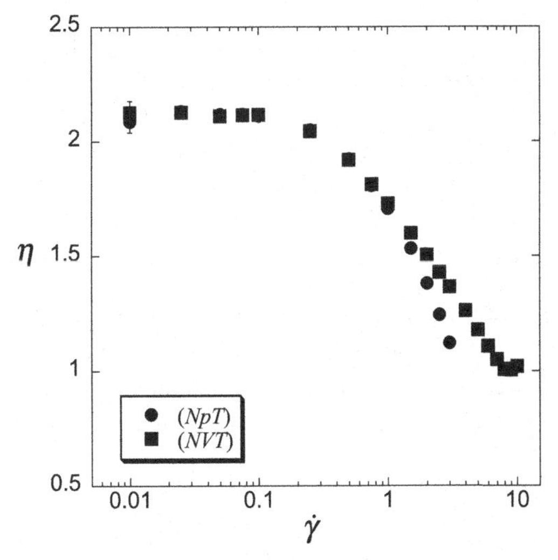

Figure 6.1 Planar shear viscosities for the isothermal-isochoric (NVT) and isothermal-isobaric (NpT) ensembles as functions of shear rate $\dot{\gamma}$. The system consists of 2048 WCA atoms at a state point of $T = 1.00$, $\rho = 0.840$, $p_0 = 7.8365 \pm 0.0001$, in reduced units. (NVT) simulations were performed using a configurational thermostat [196].

either a profile unbiased thermostat (PUT) or a configurational thermostat. We chose the latter in this case. The reason for needing a PUT or configurational thermostat is that for very high strain rates a string phase forms for strain rates at approximately $\dot{\gamma} \geq 4$ if a profile biased thermostat is used. This results in an artificially reduced shear viscosity (see Section 5.3 on thermostats). High strain rates are used in the case of the (NVT) simulations to illustrate an important point, namely the shear thickening (increase of viscosity as strain rate increases) that occurs at excessive strain rates. This happens because the pressure increases rapidly at such high strain rates, leading to increasing values of the viscosity. In many natural, industrial and experimental rheological processes however, at least one surface of the fluid is in contact with the atmosphere and thus remains in mechanical equilibrium with it. Therefore the most useful and realistic physical ensemble to use is actually the (NpT) ensemble.

As can be seen in Figure 6.1, (NpT) simulations show a steady decline in shear viscosity as a function of strain rate. At low strain rates both (NpT) and (NVT) viscosities are identical to within statistical precision, but as strain rate increases (NpT) viscosities are consistently lower than (NVT) counterparts due to shear dilatancy. (NpT) simulations maintain a constant pressure by varying the system volume, and hence density. Thus, at any given and sufficiently high strain rate, the density of an (NpT) system is *always* lower than an (NVT) system at the same strain rate due to the volume expansion needed to maintain constant pressure. Hence the viscosities are correspondingly lower.

Algorithms 6.6 and 6.7 display pseudocode for the (NpT) ensemble for shear and elongational flows. It is assumed here that the steps depicted are all implemented from *within* the numerical integrator routine. This is because the solution for the barostat multiplier occurs within the integrator routine, whereas this is not necessarily so for the thermostat, which may be computed within the force routine (e.g. the Gaussian implementation). In the case of the Nosé-Hoover or configurational thermostats these multipliers are in fact computed within the integrator routine as well (though this is not indicated in the algorithm schematic). Note however that if a Gaussian thermostat is used, then the constraint must subtract the term $\dot{\xi}$ (see Equations (6.11) and (6.14)).

6.1.3.2 Planar Elongation

Implementation of (NpT) simulations for elongating polymer melts were first published by Daivis *et al.* [197], but in this study the pressure was kept constant by adjusting the system density for each elongation rate rather than by application of Equations (6.8) and (6.9). To apply the constant pressure algorithm to planar elongational flow, Equation (6.8) reduces to

$$\begin{aligned}
\dot{\mathbf{r}}_i &= \frac{\mathbf{p}_i}{m_i} + \dot{\epsilon}(\mathbf{i}x_i - \mathbf{j}y_i) + \dot{\xi}\mathbf{r}_i \\
\dot{\mathbf{p}}_i &= \mathbf{F}_i - \dot{\epsilon}(\mathbf{i}p_{xi} - \mathbf{j}p_{yi}) - \alpha\mathbf{p}_i - \dot{\xi}\mathbf{p}_i \\
\dot{V} &= 3\dot{\xi}V.
\end{aligned} \tag{6.13}$$

Algorithm 6.6 (NpT) pseudocode for shear and elongation: calls sllod

while executing integration routine **do**
  $T_C \leftarrow 0$; {initialise temperature}
  $V(t) \leftarrow L_x L_y L_z$; {old volume}
  call neighbour; {apply neighbour list routine}
  call force; {compute forces}
  for i $\leq$ N **do** {compute current temperature}
    $T_C \leftarrow \mathbf{p}_i \cdot \mathbf{p}_i / m_i$;
  end for
  $T_C \leftarrow T_C / (3N - N_C) k_B$;
  $p \leftarrow \mathrm{Tr}(\mathbf{P})/3$; {current pressure (note $\mathbf{P}$ computed in force)}
  $\dot{\xi} \leftarrow$ integrator $((p - p_0) V_C / QNk_B T_C)$; {integrate Equation (6.9) for barostat multiplier}
  $V_T \leftarrow$ integrator $(3\dot{\xi} V_T)$ {integrate 3rd line of Equation (6.10) for new volume}
  $\lambda(t) \leftarrow [V_T / V(t)]^{1/3}$;
  if shear flow **then**
    $L_{x,y,z} \leftarrow \lambda(t) L_{x,y,z}$; {rescale simulation box via Equation (6.12)}
  else {elongational flow}
    $L_{x,y,z}(t) \leftarrow \lambda(t) L^{cube}_{x,y,z}$; {box rescaled to a cube with volume V_T}
    $\mathbf{L}_{1,2} \leftarrow \mathbf{M}^{-1}(t) \mathbf{L}'_{1,2}$; {rotate cube counterclockwise in xy plane}
    $L_{1x} \leftarrow L_{1x}(0) * \exp(\dot{\epsilon} t)$; {evolve box lattice vectors to current time}
    $L_{1y} \leftarrow L_{1y}(0) * \exp(-\dot{\epsilon} t)$;
    $L_{2x} \leftarrow L_{2x}(0) * \exp(\dot{\epsilon} t)$;
    $L_{2y} \leftarrow L_{2y}(0) * \exp(-\dot{\epsilon} t)$;
    $\theta \leftarrow \arctan(L_{1y}/L_{1x})$; {evolve box angle}
    pysum $\leftarrow 0$; {initialise sum of y-component of momenta}
  end if
  call sllod; {integrate SLLOD equations of motion, described in Algorithm 6.7}
end while

The thermostat multiplier is again obtained by taking the time derivative of the instantaneous kinetic energy and setting it to zero, giving

$$\alpha = \frac{\sum_i \frac{1}{m_i} \left[\mathbf{F}_i^{\phi} \cdot \mathbf{p}_i - \dot{\epsilon} \left(p_{xi}^2 - p_{yi}^2 \right) \right]}{\sum_i \frac{\mathbf{p}_i \cdot \mathbf{p}_i}{m_i}} - \dot{\xi}. \tag{6.14}$$

As with the case of shear flow, one solves for these equations of motion, scaling the volume by the solution of the third line in Equation (6.13) at each timestep. However, it is important that the angles of the lattice vectors $\mathbf{L}_1$ and $\mathbf{L}_2$ with respect to the x and y axes do not change from what they would normally be. This is essential because the integrity of these angles ensures that the system becomes reproducible in space and time, as discussed in Section 5.1.5. There are several ways in which this can be achieved. The simplest, though perhaps not the best, is to just rescale the z-dimension

Algorithm 6.7 sllod integrator called in Algorithm 6.6

for i $\leq$ N **do**
  if shear flow **then**
    $\mathbf{r}_i \leftarrow$ integrator $\left(\mathbf{p}_i/m_i + \mathbf{i}\dot{\gamma}y_i + \dot{\xi}\mathbf{r}_i\right)$; {integrate particle positions forward in time}
    $\mathbf{r}_i \leftarrow$ pbcsShear $(\mathbf{r}_i)$; {apply shear pbcs to particle positions}
    $\mathbf{p}_i \leftarrow$ integrator $\left(\mathbf{F}_i - \mathbf{i}\dot{\gamma}p_{yi} - \alpha\mathbf{p}_i - \dot{\xi}\mathbf{p}_i\right)$; {integrate particle momenta forward in time}
  else {elongational flow}
    $\mathbf{r}_i \leftarrow$ integrator $\left(\mathbf{p}_i/m_i + \dot{\epsilon}\left(\mathbf{i}x_i - \mathbf{j}y_i\right) + \dot{\xi}\mathbf{r}_i\right)$; {integrate particle positions forward in time}
    $\mathbf{r}_i \leftarrow$ pbcsElongation $(\mathbf{r}_i)$; {apply elongation pbcs to particle positions}
    $\mathbf{p}_i \leftarrow$ integrator $\left(\mathbf{F}_i - \dot{\epsilon}\left(\mathbf{i}p_{xi} - \mathbf{j}p_{yi}\right) - \alpha\mathbf{p}_i - \dot{\xi}\mathbf{p}_i\right)$; {integrate particle momenta forward in time}
    pysum $\leftarrow$ pysum + p_{yi}; {sum y-component of momenta}
  end if
end for
if elongational flow **then**
  pysum $\leftarrow$ pysum/N;
  for i $\leq$ N **do**
    $p_{yi} \leftarrow p_{yi}$ − pysum; {correct for momentum drift}
  end for
end if

of the simulation box while leaving $\mathbf{L}_1$ and $\mathbf{L}_2$ unchanged. However, if the elongation rate is very high then the pressure and/or density could become very high, in which case the distance between particles in the z-direction could become too small and cause catastrophic failure of the simulation [198].

The next simplest method is to apply contraction/dilation to all three lattice box dimensions. If V_T is again the updated volume that ensures that the average pressure is maintained at its target value of p_0, then, as with the case of shear flow, one again just rescales the elongated simulation box dimensions such that

$$\|\mathbf{L}_i(t)\| \to \lambda \|\mathbf{L}_i(t)\| \tag{6.15}$$

where $i = x, y, z$, and the volume of the simulation cell is thus

$$V_T = \|\mathbf{L}_1(t)\| \; \|\mathbf{L}_2(t)\| \; \|\mathbf{L}_3(t)\| \; \sin\left[\theta(t) + \varphi(t)\right] \tag{6.16}$$

where the $\|\mathbf{L}_i(t)\|$ are the rescaled box lengths (see Figure 6.2). A further variant of this method, which leads to identical results, is to not deal directly with the evolving simulation box, but rather create a "new" cubic unrotated box with volume V_T, rotate it with respect to θ as detailed in Section 5.1.5, and then use the evolution equations (Equation (5.55)) to obtain the appropriate box dimensions in the simulation frame of reference. The equations of motion for all particles are then integrated forward in time and periodic boundary conditions are applied to particle coordinates. It is a matter of

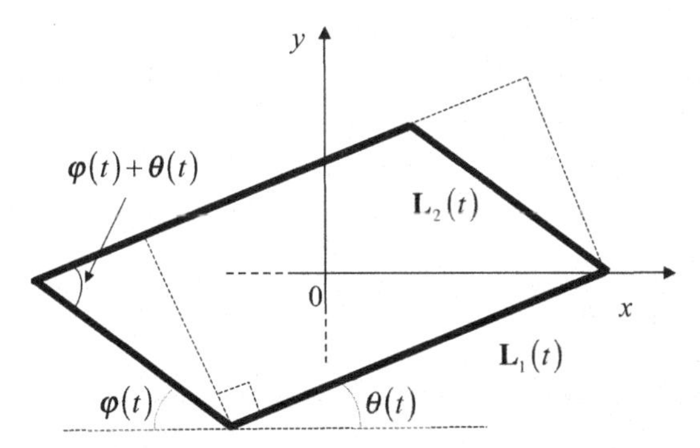

Figure 6.2 The volume of the tilted lattice at a time t is given by the area of the xy parallelogram, i.e. $\|\mathbf{L}_1(t)\| \; \|\mathbf{L}_2(t)\| \sin[\theta(t) + \varphi(t)]$, multiplied by the length of the z dimension. Multiplying $\|\mathbf{L}_i(t)\|$ by the factor λ given by Equation (6.12) gives the target value for the volume according to the last line of Equation (6.13) and preserves the angles θ and φ. Reprinted from reference [198] with the permission of AIP Publishing.

preference which method to use, and details of both methods can be found in reference [198]. Pseudocode for this latter procedure is depicted in Algorithm 6.6.

Elongational viscosities of a simple WCA atomic fluid under PEF for both the (NVT) and (NpT) ensembles are shown in Figure 6.3. Clearly, as already noted in the case of shear flow, (NpT) simulations lead to reduced viscosity at high elongation rates because the system pressure is now kept constant. As with shear, (NVT) simulations run the risk of enhancing viscosities at high rates of strain, which contravenes what would be experimentally observed when fluid samples have at least one surface in contact with the atmosphere. Similar anomalies are observed in simulations of alkanes and polymer melts and solutions (see Chapter 8).

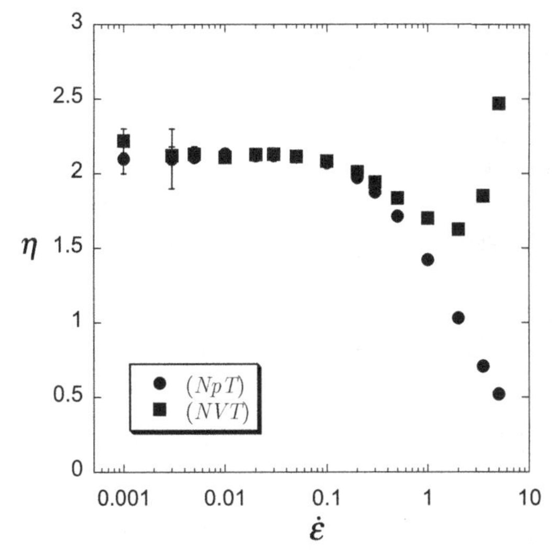

Figure 6.3 Planar elongational viscosities for the isothermal-isochoric (NVT) and isothermal-isobaric (NpT) ensembles as functions of the elongation rate $\dot{\epsilon}$. The system consists of 2048 WCA atoms at a state point of $T = 1.00$, $\rho = 0.840$, $p_0 = 7.8365 \pm 0.0001$, in reduced units. Replotted from reference [198].

6.1.3.3 Configurational Thermostat Implementation

Implementations of configurational thermostats have been described by Delhommelle and Evans [190] and Braga and Travis [192]. In the case of the former, the equations of motion for shear flow are

$$\begin{aligned}
\dot{\mathbf{r}}_i &= \frac{\mathbf{p}_i}{m_i} + \mathbf{i}\dot{\gamma} y_i + \frac{s}{T_0}\frac{\partial T_C}{\partial \mathbf{r}_i} + \dot{\xi}\mathbf{r}_i \\
\dot{\mathbf{p}}_i &= \mathbf{F}_i^{\phi} - \mathbf{i}\dot{\gamma} p_{yi} - \dot{\xi}\mathbf{p}_i \\
\dot{s} &= -Q_{T_C}\frac{(T_C - T_0)}{T_0} \\
\dot{V} &= 3\dot{\xi} V,
\end{aligned} \tag{6.17}$$

whereas for planar elongation they are

$$\begin{aligned}
\dot{\mathbf{r}}_i &= \frac{\mathbf{p}_i}{m_i} + \dot{\epsilon}(\mathbf{i}x_i - \mathbf{j}y_i) + \frac{s}{T_0}\frac{\partial T_C}{\partial \mathbf{r}_i} + \dot{\xi}\mathbf{r}_i \\
\dot{\mathbf{p}}_i &= \mathbf{F}_i^{\phi} - \dot{\epsilon}(\mathbf{i}p_{xi} - \mathbf{j}p_{yi}) - \dot{\xi}\mathbf{p}_i \\
\dot{s} &= -Q_{T_C}\frac{(T_C - T_0)}{T_0} \\
\dot{V} &= 3\dot{\xi} V.
\end{aligned} \tag{6.18}$$

Here T_0 is the target temperature, T_C is the configurational temperature and Q_{T_C} is a damping factor given as $Q_{T_C} = \sigma^2/\tau_C^2$, where τ_C is the response time of the feedback mechanism and σ is the interaction distance (effective atomic diameter). T_C is computed at each timestep (to leading order in the number of particles, N) as

$$k_B T_C = \frac{\sum_i \mathbf{F}_i^{\phi} \cdot \mathbf{F}_i^{\phi}}{-\sum_{i \neq j} \frac{\partial}{\partial \mathbf{r}_{ij}} \cdot \mathbf{F}_{ij}^{\phi}}. \tag{6.19}$$

With Braga and Travis's methodology, the governing equations of motion for planar shear, expressed this time in terms of forces rather than (the equivalent) gradients of the inter-atomic potential (cf. Equations (5.112)–(5.114) for the (NVT) ensemble), are

$$\begin{aligned}
\dot{\mathbf{r}}_i &= \frac{\mathbf{p}_i}{m_i} + \mathbf{i}\dot{\gamma} y_i + \dot{\xi}\left[\mathbf{r}_i + 3N\left(\frac{\mathbf{F}_i^{\phi}}{\Delta}\right)\right] + \zeta \mathbf{F}_i^{\phi} \\
\dot{\mathbf{p}}_i &= \mathbf{F}_i^{\phi} - \mathbf{i}\dot{\gamma} p_{yi} \\
\dot{V} &= 3V\dot{\xi} \\
\ddot{\xi} &= \frac{3V}{Q_{\xi}}[p_C - p_0] \\
\dot{\zeta} &= \frac{1}{Q_{\zeta}}\left[\sum_i \frac{1}{\Delta}\left(\mathbf{F}_i^{\phi}\right)^2 - k_B T_0\right],
\end{aligned} \tag{6.20}$$

where $\Delta = \sum_i (\partial/\partial \mathbf{r}_i) \cdot (\partial\phi/\partial \mathbf{r}_i) = -\sum_i (\partial/\partial \mathbf{r}_i) \cdot \mathbf{F}_i^{\phi}$ is the so-called configurational Laplacian and

$$p_C = \frac{1}{3V}\left[3N\sum_i \frac{1}{\Delta}\left(\mathbf{F}_i^{\phi}\right)^2 + \sum_i \mathbf{r}_i \cdot \mathbf{F}_i^{\phi}\right] \tag{6.21}$$

is the so-called "configurational pressure". The configurational pressure is a useful quantity to use instead of the standard pressure given by taking one-third of the trace of Equation (4.88), because it has no kinetic contribution. Indeed, as we noted in Section 5.3.4, one of the advantages of a configurational thermostat is that one does not need to know the streaming velocity of the fluid, which can sometimes be a difficult quantity to compute for complicated flow geometries and/or molecular fluid systems with many degrees of freedom, including the rotational streaming velocity or vibrational degrees of freedom (see Section 8.5 on molecular thermostats). For a system at equilibrium, one simply sets $\dot{\gamma} = 0$ in the above equations of motion.

For planar elongational flow the governing equations of motion are

$$\begin{aligned}
\dot{\mathbf{r}}_i &= \frac{\mathbf{p}_i}{m_i} + \dot{\epsilon}(\mathbf{i}x_i - \mathbf{j}y_i) + \dot{\xi}\left[\mathbf{r}_i + 3N\left(\frac{\mathbf{F}_i^{\phi}}{\Delta}\right)\right] + \zeta\mathbf{F}_i^{\phi} \\
\dot{\mathbf{p}}_i &= \mathbf{F}_i^{\phi} - \dot{\epsilon}(\mathbf{i}p_{xi} - \mathbf{j}p_{yi}) \\
\dot{V} &= 3V\dot{\xi} \\
\ddot{\xi} &= \frac{3V}{Q_{\xi}}\left[p_C - p_0\right] \\
\dot{\zeta} &= \frac{1}{Q_{\zeta}}\left[\sum_i \frac{1}{\Delta}\left(\mathbf{F}_i^{\phi}\right)^2 - k_B T_0\right]
\end{aligned} \tag{6.22}$$

with p_C defined as before.

6.2 General Homogeneous Flows

We have so far focussed exclusively on two types of homogeneous flow: planar shear and elongational flows. Planar shear or Couette flow is an extremely important type of flow in many industrial applications and there are countless experimental measurements of an enormous amount of different fluids under this type of flow in the literature. The need for establishing reliable NEMD methods for such flows is thus paramount. Planar elongational flow is also a very important flow, particularly in the processing of polymer melts, but also has applications in molecular biology, such as studying the conformations of stretched DNA (e.g. Perkins *et al.* [199]). We have so far devoted considerable time describing the application of SLLOD dynamics to planar elongation because until very recently it was the *only* type of elongational flow for which spatio-temporal periodic boundary conditions could be applied to enable time-independent steady-state calculations of the structural and rheological properties of dense liquids. Kraynik and Reinelt [129] proved that compatible and reproducible periodic boundary conditions could be found for planar elongation, but not for certain other flows such as uniaxial flow. However, both Dobson [200] and Hunt [201] have independently made the clever observation that lattice compatibility (i.e. particles do not come closer than their minimum allowable distances) rather than periodicity (in which the simulation box is remapped at periodic time intervals into its original shape) is all one really needs to perform indefinite NEMD simulations for *any* type of homogeneous deformation. In other words, the simulation box shape can be remapped into a variety of suitable shapes after required times, as long as the compatibility condition is maintained. This allowed Hunt to simulate uniaxial flow for indefinite times, and Dobson demonstrated the procedure for planar, uniaxial and biaxial elongational flows. We refer readers to their original papers for the details of the method but will not discuss them further in this book. Their method involves time-independent flow, but in what follows we will look at time-dependent methods to simulate planar elongation, uniaxial and biaxial flow.

As we have seen, traditional "rectangular" periodic boundary conditions are inadequate for flows that involve compression of at least one spatial dimension because the simulation must cease when the length of the simulation box reaches the minimum allowable length of twice the interaction potential radius. However, there is an alternative way in which *all* other types of compressional flows can be simulated for indefinite times. The method here is to use a time-dependent oscillatory field in the equations of motion and then take the limit as the frequency goes to zero. In doing so one can extrapolate all the time-independent properties of interest, such as viscosities, normal stresses, and the like. While the method is powerful in that it can be applied to any flow field, the most serious disadvantage is that the simulations are computationally intensive, since for each value of the strain rate one needs to simulate over a range of frequencies in order to make reliable extrapolations to zero-frequency. In what follows we will describe the methods used.

6.2.1 Time-dependent Flows in the Steady-state

In order to proceed we must first derive the SLLOD equations of motion again, only this time by including a time-dependent field. It is important to appreciate that the flow is still homogeneous in space, even though we now build in a time dependence into the governing equations of motion. Therefore all arguments about homogeneity still apply equally here. The derivation closely follows the steps in Section 5.1.2. We first write down Equation (5.7) again, which can now be interpreted as the total force experienced by a system of N particles under the influence of an external field. As we are only interested in the form of the external forces we do not yet consider any inter-atomic interactions acting between particles, so set $\mathbf{F}_i^{\phi} = 0$. Furthermore, we now explicitly consider the time-dependence in the external forces:

$$\frac{d}{dt}\sum_i m_i \mathbf{v}_i = \sum_i \mathbf{F}_i^e(t). \tag{6.23}$$

Using the same definitions as in Section 5.1.2, we have

$$\begin{aligned}\mathbf{v}_i &= \mathbf{c}_i + \mathbf{v}(\mathbf{r}_i)\,\Theta(t) \\ &= \mathbf{c}_i + \mathbf{r}_i \cdot \nabla\mathbf{v}(t)\,\Theta(t).\end{aligned} \tag{6.24}$$

Substituting Equation (6.24) into Equation (6.23), and invoking conservation of peculiar momentum gives

$$\begin{aligned}\frac{d}{dt}\sum_i m_i\mathbf{v}_i &= \frac{d}{dt}\sum_i m_i\mathbf{r}_i \cdot \nabla\mathbf{v}(t)\,\Theta(t) \\ &= \sum_i \left[m_i\frac{d\mathbf{r}_i}{dt} \cdot \nabla\mathbf{v}(t)\,\Theta(t) + m_i\mathbf{r}_i \cdot \frac{d}{dt}\left(\nabla\mathbf{v}(t)\,\Theta(t)\right)\right] \\ &= \sum_i \left[m_i\mathbf{v}_i \cdot \nabla\mathbf{v}(t)\,\Theta(t) + m_i\mathbf{r}_i \cdot \left(\frac{d\nabla\mathbf{v}(t)}{dt}\Theta(t) + \nabla\mathbf{v}(t)\,\delta(t)\right)\right] \\ &= \sum_i \left[m_i\mathbf{v}_i \cdot \nabla\mathbf{v}(t)\,\Theta(t) + m_i\mathbf{r}_i \cdot \frac{d\nabla\mathbf{v}(t)}{dt}\Theta(t) + m_i\mathbf{r}_i \cdot \nabla\mathbf{v}(t)\,\delta(t)\right].\end{aligned} \tag{6.25}$$

Substituting Equation (6.24) into Equation (6.25), noting that the product of two Heaviside functions is just the Heaviside function, and invoking the fact that spatial homogeneity demands that the form of equations of motion are the same for each particle, gives the external force acting on each particle as

$$\mathbf{F}_i^e(t) = m_i\mathbf{r}_i \cdot \nabla\mathbf{v}(t)\,\delta(t) + m_i\mathbf{r}_i \cdot \nabla\mathbf{v}(t) \cdot \nabla\mathbf{v}(t)\,\Theta(t) + m_i\mathbf{r}_i \cdot \frac{d\nabla\mathbf{v}(t)}{dt}\Theta(t). \tag{6.26}$$

Equation (6.24) is just the SLLOD equation of motion for the particle positions. To obtain the second equation of motion for the evolution of particle velocities, we have to now include the inter-atomic interaction between particles. The total force on particle i for a system of N *interacting* particles is therefore just

$$m_i\frac{d\mathbf{v}_i}{dt} = -\frac{\partial\phi}{\partial\mathbf{r}_i} + \mathbf{F}_i^e(t) = \mathbf{F}_i^\phi + \mathbf{F}_i^e(t). \tag{6.27}$$

Substituting Equation (6.24) into the left-hand side of Equation (6.27) and Equation (6.26) into the right-hand side of Equation (6.27) gives

$$\begin{aligned}m_i\frac{d}{dt}\left(\mathbf{c}_i + \mathbf{r}_i \cdot \nabla\mathbf{v}(t)\,\Theta(t)\right) = \mathbf{F}_i^\phi &+ m_i\mathbf{r}_i \cdot \nabla\mathbf{v}(t)\,\delta(t) + m_i\mathbf{r}_i \cdot \nabla\mathbf{v}(t) \cdot \nabla\mathbf{v}(t)\,\Theta(t) \\ &+ m_i\mathbf{r}_i \cdot \frac{d\nabla\mathbf{v}(t)}{dt}\Theta(t).\end{aligned} \tag{6.28}$$

Straightforward manipulation of this equation results in the desired equation for the evolution of particle peculiar velocities:

$$m_i\frac{d\mathbf{c}_i}{dt} = \mathbf{F}_i^\phi - m_i\mathbf{c}_i \cdot \nabla\mathbf{v}(t)\,\Theta(t). \tag{6.29}$$

Equations (6.24) and (6.29) thus constitute the SLLOD equations of motion for arbitrary time-dependent fields, and can be written in the standard form for $t > 0$ as:

$$\dot{\mathbf{r}}_i = \frac{\mathbf{p}_i}{m_i} + \mathbf{r}_i \cdot \nabla\mathbf{v}(t)$$
$$\dot{\mathbf{p}}_i = \mathbf{F}_i^{\phi} - \mathbf{p}_i \cdot \nabla\mathbf{v}(t), \tag{6.30}$$

where $\mathbf{p}_i$ is the peculiar momentum of particle i. By comparing Equations (6.30) with Equations (5.18) we see that the time-dependent and time-independent SLLOD equations of motion are identical. This is convenient because once the equations have been coded up for use in, say, time-independent flow, it is a very simple matter of coding the time-dependence of the external field into the equations of motion. Typically this time-dependence will involve a relatively simple functional form, such as an oscillatory field, as we will demonstrate in the sections that follow. Furthermore, an examination of the form of the external forces in Equation (6.26) demonstrates that in the case of a time-dependent external force at $t > 0$ there is *no case at all* in which the SLLOD equations of motion can be generated from just an impulse term at $t = 0$, as was the case for time-independent shear flow. The third term in this equation, $m_i\mathbf{r}_i \cdot (d\nabla\mathbf{v}(t)/dt)\,\Theta(t)$, shows that at all times a time-dependent force proportional to the time derivative of the velocity gradient tensor exists.

6.2.2 Oscillatory Shear Flow

From Equation (6.30), the time-dependent SLLOD equations for thermostatted shear flow ((NVT) ensemble) may now be written down as

$$\dot{\mathbf{r}}_i = \frac{\mathbf{p}_i}{m_i} + \mathbf{i}\dot{\gamma}(t)\,y_i$$
$$\dot{\mathbf{p}}_i = \mathbf{F}_i^{\phi} - \mathbf{i}\dot{\gamma}(t)\,p_{yi} - \alpha\mathbf{p}_i, \tag{6.31}$$

where, for the case of a Gaussian thermostat, the thermostat multiplier is given by

$$\alpha = \frac{\sum_i \frac{1}{m_i}\left(\mathbf{F}_i^{\phi} \cdot \mathbf{p}_i - \dot{\gamma}(t)\,p_{xi}p_{yi}\right)}{\sum_i \frac{\mathbf{p}_i \cdot \mathbf{p}_i}{m_i}}. \tag{6.32}$$

Implementations involving other thermostats, such as Nosé-Hoover or configurational, can be straightforwardly formulated (see Section 5.3). Similarly, applications to other ensembles, such as (NVU) and (NpT) for example, can also be performed following analogous steps to those in Sections 6.1.1 and 6.1.3 above. In what follows, for the purposes of brevity, we will only consider time-dependent flows in the (NVT) ensemble under the influence of a Gaussian thermostat.

As mentioned above, the most useful time-dependent field is the oscillatory field. For shear flow it is particularly relevant for rheometric experiments in which, typically, a polymeric fluid is sandwiched between plates that move in opposite directions with respect to each other in an oscillatory manner [202]. In this way the linear viscoelastic frequency-dependent properties of the fluid can be studied. If we consider a velocity gradient tensor with the form $\nabla\mathbf{v}(t) = \nabla\mathbf{v}\cos(\omega t)$, where ω is the frequency defined as $\omega = 2\pi/\tau$ and τ is the period of oscillation, then the equations of motion become

$$\begin{aligned}\dot{\mathbf{r}}_i &= \frac{\mathbf{p}_i}{m_i} + \mathbf{i}\dot{\gamma}\cos(\omega t)\, y_i \\ \dot{\mathbf{p}}_i &= \mathbf{F}_i^{\phi} - \mathbf{i}\dot{\gamma}\cos(\omega t)\, p_{yi} - \alpha\mathbf{p}_i,\end{aligned} \tag{6.33}$$

with

$$\alpha = \frac{\sum_i \frac{1}{m_i}\left(\mathbf{F}_i^{\phi}\cdot\mathbf{p}_i - \dot{\gamma}\cos(\omega t)\, p_{xi}p_{yi}\right)}{\sum_i \frac{\mathbf{p}_i\cdot\mathbf{p}_i}{m_i}}. \tag{6.34}$$

Here $\dot{\gamma}$ is the amplitude of the velocity gradient field. The simulation box lattice vectors are found by solving Equation (5.47), which for this flow geometry leads to

$$L_x(t) = L_y(0)\,\dot{\gamma}\frac{\sin(\omega t)}{\omega} + L_x(0). \tag{6.35}$$

The box lengths in either the y or z directions do not change in time, as is the case for time-independent shear. If, as is usually the case, we choose the simulation box such that at $t = 0$, $L_x(0) = L_y(0) = L_z(0) = L(0)$, then Equation (6.35) becomes

$$L_x(t) = L(0)\left[1 + \dot{\gamma}\frac{\sin(\omega t)}{\omega}\right]. \tag{6.36}$$

The strain and shear stress will be out of phase by some phase angle δ, so

$$\sigma_{xy}(t) = -P_{xy}(t) = \sigma_{xy}^0 \sin(\omega t + \delta), \tag{6.37}$$

where σ_{xy}^0 is the shear stress amplitude and we represent the strain as $\gamma = \gamma_0\sin(\omega t)$, with γ_0 being the maximum strain amplitude, given as $\gamma_0 = \dot{\gamma}\tau/2$, where τ is the period of oscillation. The shear stress (negative of the xy component of the pressure tensor) can be computed directly from the NEMD simulation data via Equation (4.88).

What is of interest in such simulations is the elastic (or storage) modulus, G', the viscous (or loss) modulus, G'', and the complex, or dynamic, viscosity, defined as

$$\eta^* \equiv -iG^*/\omega, \tag{6.38}$$

where

$$G^* = G' + iG''. \tag{6.39}$$

At sufficiently low frequencies the fluid has time to respond to the oscillatory field and one has liquid-like behaviour. Therefore the steady-state dynamic modulii can be computed as

$$\begin{aligned}G' &= \left(\sigma_{xy}^0/\gamma_0\right)\cos\delta \\ G'' &= G'\tan\delta.\end{aligned} \tag{6.40}$$

Usually such methods would only be applied to more complex molecular fluids, rather than simple atomic fluids, such as the simulation study performed by Cifre *et al.* [202], in which the details of the method can also be found. However, we note that Cifre *et al.* did not use the SLLOD equations of motion, but rather drove the flow entirely by oscillatory modulation of the Lees-Edwards boundary conditions.

6.2.3 Uniaxial, Biaxial and Planar Elongational Viscosities

There are three different types of elongational (or extensional) flows possible. If we write the strain rate tensor as

$$\nabla \mathbf{v} = \begin{pmatrix} \dot{\epsilon}_{xx} & \dot{\gamma}_{yx} & \dot{\gamma}_{zx} \\ \dot{\gamma}_{xy} & \dot{\epsilon}_{yy} & \dot{\gamma}_{zy} \\ \dot{\gamma}_{xz} & \dot{\gamma}_{yz} & \dot{\epsilon}_{zz} \end{pmatrix} \tag{6.41}$$

and, as before, assume that elongation is parallel to the x, y or z directions, then its form for shear-free flows becomes

$$\nabla \mathbf{v} = \begin{pmatrix} \dot{\epsilon}_{xx} & 0 & 0 \\ 0 & \dot{\epsilon}_{yy} & 0 \\ 0 & 0 & \dot{\epsilon}_{zz} \end{pmatrix}. \tag{6.42}$$

Noting that compression/expansion can take place in any of the x, y or z directions, and imposing the constraint of constant volume flow (i.e. incompressible) we obtain the following possible types of pure elongational flow:

Planar elongational flow (PEF): $\dot{\epsilon}_{xx} = \dot{\epsilon}, \quad \dot{\epsilon}_{yy} = -\dot{\epsilon}, \quad \dot{\epsilon}_{zz} = 0$

Uniaxial stretching flow (USF): $\dot{\epsilon}_{xx} = -\dot{\epsilon}/2, \quad \dot{\epsilon}_{yy} = \dot{\epsilon}, \quad \dot{\epsilon}_{zz} = -\dot{\epsilon}/2$

Biaxial stretching flow (BSF): $\dot{\epsilon}_{xx} = \dot{\epsilon}/2, \quad \dot{\epsilon}_{yy} = -\dot{\epsilon}, \quad \dot{\epsilon}_{zz} = \dot{\epsilon}/2.$

In what follows we will only consider conventional rectangular periodic boundary conditions rather than the K-R scheme, which after all is only applicable for PEF and can not be implemented for USF or BSF. In the case of oscillatory PEF, there is absolutely no advantage in the K-R scheme of pbcs, as we will proceed to demonstrate. Bearing in mind that for time-oscillatory flow USF and BSF are actually the same flow (with the flow being USF for half a cycle and BSF for the other half cycle), the thermostatted SLLOD equations of motion become [147]:

PEF

$$\begin{aligned} \dot{\mathbf{r}}_i &= \frac{\mathbf{p}_i}{m_i} + \dot{\epsilon}\left(\mathbf{i}x_i - \mathbf{j}y_i\right)\cos(\omega t) \\ \dot{\mathbf{p}}_i &= \mathbf{F}_i^{\phi} - \dot{\epsilon}\left(\mathbf{i}p_{xi} - \mathbf{j}p_{yi}\right)\cos(\omega t) - \alpha \mathbf{p}_i \end{aligned} \tag{6.43}$$

with

$$\alpha = \frac{\sum_i \frac{1}{m_i}\left\{\mathbf{F}_i^{\phi}\cdot\mathbf{p}_i - \dot{\epsilon}\left[p_{xi}^2 - p_{yi}^2\right]\cos(\omega t)\right\}}{\sum_i \frac{\mathbf{p}_i\cdot\mathbf{p}_i}{m_i}} \tag{6.44}$$

and

USF/BSF

$$\dot{\mathbf{r}}_i = \frac{\mathbf{p}_i}{m_i} + \dot{\epsilon}\left[\frac{1}{2}\left(\mathbf{i}x_i + \mathbf{k}z_i\right) - \mathbf{j}y_i\right]\cos\left(\omega t\right)$$
$$\dot{\mathbf{p}}_i = \mathbf{F}_i^{\phi} - \dot{\epsilon}\left[\frac{1}{2}\left(\mathbf{i}p_{xi} + \mathbf{k}p_{zi}\right) - \mathbf{j}p_{yi}\right]\cos\left(\omega t\right) - \alpha\mathbf{p}_i, \tag{6.45}$$

where

$$\alpha = \frac{\sum_i \frac{1}{m_i}\left\{\mathbf{F}_i^{\phi}\cdot\mathbf{p}_i - \dot{\epsilon}\left[\frac{1}{2}\left(p_{xi}^2 + p_{zi}^2\right) - p_{yi}^2\right]\cos\left(\omega t\right)\right\}}{\sum_i \frac{\mathbf{p}_i\cdot\mathbf{p}_i}{m_i}}. \tag{6.46}$$

In the process of the simulation either or all of $L_x(t)$, $L_y(t)$, or $L_z(t)$ will oscillate in time. The solution of Equation (5.47) shows that

$$L_i(t) = L_i(0)\exp\left[\frac{\dot{\epsilon}_{ii}\sin\left(\omega t\right)}{\omega}\right] \tag{6.47}$$

where $i = x, y, z$. Note however that there is a limit on the frequency which is set by the minimum allowable length of any box side when that side is compressing. As mentioned in Section 5.1.5, for time-independent flow the simulation can only run for a maximum time of $t_{\max} = \dot{\epsilon}_{ii}^{-1}\ln(2r_c/L_i(0))$ for rectangular pbcs, where i is the direction of contraction and $\dot{\epsilon}$ is negative. For oscillatory flow, Equation (6.47) demands that the frequency must not be below a minimum, set by

$$\omega_{\min} = \frac{\dot{\epsilon}_{ii}}{\ln\left[2r_c/L_i(0)\right]}. \tag{6.48}$$

$\omega < \omega_{\min} \Rightarrow L_i(t) < 2r_c$, which is not allowable as it violates the minimum image convention.

As the field is periodic in time the system energy and normal pressures are also periodic, but slightly phase-shifted by an amount δ_n to account for the finite response time of the fluid to the field. They can be expressed as higher-order harmonics of the field [147]

$$U(t) = \sum_n U(\omega_n)\cos\left(\omega t + \delta_n\right)$$
$$P_{ii}(t) = \sum_n P_{ii}(\omega_n)\cos\left(\omega t + \delta_n\right), \tag{6.49}$$

where $n = 1, 2, 3, \ldots, \omega_n = n\omega = 2\pi n/\tau$ and τ is the period. Here the energy is defined to be the total internal energy per particle and the coefficients are just the amplitudes of the energy or pressure response per harmonic. We note here that for moderate frequencies one only need consider harmonics up to $n \approx 2$ for a simple atomic fluid [147]. As $\omega \rightarrow 0$ higher-order harmonics do become manifest and the shape of the energy/pressure responses can become quite complex. In general, it is sensible to choose frequencies such that higher-order harmonics are minimised.

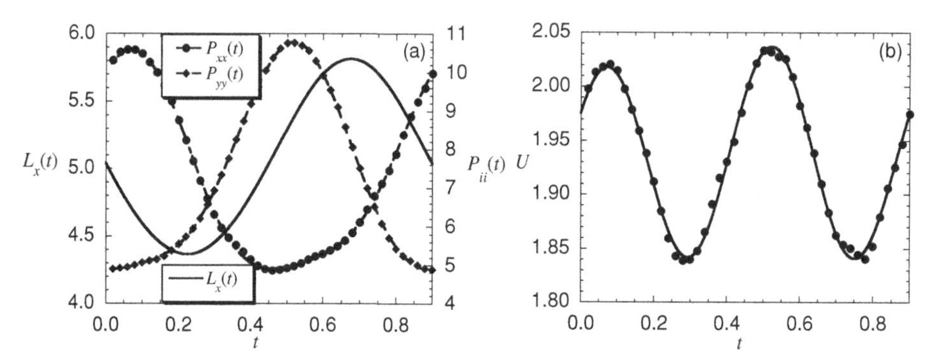

Figure 6.4 Oscillatory PEF for a system of 108 WCA atoms at reduced density and temperature of 0.8442 and 0.722 respectively. $\dot{\epsilon} = 0.1$ and $\omega = 6.981$. (a) Normal pressure components and $L_x(t)$; (b) total internal energy per particle. The data are plotted over one cycle, though each data point is an average over 200 cycles. Symbols are NEMD data and curved lines are fits to Equations (6.47) and (6.49). Replotted from reference [147].

In Figure 6.4, we plot the dimensions of an oscillatory PEF system with $\dot{\epsilon} = 0.1$ and $\omega = 6.981$ along with the normal pressures in the contracting/expanding directions over one period, averaged over 200 cycles. Also shown is the total internal energy per particle. Clearly we can see that the extrema in the normal pressures are out of phase by some angle $\delta\theta$ with respect to the extremum in $L_i(t)$, which as we noted is due to the finite response time of the fluid to the imposed time-dependent strain rate. The normal pressures are also out of phase by π with respect to each other, as one would expect for this type of symmetrical flow.

For PEF let us now define $P_{yy}(\omega)$ and $P_{xx}(\omega)$ to be the maximum and minimum normal stresses in the contracting and expanding directions respectively, i.e. $P_{yy}(\omega) \equiv \max[P_{yy}(\omega, t)]$ and $P_{xx}(\omega) \equiv \min[P_{xx}(\omega, t)]$. Similarly for BSF we define $P_{yy}(\omega) \equiv \max[P_{yy}(\omega, t)]$ and $P_{xx}(\omega) = P_{zz}(\omega) \equiv \min[P_{xx}(\omega, t)] = \min[P_{zz}(\omega, t)]$, whereas for USF we have $P_{yy}(\omega) \equiv \min[P_{yy}(\omega, t)]$ and $P_{xx}(\omega) = P_{zz}(\omega) \equiv \max[P_{xx}(\omega, t)] = \max[P_{zz}(\omega, t)]$. Using these definitions we can now define the steady-state, time-independent elongational viscosities as:

$$\eta_E^{PEF}(\dot{\epsilon}) = \lim_{\omega \to 0} \frac{P_{yy}(\omega) - P_{xx}(\omega)}{4\dot{\epsilon}} \tag{6.50}$$

$$\eta_E^{BSF}(\dot{\epsilon}) = \lim_{\omega \to 0} \frac{P_{yy}(\omega) - \frac{1}{2}[P_{xx}(\omega) + P_{zz}(\omega)]}{3\dot{\epsilon}} \tag{6.51}$$

$$\eta_E^{USF}(\dot{\epsilon}) = \lim_{\omega \to 0} \frac{\frac{1}{2}[P_{xx}(\omega) + P_{zz}(\omega)] - P_{yy}(\omega)}{3\dot{\epsilon}}. \tag{6.52}$$

Equations (6.50)–(6.52) are standard rheological definitions for these types of flow and ensure that the zero-strain rate viscosity for all flows are equal, i.e.

$$\eta_0 \equiv \lim_{\nabla \mathbf{v} \to 0} \eta(\nabla \mathbf{v}) = \lim_{\dot{\gamma} \to 0} \eta^{PSF}(\dot{\gamma}) = \lim_{\dot{\epsilon} \to 0} \eta^{PEF}(\dot{\epsilon}) = \lim_{\dot{\epsilon} \to 0} \eta^{USF/BSF}(\dot{\epsilon}). \tag{6.53}$$

In Figure 6.5, we plot PEF elongational viscosities as a function of $\dot{\epsilon}^{1/2}$ for the system depicted in Figure 6.4, along with the zero-shear viscosity (2.3 ± 0.1) calculated by the

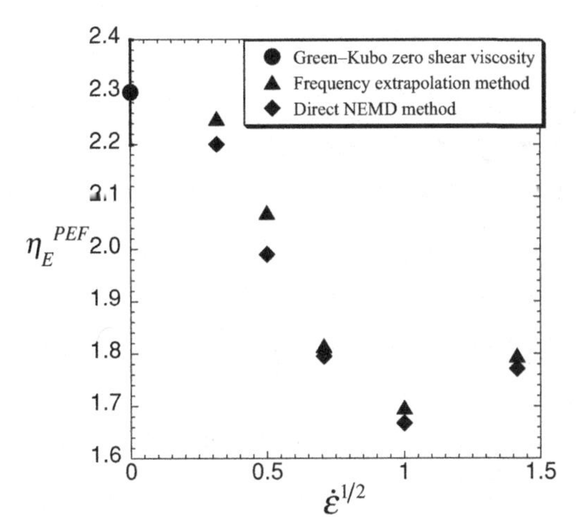

Figure 6.5 Comparison of PEF viscosities for direct steady-state time-independent flow and the frequency extrapolation method for a simple WCA fluid as described in Figure 6.4. Replotted from reference [147].

Green–Kubo equilibrium method. PEF viscosities are compared from direct NEMD simulations of steady-state time-independent flow against those of the extrapolation method described above. The zero-frequency extrapolated viscosities agree with the direct NEMD viscosities to within 4%, and are consistent when extrapolated to zero elongation rate, where they match with the reliable Green–Kubo result. Similar agreement is also found for USF/BSF viscosities [147]. The method is thus seen to be a viable alternative to the K-R scheme for USF/BSF flows, where K-R boundary conditions can not be implemented. Thus far the frequency extrapolation method has only been used for simple atomic systems. A potentially very useful application would be to apply the method to molecular fluids such as alkanes or polymer melts; however, it is not yet known how the normal stresses will extrapolate to zero frequency.

Note also that at high elongation rates the viscosity increases, a phenomenon known as strain hardening (or "thickening"). This is also seen in simulations of chain molecules such as polymer melts and will be discussed later in Chapter 8. However, the physical process through which this phenomenon is observed for simple fluids is distinct from the analogous molecular fluid situation. In the case of molecular fluids strain hardening occurs due to competitive entropic processes associated with chain stretching and flow alignment. For simple atomic fluids the mechanism is due to distortions of the pair distribution function, which in fact is the same mechanism through which shear thinning for simple atomic fluids occurs [132]. Thickening also results from dilatancy effects in simple liquids.

The data extracted from the oscillatory elongation simulations actually is very rich. In addition to just extracting the zero-frequency viscosities one can also obtain linear viscoelastic data by looking at the zero-strain rate, nonzero frequency case. To see this, we can write the frequency-dependent viscosity as a complex quantity [203]:

$$\eta(\omega) = \eta'(\omega) - i\eta''(\omega), \tag{6.54}$$

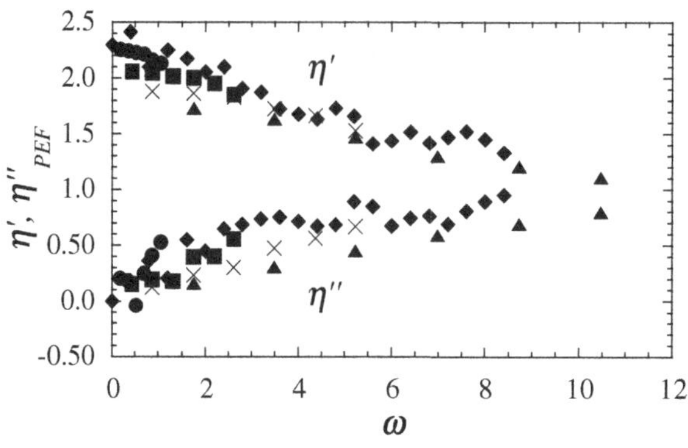

Figure 6.6 Real (upper points) and imaginary (lower points) parts of the planar elongational viscosity defined by Equation (6.57) plotted against frequency for four different elongation rates: $\dot{\epsilon} = 0.1$ (circles), $\dot{\epsilon} = 0.25$ (squares), $\dot{\epsilon} = 0.5$ (crosses), and $\dot{\epsilon} = 1.0$ (triangles). Also plotted are the real (upper) and imaginary (lower) parts of the linear shear viscosity obtained via the Green–Kubo method (diamonds). The convergence as $\dot{\epsilon}$ decreases indicates approach to the linear limit. Replotted from reference [172].

where the components of the frequency-dependent viscosity are characterised by

$$\lim_{\omega \to 0} \eta'(\omega) = \eta$$
$$\lim_{\omega \to 0} \eta''(\omega) = 0. \tag{6.55}$$

If we now express the pressure tensor in the form

$$\mathbf{P}(t) = \mathbf{P} \exp\left[i(\omega t + \delta)\right], \tag{6.56}$$

where δ is the phase shift between the strain rate and pressure tensors in the linear limit, we are then able to define the real and imaginary components of the effective frequency-dependent viscosity. In the case of PEF, we thus have [172]

$$\begin{aligned} \eta'_{PEF}(\omega, \dot{\epsilon}) &= \frac{P_{yy} - P_{xx}}{4\dot{\epsilon}} \cos(\delta) \\ \eta''_{PEF}(\omega, \dot{\epsilon}) &= \frac{P_{xx} - P_{yy}}{4\dot{\epsilon}} \sin(\delta). \end{aligned} \tag{6.57}$$

Equations (6.57) are only strictly valid in the zero strain rate limit. Figure 6.6 shows the real and imaginary viscosity components plotted against the four lowest strain rates, along with the real part of the Green–Kubo viscosity. Clearly, as the strain rate decreases the results converge to the Green–Kubo results, as anticipated. Note also that the agreement improves at higher frequency for any given value of $\dot{\epsilon}$. This is because elasticity is important at nonzero frequency, and the strain, rather than the strain rate, becomes the important parameter for the onset of nonlinear elastic behaviour. Therefore, at a given strain rate, the strain amplitude decreases with increasing frequency and the linear limit for elasticity is approached. In a similar way, the disparity between results for different strain rates at a given frequency increases for decreasing frequency, as seen in Figure 6.6.

6.2.4 Bulk Viscosity

The bulk viscosity can, at least in principle, be computed in a similar manner as the elongational viscosities described in the section above. The strain rate tensor is again given by Equation (6.42), except now we relax the condition of incompressibility. The simulation cell is thus now able to change volume during the simulation, in an oscillatory manner, but with each box direction changing at the same rate. Thus $\dot{\epsilon}_{xx} = \dot{\epsilon}_{yy} = \dot{\epsilon}_{zz} = \dot{\epsilon}\cos(\omega t)$. With this flow geometry, the governing SLLOD equations of motion are

$$\begin{aligned} \dot{\mathbf{r}}_i &= \frac{\mathbf{p}_i}{m_i} + \dot{\epsilon}\left(\mathbf{i}x_i + \mathbf{j}y_i + \mathbf{k}z_i\right)\cos(\omega t) \\ \dot{\mathbf{p}}_i &= \mathbf{F}_i^{\phi} - \dot{\epsilon}\left(\mathbf{i}p_{xi} + \mathbf{j}p_{yi} + \mathbf{k}p_{zi}\right)\cos(\omega t) - \alpha\mathbf{p}_i \end{aligned} \tag{6.58}$$

with

$$\alpha = \frac{\sum_i \frac{1}{m_i}\left\{\mathbf{F}_i^{\phi}\cdot\mathbf{p}_i - \dot{\epsilon}\left[p_{xi}^2 + p_{yi}^2 + p_{zi}^2\right]\cos(\omega t)\right\}}{\sum_i \frac{\mathbf{p}_i\cdot\mathbf{p}_i}{m_i}}. \tag{6.59}$$

Note again that, as must always be the case with the SLLOD equations, the boundaries of the simulation box must not exert any forces on the system, and so the box lattice vectors must evolve according to Equation (5.47). In the case of equal compression/dilation this implies

$$L_i(t) = L_i(0)\exp\left[\frac{\dot{\epsilon}\sin(\omega t)}{\omega}\right], \tag{6.60}$$

where $i = x, y$ or z. Once again, this implies a minimum allowable frequency, given by

$$\omega_{\min} = \frac{\dot{\epsilon}}{\ln\left[2r_c/L_i(0)\right]}. \tag{6.61}$$

The constitutive equation for the bulk viscosity is given as

$$\Pi = -\eta_v \nabla\cdot\mathbf{v} \tag{6.62}$$

with

$$\Pi = \frac{1}{3}\mathrm{Tr}(\mathbf{\Pi}) = \frac{1}{3}\mathrm{Tr}\left[\mathbf{P} - p_0\mathbf{I}\right]. \tag{6.63}$$

Here p_0 is the equilibrium pressure. Substituting Equation (6.63) into Equation (6.62), and noting that for bulk compression we have $\nabla\cdot\mathbf{v} = 3\dot{\epsilon}$, with $\dot{\epsilon}$ being negative, gives

$$\eta_v = -\frac{\left(\Pi_{xx} + \Pi_{yy} + \Pi_{zz}\right)}{9\dot{\epsilon}} \tag{6.64}$$

where Π_{ii} here represents the nonequilbrium part of the ii component of the pressure tensor taken at maximum compression of the simulation box. For the other half of the cycle, in which expansion occurs, the sign of Equation (6.64) is reversed and the pressures are evaluated at maximum expansion.

Equation (6.64) is however not easy to use. For one thing, in pure compression or dilation the volume is constantly changing. Thus, the state point is constantly evolving (density always increasing for compression or decreasing for dilation), even though the system is thermostatted instantaneously (if using a Gaussian thermostat, for example). Furthermore, if one were to use the frequency dependent equations of motion given by Equation (6.58) the bulk viscosity becomes both state-point *and* frequency dependent. Thus one would have to perform many simulations in which the bulk viscosity corresponding to a particular volume V (density) and frequency ω is computed by Equation (6.64), perform many such simulations for a variety of frequencies, and then extrapolate to zero-frequency to obtain the required zero-frequency viscosity. In addition, this must be done for each strain rate $\dot{\epsilon}$ if one wishes to compute the strain rate dependence. Such a method is clearly tedious and not advisable.

An alternative method was devised by Hoover *et al.* [20], in which rather than trying to compute the bulk viscosity directly from the constitutive relation (Equation (6.62)) they do it indirectly via the relation for the work over a cycle of expansion/compression. If $\bar{W}$ is the average work done over one cycle with frequency ω, then it is related to the bulk viscosity by

$$\bar{W} = \int dW = 9\pi \xi^2 V_0 \, \omega \, \eta_v, \tag{6.65}$$

where ξ is the maximum strain amplitude and V_0 is the volume amplitude. Clearly one can obtain the frequency-dependent bulk viscosity through this method and extrapolate to zero frequency.

6.3 Mixed Shear and Planar Elongational Flows

6.3.1 Periodic Boundary Conditions

NEMD simulations of mixed flows of planar elongation and shear for indefinite time have recently been made possible by ideas derived from the Kraynik-Reinelt method of pbc implementation, and references therein. We do not go into specific details of the theoretical development or implementation of the algorithm, and will only sketch the procedure here. Readers can find specific details in Hunt *et al.* [134]. The basic algorithm applicable to the implementation of the pbc scheme will be valid for *any* simulation scheme and is not limited to just NEMD (e.g. semi-dilute Brownian dynamics simulations of mixed shear and elongation [204] etc. are all feasible using this method).

The basic idea stems from the theory of lattices [205, 206], in which a general lattice can be expressed as a set of points $\mathbf{L}_i(t)$ as defined in Equation (5.52), which evolves in time as[1]

$$\mathbf{L}_i(t) = \mathbf{L}_i(0) \cdot \exp(\nabla \mathbf{v}\, t) \equiv \mathbf{L}_i(0) \cdot \mathbf{\Lambda}. \tag{6.66}$$

As with the discussion of indefinite planar elongational flow in Section 5.1.5, the lattice is reproducible in time if

$$\mathbf{L}_i(t = \tau_p) = N_{i1}\mathbf{L}_1(0) + N_{i2}\mathbf{L}_2(0) + N_{i3}\mathbf{L}_3(0), \tag{6.67}$$

where N_{ij} are integers and τ_p is the reproducibility time. Kraynik and Reinelt [129] noticed that the velocity gradient tensor may be replaced by any diagonalizable constant matrix with zero trace and real eigenvalues. This in turn leads to a new basis set applicable for mixed flows in which

$$\mathbf{L}'_i(t = \tau_p) = N_{i1}\mathbf{L}'_1(0) + N_{i2}\mathbf{L}'_2(0) + N_{i3}\mathbf{L}'_3(0), \tag{6.68}$$

where the new basis vectors $\mathbf{L}'_i(0)$ are given by the transformation operation:

$$\mathbf{L}'_i(0) = \mathbf{L}_i(0) \cdot \mathbf{S}^{-1}. \tag{6.69}$$

If we consider a mixed flow velocity gradient tensor such as

$$\nabla \mathbf{v} = \begin{pmatrix} \dot{\epsilon} & 0 & 0 \\ \dot{\gamma} & -\dot{\epsilon} & 0 \\ 0 & 0 & 0 \end{pmatrix} \tag{6.70}$$

$$= \begin{pmatrix} 1 & 0 & 0 \\ \frac{\dot{\gamma}}{2\dot{\epsilon}} & 1 & 0 \\ 0 & 0 & 1 \end{pmatrix} \begin{pmatrix} \dot{\epsilon} & 0 & 0 \\ 0 & -\dot{\epsilon} & 0 \\ 0 & 0 & 0 \end{pmatrix} \begin{pmatrix} 1 & 0 & 0 \\ -\frac{\dot{\gamma}}{2\dot{\epsilon}} & 1 & 0 \\ 0 & 0 & 1 \end{pmatrix}$$

$$= \mathbf{S} \cdot \mathbf{D} \cdot \mathbf{S}^{-1} \tag{6.71}$$

then the diagonal matrix $\mathbf{D}$ governs the reproducibility of the lattice and is just the elongational flow velocity gradient tensor. Thus, the mapping operation applicable to planar elongational flow pbcs (see Section 5.1.5) is *exactly* the same mapping operation applicable to mixed flow simulations. The matrices $\mathbf{S}$ and $\mathbf{S}^{-1}$ are the transformation matrices that map the lattice vectors $\mathbf{L}_i(0)$ into their initial states $\mathbf{L}'_i(0)$ via Equation (6.69). Note that $\mathbf{L}_i(0)$ are the initial lattice vectors of the Kraynik-Reinelt lattice required to simulate planar elongational flow for indefinite times.

The transformation operation given by Equation (6.69) results in initial lattice vectors that are *not orthogonal* in general. $\mathbf{S}$ and $\mathbf{S}^{-1}$ determine the magnitude and orientation direction of the initial lattice vectors $\mathbf{L}'_i(0)$, and clearly depend on both the shear and elongational strain rates. In contrast, $\mathbf{D}$ determines the reproducibility of the lattice and angle of orientation of the K-R lattice with respect to the x-axis. This angle is just the angle θ required to rotate a square simulation box to ensure lattice reproducibility in planar elongational flow (see Section 5.1.5). However, it is *not* the angle in which the

[1] We assume here that the lattice vectors are expressed as row rather than column vectors, to be compatible with the discussion in reference [134].

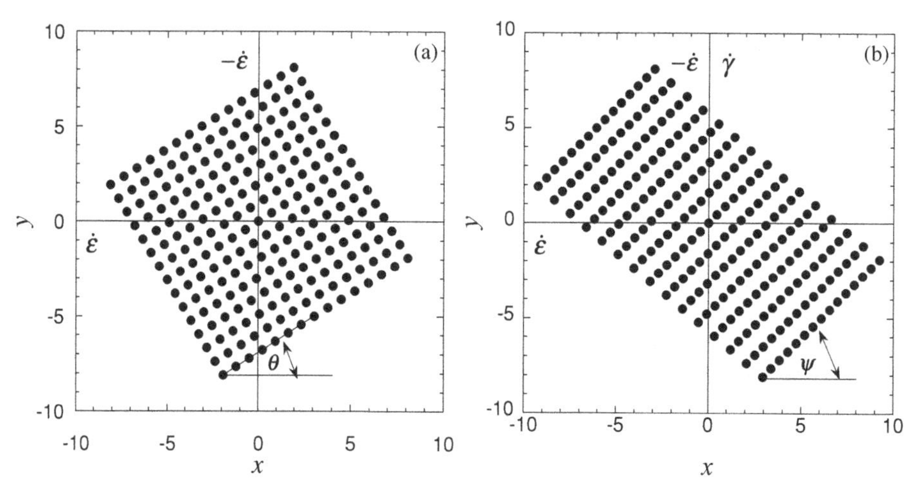

Figure 6.7 Initial lattice set up for (a) planar elongational flow, and (b) planar mixed flow. (b) is obtained by mapping the lattice (a) via the transformation Equation (6.69), where we have used values $\dot{\gamma} = 1.2\dot{\epsilon}$. Note that the initial lattice for mixed flow is no longer square but rhomboid for this particular choice of $\dot{\gamma}$ and $\dot{\epsilon}$ and the lattice vectors are not orthogonal (in this example, they are separated by $\sim 96^\circ$). Even though the elongation (expansion in x, contraction in y) and shear (flow in x with gradient in y) fields act in the x and y directions, the result of coupling of these fields means that the direction of pure contraction is *not* along the y-axis and hence not orthogonal to the direction of expansion (along the x-axis) and can be obtained from the eigenvectors of the velocity gradient tensor. Note also that θ ($\sim 37.1^\circ$) $\neq \psi$ ($\sim 44.5^\circ$). See reference [134] for further details.

transformed simulation box will be aligned for the mixed flow simulation. This angle, ψ, is shown in Figure 6.7, along with initial lattices for both elongation and mixed flows. The fact that the stretching and compressing axes are now no longer orthogonal in general is due to the shearing of the fluid which also stretches the simulation box in the x-direction (assuming the shear flow velocity is in the x-direction with gradient in the y-direction). For the purposes of convenient simulation, one can in fact commence the simulation from orthogonal lattice vectors by noting that the reproducibility time (or equivalently, Hencky strain) is independent of the choice of time origin. Thus, if t_0 is a time at which the basis vectors $\mathbf{L}_1'$ and $\mathbf{L}_2'$ are orthogonal (which may be some time less than zero), then reproducibility will be obtained when $\mathbf{L}_i'(t_0 + \tau_p) = \mathbf{L}_i'(t_0)$. The time t_0 is readily obtained from the orthogonality condition $\mathbf{L}_1'(t_0) \cdot \mathbf{L}_2'(t_0) = 0$ and solving for t_0. Note also that the transformation matrices $\mathbf{S}$ and $\mathbf{S}^{-1}$ are not unique, and so one chooses in practice whatever is the simplest geometry for simulation purposes. It is also possible to perform pure rotational and elliptical flows with this implementation, as both are examples of mixed shear and elongational flow [134].

A final observation worth making is that the reproducibility time of a mixed flow simulation depends only on the elongational component of the flow. As we have seen in Section 5.1.5, one could in principle perform pure shear flow simulations without any mapping of the simulation box because shear only stretches the box in the direction of flow, but no compression occurs. Therefore the simulation never needs to cease. In practice, one does map the shearing box back into its original shape in order to limit floating point precision errors that would result in very large numbers for particle coordinates and pbc translations and minimum image computations. It should not therefore be surprising that the reproducibility time for mixed flow simulations depends only on the elongation rate $\dot{\epsilon}$.

6.3.2 Equations of Motion

The SLLOD equations of motion for mixed flows in the (NVT) ensemble are simply obtained by substituting the velocity gradient tensor given in Equation (6.70) into Equation (5.82), leading to

$$\begin{aligned}\dot{\mathbf{r}}_i &= \frac{\mathbf{p}_i}{m} + \dot{\epsilon}(\mathbf{i}x_i - \mathbf{j}y_i) + \dot{\gamma}\mathbf{i}y_i \\ \dot{\mathbf{p}}_i &= \mathbf{F}_i^{\phi} - \dot{\epsilon}(\mathbf{i}p_{xi} - \mathbf{j}p_{yi}) - \dot{\gamma}\mathbf{i}p_{yi} - \alpha\mathbf{p}_i.\end{aligned} \tag{6.72}$$

Once the initial simulation box has been set up as described in Section 6.3.1, the implementation of pbcs is similar to that already outlined in Section 5.1.5, either by the rotation of the simulation box or by direct application of lattice displacement vectors. Again we note that the mapping of the fully stretched simulation box back into its $t = 0$ configuration is determined entirely from the elongational component of the flow, i.e. when the maximum Hencky strain is reached, as discussed in Section 5.1.5. The box boundaries themselves evolve in time via the velocity streamlines

$$\begin{aligned}L_x(t) &= \frac{\dot{\gamma}}{\dot{\epsilon}} L_y(0) \sinh(\dot{\epsilon}t) + L_x(0) e^{\dot{\epsilon}t} \\ L_y(t) &= L_y(0) e^{-\dot{\epsilon}t} \\ L_z(t) &= L_z(0),\end{aligned} \tag{6.73}$$

where L_x, L_y and L_z are the vertices of the box in the x, y and z directions, respectively.

Specific details of the algorithm can be found in reference [134]. One could also apply the NpT algorithm to this type of flow by implementing the procedure described in Section 6.1.3.

6.4 Thermodynamic, Rheological and Structural Results for Simple Fluids under Shear and Extensional Flows

It is a relatively straightforward task to compute the properties of interest once the equations of motion and suitable boundary conditions are implemented into code. In the 1980s, considerable effort was expended in computing a number of properties of simple atomic fluids under shear flow, such as shear viscosities, normal stresses, and corresponding correlation functions to test out existing theories of "long time tails". The mode-coupling theory of Kawasaki and Gunton [47] was of particular interest, and

NEMD simulations of LJ fluids close to the triple point, at that time, strongly supported the view that energy, pressure, viscosity and the normal stresses were all nonanalytic functions of strain rate in the zero strain rate limit. Within the range of strain rates that was then accessible, the pressure and energy were confirmed to vary as $\dot{\gamma}^{3/2}$, while the shear viscosity varied as $\dot{\gamma}^{1/2}$. Subsequently, it was then often assumed that this is true for all simple fluids. However, recent work since 2000 has strongly suggested otherwise. Marcelli *et al.* [99] performed NEMD simulations of atomic fluids including the effects of 3-body forces and found that pressure and energy tended to vary as $\dot{\gamma}^2$, which in turn suggested they behaved analytically, i.e. they could be expanded as a Taylor series in the strain rate. Ge *et al.* [207–209] then performed simulations of LJ fluids and again observed power law behaviour of the pressure and internal energy over a wide range of strain rates, but the mode-coupling predictions of the exponents were only true close to the triple point. Away from the triple point, the energy and pressure behaved as a power law with thermodynamic state dependent exponent μ in $\dot{\gamma}$. This result was later confirmed for the shear viscosity as well [210]. More recent work has demonstrated that it is also valid for liquid metals [211].

Studies that focussed on the limiting low strain rate behaviour of the internal energy, pressure and viscosity for simple fluids have found that these properties do not follow a simple, single power law in this limit. Travis *et al.* [212] concluded that it was necessary to introduce a second region of $\dot{\gamma}^{1/2}$ viscosity dependence at strain rates below 0.16 to adequately fit their low strain rate data for a WCA fluid at the LJ triple point. Ferrario *et al.* [213] found that the low strain rate data for a LJ fluid at its triple point were well described by a quadratic dependence on strain rate. They also found that the limiting value obtained by extrapolating the quadratic fit to zero strain rate was consistent with the (equilibrium) Green–Kubo result. TTCF computations by Borzsák *et al.* [141] again showed that the low strain rate viscosity data could not be described by a single square root dependence and, at the very lowest strain rates accessible with the TTCF technique, the slope of the square root dependence was almost indistinguishable from zero (i.e. the viscosity is essentially constant). Their data was very well described by the Cross equation, with a strain rate exponent very close to 2.

From a practical point of view, it seems that excellent agreement between low strain rate NEMD results and equilibrium Green–Kubo results for the viscosity can be obtained by assuming quadratic dependence at the lowest strain rates, followed by power law behaviour at higher strain rates. This also appears to be the case for the pressure [214] and internal energy.

6.4.1 Energy, Pressure and Viscosity

6.4.1.1 Energy and Pressure

If energy and pressure are analytic in the strain rate, then they can be expanded as a Taylor series in $\dot{\gamma}$. In the following derivation we consider only systems under shear flow, though it would be straightforward to perform the derivation for other flows such as the various elongational flows. For the internal energy U expanded to second order in $\nabla \mathbf{v}$ we have [207, 208]:

$$U = U(\nabla \mathbf{v})$$
$$= U(0) + \nabla \mathbf{v} : \left.\frac{\partial U}{\partial (\nabla \mathbf{v})}\right|_{\nabla \mathbf{v}=0} + \frac{1}{2} (\nabla \mathbf{v})(\nabla \mathbf{v}) :: \left.\frac{\partial^2 U}{\partial (\nabla \mathbf{v}) \partial (\nabla \mathbf{v})}\right|_{\nabla \mathbf{v}=0} + \cdots \quad (6.74)$$

where the notation :: stands for a fourth-order contraction between the two fourth-rank tensors $(\nabla \mathbf{v})(\nabla \mathbf{v})$ and $\left.\frac{\partial^2 U}{\partial (\nabla \mathbf{v}) \partial (\nabla \mathbf{v})}\right|_{\nabla \mathbf{v}=0}$. The two partial derivatives in Equation (6.74) are both evaluated at zero applied thermodynamic force, i.e. at equilibrium. If we assume spatial isotropy we can simplify both partial derivatives for shear flow. We can express the second term in Equation (6.74) as

$$\begin{aligned} U^{(2)} &\equiv \nabla \mathbf{v} : \left.\frac{\partial U}{\partial (\nabla \mathbf{v})}\right|_{\nabla \mathbf{v}=0} \\ &= \nabla_\beta v_\alpha \, a \, \delta_{\alpha\beta} \\ &= a \nabla_\beta v_\beta \\ &= 0, \end{aligned} \quad (6.75)$$

where Einstein notation is invoked and we have used

$$\left.\frac{\partial U}{\partial (\nabla \mathbf{v})}\right|_{\nabla \mathbf{v}=0} = a \delta_{\alpha\beta} \quad (6.76)$$

where a is a constant.

The second-order partial derivative in Equation (6.74) is a fourth-rank isotropic tensor, which we expand as a linear combination of the three isotropic fourth-rank polar tensors [2, 215, 216]:

$$\begin{aligned} \left.\frac{\partial^2 U}{\partial (\nabla \mathbf{v}) \partial (\nabla \mathbf{v})}\right|_{\nabla \mathbf{v}=0} &\equiv B_{\alpha\beta\gamma\delta} \\ &= b_1 \delta_{\alpha\beta} \delta_{\gamma\delta} + b_2 \delta_{\alpha\gamma} \delta_{\beta\delta} + b_3 \delta_{\alpha\delta} \delta_{\gamma\beta}. \end{aligned} \quad (6.77)$$

This leads to the third term in Equation (6.74) simplifying to

$$\begin{aligned} U^{(3)} &\equiv \frac{1}{2} (\nabla \mathbf{v})(\nabla \mathbf{v}) :: \left.\frac{\partial^2 U}{\partial (\nabla \mathbf{v}) \partial (\nabla \mathbf{v})}\right|_{\nabla \mathbf{v}=0} \\ &= \frac{1}{2} \left(\nabla_\delta v_\gamma\right) \left(\nabla_\beta v_\alpha\right) B_{\alpha\beta\gamma\delta}. \end{aligned} \quad (6.78)$$

Substitution of Equation (6.77) into Equation (6.78) gives $U^{(3)} = \frac{1}{2} b_2 \dot{\gamma}^2$ which, along with Equation (6.75), when substituted back into Equation (6.74), gives the result

$$U(\dot{\gamma}) = U(0) + \frac{1}{2} b_2 \dot{\gamma}^2. \quad (6.79)$$

Similarly for the pressure, we expand the full pressure tensor about $\nabla\mathbf{v}$ to second order:

$$\begin{aligned}
\mathbf{P} &= \mathbf{P}(\nabla\mathbf{v}) \\
&= \mathbf{P}(0) + \nabla\mathbf{v} : \left.\frac{\partial\mathbf{P}}{\partial(\nabla\mathbf{v})}\right|_{\nabla\mathbf{v}=0} + \frac{1}{2}(\nabla\mathbf{v})(\nabla\mathbf{v}) :: \left.\frac{\partial^2\mathbf{P}}{\partial(\nabla\mathbf{v})\,\partial(\nabla\mathbf{v})}\right|_{\nabla\mathbf{v}=0} + \cdots \\
&= p_0\mathbf{1} + \nabla\mathbf{v} : \left.\frac{\partial\mathbf{P}}{\partial(\nabla\mathbf{v})}\right|_{\nabla\mathbf{v}=0} + \frac{1}{2}(\nabla\mathbf{v})(\nabla\mathbf{v}) :: \left.\frac{\partial^2\mathbf{P}}{\partial(\nabla\mathbf{v})\,\partial(\nabla\mathbf{v})}\right|_{\nabla\mathbf{v}=0} + \cdots
\end{aligned} \tag{6.80}$$

where $\mathbf{1}$ is the unit tensor and p_0 is the equilibrium pressure $(1/3\mathrm{Tr}(\mathbf{P}(0)))$. The second term is a second-order contraction of the second-rank tensor $\nabla\mathbf{v}$ with the fourth-rank tensor, $\partial\mathbf{P}/\partial(\nabla\mathbf{v})|_{\nabla\mathbf{v}=0}$, while the third term is a fourth-order contraction of the fourth-rank tensor $(\nabla\mathbf{v})(\nabla\mathbf{v})$ with the sixth-rank tensor, $\partial^2\mathbf{P}/\partial(\nabla\mathbf{v})\,\partial(\nabla\mathbf{v})|_{\nabla\mathbf{v}=0}$. Both operations result in second-rank tensors. If we define the nonequilibrium part of the pressure tensor as $\mathbf{\Pi} \equiv \mathbf{P} - p_0\mathbf{1}$ then

$$\begin{aligned}
\mathbf{\Pi} &= \nabla\mathbf{v} : \left.\frac{\partial\mathbf{P}}{\partial(\nabla\mathbf{v})}\right|_{\nabla\mathbf{v}=0} + \frac{1}{2}(\nabla\mathbf{v})(\nabla\mathbf{v}) :: \left.\frac{\partial^2\mathbf{P}}{\partial(\nabla\mathbf{v})\,\partial(\nabla\mathbf{v})}\right|_{\nabla\mathbf{v}=0} + \cdots \\
&\equiv \mathbf{\Pi}^{(1)} + \ \mathbf{\Pi}^{(2)} + \cdots .
\end{aligned} \tag{6.81}$$

The first-order partial derivative is a fourth-order isotropic tensor, which can be represented as a linear combination of the three isotropic fourth-rank polar tensors,

$$\begin{aligned}
\left.\frac{\partial\mathbf{P}}{\partial(\nabla\mathbf{v})}\right|_{\nabla\mathbf{v}=0} &\equiv B_{\alpha\beta\gamma\delta} \\
&= b_1\delta_{\alpha\beta}\delta_{\gamma\delta} + b_2\delta_{\alpha\gamma}\delta_{\beta\delta} + b_3\delta_{\alpha\delta}\delta_{\gamma\beta}.
\end{aligned} \tag{6.82}$$

Substitution of Equation (6.82) into Equation (6.81) and simplifying leads to

$$\mathbf{\Pi}^{(1)} - \dot{\gamma}\begin{pmatrix} 0 & b_2 & 0 \\ b_3 & 0 & 0 \\ 0 & 0 & 0 \end{pmatrix} \tag{6.83}$$

for the first term in Equation (6.81).

The second-order partial derivative in Equation (6.81) is a sixth-rank isotropic tensor, which is expressible as a linear combination of the 15 independent isotropic sixth-rank tensors:

$$\begin{aligned}
\left.\frac{\partial^2\mathbf{P}}{\partial(\nabla\mathbf{v})\,\partial(\nabla\mathbf{v})}\right|_{\nabla\mathbf{v}=0} &\equiv E_{\alpha\beta\gamma\delta\varepsilon\zeta} \\
&= e_1\delta_{\alpha\beta}\delta_{\gamma\varepsilon}\delta_{\delta\zeta} + e_2\delta_{\alpha\beta}\delta_{\gamma\delta}\delta_{\varepsilon\zeta} + e_3\delta_{\alpha\beta}\delta_{\gamma\zeta}\delta_{\delta\varepsilon} \\
&\quad + e_4\delta_{\alpha\gamma}\delta_{\beta\delta}\delta_{\varepsilon\zeta} + e_5\delta_{\alpha\gamma}\delta_{\beta\varepsilon}\delta_{\delta\zeta} + e_6\delta_{\alpha\gamma}\delta_{\beta\zeta}\delta_{\varepsilon\delta} \\
&\quad + e_7\delta_{\alpha\delta}\delta_{\beta\gamma}\delta_{\varepsilon\zeta} + e_8\delta_{\alpha\delta}\delta_{\beta\varepsilon}\delta_{\gamma\zeta} + e_9\delta_{\alpha\delta}\delta_{\beta\zeta}\delta_{\varepsilon\gamma} \\
&\quad + e_{10}\delta_{\alpha\varepsilon}\delta_{\gamma\delta}\delta_{\beta\zeta} + e_{11}\delta_{\alpha\varepsilon}\delta_{\beta\gamma}\delta_{\delta\zeta} + e_{12}\delta_{\alpha\varepsilon}\delta_{\gamma\zeta}\delta_{\beta\delta} \\
&\quad + e_{13}\delta_{\alpha\zeta}\delta_{\gamma\delta}\delta_{\varepsilon\beta} + e_{14}\delta_{\alpha\zeta}\delta_{\beta\gamma}\delta_{\delta\varepsilon} + e_{15}\delta_{\alpha\zeta}\delta_{\gamma\varepsilon}\delta_{\beta\delta}.
\end{aligned} \tag{6.84}$$

Substitution of Equation (6.84) into Equation (6.81) leads to

$$\mathbf{\Pi}^{(2)} = \frac{1}{2}\dot{\gamma}^2 \begin{pmatrix} e_4 + e_{12} + e_{15} & 0 & 0 \\ 0 & e_4 + e_5 + e_6 & 0 \\ 0 & 0 & e_4 \end{pmatrix}. \tag{6.85}$$

Substituting Equations (6.83) and (6.85) into Equation (6.81) gives

$$\mathbf{\Pi} = \dot{\gamma} \begin{pmatrix} 0 & b_2 & 0 \\ b_3 & 0 & 0 \\ 0 & 0 & 0 \end{pmatrix} + \frac{1}{2}\dot{\gamma}^2 \begin{pmatrix} e_4 + e_{12} + e_{15} & 0 & 0 \\ 0 & e_4 + e_5 + e_6 & 0 \\ 0 & 0 & e_4 \end{pmatrix}. \tag{6.86}$$

Finally, the isotropic pressure is given as

$$\begin{aligned} p(\dot{\gamma}) &\equiv \frac{1}{3}\mathrm{Tr}(\mathbf{P}) \\ &= p_0 + \frac{1}{3}\mathrm{Tr}(\mathbf{\Pi}) \\ &= p_0 + \frac{1}{6}\left[3e_4 + e_5 + e_6 + e_{12} + e_{15}\right]\dot{\gamma}^2 \\ &= p_0 + p_2\dot{\gamma}^2, \end{aligned} \tag{6.87}$$

where we have grouped all coefficients e_n and the factor $1/6$ in the third line of Equation (6.87) into the one term p_2 in the last line.

Equations (6.79) and (6.87) tell us that if the energy and pressure are analytic in powers of the strain rate, then their leading terms must be quadratic in the strain rate. While there are terms involving $\dot{\gamma}$, these are off-diagonal and relate to the shear stress. b_2 and b_3 are constants equivalent and equal in magnitude to the shear viscosity [207], so only diagonal terms contribute to the isotropic pressure.

As already noted, Daivis *et al.* [217] observed this leading $\dot{\gamma}^2$ behaviour for linear polymer melts at low strain rates and related it to the retarded motion expansion of a simple viscoelastic fluid [28, 31]. However, over a wider range of strain rates that are typical for NEMD simulations, Ge *et al.* [209] observed that energy and pressure were not analytic in $\dot{\gamma}$, but rather followed a power-law relationship. Nevertheless, they found that the exponents were in fact simple linear functions of the thermodynamic state point. If the energy and pressure are written as

$$\begin{aligned} U(\dot{\gamma}) &= U_0 + a\dot{\gamma}^{\mu} \\ p(\dot{\gamma}) &= p_0 + b\dot{\gamma}^{\mu}, \end{aligned} \tag{6.88}$$

then, curiously, the exponents for *both* energy and pressure can be simply expressed as

$$\mu(T, \rho) = A + BT - C\rho, \tag{6.89}$$

where T and ρ are the temperature and density, respectively. For the Lennard-Jones fluid, the coefficients were determined as $A = 3.67 \pm 0.04$, $B = 0.69 \pm 0.03$, $C = 3.35 \pm 0.03$. The exponent is plotted as a function of density for three temperatures for the Lennard-Jones fluid in Figure 6.8. This simple linear relationship has also been found to hold true for liquid metals, in which many-body potentials were used [211].

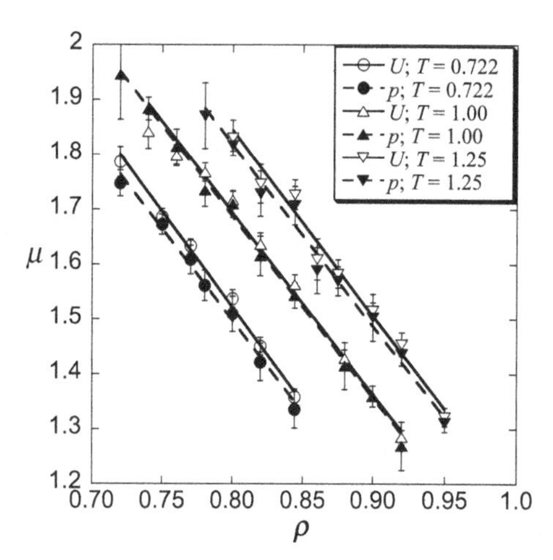

Figure 6.8 Pressure and energy exponent μ as a function of density for different temperatures replotted from reference [209]. Open symbols represent fits determined from the potential energy profile, whereas solid symbols refer to fits determined from the pressure profile. The simulations were run at constant temperature so only the potential energy need be plotted. Lines are linear fits to the data weighted by the errors.

Furthermore, it has been shown to be a useful alternative method to determine the liquid-solid phase transition line for equilibrium fluids [218].

6.4.1.2 Shear Viscosity

Viscosity was found to have a similar behavior to pressure and energy, in that the exponent of the strain rate is also a linear function of temperature and density [210]. However, it again has to be stressed that these computations were performed via standard NEMD simulations in which the strain rates need to be sufficiently high to obtain reasonable signal to noise ratios. For sufficiently low strain rates in the range 0–0.14, Daivis *et al.* [170] also found viscosity to vary quadratically with the strain rate, in accord with the retarded motion expansion.

Todd [210] found that for any state point, the viscosity computed over a wider range of strain rates, typically 0.1–0.9, the viscosity was found to vary as

$$\eta = \eta_0 - \eta_1 \dot{\gamma}^{\nu} \tag{6.90}$$

where η_0 is the zero-shear viscosity and η_1 is a constant and both quantities are functions of density and temperature. As with pressure and energy, the dependence of the exponent ν on temperature and density is linear, given by

$$\nu(T, \rho) = a + bT - c\rho \tag{6.91}$$

where the coefficients were found to be $a = 3.9 \pm 0.2$, $b = 1.04 \pm 0.06$ and $c = 5.1 \pm 0.2$ for the Lennard-Jones fluid. Figure 6.9 depicts this dependence for three temperatures. It is again seen that the widely accepted $\dot{\gamma}^{1/2}$ dependence predicted by

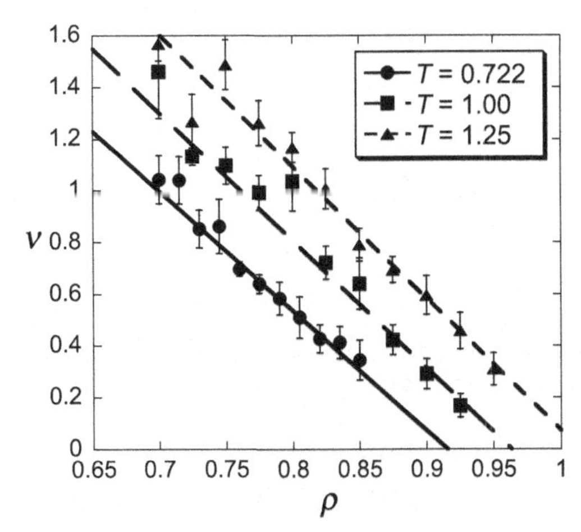

Figure 6.9 Shear viscosity exponents ν for the Lennard-Jones fluid as a function of density for different temperatures. Lines are linear fits to the data weighted by the errors. Replotted from reference [210].

mode-coupling theory is valid only around the triple point, $(T, \rho) \sim (0.722, 0.844)$. Beyond this region the exponent can take on any value between around 0.2 to 1.6.

The values of η_0 and η_1 can also be determined by plotting each value for each temperature and density. For each temperature, the values of η_0 and η_1 increase rapidly as density increases. This increase is so rapid that a stretched exponential of the form

$$\eta_i = k_i \exp\left(m_i \rho^{q_i}\right) \tag{6.92}$$

fits the data well. Here $i = 0$ or 1 labels the parameter η_0 or η_1 respectively, and k_i, m_i, and q_i are constants for each particular temperature. Equation (6.92) gives a better fit to the data than a simple power law expression. Values of all fitting parameters k_i, m_i and q_i for each of the three temperatures can be found in reference [210]. That η_0 or η_1 are well represented by a stretched exponential reflects the rapidity at which they diverge as functions of density. This divergence is rapid as the fluid approaches the liquid-solid phase transition, but its actual behaviour in this coexistence phase is not yet known. Evidence suggests [219] that the simple linear scaling of the pressure and energy exponents as functions of temperature and density no longer applies in the liquid-solid or vapour-liquid coexistence phase regions, so care should be taken in not extrapolating beyond the data set that spans only the pure liquid phase.

6.4.2 Normal Stress Differences

The traceless symmetric pressure tensor has two independent diagonal components, so we can define two normal pressure differences. These are more conventionally defined in terms of the stresses and called normal stress differences, the term that we will also adopt.

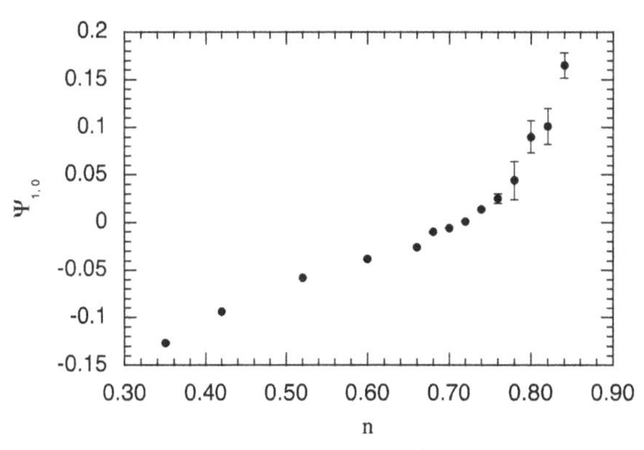

Figure 6.10 Number density dependence of the first normal stress coefficient at zero strain rate for the WCA fluid at a reduced temperature $T = 1.0$. Replotted from reference [214].

Over the range of strain rates that are accessible in experiments, nonzero normal stress differences are only ever observed for strongly non-Newtonian fluids like polymer melts and solutions. At the extremely high shear rates that are easily accessible in nonequilibrium molecular dynamics simulations, they are also measurable for simple and low molecular weight liquids. The normal stress differences are generally quite small and difficult to compute accurately for simple liquids. At sufficiently low strain rates the normal stress differences vary quadratically with the strain rate and it is conventional to define the normal stress *coefficients* as

$$\Psi_1(\dot{\gamma}) = \frac{P_{yy} - P_{xx}}{\dot{\gamma}^2} \tag{6.93}$$

$$\Psi_2(\dot{\gamma}) = \frac{P_{zz} - P_{yy}}{\dot{\gamma}^2}. \tag{6.94}$$

The zero shear rate normal stress coefficients can be obtained accurately by extrapolating a plot of Ψ_1 versus $\dot{\gamma}^2$ to zero strain rate. Figure 6.10 shows the density dependence of the first normal stress coefficient obtained in this way for the WCA fluid. We see that the first normal stress coefficient varies strongly with density, increasing rapidly when the number density is greater than about 0.8. The first normal stress coefficient is often associated with the elasticity of a liquid. The results confirm that a simple WCA fluid is indeed almost completely inelastic. It is well established that when a simple liquid is subjected to planar shear using the SLLOD algorithm, application of the usual isokinetic thermostat results in unequal kinetic temperatures calculated using the x, y and z components of the thermal momentum. The configurational temperature may also be anisotropically distributed. None of these is necessarily equal to the temperature of a small subsystem (i.e. a thermometer) in thermal contact with the system of interest [84].

Despite this, it has been shown that the values of linear transport coefficients such as the viscosity are independent of the details of the thermostat, provided that the underlying equilibrium thermodynamic state remains the same [2]. For example, constant energy simulations and constant temperature simulations should give the same results for the linear shear viscosity, as long as the temperature and density of the limiting equilibrium state are the same for both sets of simulations.

Because the normal stress coefficients enter at second order in the strain rate, they are more sensitive to the details of the thermostat. Very little work has been done to determine the sensitivity of the normal stress coefficients to the thermostatting mechanism and implementation details. To investigate this question, Daivis, Dalton and Morishita conducted a series of homogeneous shear simulations with the SLLOD algorithm for a simple fluid subjected to a variety of thermostats [170]. The first thermostat was the usual kinetic thermostat, which keeps the total thermal kinetic energy constant, and the second was a Braga-Travis type configurational thermostat. The equations of motion and thermostat multiplier for these thermostats have been given previously. A directional kinetic thermostat, which imposes equal temperatures on the x, y and z components of the kinetic energy was also applied. To control all of the directional components of the kinetic temperature individually, we must use the equations of motion

$$\begin{aligned}\dot{\mathbf{r}}_i &= \frac{\mathbf{p}_i}{m_i} + \mathbf{r}_i \cdot \nabla\mathbf{v} \\ \dot{\mathbf{p}}_i &= \mathbf{F}_i - \mathbf{p}_i \cdot \nabla\mathbf{v} - \alpha_x p_{ix}\mathbf{i} - \alpha_y p_{iy}\mathbf{j} - \alpha_z p_{iz}\mathbf{k}\end{aligned} \tag{6.95}$$

with the thermostat multiplier for the x-kinetic thermostat equal to

$$\alpha_x = \frac{\sum\limits_{i=1}^{N} p_{ix}\left(F_{ix} - (\mathbf{p}_i \cdot \nabla\mathbf{v})_x\right)/m_i}{\sum\limits_{i=1}^{N} p_{ix}^2/m_i} \tag{6.96}$$

and similar expressions for the y and z multipliers.

A directional configurational thermostat which acts in a similar way on the configurational degrees of freedom was also applied. The equations of motion used for the directional configurational thermostats were

$$\begin{aligned}\dot{\mathbf{r}}_i &= \frac{\mathbf{p}_i}{m_i} + \mathbf{r}_i \cdot \nabla\mathbf{v} + \zeta_x F_{ix}\mathbf{i} + \zeta_y F_{iy}\mathbf{j} + \zeta_z F_{iz}\mathbf{k} \\ \dot{\mathbf{p}}_i &= \mathbf{F}_i - \mathbf{p}_i \cdot \nabla\mathbf{v}\end{aligned} \tag{6.97}$$

and the equations of motion for the thermostat variables were

$$\begin{aligned}\dot{\zeta}_x &= \frac{1}{Q_\zeta}\left(\sum_{i=1}^{N} F_{ix}^2 - k_B T \sum_{i=1}^{N} \frac{\partial F_{ix}}{\partial x_i}\right) \\ \dot{\zeta}_y &= \frac{1}{Q_\zeta}\left(\sum_{i=1}^{N} F_{iy}^2 - k_B T \sum_{i=1}^{N} \frac{\partial F_{iy}}{\partial y_i}\right) \\ \dot{\zeta}_z &= \frac{1}{Q_\zeta}\left(\sum_{i=1}^{N} F_{iz}^2 - k_B T \sum_{i=1}^{N} \frac{\partial F_{iz}}{\partial z_i}\right).\end{aligned} \tag{6.98}$$

Lastly, a thermostat that simultaneously fixes all directional components of the kinetic temperature and the configurational temperature was also applied.

Table 6.1 Zero shear rate viscosities and first normal stress coefficients calculated by NEMD for various homogeneous synthetic thermostats.

Thermostat type	η_0	Error	$\Psi_{1,0}$	Error
T_K	2.119	0.001	0.206	0.004
T_C	2.118	0.002	0.193	0.009
$T_{K\alpha}, \alpha = x, y, z$	2.124	0.006	0.27	0.03
$T_{C\alpha}, \alpha = x, y, z$	2.115	0.002	0.205	0.009
$T_{K\alpha}, T_{C\alpha}, \alpha = x, y, z$	2.120	0.003	0.24	0.02

The results showed that the linear transport coefficient – in this case the viscosity – was insensitive to the thermostat details while the first normal stress coefficient was measurably affected by the thermostat. Table 6.1 summarises these results.

The conclusions are in broad agreement with those reported by Hoover *et al.* [126]. Work on this topic is ongoing, and it remains to be seen whether a synthetic, homogeneous thermostat can be devised that will reproduce the nonlinear rheological properties of systems undergoing naturally thermostatted, boundary-driven shear.

6.4.3 Fluid Structure

The shear stress can also be calculated by another independent method, namely by integrating over the total nonequilibrium pair distribution function. The nonequilibrium pair distribution function gives information about the distortion from isotropy of the fluid. Such a distortion is what leads to shear thinning in simple fluids, even though an atomic fluid has no intrinsic structure so that flow alignment or rotational effects do not exist. We can expand the nonequilibrium pair distribution function (denoted here as g_{ne}) as a Taylor series, and consider only terms that are first order in the gradient of the streaming velocity [220]:

$$g_{ne}(\mathbf{r}, \nabla\mathbf{v}(\mathbf{r})) = g(\mathbf{r}) + \nu(\mathbf{r})\left[\frac{\mathbf{rr}}{r^2}\right] : \nabla\mathbf{v}(\mathbf{r}) + \left[\nu_0(\mathbf{r}) - \frac{1}{3}\nu(\mathbf{r})\right] \nabla\cdot\mathbf{v}(\mathbf{r}) + \cdots \tag{6.99}$$

where $g(\mathbf{r})$ is the equilibrium pair distribution function and $\nu(\mathbf{r})$ represents the radial part of the distortion from spherical symmetry of the nonquilibrium pair distribution function. For constant volume deformation, $\nabla\cdot\mathbf{v}(\mathbf{r}) = 0$ and so the last term vanishes. Following the procedure outlined by Pryde [17, 221, 222] one can define a strain rate dependent ν that approaches $\nu(r)$ in the zero strain rate limit $\nu(r, \nabla\mathbf{v}(\mathbf{r}))$ for a three-dimensional fluid shearing in the x-y plane as

$$\nu(r, \dot{\gamma}) = (15/\dot{\gamma})\frac{\left\langle x_{ij}y_{ij}/r_{ij}^2\right\rangle}{4\pi\rho r^2 dr} \tag{6.100}$$

and the averaging is performed in a small volume of the fluid between r and $r + dr$. Clearly the denominator $4\pi \rho r^2 dr$ is just the number of particles in a shell of volume $4\pi r^2 dr$. Also note that the function is undefined for zero strain rate, since it is only valid for nonzero shear. The configurational component of the pressure may be calculated from the virial as

$$p^{\phi} = -(2\pi/3)\rho^2 \int_0^{\infty} \frac{\partial \phi(r)}{\partial r} g(r, \dot{\gamma}) r^3 \, dr, \tag{6.101}$$

whereas the corresponding shear stress component is

$$P_{xy}^{\phi} = -(2\pi/15)\dot{\gamma}\rho^2 \int_0^{\infty} \frac{\partial \phi(r)}{\partial r} \nu(r, \dot{\gamma}) r^3 dr. \tag{6.102}$$

Observe that $\nu(r, \dot{\gamma})$ only contributes to the off-diagonal elements of the pressure tensor, appearing in Equation (6.102) but not Equation (6.101). The potential (i.e. structural) contribution to the shear viscosity, using the constitutive relation $P_{xy} = -\eta\dot{\gamma}$, is thus

$$\eta^{\phi} = (2\pi/15)\rho^2 \int_0^{\infty} \frac{\partial \phi(r)}{\partial r} \nu(r, \dot{\gamma}) r^3 \, dr. \tag{6.103}$$

Note the correct limiting behaviour of the shear stress and viscosity in the weak field limit as $\dot{\gamma} \to 0$. In this case $\nu(r, \dot{\gamma}) \to \nu(r)$, so the shear stress must be linear in $\dot{\gamma}$ whereas the shear viscosity becomes explicitly independent of strain rate. In other words, these expressions in the weak-field limit are descriptive of Newtonian flow. Note also that even though the pressure p^{ϕ} does not have any explicit dependence on $\dot{\gamma}$, there is still an implicit strain rate dependence in the pair distribution function $g(r, \dot{\gamma})$. In the weak field limit $g(r, \dot{\gamma}) \to g(r)$, so we obtain the equilibrium result. At higher strain rates it is well known that $g(\mathbf{r}, \nabla\mathbf{v})$ is no longer spherically symmetric but becomes distorted at an angle of 45 degrees to the fluid velocity streamlines [2].

In Figure 6.11 we show $g(r, \dot{\gamma})$ and $\nu(r, \dot{\gamma})$ for a shearing atomic fluid. Also shown is $g(r, \dot{\gamma} = 0)$ at equilibrium for comparison purposes. The difference between $g(r, \dot{\gamma})$ for $\dot{\gamma} = 0$ and nonzero values reflects the change in the fluid structure with imposed strain rate, which is to be expected. Comparisons between the direct NEMD configurational components of the pressure, shear stress and shear viscosity and those computed via Equations (6.101)–(6.103) are displayed in Table 6.2, showing excellent agreement. Equations (6.101)–(6.103) can be used as a useful check to the validity of direct NEMD evaluation of pressures and viscosity and vice versa. Caution must be used for higher field strengths as the expansion given by Equation (6.99) is only valid up to first order in the velocity gradient (weak to moderate field strengths).

One can follow the same procedure for planar elongational flow and obtain the following expressions for the configurational normal pressures and planar elongational

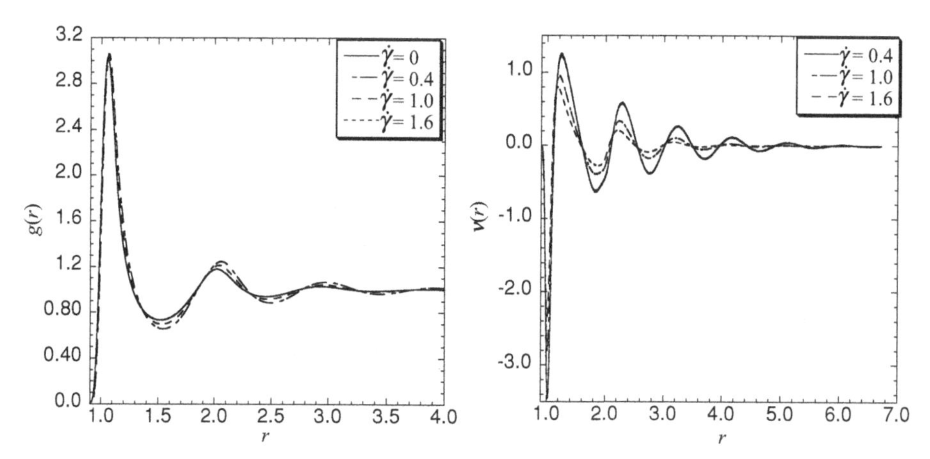

Figure 6.11 $g(r, \dot{\gamma})$ and $\nu(r, \dot{\gamma})$ for a shearing WCA fluid of 2048 atoms at several shear rates. $(\rho, T) = (0.8442, 0.722)$.

viscosity, defined as $\eta_E^{\phi} = (P_{yy}^{\phi} - P_{xx}^{\phi})/4\dot{\epsilon}$:

$$P_{xx}^{\phi} = -(2\pi/15)\rho^2 \int_0^{\infty} \frac{\partial\phi(r)}{\partial r} [5g(r, \dot{\epsilon}) + 2\dot{\epsilon}\nu(r, \dot{\epsilon})] r^3 \, dr \tag{6.104}$$

$$P_{yy}^{\phi} = -(2\pi/15)\rho^2 \int_0^{\infty} \frac{\partial\phi(r)}{\partial r} [5g(r, \dot{\epsilon}) - 2\dot{\epsilon}\nu(r, \dot{\epsilon})] r^3 \, dr \tag{6.105}$$

$$P_{zz}^{\phi} = -(2\pi/3)\rho^2 \int_0^{\infty} \frac{\partial\phi(r)}{\partial r} g(r, \dot{\epsilon}) r^3 \, dr \tag{6.106}$$

$$p_E^{\phi} \equiv \frac{1}{3}\mathrm{Tr}(\mathbf{P}) = -(2\pi/3)\,\rho^2 \int_0^{\infty} \frac{\partial\phi(r)}{\partial r} g(r, \dot{\epsilon}) r^3 \, dr \tag{6.107}$$

$$\eta_E^{\phi} = (2\pi/15)\,\rho^2 \int_0^{\infty} \frac{\partial\phi(r)}{\partial r} \nu(r, \dot{\epsilon}) r^3 \, dr \tag{6.108}$$

Table 6.2 Configurational components of the isotropic pressure, shear stress and shear viscosity for a shearing fluid at various strain rates (and at equilibrium) computed by direct NEMD and via Equations (6.101)–(6.103). Bins used to accumulate data were of width 0.001 reduced units and total simulation time after steady-state was achieved was 1 million timesteps, with $\Delta t = 0.001$.

	Direct NEMD			via Equations (6.101)–(6.103)		
$\dot{\gamma}$	p^{ϕ}	P_{xy}^{ϕ}	η^{ϕ}	p^{ϕ}	P_{xy}^{ϕ}	η^{ϕ}
0.0	5.777	0.000	–	5.718	–	–
0.4	5.968	−0.789	1.973	5.912	−0.783	1.957
1.0	6.520	−1.721	1.721	6.459	−1.707	1.707
1.6	7.256	−2.527	1.579	7.191	−2.507	1.567

with

$$\nu(r, \dot{\epsilon}) = (15/4\dot{\epsilon}) \frac{\langle x_{ij}^2/r_{ij}^2 \rangle - \langle y_{ij}^2/r_{ij}^2 \rangle}{4\pi \rho r^2 \, dr}. \tag{6.109}$$

Once again Equations (6.104)–(6.108) have the right limiting behaviour as $\dot{\epsilon} \to 0$. Note also that the configurational shear and elongational viscosities given by Equations (6.103) and (6.108) are identical in the weak field limit. This is because in the Newtonian limit both viscosities are defined such that they converge to the zero-shear viscosity. Similarly, the isotropic pressure p_E^{ϕ} is equivalent to that for shear given by Equation (6.101). We note that in the limit as shear or elongation rate go to zero both expressions yield the expected equilibrium result. However, as already mentioned, at higher strain rates the pair distribution function implicity depends on both r and either $\dot{\gamma}$ or $\dot{\epsilon}$, depending on the type of flow. Thus the distortions to this function will differ for the two different flows and so the pressures will not be identical as shear/elongation rate increases. Furthermore, the form of the dependence of $g(r, \dot{\gamma})$ or $g(r, \dot{\epsilon})$ on applied strain rate can be obtained by expanding the distribution function about $\dot{\gamma}$ or $\dot{\epsilon}$, but we will not perform the expansion here. Interested readers can refer to studies performed by Hess [223, 224] and Kalyuzhnyi *et al.* [225] for more detailed analyses of angle dependence of the distribution functions of simple fluids under shear flow and also to Gan and Eu for further theoretical discussion [226, 227].

We plot $g(r, \dot{\epsilon})$ and $\nu(r, \dot{\epsilon})$ as functions of r in Figure 6.12 for three values of the elongation rate to compare how the function distorts with increasing $\dot{\epsilon}$. From a computational perspective, there is one word of caution to be made: the maximum extent of either $g(r, \dot{\epsilon})$ or $\nu(r, \dot{\epsilon})$ must be the dimension of minimum extension of the simulation box. For shear flow, there is no problem as this dimension does not change as time evolves (see Section 5.1.5). This is *not* the case for elongation, in which the length

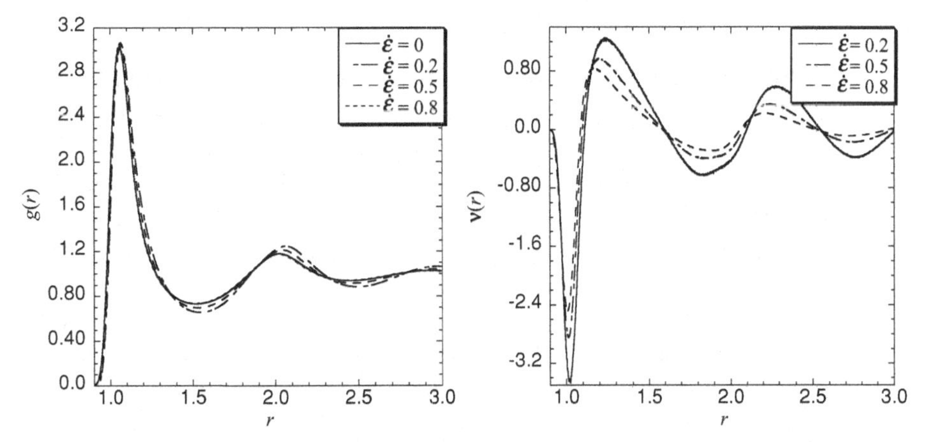

Figure 6.12 $g(r, \dot{\epsilon})$ and $\nu(r, \dot{\epsilon})$ for a WCA fluid of 2048 atoms undergoing planar elongation at several elongation rates. $(\rho, T) = (0.8442, 0.722)$.

Table 6.3 Configurational components of normal stresses, isotropic pressure and elongational viscosity for a WCA fluid at various elongation rates computed by direct NEMD and via Equations (6.104)–(6.108). Bins used to accumulate data were of width 0.001 reduced units and total simulation time after steady-state was achieved was 1 million timesteps, with $\Delta t = 0.001$.

	Direct NEMD					via Equations (6.104)–(6.108)				
$\dot{\epsilon}$	P^{ϕ}_{xx}	P^{ϕ}_{yy}	P^{ϕ}_{zz}	p^{ϕ}_{E}	η^{ϕ}_{E}	P^{ϕ}_{xx}	P^{ϕ}_{yy}	P^{ϕ}_{zz}	p^{ϕ}_{E}	η^{ϕ}_{E}
0.2	5.205	6.786	5.926	5.972	1.976	5.128	6.696	5.913	5.913	1.960
0.5	4.894	8.383	6.327	6.535	1.745	4.739	8.201	6.470	6.470	1.731
0.8	4.896	10.195	6.914	7.335	1.656	4.630	9.890	7.260	7.260	1.644

of the simulation box in the direction of contraction (in our case the y-direction) is decreasing in time. Thus the maximum extension in space for $g(r, \dot{\epsilon})$ or $v(r, \dot{\epsilon})$ is the *minimum* box length in the y-direction, which can be easily computed at the beginning of the simulation (see Section 5.1.5). Figure 6.12 therefore shows r extending out to about half the distance that it does for shear or equilibrium computations displayed in Figure 6.11.

As was the case for shear, the expressions for the pressure, normal stresses and elongational viscosities presented above are useful checks for the correct execution of code *in the weak to moderate field limit.* They can be compared with NEMD values computed directly from the configurational component of the pressure tensor, as is seen in Table 6.3. The full pressure tensor is of course obtained trivially by adding on the kinetic contribution, which for a homogeneous fluid is $(1/V)\langle\sum_i m_i \mathbf{c}_i \mathbf{c}_i\rangle$. At equilibrium or weak fields this is just the kinetic contribution to the isotropic pressure, $\rho k_B T$.

6.5 TTCF Algorithms for Shear and Elongational Flows

In Chapter 3 we derived the generalised TTCF expressions for arbitrary homogeneous flow fields. In this section we present specific results for shear and elongational flows for atomic fluids. We will only present results for the relevant phase variables of interest to us for thermostatted viscous flow, namely the shear stress and normal pressures and, in the case of shear flow, viscosity as well (see for example Borzsák *et al.* [141]). Our purpose here is to succinctly present the relevant TTCF expressions for the various flow fields and to show how the phase variables computed by them compare to direct NEMD time-averaged phase variables. While we restrict ourselves to the various elements of the pressure tensor, the TTCF expressions can be used to compute the nonlinear response of *any* phase variable (see for example "colour" conductivity [54, 55] or energy for a shearing fluid [228]). We also note that the TTCF methodology has been used to study the transport properties of liquid metals [229, 230], confined atomic fluids [135, 136,

231, 232], the rheology of shearing molecular fluids [233] and even for mixed shear and elongation flows [234].

6.5.1 Time-independent Flow

6.5.1.1 Shear

In Chapter 3 (Section 3.4.3) we derived the general TTCF expression for the time evolution of any phase variable $B(t)$ (given by Equation (3.62)), which we again write here as:

$$\langle B(t)\rangle = \langle B(0)\rangle - \frac{1}{k_B T}\sum_{i,j} F^e_{ij}\int_0^t ds\left\langle B(s) J_{ji}(0)\right\rangle \tag{6.110}$$

where F^e_{ij} is the ij component of some generalised tensorial field and $J_{ji}(0)$ is the conjugate thermodynamic flux evaluated at equilibrium. We again stress that the expression $\langle\ldots\rangle$ implies that time averaging is performed over the nonequilibrium system, i.e. the phase variable $B(s)$ is a *nonequilibrium* quantity sampled from within a NEMD simulation, whereas the flux $J_{ji}(0)$ is sampled at $t = 0$, the instant just before the field is switched on (i.e. still at equilibrium). $B(0)$ is likewise the equilibrium value of the phase variable at the moment just before the field is switched on. The resulting $\langle B(t)\rangle$ is thus the nonlinear response of the phase variable to the imposed field F^e_{ij}.

For planar shear, the generalised tensorial force is the strain rate tensor given by Equation (5.19), while the dissipative flux is just the product of the shear stress and system volume, so Equation (6.110) reduces to

$$\langle B(t)\rangle = \langle B(0)\rangle - \beta\dot{\gamma}V\int_0^t ds\left\langle B(s) P_{xy}(0)\right\rangle, \tag{6.111}$$

where $\beta \equiv 1/(k_B T)$.

Equation (6.111) is very simple to use: one just forms the relevant time correlation function and then integrates it to obtain the nonlinear response of the phase variable $B(t)$ of interest. However, there are some important and subtle points to take note of, which we will now describe.

As already mentioned, a TTCF simulation involves the correlation of two quantities in *different* ensembles. The dissipative flux, $J_{ji}(0)$, is computed at equilibrium, whereas the phase variable of interest, $B(s)$, is evaluated in the nonequilibrium ensemble. In practice what one does is evolve one equilibrium "mother" trajectory, from which spawn many nonequilibrium "daughter" trajectories, as depicted in Figure 6.13(a) below. For convenience, the daughters are separated by a time interval Δt_D such that Δt_D is long enough so that all of the $J_{ji}(0)$ at each time origin are statistically uncorrelated. The time "0" in the argument of J_{ji} thus represents the time origin of the nonequilibrium daughter trajectory. It is not to be confused with the time origin of the equilibrium mother. While it is convenient and simplest to separate daughters by a constant Δt_D, there is no theoretical reason why Δt_D can't vary between daughter trajectories.

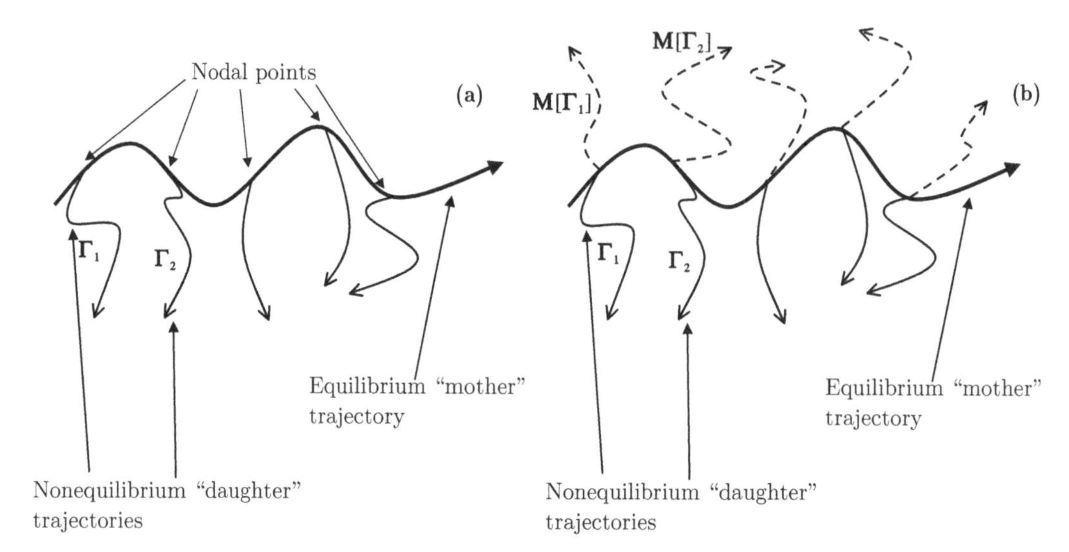

Figure 6.13 Schematic diagram of TTCF algorithm. A "mother" phase space trajectory evolves under equilibrium conditions. At time intervals Δt_D, "daughter" nonequilibrium trajectories are spawned from "nodal points" on the mother. In (a) daughters only are spawned, whereas in (b) daughters and their phase space mappings are spawned, where mapped trajectories are depicted by dashed lines. Arrow heads on phase space trajectories depict their evolution in time.

The overall statistics of the phase variable B can be increased substantially by ensemble averaging over all daughter trajectories. The number of such daughters can, in principle, be many. Morriss and Evans [2, 52] demonstrated that for planar shear simulations, a substantial improvement in the signal-to-noise ratio at long times can be obtained through a judicious choice of phase space symmetry mappings that rely on mixing properties of the correlation variables. If two quantities "mix" (in the context of statistical mechanics), then they become statistically de-correlated at sufficiently long times. If mixing occurs in the limit as $t \to \infty$, then $\langle B(t) P_{xy}(0) \rangle \to \langle B(t) \rangle \langle P_{xy}(0) \rangle$. In general, for a nonequilibrium phase variable, $\langle B(t) \rangle \neq 0$. Morriss and Evans showed that for planar shear flow, if one could generate an ensemble of initial phases such that $\sum_i P_{xy}(\mathbf{\Gamma}_i(0)) = 0$, then as a consequence of mixing, the statistical uncertainties at long time that are associated with small nonzero fluctuations around $P_{xy}(0)$ are eliminated. In other words, in the limit as $t \to \infty$ the sum of several correlation functions of the form $\langle B(t) P_{xy}(0) \rangle$ will tend to zero. For example, if $\mathbf{\Gamma}_1$ is a daughter trajectory and $\mathbf{\Gamma}_2$ its phase-space mapping "sister", then we would need the property that $P_{xy}(\mathbf{\Gamma}_1(0)) + P_{xy}(\mathbf{\Gamma}_2(0)) = 0$. The so-called y-reflection map, $(x, y, z, p_x, p_y, p_z) \to (x, -y, z, p_x, -p_y, p_z)$, is an example of one sort of mapping that fulfils this requirement, but others exist and none is unique [2].

This insight is particularly useful; without it one finds that the statistical fluctuations in the correlation functions of a single daughter trajectory are large in the long-time limit. One can eliminate this of course by averaging over many daughters, each separated by Δt_D. However, this is computationally very intensive and highly inefficient. The use of phase space mapping symmetry significantly reduces this computational cost.

Algorithm 6.8 Pseudocode for time-independent TTCF for shear flow, calling daughter routine described in Algorithm 6.9

while $t < t_{max}$ **do** {within mother equilibrium MD trajectory}
  if mod(t, Δt_D) = 0 **then**
    $B(0) \leftarrow B(t)$;
    $P_{xy}(0) \leftarrow P_{xy}(t)$; {zero time values computed within mother equilibrium trajectory at multiple daughter time-origins $t_D = 0$}
    call daughter$\left(B(0), P_{xy}(0), \mathbf{\Gamma}(0)\right)$; {call daughter routine; $\mathbf{\Gamma}(0)$ is the mother equilibrium trajectory phase space vector at t}
    call averageTTCF($B(0)$, TTCF, etc); {pass $B(0)$ value, TTCF array plus other relevant parameters to averaging routine to compute $\langle B(0)\rangle$ and $\left\langle B(t) P_{xy}(0)\right\rangle$}
    call phaseMap($\mathbf{\Gamma}(0)$); {call phase mapping routine}
    $\mathbf{\Gamma}(0) \leftarrow \mathbf{\Gamma}^M(0)$; {initialise new time origin daughter phase trajectory to mapped trajectory}
    call daughter$\left(B(0), P_{xy}(0), \mathbf{\Gamma}(0)\right)$;
    call averageTTCF($B(0)$, TTCF, etc);
    call phaseMap($\mathbf{\Gamma}(0)$); {call phase mapping routine repeatedly as required for each new mapping and repeat above procedure for each TTCF mapping trajectory}
    ... etc;
    ... etc;
  end if
end while
$\langle B(t_D)\rangle \leftarrow \langle B(0)\rangle - \beta\dot{\gamma}V \int_0^{t_D} ds \left\langle B(s) P_{xy}(0)\right\rangle$; {numerically integrate time averaged time-correlation function}

Thus, instead of spawning just one daughter at each nodal point on the mother trajectory, as was seen in Figure 6.13(a), we can now evolve several daughters from each node, as now depicted in Figure 6.13(b). The mapped daughter trajectories need to have only three characteristics: (1) that the sum of the stresses of *all* daughters at $t = 0$ are zero, as explained above; (2) that they preserve the total internal energy of the equilibrium mother system at $t = 0$ (i.e. for all daughter trajectories $\mathbf{\Gamma}_i^d$, $U(\mathbf{\Gamma}_i^d(0)) = U(\mathbf{\Gamma}^m(t_n))$, where d stands for daughter, m for mother and n for a nodal point in the mother's time coordinate system); and (3) that the mapped daughters evolve along distinct trajectories to the original daughter. Recall that the phase space vector of the original daughter trajectory at $t = 0$ is just the phase space vector that the mother trajectory has at the nodal point t_n where the daughter is spawned. Point (3) ensures that different daughter trajectories are completely de-correlated, and is easily accomplished by a symmetry mapping. This is important because the final result is obtained by averaging over *all* daughter trajectories and their phase space mappings.

Algorithms for TTCF calculations for time-independent shear flow are outlined in Algorithms 6.8 and 6.9.

Algorithm 6.9 Pseudocode for daughter routine called in Algorithm 6.8

```
while in daughter routine do
  t_D ← 0; {initialise daughter time variable}
  for t_D = 0 → t_Dmax do
    TTCF [t_D] ← 0; {initialise TTCF accumulation arrays}
  end for
  for t_D = 0 → t_Dmax do
    call molecularDynamics(...); {call molecular dynamics routine for general flow
    field, passing in required parameters}
    TTCF [t_D] ← B (t_D) P_xy (0);
  end for
end while
```

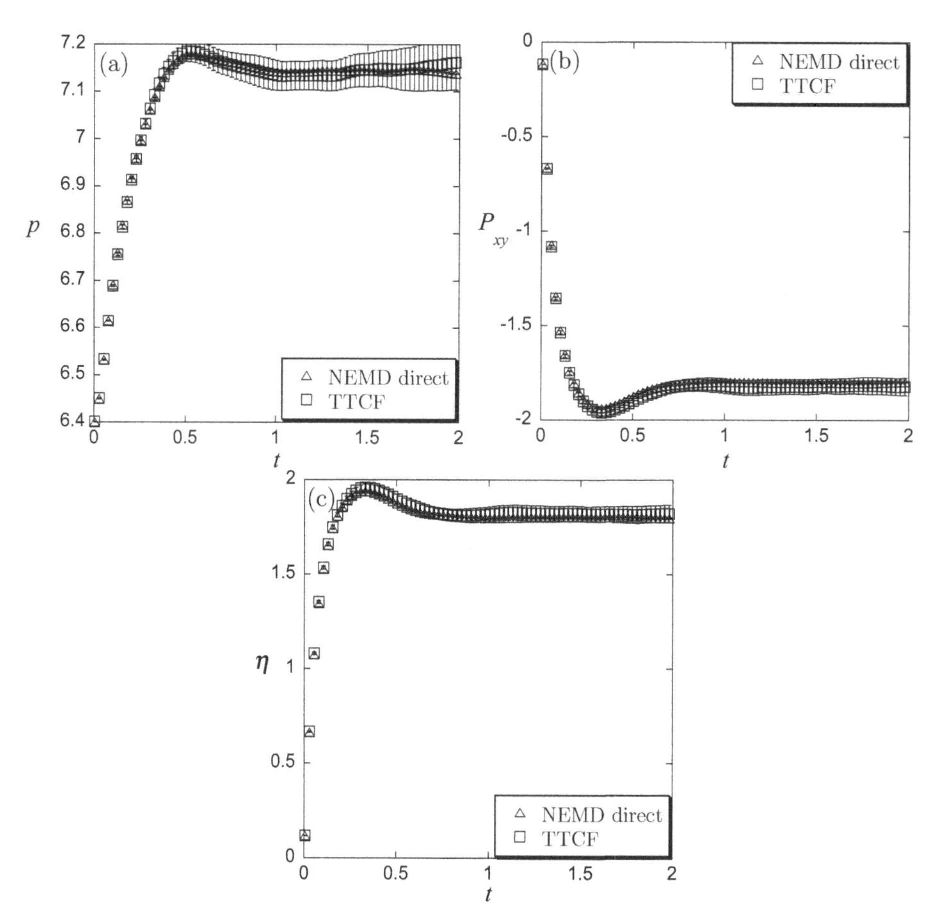

Figure 6.14 Comparison of direct NEMD time averages and TTCF for a system of 256 WCA atoms at a state point of $(T, \rho) = (0.722, 0.844)$ under planar shear at a strain rate of $\dot{\gamma} = 1.0$. (a) pressure, (b) negative shear stress (P_{xy}), and (c) shear viscosity. Error bars of direct NEMD data are smaller than plotting symbol size. Data are replotted from reference [228].

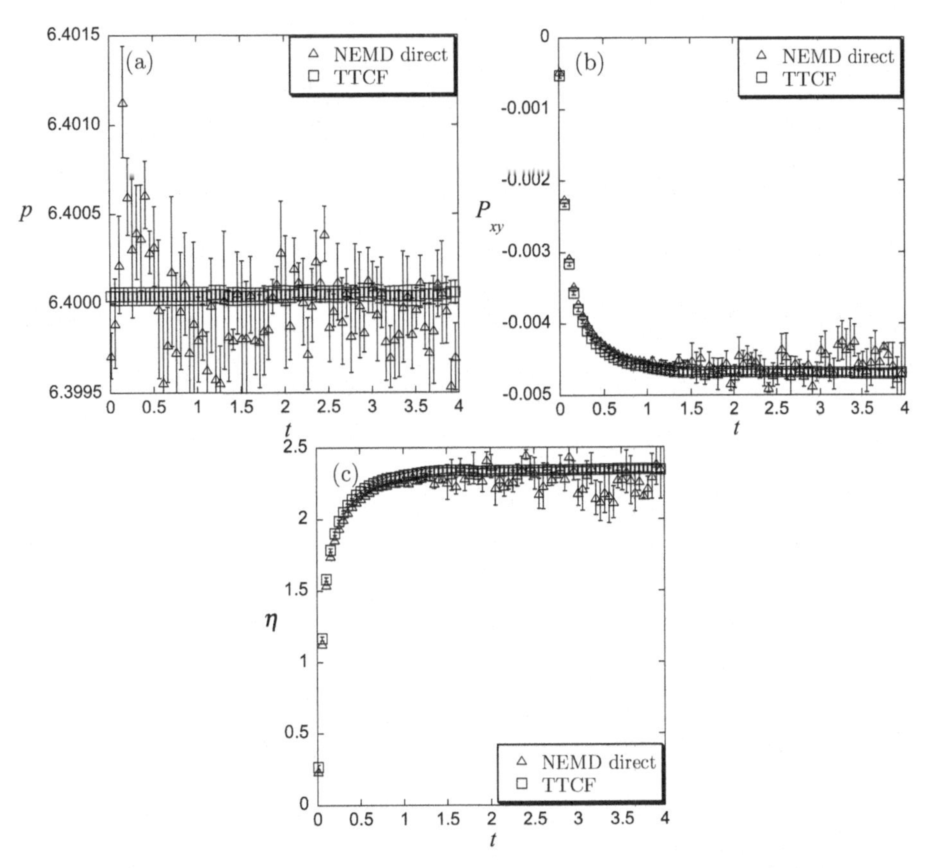

Figure 6.15 Comparison of direct NEMD time averages and TTCF for a system of 500 WCA atoms at a state point of $(T, \rho) = (0.722, 0.844)$ under planar shear at a strain rate of $\dot{\gamma} = 0.002$. (a) pressure, (b) negative shear stress (P_{xy}), and (c) shear viscosity. Data are replotted from reference [228].

In Figures 6.14 and 6.15 we display some typical TTCF results for a system of WCA atoms under planar shear flow (SLLOD dynamics with a Gaussian thermostat). In Figure 6.14 we show results for a strain rate of $\dot{\gamma} = 1.0$. The pressure, shear stress and viscosity are displayed as functions of time for both direct NEMD time averaging and TTCF calculations. The data are generated from 10 equilibrium trajectories of 25 million time steps, each from different initial conditions, where the integration time step is 0.001. From every equilibrium trajectory $2 \times 10\,000$ nonequilibrium trajectories are initiated and each nonequilibrium trajectory is 2000 time steps long. Thus, each nonequilibrium daughter trajectory is spawned every $\Delta t_D = 2500$ time steps from the equilibrium mother. Note that at each nodal point two trajectories are spawned: the daughter plus its symmetry mapped sister, where the mapping used was the y-reflection map. The final results are obtained by averaging over all the nonequilibrium trajectories.

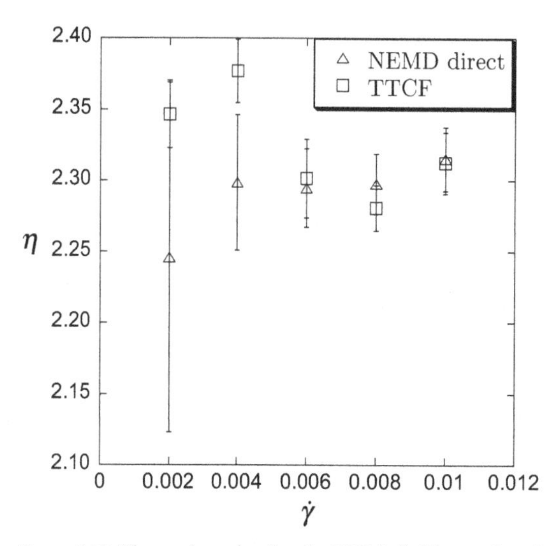

Figure 6.16 Shear viscosity for the WCA fluid as a function of strain rate for direct NEMD and TTCF calculations.

The shear rate used for the system described in Figure 6.14 is large. At such high strain rates direct NEMD is highly efficient and is superior to TTCF, as can be observed from the error bars of both sets of data. However, the situation is completely reversed for low strain rates, as depicted in Figure 6.15, where now the strain rate used was $\dot{\gamma} = 0.002$, and the state point is identical. In order to obtain these results, 4 equilibrium trajectories of 250 million time steps each were run. Once again nonequilibrium daughter trajectories (daughter plus symmetry mapping sister) are spawned every 2500 time steps (total of $2 \times 100\ 000$ daughters), and each daughter is of length 4000 time steps. Clearly now the statistics for the TTCF calculations are far superior to those of direct NEMD averaging. Most importantly, the shear viscosity data and normal stresses (not shown here) are statistically much more reliable. Thus the Newtonian limit is able to be probed to much greater precision than would otherwise be obtainable through direct NEMD averaging. This is again clearly demonstrated in Figure 6.16, in which direct and TTCF viscosities are plotted for several values of strain rate. At low $\dot{\gamma}$ the TTCF viscosities are much better than the direct NEMD values, whereas as $\dot{\gamma}$ increases the trend is reversed.

6.5.1.2 Elongation

In the case of elongational flow under adiabatic SLLOD dynamics, the rate of energy dissipation is given by

$$\begin{aligned}\dot{E}_0 &= -\dot{\epsilon}_{xx}\sum_i\left[\frac{p_{xi}^2}{m_i}+F_{xi}x_i\right]-\dot{\epsilon}_{yy}\sum_i\left[\frac{p_{yi}^2}{m_i}+F_{yi}y_i\right]-\dot{\epsilon}_{zz}\sum_i\left[\frac{p_{zi}^2}{m_i}+F_{zi}z_i\right]\\ &= -\left[\dot{\epsilon}_{xx}P_{xx}+\dot{\epsilon}_{yy}P_{yy}+\dot{\epsilon}_{zz}P_{zz}\right]V = -\left[V\,\mathbf{P}:\nabla\mathbf{v}\right] \equiv -\mathbf{J}:\mathbf{F}^e.\end{aligned} \tag{6.112}$$

Substituting the values of **J** and $\mathbf{F}^e$ into Equation (6.110) leads to the TTCF expression for elongational flows [53]:

$$\langle B(t)\rangle = \langle B(0)\rangle - \beta V \left[\dot{\epsilon}_{xx} \int_0^t ds \, \langle B(s)P_{xx}(0)\rangle + \dot{\epsilon}_{yy} \int_0^t ds \, \langle B(s)P_{yy}(0)\rangle \right.$$
$$\left. + \, \dot{\epsilon}_{zz} \int_0^t ds \, \langle B(s)P_{zz}(0)\rangle \right]. \tag{6.113}$$

As with shear, one can find appropriate symmetry mappings to reduce statistical noise in the limit as $t \to \infty$. In the case of elongational flow, we require $\sum_n \sum_i \dot{\epsilon}_{ii} P_{ii}(\mathbf{\Gamma}_n(0)) = 0$. An appropriate phase space mapping which can achieve this for PEF is $\mathbf{\Gamma}_1 \to \mathbf{\Gamma}_2$, where $\mathbf{\Gamma}_1 = (x_i, y_i, z_i, p_{xi}, p_{yi}, p_{zi})$ and $\mathbf{\Gamma}_2 = (y_i, x_i, z_i, p_{yi}, p_{xi}, p_{zi})$. Thus $P_{xx}(\mathbf{\Gamma}_2(0)) = P_{yy}(\mathbf{\Gamma}_1(0))$ and $P_{yy}(\mathbf{\Gamma}_2(0)) = P_{xx}(\mathbf{\Gamma}_1(0))$. For USF and BSF flows two initial equilibrium phases are insufficient for zeroing the long time fluctuations. Now we require an additional phase space mapping. One such map is $\mathbf{\Gamma}_3 = (z_i, y_i, x_i, p_{zi}, p_{yi}, p_{xi})$. Once again all these mapping schemes are not unique and others can be found [53].

The algorithm for planar elongational flow is very similar to that for shear depicted in Algorithms 6.8 and 6.9, except now one forms correlation functions of a set of three time-correlation functions, averages over these, integrates and finally sums them as given by Equation (6.113). TTCF calculations have also been performed for planar mixed flow [234].

In Figure 6.17(a) we display the diagonal elements of the pressure tensor for a system of 864 WCA atoms undergoing PEF, with an elongational strain rate of 0.1. For comparison purposes both the TTCF and direct NEMD pressures are displayed. Once more, for this relatively high field strength, the direct method is far more efficient. In contrast, when the strain rate is reduced significantly to 0.001, the TTCF method is seen to be

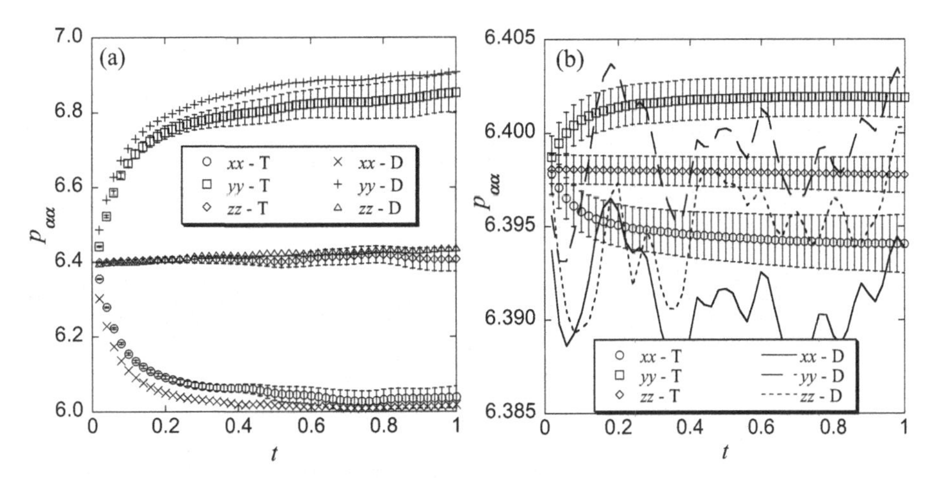

Figure 6.17 Comparison of TTCF (T) and direct (D) NEMD pressures for a system of 864 WCA atoms under PEF. (a) $\dot{\epsilon} = 0.1$. Simulations were averaged over a total of 5×2000 independent trajectories, each of length 1000 timesteps ($\Delta t = 0.001$), at a state point of $T = 0.722$, $\rho = 0.8442$. Error bars for direct pressures are smaller than their symbol sizes. (b) As with (a), but with $\dot{\epsilon} = 0.001$. Error bars for direct pressures are typically 2–3 times greater than corresponding TTCF errors.

superior, as clearly seen in Figure 6.17(b). Similar results have also been observed for systems undergoing USF and BSF [53].

6.5.2 Time-dependent Flow

In Chapter 3, Section 3.4.3, we sketched a derivation of the general TTCF expression for arbitrary homogeneous time-oscillatory flow. We now demonstrate the validity of this expression by applying it to an atomic fluid undergoing oscillatory PEF and BSF/USF (recall from Section 6.2.3 that oscillatory BSF is the same as oscillatory USF, each swapping its flow role every half-cycle). Results for shear flow are given in references [54–57] and pseudocode for general elongation is displayed in Algorithms 6.10 and 6.11.

For oscillatory elongational flow, $\mathbf{F}^e$ can be expressed as an applied time-dependent strain rate tensor,

$$\mathbf{F}^e(\varphi) \equiv \nabla \mathbf{v}(\varphi) = \begin{pmatrix} \dot{\epsilon}_{xx} & 0 & 0 \\ 0 & \dot{\epsilon}_{yy} & 0 \\ 0 & 0 & \dot{\epsilon}_{zz} \end{pmatrix} \cos(\varphi) \equiv \dot{\boldsymbol{\epsilon}} \cos(\varphi). \tag{6.114}$$

Here the explicit time dependence of the field has been transformed into a dependence on an extended phase-space variable, namely the phase angle, $\varphi = \omega t + \varphi_0$, where ω is the frequency of the driving field. Assuming unthermostatted *extended* SLLOD equations of motion for the particle dynamics [60],

$$\begin{aligned} \dot{\mathbf{r}}_i &= \frac{\mathbf{p}_i}{m_i} + \mathbf{r}_i \cdot \nabla \mathbf{v}(\varphi) \\ \dot{\mathbf{p}}_i &= \mathbf{F}_i - \mathbf{p}_i \cdot \nabla \mathbf{v}(\varphi) \\ \dot{\varphi} &= \omega, \end{aligned} \tag{6.115}$$

the adiabatic time derivative of the total internal energy is

$$\begin{aligned} \dot{U}_0 &= -\left[\dot{\epsilon}_{xx} P_{xx} + \dot{\epsilon}_{yy} P_{yy} + \dot{\epsilon}_{zz} P_{zz}\right] \cos(\varphi) V \\ &= -\left[V\, \mathbf{P} : \nabla \mathbf{v}(\varphi)\right] \equiv -\mathbf{J} : \mathbf{F}^e(\varphi). \end{aligned} \tag{6.116}$$

Substitution of these values of $\mathbf{J}$ and $\mathbf{F}^e$ into Equation (3.63) gives the time-dependent TTCF expression for the time evolution of any arbitrary phase variable $B(t)$ at a constant value of the phase angle, φ_p [60]:

$$\begin{aligned} \langle B[\boldsymbol{\Gamma}(t;\, \varphi(t) = \varphi_p)]\rangle &= \langle B[\boldsymbol{\Gamma}(0;\, \varphi(0) = \varphi_p)]\rangle - \beta V \Big[\dot{\varepsilon}_{xx} \int_0^t ds \, \cos(\varphi_p - \omega s) \\ &\quad \times \langle B[\boldsymbol{\Gamma}(s;\, \varphi(s) = \varphi_p)]\, P_{xx}[\boldsymbol{\Gamma}(0;\, \varphi(0) = \varphi_p - \omega s)]\rangle \\ &\quad + \dot{\varepsilon}_{yy} \int_0^t ds \, \cos(\varphi_p - \omega s) \\ &\quad \times \langle B[\boldsymbol{\Gamma}(s;\, \varphi(s) = \varphi_p)]\, P_{yy}[\boldsymbol{\Gamma}(0;\, \varphi(0) = \varphi_p - \omega s)]\rangle \\ &\quad + \dot{\varepsilon}_{zz} \int_0^t ds \, \cos(\varphi_p - \omega s) \\ &\quad \times \langle B[\boldsymbol{\Gamma}(s;\, \varphi(s) = \varphi_p)]\, P_{zz}[\boldsymbol{\Gamma}(0;\, \varphi(0) = \varphi_p - \omega s)]\rangle \Big]. \end{aligned} \tag{6.117}$$

Algorithm 6.10 Pseudocode for time-dependent TTCF for general elongational flows, calling daughter routine described in Algorithm 6.11

while $t < t_{max}$ **do** {within mother equilibrium trajectory}
  if mod(t, Δt_D) = 0 **then** {spawn daughter trajectories every Δt_D timesteps}
    for $\varphi_p = 0, 2\pi$ **do** {loop over φ_p in increments of $\Delta\varphi_p$}
      $B\left(0, \varphi_p\right) \leftarrow B\left[\mathbf{\Gamma}\left(t;\ \varphi(0) = \varphi_p\right)\right]$;
      {$B\left[\mathbf{\Gamma}\left(0;\ \varphi(0) = \varphi_p\right)\right]$ is computed from within the mother equilibrium trajectory at multiple daughter time-origins $t_D = 0$}
      call daughter$\left(B\left(0, \varphi_p\right), P_{xx}(0), P_{yy}(0), P_{zz}(0), \mathbf{\Gamma}\left(0;\ \varphi(0) = \varphi_p\right)\right)$; {call daughter routine}
      call averageTTCF$\left(\varphi_p, B\left(0, \varphi_p\right)\right.$, TTCFX, TTCFY, TTCFZ, etc$\left.\right)$;
      {pass current phase angle, $B\left(0, \varphi_p\right)$ value, TTCF arrays plus other relevant parameters to averaging routine to compute time averages of relevant phase variables and time correlation functions.}
    end for
  end if
end while
for $\varphi_p = 0, 2\pi$ **do** {loop over φ_p in increments of $\Delta\varphi_p$}
  for $t_D = 0 \rightarrow t_{D_{max}}$ **do** {loop over daughter trajectory times}
    $\mathrm{IX}\left[t_D, \varphi_p\right] \leftarrow \int_0^{t_D} \left\langle \mathrm{TTCFX}\left[t_D, \varphi_p - \omega s, \varphi_p\right]\right\rangle ds$;
    $\mathrm{IY}\left[t_D, \varphi_p\right] \leftarrow \int_0^{t_D} \left\langle \mathrm{TTCFY}\left[t_D, \varphi_p - \omega s, \varphi_p\right]\right\rangle ds$;
    $\mathrm{IZ}\left[t_D, \varphi_p\right] \leftarrow \int_0^{t_D} \left\langle \mathrm{TTCFZ}\left[t_D, \varphi_p - \omega s, \varphi_p\right]\right\rangle ds$;
    {numerically integrate time-correlation functions, i.e. down shaded columns in Figure 3.3, representing the integrals in Equations (3.63) and (6.117). See also Figures 3.2 and 6.18(a).}
  end for
  for $t_D = 0 \rightarrow t_{D_{max}}$ **do** {sum over all integrals}
    $\left\langle B\left[\Gamma\left(t_D;\ \varphi\left(t_D\right) = \varphi_p\right)\right]\right\rangle \leftarrow \left\langle B\left(0, \varphi\left(0\right) = \varphi_p\right)\right\rangle -$
    $\beta V\left(\dot{\epsilon}_{xx}\mathrm{IX}\left[t_D, \varphi_p\right] + \dot{\epsilon}_{yy}\mathrm{IY}\left[t_D, \varphi_p\right] + \dot{\epsilon}_{zz}\mathrm{IZ}\left[t_D, \varphi_p\right]\right)$;
  end for
end for
Store $\lim_{t_D \rightarrow \infty} \left\langle B\left(t_D, \varphi_p\right)\right\rangle \forall\varphi_p$ in array to obtain $\langle B\left(t\right)\rangle$;
{representing, for example, the time averaged, time dependent, steady-state normal pressures in Figure 6.18(b).}

For elongation the only nonzero elements of the pressure tensor are in the diagonal, P_{ii}. In this example, we restrict ourselves to the computation of normal pressures, thus we replace B with P_{ii} in Equation (6.117). These elements are used to compute the elongational viscosity (see Equations (6.50)–(6.52)).

In Figure 6.18 we show the results of an oscillatory PEF simulation with $\dot{\epsilon}_{xx} = 0.5$, $\dot{\epsilon}_{yy} = -0.5$, $\dot{\epsilon}_{zz} = 0$ [60]. The system consists of 108 thermostatted WCA atoms at a state point $(\rho\,, T) = (0.844\,, 0.722)$. These results consisted of data averaged over a total of $10 \times 2 \times 200 \times 50$ NEMD trajectories (i.e. 10 separate runs of 2 phase-space mappings of 200 nodal points of 50 discrete phase angles spanning the range

Algorithm 6.11 Pseudocode for daughter routine called in Algorithm 6.10

```
while in daughter NEMD routine do
  t_D ← 0; {initialise daughter time variable}
  for t_D = 0 → t_Dmax do {initialise TTCF accumulation arrays}
    TTCFX[t_D, φ(t_D), φ_p] ← 0;
    TTCFY[t_D, φ(t_D), φ_p] ← 0;
    TTCFZ[t_D, φ(t_D), φ_p] ← 0;
    IX[t_D, φ_p] ← 0;
    IY[t_D, φ_p] ← 0;
    IZ[t_D, φ_p] ← 0;
  end for
  for t_D = 0 → t_Dmax do {Form diagonal matrix elements in Figure 3.3.}
    φ(t_D) ← ωt_D + φ_p;
    call molecularDynamics(...); {call molecular dynamics routine for general flow
    field, passing in required parameters}
    TTCFX[t_D, φ(t_D), φ_p] ← cos[φ_p] × B[φ(t_D)] × P_xx[φ_p];
    TTCFY[t_D, φ(t_D), φ_p] ← cos[φ_p] × B[φ(t_D)] × P_yy[φ_p];
    TTCFZ[t_D, φ(t_D), φ_p] ← cos[φ_p] × B[φ(t_D)] × P_zz[φ_p];
  end for
end while
```

$\varphi \in [0, 2\pi]$). The frequency of the applied strain rate was $\omega = 12.566$, corresponding to a period of $\tau = 0.5$ reduced time units, and each NEMD trajectory was run for a total length of 3τ timesteps.

Figure 6.18(a) shows the normal pressures as functions of time at a constant phase angle of $\varphi_p = 0$ for both direct and TTCF calculations. The normal stresses at this fixed

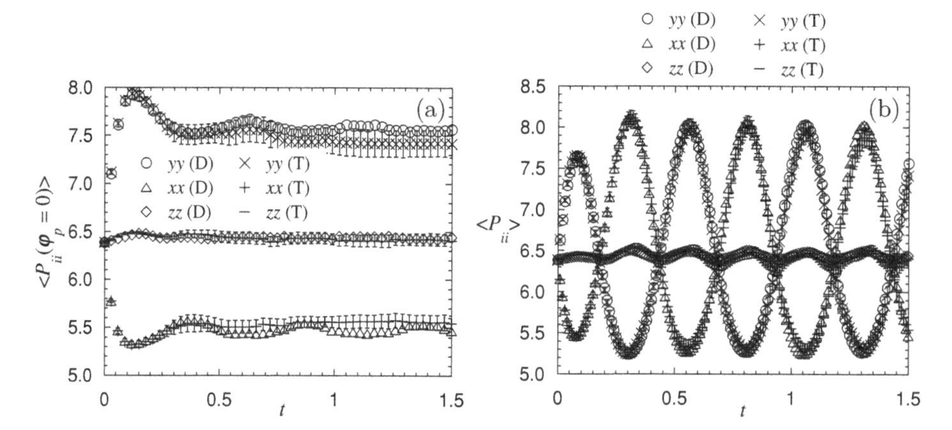

Figure 6.18 Normal pressures for an atomic WCA fluid undergoing oscillatory PEF with $\dot{\epsilon}_{xx} = 0.5$, $\dot{\epsilon}_{yy} = -0.5$, $\dot{\epsilon}_{zz} = 0$ and frequency $\omega = 12.566$. Direct NEMD averaging and TTCF results are labeled (D) and (T), respectively. Replotted from reference [60].

value of the phase angle are seen to relax to a nonequilibrium steady-state after some time. As the strain rate is high, the direct stresses are statistically superior to those computed via the TTCF formalism.

Of further interest in Figure 6.18(a) is the similarity of $P_{ii}(\varphi = \varphi_p)$ with P_{ii} in the time-independent TTCF studies of elongational flow presented in the previous section. A comparison of both results in fact stresses the point that P_{ii} evaluated at a constant phase angle for a time-dependent simulation evolves in time in much the same manner as P_{ii} would for a time-independent simulation. As stressed in Chapter 3, Section 3.4.3, it is *precisely* this similarity which enables the relatively simple time-dependent TTCF expressions in Equations (3.63) and (6.117) to be derived.

The normal pressures for both the direct and TTCF methods are plotted in Figure 6.18(b). In this case the TTCF data represents the shaded cells in Figure 3.3 (i.e. each shaded column represents a point on Figure 6.18(b)). P_{xx} and P_{yy} have a period of τ and are out of phase by $\pi/2$, while P_{zz} is weakly oscillatory with period 2τ, even though $\dot{\epsilon}_{zz} = 0$, which represents a nonlinear coupling of the field to phase variables in the field-independent direction if the field strength is sufficiently large.

In contrast, Figure 6.19 displays the same data, but this time for weak flow parameters $\dot{\epsilon}_{xx} = 0.002$, $\dot{\epsilon}_{yy} = -0.002$, $\dot{\epsilon}_{zz} = 0$. Now the TTCF calculations for normal pressures are vastly better than direct averaging. The frequency is the same as before, but the total simulation time per NEMD trajectory is 2τ. On the magnified scale of Figure 6.19(a), $\langle P_{xx}(0)\rangle = \langle P_{yy}(0)\rangle \neq \langle P_{zz}(0)\rangle$ due to averaging over the phase space mappings used. These zero-time (equilibrium) pressures are nevertheless within error bounds of each other.

Results for a BSF/USF system with $\dot{\epsilon}_{xx} = 0.25$, $\dot{\epsilon}_{yy} = -0.5$, $\dot{\epsilon}_{zz} = 0.25$ and same frequency as the PEF simulation are displayed in Figure 6.20. Each NEMD trajectory was run for a total simulation time of 3τ and a total of $10 \times 3 \times 140 \times 50$ trajectories

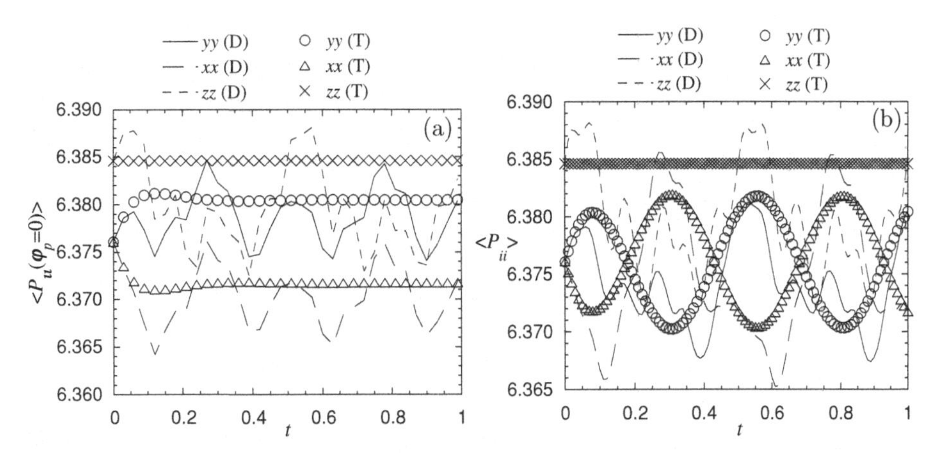

Figure 6.19 Normal pressures for an atomic WCA fluid undergoing oscillatory PEF with $\dot{\epsilon}_{xx} = 0.002$, $\dot{\epsilon}_{yy} = -0.002$, $\dot{\epsilon}_{zz} = 0$ and frequency $\omega = 12.566$. Direct NEMD averaging and TTCF results are labeled (D) and (T), respectively. Replotted from reference [60].

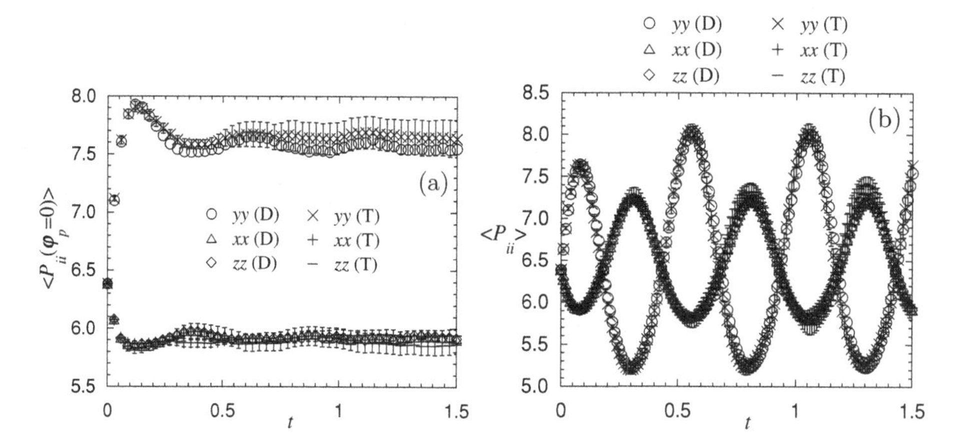

Figure 6.20 Normal pressures for an atomic WCA fluid undergoing oscillatory USF/BSF with $\dot{\epsilon}_{xx} = 0.25$, $\dot{\epsilon}_{yy} = -0.5$, $\dot{\epsilon}_{zz} = 0.25$ and frequency $\omega = 12.566$. Direct NEMD averaging and TTCF results are labeled (D) and (T), respectively. Replotted from reference [60].

were run [60]. Once again, agreement between the direct and TTCF methods is excellent. Also, as $\dot{\epsilon}_{xx} = \dot{\epsilon}_{zz}$, P_{xx} and P_{zz} are identical to within statistical errors.

We note here that as TTCF involves time integrations of correlation functions, statistical errors will propagate as time advances. As these calculations are computationally intensive (e.g. a total of 3.15×10^8 time steps for the results in Figure 6.19), significant computational resources are required for large system sizes or systems of molecules. Although the data presented in Figures 6.18–6.20 were obtained using only a single processor, as the TTCF method involves nonequilibrium trajectories that span the entire set of extended phase space, it is ideally suited to parallelisation. For maximum efficiency, at each nodal point one processor could be used for each trajectory which spans $\varphi \in [0, 2\pi]$. In such a way the time-dependent TTCF calculation would then become equally as efficient as the time-independent TTCF calculation running on a single processor, in that a larger number of nodes are possible, which in turn would improve the overall statistical accuracy of the results. In turn the time-independent TTCF calculation itself is amenable to parallel computation in that each daughter can be simulated on a separate processor.

7 Homogeneous Heat and Mass Transport

As we have shown in Chapter 2, the transport coefficients appearing in the entropy production of a simple, one component fluid are the bulk and shear viscosities and the thermal conductivity. We have already devoted considerable attention to the shear viscosity, because of its importance as a practical transport coefficient and as a test property for nonequilibrium statistical mechanics. The determination of the thermal conductivity by homogeneous nonequilibrium molecular dynamics techniques is also an interesting and subtle subject, leading to fundamental questions regarding the relationship between the synthetic heat field appearing in nonequilibrium molecular dynamics simulations and the temperature gradient, and the existence of "heat waves". When binary and multicomponent systems are considered, new transport coefficients associated with diffusion and the Soret and Dufour effects appear, and additional questions regarding the application of linear response theory to equations of motion with phase-space compression arise.

The thermal conductivity can be determined by equilibrium molecular dynamics simulations using the Green–Kubo formula, by homogeneous molecular dynamics methods that introduce a synthetic "heat field", or by inhomogeneous molecular dynamics methods that examine the heat flow between thermal reservoirs. Here, we will focus on the homogeneous nonequilibrium molecular dynamics methods, because the equilibrium methods have been described in detail elsewhere, and the reservoir methods suffer from the same problems as the "boundary-driven" methods for the determination of the shear viscosity. Reservoir methods work by directly inducing a temperature gradient which must be large due to the poor signal to noise ratio of the molecular dynamics methods. This large temperature gradient also inevitably leads to inhomogeneity of the density. In this case the value of the thermal conductivity obtained from the simulation is the average over a range of temperatures and densities. This is an even greater limitation if we wish to compute the thermal conductivity near the critical point or near a phase transition. Similar considerations apply to the determination of the diffusion coefficient of the binary fluid, and the Soret and Dufour coefficients. Fortunately, in all cases, there exist homogeneous nonequilibrium molecular dynamics algorithms for the determination of these transport coefficients. The equations of motion used in these algorithms generate the desired fluxes in homogeneous systems. In each case, there is a statistical-mechanical proof that the equations of motion lead to the correct determination of the corresponding linear transport coefficient. Nonlinear heat flow and diffusion are less well understood than nonlinear shear flow, and many interesting questions in this area remain unanswered.

7.1 Single Component Heat Transport

7.1.1 Equations of Motion

A nonequilibrium molecular dynamics algorithm for heat flow in a single component fluid can be derived by beginning with the assumption that it should be possible to devise an external field that will drive a heat flux. By applying linear response theory, we can obtain the relationship between this external field and the steady state average of the heat flux. Furthermore, we will assume (and later verify) that the equations of motion satisfy AIΓ (see Section 3.4.2). This condition requires the phase space compression factor to be zero when the field is applied in the absence of a thermostat. Altogether, this means that we want the rate of change of the instantaneous internal energy to satisfy

$$\frac{dU(t)}{dt} = V\mathbf{J}_q \cdot \mathbf{F}_q, \tag{7.1}$$

where $\mathbf{F}_q$ is the synthetic heat field. The total internal energy is the zero wavevector component of the Fourier transform of the instantaneous internal energy density, which obeys the internal energy balance equation

$$\frac{\partial \widetilde{\rho u}(\mathbf{k}, t)}{\partial t} = i\mathbf{k} \cdot \widetilde{\mathbf{J}}_q(\mathbf{k}, t) + \widetilde{W}(\mathbf{k}, t). \tag{7.2}$$

Here, W represents the rate of change of the internal energy due to the synthetic "heat field". Comparing these two equations, we see that the required zero-wavevector internal energy derivative can only be provided by a source term in the internal energy balance equation, and not by a heat flux due to temperature gradients. For heat flow and diffusion, we want equations of motion that do not generate streaming flow of the bulk fluid, so we will assume that the required equations of motion must have the form

$$\dot{\mathbf{r}}_i = \frac{\mathbf{p}_i}{m_i} \tag{7.3a}$$

$$\dot{\mathbf{p}}_i = \mathbf{F}_i + \mathbf{D}_i \cdot \mathbf{F}_q, \tag{7.3b}$$

where $\mathbf{F}_i$ represents the force on atom i due to interatomic forces, which we assume here consist entirely of two body interactions. The task now is to find the form of the coupling between the heat field $\mathbf{F}_q$ and the atomic dynamics that will produce a heat flux. The derivative of the microscopic internal energy of the system is

$$\frac{dU(t)}{dt} = \frac{d}{dt}\sum_{i=1}^{N} u_i = \frac{d}{dt}\left(\sum_{i=1}^{N} \frac{\mathbf{p}_i^2}{2m_i} + \frac{1}{2}\sum_{i=1}^{N}\sum_{j \neq i}^{N} \phi_{ij}\right), \tag{7.4}$$

where u_i is the internal energy of atom i. Taking the derivatives of the kinetic and potential energies, and substituting the equations of motion into the result, we see that only the kinetic energy derivative involves the heat field term, and the other terms cancel, giving

$$\frac{dU(t)}{dt} = \sum_{i=1}^{N} \frac{\mathbf{p}_i}{m_i} \cdot \mathbf{D}_i \cdot \mathbf{F}_q. \tag{7.5}$$

This can be compared with the right-hand side of Equation (7.1) after inserting the microscopic expression for the heat flux vector,

$$\begin{aligned} V\mathbf{J}_q(t) &= \sum_{i=1}^{N} u_i \frac{\mathbf{p}_i}{m_i} - \frac{1}{2} \sum_{i=1}^{N} \sum_{j \neq i}^{N} \mathbf{r}_{ij} \mathbf{F}_{ij} \cdot \frac{\mathbf{p}_i}{m_i} \\ &= \sum_{i=1}^{N} \frac{\mathbf{p}_i}{m_i} \cdot \mathbf{S}_i, \end{aligned} \tag{7.6}$$

where we have defined

$$\mathbf{S}_i = u_i \mathbf{1} - \frac{1}{2} \sum_{j \neq i}^{N} \mathbf{F}_{ij} \mathbf{r}_{ij}. \tag{7.7}$$

Equations (7.1) and (7.5) together with (7.6) and (7.7) show that we must have

$$\sum_{i=1}^{N} \frac{\mathbf{p}_i}{m_i} \cdot \mathbf{D}_i \cdot \mathbf{F}_q = \sum_{i=1}^{N} \frac{\mathbf{p}_i}{m_i} \cdot \mathbf{S}_i \cdot \mathbf{F}_q. \tag{7.8}$$

For a homogeneous nonequilibrium molecular dynamics algorithm, we require the equations of motion to be the same for all particles, so they must be identical for every value of the particle index i. This can be achieved by defining

$$\mathbf{D}_i = \mathbf{S}_i. \tag{7.9}$$

However, it is easily verified that the equations of motion that we obtain with this expression for $\mathbf{D}_i$ do not conserve momentum. If we take the time derivative of the total system momentum and substitute Equation (7.3b), we find

$$\frac{d}{dt} \sum_{i=1}^{N} \mathbf{p}_i = \sum_{i=1}^{N} \mathbf{D}_i \cdot \mathbf{F}_q, \tag{7.10}$$

which will only be (nontrivially) zero if the sum of $\mathbf{D}_i$ over all particles is zero. We can satisfy this condition without changing the result for the internal energy derivative if the particles all have the same mass and the initial momentum is set to zero by redefining $\mathbf{D}_i$ as

$$\mathbf{D}_i = \mathbf{S}_i - \frac{1}{N} \sum_{i=1}^{N} \mathbf{S}_i = \mathbf{S}_i - \bar{\mathbf{S}}. \tag{7.11}$$

Thus, the equations of motion for the Evans heat flow algorithm (in the absence of a thermostat) are [235]:

$$\dot{\mathbf{r}}_i = \frac{\mathbf{p}_i}{m_i} \tag{7.12a}$$

$$\dot{\mathbf{p}}_i = \mathbf{F}_i + \left(\mathbf{S}_i - \bar{\mathbf{S}}\right) \cdot \mathbf{F}_q. \tag{7.12b}$$

It is important to check whether these equations lead to incompressible phase space dynamics. For equations of motion of the form given above, the phase space compression factor is given by

$$\begin{aligned}\Lambda &= \frac{\partial}{\partial \boldsymbol{\Gamma}} \cdot \dot{\boldsymbol{\Gamma}} \\ &= \sum_i \left(\frac{\partial}{\partial \mathbf{r}_i} \cdot \dot{\mathbf{r}}_i + \frac{\partial}{\partial \mathbf{p}_i} \cdot \dot{\mathbf{p}}_i \right) \\ &= \sum_i \frac{\partial}{\partial \mathbf{p}_i} \cdot \left(\mathbf{D}_i \cdot \mathbf{F}_q\right) \\ &= \left(1 - \frac{1}{N}\right) \sum_i \frac{\mathbf{p}_i}{m_i} \cdot \mathbf{F}_q. \end{aligned} \tag{7.13}$$

For a single component system where the masses of all molecules are equal, this is equal to zero if the total momentum is initialised to zero, since the equations of motion conserve the total momentum of the system. For a multicomponent system this is no longer true. We will discuss the multicomponent case in Section 7.3.

The synthetic heat field does work on the system, increasing its internal energy, so these equations of motion must be supplemented by a thermostat to keep the temperature constant and achieve a steady state. The simplest thermostat to apply to this case is the Gaussian isokinetic thermostat which was discussed in general terms in Chapter 5. The equations of motion for the Evans heat flow algorithm are given by the general form of Equation (3.48), except that the $\mathbf{C}_i$ term is zero and F^e is now the heat field, $\mathbf{F}_q$. All that is required in order to obtain the equation for the thermostat multiplier is to substitute the appropriate expression for $\mathbf{D}_i$, i.e.

$$\alpha = \frac{\sum_{i=1}^{N} \frac{\mathbf{p}_i}{m_i} \cdot \left(\mathbf{F}_i + \mathbf{D}_i \cdot \mathbf{F}_q\right)}{\sum_{i=1}^{N} \frac{\mathbf{p}_i^2}{m_i}}, \tag{7.14}$$

where $\mathbf{D}_i$ is given by Equation (7.11).

Linear response theory provides the link between the steady state flux that results from the application of the heat field and the corresponding transport coefficient – the thermal conductivity. From Equation (3.70), we see that the ensemble averaged, linear response of any phase variable B after the application of a time-independent external field can be written as

$$\langle B(t) \rangle = \langle B(0) \rangle - \frac{F^e}{k_B T} \int_0^t \langle B(s) J(0) \rangle \, ds. \tag{7.15}$$

If we now take the external field to be given by the x-component of the heat field $\mathbf{F}_q$ and the dissipative flux to be the corresponding component of $-V\mathbf{J}_q$ (which has zero mean in the equilibrium state), the ensemble averaged value of the steady state heat flux in the linear limit is given by

$$\langle J_{qx} \rangle = \frac{F_q V}{k_B T} \int_0^\infty \langle J_{qx}(t) J_{qx}(0) \rangle \, dt, \tag{7.16}$$

where we have used Equation (3.56) to define the dissipative flux. It is already known from the theory describing the relaxation of hydrodynamic fluctuations in equilibrium systems that the equilibrium time autocorrelation function of the heat flux vector is related to the phenomenological coefficient L_{qq} defined in Equations (2.54) by the Green–Kubo relation [236]

$$L_{qq} = \frac{V}{k_B} \int_0^\infty \langle J_{qx}(t) J_{qx}(0) \rangle \, dt. \tag{7.17}$$

By comparing this equation with the linear response equation for the heat flux and Equation (2.54) we see that the thermal conductivity can be computed from the Evans heat flow algorithm in the linear response (zero field) limit as

$$\lambda = \frac{L_{qq}}{T^2} = \lim_{F_q \to 0} \frac{\langle J_{qx} \rangle}{T F_q}. \tag{7.18}$$

Implementation of the Evans heat flow algorithm is illustrated in Algorithms 7.1 and 7.2.

When the Evans heat flow algorithm is applied to large systems, the diffusive transport of energy becomes unstable and an alternative mechanism for heat transport becomes evident [237]. Transport of heat via shock waves occurs above a critical value of the heat field, which decreases as the system size grows. For simple atomic systems in 2 dimensions of more than 896 particles, the shock waves occur at values of the

Algorithm 7.1 Pseudocode for Evans heat flow algorithm for simple liquids – integration of equations of motion

for i $\leq$ N **do**
 $\mathbf{r}_i \leftarrow$ integrator $(\mathbf{p}_i/m_i)$;
 $\mathbf{r}_i \leftarrow$ pbcs $(\mathbf{r}_i)$; {apply pbcs to particle positions}
 $D_{ixx} \leftarrow \left(S_{ixx} - \bar{S}_{xx}\right)$;
 $D_{iyx} \leftarrow \left(S_{iyx} - \bar{S}_{yx}\right)$;
 $D_{izx} \leftarrow \left(S_{izx} - \bar{S}_{zx}\right)$;
 $\mathbf{p}_i \leftarrow$ integrator $\left(\mathbf{F}_i + \mathbf{i} * D_{ixx} * F_{qx} + \mathbf{j} * D_{iyx} * F_{qx} + \mathbf{k} * D_{izx} * F_{qx} - \alpha \mathbf{p}_i\right)$;
end for

Algorithm 7.2 Pseudocode for Evans heat flow algorithm for simple liquids - calculation of **S** tensor and heat flux vector

call neighbour; {form neighbour list}
call force; {calculate minimum image pair separations $\mathbf{r}_{ij}$ and pair forces $\mathbf{F}_{ij}$}
$\bar{S}_{xx} \leftarrow 0$; {initialise sum of S_{ixx}}
$\bar{S}_{yx} \leftarrow 0$; {initialise sum of S_{iyx}}
$\bar{S}_{zx} \leftarrow 0$; {initialise sum of S_{izx}}
for i $\leq$ N **do**
  $u_i \leftarrow 0$; {initialise atomic internal energy}
end for
{loop over all interacting pairs}
for ipair $\leq$ npair **do**
  $i \leftarrow$ pairlist(ipair, 1); {get index for particle i of pair ipair from pair list}
  $j \leftarrow$ pairlist(ipair, 2); {get index for particle j of pair ipair from pair list}
  $u_i \leftarrow u_i + 0.5 * u_{ij}(\text{ipair})$;{add potential energy to internal energy}
  $u_j \leftarrow u_j + 0.5 * u_{ij}(\text{ipair})$;{add potential energy to internal energy}
  $S_{ixx} \leftarrow S_{ixx} - 0.5 * F_{ijx}(\text{ipair}) * r_{ijx}(\text{ipair})$;
  $S_{iyx} \leftarrow S_{iyx} - 0.5 * F_{ijy}(\text{ipair}) * r_{ijx}(\text{ipair})$;
  $S_{izx} \leftarrow S_{izx} - 0.5 * F_{ijz}(\text{ipair}) * r_{ijx}(\text{ipair})$;
  $S_{jxx} \leftarrow S_{jxx} - 0.5 * F_{ijx}(\text{ipair}) * r_{ijx}(\text{ipair})$;
  $S_{jyx} \leftarrow S_{jyx} - 0.5 * F_{ijy}(\text{ipair}) * r_{ijx}(\text{ipair})$;
  $S_{jzx} \leftarrow S_{jzx} - 0.5 * F_{ijz}(\text{ipair}) * r_{ijx}(\text{ipair})$;
end for
for i $\leq$ N **do**
  $u_i \leftarrow u_i + 0.5 * \mathbf{p}_i^2/m_i$; {add kinetic energy to internal energy}
  $S_{ixx} \leftarrow S_{ixx} + u_i$;
  $\bar{S}_{xx} \leftarrow \bar{S}_{xx} + S_{ixx}$;
  $\bar{S}_{yx} \leftarrow \bar{S}_{yx} + S_{iyx}$;
  $\bar{S}_{zx} \leftarrow \bar{S}_{zx} + S_{izx}$;
  $J_{qx} \leftarrow p_{ix} * S_{ixx} + p_{iy} * S_{iyx} + p_{iz} * S_{izx}$; {calculate heat flux vector}
end for
$\bar{S}_{xx} \leftarrow \bar{S}_{xx}/N$; {calculate average value of S_{ixx}}
$\bar{S}_{yx} \leftarrow \bar{S}_{yx}/N$;
$\bar{S}_{zx} \leftarrow \bar{S}_{zx}/N$;
$J_{qx} \leftarrow J_{qx}/V$; {calculate (intensive) heat flux vector density}

heat field as low as 0.04 (in reduced units). The shock waves occur due to a feedback mechanism inherent in the Evans heat flow algorithm. Examination of the equations of motion shows that if the energy of a particle is greater than the average value, it will experience a greater force due to the heat field. This term can grow rapidly, leading to the formation of regions of heat flow that are selectively driven by the heat field. To eliminate this artefact, the average energy can be calculated over a small region instead of the entire system, so it is the energy of a particle relative to the average energy in its local region that couples to the heat field [238].

7.1.2 Results for Simple Fluids

The Evans heat flow algorithm has been used to compute the thermal conductivity of liquid Argon over a wide range of temperatures and densities including states near the triple point, supercritical states and states on the critical isotherm [239]. Except for states very near the critical point, excellent agreement between the nonequilibrium molecular dynamics results and experimental values was found.

Figure 7.1 shows an example of thermal conductivity data for a simple liquid obtained with the Evans heat flow algorithm. The system consisted of $N = 500$ WCA particles at reduced temperature $T = 0.722$ and reduced number density $n = 0.8442$. Each point was obtained by averaging over three production runs of 10^6 steps with a reduced timestep of $\delta t = 0.002$. This corresponds to a total run time of 1.2×10^7 steps. The heat-field dependent thermal conductivity was extrapolated to zero field by fitting a quadratic polynomial to the data when plotted against the heat-field squared. The statistical uncertainty of each point is clearly very small, typically less than ± 0.02, or around 0.3%. The zero field value obtained from the fit is 6.71. Also shown is the value obtained from equilibrium simulations using the Green–Kubo relation, 6.65 with an uncertainty based on one standard error in the mean of ± 0.05 evaluated using a total run time of 2.4×10^7 steps. The two results agree, but the nonequilibrium method is clearly statistically superior under these conditions.

The Evans heat flow algorithm has also been applied to the molten salts NaCl and KCl [240]. In these simulations, the molten salts were treated as single component fluids due to the difficulty of separately evaluating the partial enthalpies of the individual

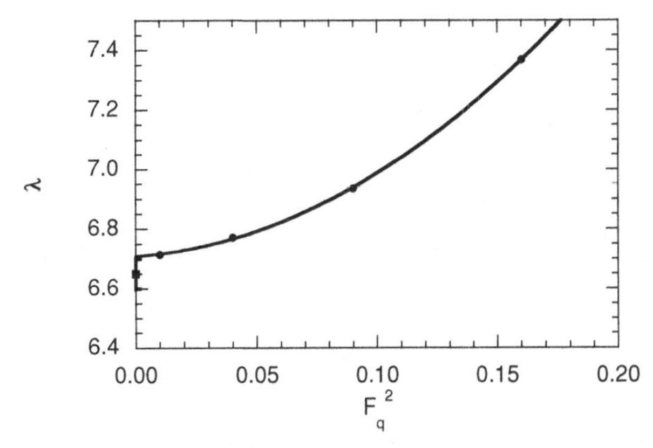

Figure 7.1 Heat field dependence of the thermal conductivity calculated from the Evans heat flow algorithm (circles) for a WCA fluid at reduced temperature $T = 0.722$ and reduced number density $n = 0.8442$. The value shown at zero field (square) was obtained from equilibrium simulations using the Green–Kubo relation. The two results agree to within one standard error in the mean of the Green–Kubo result.

components. Although the electrostatic contribution to the heat flux vector was neglected in the nonequilibrium computations, the results were in reasonable agreement with those of equilibrium simulations in which the full heat flux vector including electrostatic contributions was used.

Heat flow algorithms for molecular systems are discussed in Section 7.5.

7.2 Diffusion

7.2.1 Equations of Motion

Using similar ideas, we can derive an algorithm to simulate diffusive flow of the species in a solution of two or more components. This transport process is sometimes called interdiffusion, or in a two component system, mutual diffusion [241].

We will assume that the unthermostatted equations of motion are of the form

$$\dot{\mathbf{r}}_{\nu i} = \frac{\mathbf{p}_{\nu i}}{m_\nu} \tag{7.19a}$$

$$\dot{\mathbf{p}}_{\nu i} = \mathbf{F}_{\nu i} + \mathbf{D}_{\nu i} \cdot \mathbf{F}_\nu^e, \tag{7.19b}$$

where the additional index ν enumerates the components of a multicomponent system. At this stage, it is still assumed that the components are simple atomic fluids with two body interactions. It is also assumed that no macroscopic streaming velocity is induced by the external field. The macroscopic expression for the zero wavevector component of the time derivative of the internal energy for an r-component system with homogeneous diffusive mass fluxes driven by external fields in the absence of a thermostat is given by

$$\frac{dU(t)}{dt} = \sum_{\nu=1}^{r} \tilde{\mathbf{J}}_\nu (\mathbf{k} = 0, t) \cdot \mathbf{F}_\nu^e. \tag{7.20}$$

The Fourier transform of the diffusive flux is defined by

$$\tilde{\mathbf{J}}_\nu (\mathbf{k}, t) = \widetilde{\rho_\nu \mathbf{v}_\nu} - \widetilde{\rho_\nu \mathbf{v}} = \widetilde{\rho_\nu \mathbf{v}_\nu} = \sum_{i=1}^{N_\nu} m_\nu \mathbf{v}_{\nu i} e^{i\mathbf{k}\cdot\mathbf{r}_{\nu i}}, \tag{7.21}$$

where the second equality is assumed to be true because the equations of motion will ensure that the streaming velocity is zero. The zero wavevector component of this homogeneous diffusive flux is then

$$V\mathbf{J}_\nu (t) = \widetilde{\rho_\nu \mathbf{v}_\nu} (0, t) = \sum_{i}^{N_\nu} m_\nu \mathbf{v}_{\nu i}. \tag{7.22}$$

The corresponding microscopic expression for the internal energy derivative that results from the equations of motion given above is

$$\frac{dU(t)}{dt} = \sum_{\nu=1}^{r} \sum_{i}^{N_\nu} \frac{\mathbf{p}_{\nu i}}{m_\nu} \cdot \mathbf{D}_{\nu i} \cdot \mathbf{F}_\nu^e. \tag{7.23}$$

Since Equations (7.20) and (7.23) for the internal energy derivative must be equal and the equations of motion are required to be homogeneous (the same for every particle), the equations of motion must satisfy

$$\frac{\mathbf{p}_{\nu i}}{m_\nu} \cdot \mathbf{D}_{\nu i} \cdot \mathbf{F}_\nu^e = m_\nu \mathbf{v}_{\nu i} \cdot \mathbf{F}_\nu^e, \tag{7.24}$$

which implies that

$$\mathbf{D}_{\nu i} = m_\nu \mathbf{1}. \tag{7.25}$$

Momentum conservation also imposes the condition

$$\sum_{\nu=1}^{r} \sum_{i=1}^{N_\nu} \dot{\mathbf{p}}_{\nu i} = \sum_{\nu=1}^{r} \sum_{i=1}^{N_\nu} \left(\mathbf{F}_{\nu i} + m_\nu \mathbf{F}_\nu^e\right) = \sum_{\nu=1}^{r} N_\nu m_\nu \mathbf{F}_\nu^e = 0. \tag{7.26}$$

Now let us focus on the two-component case for simplicity. Momentum conservation then requires

$$\mathbf{F}_2^e = -\frac{N_1 m_1}{N_2 m_2} \mathbf{F}_1^e. \tag{7.27}$$

The implementation of this algorithm is simplified if we introduce a single parameter that fixes the magnitude of the external field for both species. This is sometimes called the "colour field" $\mathbf{F}_c$. The coupling of each species to this field is described by a new parameter called the colour charge, which we denote here by c_ν. Then in the two component case, we must have

$$m_1 \mathbf{F}_1^e = c_1 \mathbf{F}_c \tag{7.28a}$$

$$m_2 \mathbf{F}_2^e = -\frac{N_1 m_1}{N_2} \mathbf{F}_1^e = c_2 \mathbf{F}_c. \tag{7.28b}$$

These equations are satisfied with a pleasing symmetric form if we define

$$c_1 = \frac{N}{N_1} \tag{7.29a}$$

$$c_2 = -\frac{N}{N_2} \tag{7.29b}$$

$$\mathbf{F}_c = \frac{N_1}{N} m_1 \mathbf{F}_1, \tag{7.29c}$$

which are identical to the definitions originally given by MacGowan and Evans [242].[1] The rate of change of the internal energy for these equations of motion is given by

$$\frac{dU}{dt} = N\left(\frac{1}{N_1 m_1} + \frac{1}{N_2 m_2}\right) V \mathbf{J}_1 \cdot \mathbf{F}_c. \tag{7.30}$$

[1] Note that the sign of c_1 is wrong in Sarman *et al.* [243], and the definitions of c_1 and c_2 given by Sarman and Evans [244, 245] seem to have been interchanged due to typographical errors.

For a system consisting of r components, the external forces must again satisfy Equation (7.26). This will be achieved if we define the external body forces in such a way that

$$\mathbf{F}_{\nu}^{e} = -\frac{N_1 m_1 \mathbf{F}_1^{e}}{\sum_{\mu=2}^{r} N_{\mu} m_{\mu}}. \tag{7.31}$$

If we choose the colour charge and colour force on species 1 to be the same as in the 2-component case, i.e.

$$c_1 = \frac{N}{N_1} \tag{7.32a}$$

$$\mathbf{F}_c = \frac{N_1}{N} m_1 \mathbf{F}_1^{e}, \tag{7.32b}$$

then we find that the colour charges on the other species are given by

$$c_{\nu} = -\frac{N m_{\nu}}{\sum_{\mu=2}^{r} N_{\mu} m_{\mu}}. \tag{7.33}$$

This is the definition first proposed by Sarman, Evans and Cummings [243] and it reduces to the MacGowan-Evans result for a 2-component system. The rate of change of the internal energy due to the external field for this choice of colour charges and forces is

$$\frac{dU}{dt} = N \left(\frac{1}{N_1 m_1} + \frac{1}{\sum_{\mu=2}^{r} N_{\mu} m_{\mu}} \right) V \mathbf{J}_1 \cdot \mathbf{F}_c. \tag{7.34}$$

The colour conductivity algorithm generates diffusive fluxes of the components, but it conserves the total momentum of the system, so the average streaming velocity of the whole system remains zero after it has been initialised to that value. Applying a thermostat to the momenta of the particles relative to the total streaming velocity would thermalise the diffusive currents, so it is more efficient to define the thermal velocity of each component relative to the average velocity of its diffusive flux. Therefore, the thermal kinetic energy is defined as

$$K = \sum_{\nu}^{r} \sum_{i}^{N_{\nu}} \frac{(\mathbf{p}_{\nu i} - \bar{\mathbf{p}}_{\nu})^2}{2m_{\nu}}. \tag{7.35}$$

The thermostat multiplier for a Gaussian constraint thermostat is obtained by setting the time derivative of this to zero. The equations of motion including the thermostat are then

$$\dot{\mathbf{r}}_{\nu i} = \frac{\mathbf{p}_{\nu i}}{m_\nu} \tag{7.36a}$$

$$\dot{\mathbf{p}}_{\nu i} = \mathbf{F}_{\nu i}^{\phi} + c_\nu \mathbf{F}_c - \alpha \left(\mathbf{p}_{\nu i} - \bar{\mathbf{p}}_\nu \right) \tag{7.36b}$$

$$\alpha = \frac{\sum_{\nu}^{r} \sum_{i}^{N_\nu} \frac{1}{m_\nu} \left(\mathbf{p}_{\nu i} - \bar{\mathbf{p}}_\nu \right) \cdot \mathbf{F}_i}{\sum_{\nu}^{r} \sum_{i}^{N_\nu} \frac{1}{m_\nu} \left(\mathbf{p}_{\nu i} - \bar{\mathbf{p}}_\nu \right)^2}. \tag{7.36c}$$

We can now write down the steady state linear response for the diffusive flux generated by these equations of motion. Choosing the field to be in the positive x direction, we find that the diffusive flux of component 1 is

$$\langle J_{1x}(t) \rangle = N \left(\frac{1}{N_1 m_1} + \frac{1}{\sum_{\mu=2}^{r} N_\mu m_\mu} \right) \frac{F_c V}{k_B T} \int_0^t \langle J_{1x}(s) J_{1x}(0) \rangle \, ds. \tag{7.37}$$

Comparing this to the Green–Kubo relation for the phenomenological coefficient for the diffusive flux, [236]

$$L_{11} = \frac{V}{k_B} \int_0^\infty \langle J_{1x}(t) J_{1x}(0) \rangle \, dt \tag{7.38}$$

shows that in the limit of zero field, the phenomenological coefficient can be calculated from the ratio of the steady state flux to the colour field

$$L_{11} = \lim_{F_c \to 0} \left(\frac{N}{N_1 m_1} + \frac{N}{\sum_{\mu=2}^{r} N_\mu m_\mu} \right)^{-1} \frac{T \langle J_{1x} \rangle}{F_c}. \tag{7.39}$$

The heat flux vector and the diffusive mass flux are both polar vectors, and they both have the same parity with respect to time reversal, so the cross-correlation function of the heat flux vector with the diffusive flux is also nonzero. Linear response theory predicts that the colour field also generates a heat flux,

$$\langle J_{qx}(t) \rangle = N \left(\frac{1}{N_1 m_1} + \frac{1}{\sum_{\mu=2}^{r} N_\mu m_\mu} \right) \frac{F_c V}{k_B T} \int_0^t \langle J_{qx}(s) J_{1x}(0) \rangle \, ds \tag{7.40}$$

which enables us to determine the phenomenological coefficient L_{q1} associated with the flow of heat due to a concentration gradient

$$L_{q1} = \lim_{F_c \to 0} \left(\frac{N}{N_1 m_1} + \frac{N}{\sum\limits_{\mu=2}^{r} N_\mu m_\mu} \right)^{-1} \frac{T \langle J_{qx} \rangle}{F_c}. \tag{7.41}$$

7.2.2 Results for Simple Fluids

The colour diffusion algorithm was used by MacGowan and Evans [242] and Sarman and Evans [244, 245] to compute the L_{11} and L_{q1} phenomenological coefficients for equimolar Argon-Krypton mixtures. Very good agreement between the values of these quantities obtained by nonequilibrium and equilibrium (Green–Kubo) simulations was found. In addition, the value of L_{q1} was in very good agreement with that of the L_{1q} coefficient obtained by both nonequilibrium (Evans-Cummings heat flow algorithm) and equilibrium (Green–Kubo) methods. This result is consistent with the Onsager reciprocal relation between the two phenomenological coefficients.

The colour diffusion algorithm has also been used [246] to compute the "transport diffusivity" for permeation of a species in a porous medium. Wheeler and Newman [247] have applied a similar NEMD colour field method to compute the diffusion coefficient of an electrolyte solution.

7.3 Multicomponent Heat Transport

7.3.1 Equations of Motion

A heat flow algorithm for multicomponent fluids was first proposed by MacGowan and Evans [242, 248, 249]. Their original algorithm is formulated so as to generate a heat flux that is different from the standard Irving-Kirkwood heat flux. Thus, the Green–Kubo formula derived with this heat flux vector does not correspond to the usual definition of the thermal conductivity or the heat flow phenomenological coefficient. An alternative approach was described by Evans and Cummings [250]. Here we will discuss a route to the Evans-Cummings algorithm while also reviewing the algorithms for multicomponent heat flow proposed by Perronace *et al.* [251] and Mandadapu *et al.* [252].

The Irving-Kirkwood heat flux vector for a multicomponent solution can be written as

$$V\mathbf{J}_q(t) = \sum_{\nu=1}^{r} \sum_{i=1}^{N_\nu} \frac{\mathbf{p}_{\nu i}}{m_\nu} \cdot \mathbf{S}_{\nu i}, \tag{7.42}$$

where

$$\mathbf{S}_{\nu i} = u_{\nu i}\mathbf{1} - \frac{1}{2}\sum_{\mu=1}^{r}\sum_{j=1}^{N_\mu}\mathbf{F}_{\nu i\mu j}\mathbf{r}_{\nu i\mu j} \tag{7.43}$$

and it is implicit that the $\nu i = \mu j$ term is omitted from the double summation. When the masses of all particles are allowed to differ, the choice for $\mathbf{D}_{\nu i}$ previously given, i.e.

$$\mathbf{D}_{\nu i} = \mathbf{S}_{\nu i} - \bar{\mathbf{S}} \tag{7.44}$$

leads to a rate of change of the internal energy equal to

$$\begin{aligned}\frac{dU}{dt} &= \sum_{\nu}^{r}\sum_{i}^{N_\nu}\frac{\mathbf{p}_{\nu i}}{m_\nu}\cdot\mathbf{D}_{\nu i}\cdot\mathbf{F}_q \\ &= \sum_{\nu}^{r}\sum_{i}^{N_\nu}\frac{\mathbf{p}_{\nu i}}{m_\nu}\cdot\left(\mathbf{S}_{\nu i} - \bar{\mathbf{S}}\right)\cdot\mathbf{F}_q \\ &= V\mathbf{J}_q\cdot\mathbf{F}_q - \bar{\mathbf{S}}:\mathbf{F}_q\sum_{\nu}^{r}\sum_{i}^{N_\nu}\frac{\mathbf{p}_{\nu i}}{m_\nu},\end{aligned} \tag{7.45}$$

which is not the desired result. The desired result for the internal energy derivative can be recovered by redefining

$$\mathbf{D}_{\nu i} = \mathbf{S}_{\nu i} - \frac{m_\nu}{M}\bar{\mathbf{S}}, \tag{7.46}$$

where $M = \sum_{\nu} N_\nu m_\nu$ as suggested by Perronace *et al.* [253] but the phase space compression factor remains nonzero,

$$\begin{aligned}\Lambda &= \sum_{\nu}^{r}\sum_{i}^{N_\nu}\frac{\partial}{\partial\mathbf{p}_{\nu i}}\cdot\mathbf{D}_{\nu i}\cdot\mathbf{F}_q \\ &= \sum_{\nu}^{r}\sum_{i}^{N_\nu}\frac{\mathbf{p}_{\nu i}}{m_\nu}\cdot\mathbf{F}_q - \frac{1}{N}\sum_{\nu}^{r}\sum_{i}^{N_\nu}\frac{m_\nu}{M}\frac{\mathbf{p}_{\nu i}}{m_\nu}\cdot\mathbf{F}_q \\ &= \sum_{\nu}^{r}\sum_{i}^{N_\nu}\frac{\mathbf{p}_{\nu i}}{m_\nu}\cdot\mathbf{F}_q,\end{aligned} \tag{7.47}$$

as noted by Mandadapu *et al.* [252], who suggested further modifications which are apparently incompatible with periodic boundary conditions.

A different solution was previously proposed by Evans and Cummings [250] and implemented by Sarman and Evans [244, 245]. Their result can be obtained by the following method. Returning to the original specification of the problem, we observe that if the masses of the molecules in the system are not equal, the imposition of momentum conservation leads to an extra term in the rate of change of the internal energy. However, the rate of change of the internal energy only needs to be directly proportional to the heat flux if we insist on incompressible phase space dynamics. If we allow compressible phase space dynamics, we must find equations of motion that satisfy the generalised response theorem for compressible phase space dynamics given previously in Equation (3.49)

$$\frac{dU}{dt} - k_B T \Lambda = V \mathbf{J}_q \cdot \mathbf{F}_q. \tag{7.48}$$

This means that we must seek equations of motion that satisfy

$$\sum_{\nu}^{r} \sum_{i}^{N_\nu} \frac{\mathbf{p}_{\nu i}}{m_\nu} \cdot \mathbf{D}_{\nu i} \cdot \mathbf{F}_q - k_B T \sum_{\nu}^{r} \sum_{i}^{N_\nu} \frac{\partial}{\partial \mathbf{p}_{\nu i}} \cdot \mathbf{D}_{\nu i} \cdot \mathbf{F}_q = \sum_{i}^{N} \frac{\mathbf{p}_{\nu i}}{m_\nu} \cdot \mathbf{S}_{\nu i} \cdot \mathbf{F}_q. \tag{7.49}$$

The equations of motion must also conserve momentum, i.e.

$$\sum_{\nu}^{r} \sum_{i}^{N_\nu} \mathbf{D}_{\nu i} \cdot \mathbf{F}_q = 0. \tag{7.50}$$

By adding to $\mathbf{D}$ a constant term that is proportional to $k_B T$, we can ensure cancellation of the term due to phase space compression. By also adding a term proportional to $\bar{\mathbf{S}}$, we can null the corresponding term in the rate of change of the internal energy. To allow sufficient freedom for the coefficients of these terms while also maintaining momentum conservation, we may multiply each one by a coefficient whose sum over all particles is equal to zero. This is similar to the role played by the colour charge in the interdiffusion algorithm. We will call this coefficient the "heat charge", since it couples to the heat field.[2] Thus, Evans and Cummings proposed [250]

$$\mathbf{D}_{\nu i} = \mathbf{S}_{\nu i} - \bar{\mathbf{S}} + \chi_\nu \bar{\mathbf{S}} + \chi_\nu k_B T \mathbf{1}, \tag{7.51}$$

where momentum conservation is ensured by applying the condition

$$\sum_{\nu=1}^{r} N_\nu \chi_\nu = 0. \tag{7.52}$$

All that is required now is to find the values of the heat charges χ_i that will ensure that the response theorem is satisfied. We require

$$\sum_{\nu}^{r} \sum_{i}^{N_\nu} \frac{\mathbf{p}_{\nu i}}{m_\nu} \cdot \mathbf{D}_{\nu i} \cdot \mathbf{F}_q - k_B T \sum_{\nu}^{r} \sum_{i}^{N_\nu} \frac{\partial}{\partial \mathbf{p}_{\nu i}} \cdot \mathbf{D}_{\nu i} \cdot \mathbf{F}_q = \sum_{\nu}^{r} \sum_{i}^{N_\nu} \frac{\mathbf{p}_{\nu i}}{m_\nu} \cdot \mathbf{S}_{\nu i} \cdot \mathbf{F}_q, \tag{7.53}$$

with $\mathbf{D}_{\nu i}$ defined by Equation (7.51). Substituting $\mathbf{D}_{\nu i}$ into the first term on the left-hand

[2] Note that Evans and Cummings originally also called this factor a colour charge, but because it couples to a heat field, not the colour field, we have given it a different name.

side, we find

$$\sum_{\nu}^{r}\sum_{i}^{N}\frac{\mathbf{p}_{\nu i}}{m_\nu}\cdot\mathbf{D}_{\nu i}\cdot\mathbf{F}_q$$
$$=\sum_{\nu}^{r}\sum_{i}^{N_\nu}\frac{\mathbf{p}_{\nu i}}{m_\nu}\cdot\mathbf{S}_{\nu i}\cdot\mathbf{F}_q-\bar{\mathbf{S}}:\mathbf{F}_q\sum_{\nu}^{r}\sum_{i}^{N_\nu}(1-\chi_\nu)\frac{\mathbf{p}_{\nu i}}{m_\nu}+k_BT\mathbf{F}_q\cdot\sum_{\nu}^{r}\sum_{i}^{N_\nu}\chi_\nu\frac{\mathbf{p}_{\nu i}}{m_\nu}. \tag{7.54}$$

For the second term, we find

$$k_BT\sum_{\nu}^{r}\sum_{i}^{N_\nu}\frac{\partial}{\partial\mathbf{p}_{\nu i}}\cdot\mathbf{D}_{\nu i}\cdot\mathbf{F}_q=k_BT\mathbf{F}_q\cdot\sum_{\nu}^{r}\sum_{i}^{N_\nu}\frac{\mathbf{p}_{\nu i}}{m_\nu}\left(1-\frac{1}{N}+\frac{\chi_\nu}{N}\right). \tag{7.55}$$

For the sum of terms on the left-hand side to be equal to the right-hand side of Equation (7.53), we therefore require both

$$\sum_{\nu}^{r}\sum_{i}^{N_\nu}(1-\chi_\nu)\frac{\mathbf{p}_{\nu i}}{m_\nu}=0 \tag{7.56}$$

and

$$\sum_{\nu}^{r}\sum_{i}^{N_\nu}\frac{\mathbf{p}_{\nu i}}{m_\nu}(1-\chi_\nu)-\frac{1}{N}\sum_{\nu}^{r}\sum_{i}^{N_\nu}\frac{\mathbf{p}_{\nu i}}{m_\nu}(1-\chi_\nu)=0. \tag{7.57}$$

These are satisfied provided that

$$\sum_{\nu}^{r}\sum_{i}^{N_\nu}(1-\chi_\nu)\frac{\mathbf{p}_{\nu i}}{m_\nu}=V\sum_{\nu}^{r}(1-\chi_\nu)\frac{\mathbf{J}_\nu}{m_\nu}=0, \tag{7.58}$$

where the diffusive flux of component ν, $\mathbf{J}_\nu$ is defined by

$$V\mathbf{J}_\nu=\sum_{i}^{N_\nu}\mathbf{p}_{\nu i}. \tag{7.59}$$

The diffusive fluxes are defined so that they must obey

$$\sum_{\nu}^{r}\mathbf{J}_\nu=0, \tag{7.60}$$

which allows us to write

$$\sum_{\nu}^{r}(1-\chi_\nu)\frac{\mathbf{J}_\nu}{m_\nu}=\sum_{\nu}^{r-1}\mathbf{J}_\nu\left(\frac{1}{m_\nu}-\frac{\chi_\nu}{m_\nu}\right)-\sum_{\nu}^{r-1}\mathbf{J}_\nu\left(\frac{1}{m_r}-\frac{\chi_r}{m_r}\right)$$
$$=\sum_{\nu}^{r-1}\mathbf{J}_\nu\left(\frac{1}{m_\nu}-\frac{1}{m_r}\right)-\sum_{\nu}^{r-1}\mathbf{J}_\nu\left(\frac{\chi_\nu}{m_\nu}-\frac{\chi_r}{m_r}\right)$$
$$=0. \tag{7.61}$$

One way to satisfy this relation is to require

$$\left(\frac{1}{m_\nu}-\frac{1}{m_r}\right)=\left(\frac{\chi_\nu}{m_\nu}-\frac{\chi_r}{m_r}\right) \tag{7.62}$$

for $1 \leqslant \nu \leqslant r-1$. Multiplying both sides by $m_\nu m_r N_\nu$ and summing each term over ν up to $r-1$, we find that the heat charges are given by

$$\chi_\nu=\frac{M-Nm_\nu}{M}, \tag{7.63}$$

where the total mass M is defined as $M=\sum_\nu^r N_\nu m_\nu$. Substituting Equation (7.63) into Equation (7.52) for the sum of $N_\nu\chi_\nu$ to obtain χ_r, we find that Equation (7.63) also applies to component $\nu=r$, i.e. it is true for all $\nu=1,\ldots,r$.

This assignment of the "heat charges" satisfies the response theorem,

$$\begin{aligned}\frac{dU}{dt}-k_BT\Lambda &= \sum_\nu^r\sum_i^{N_\nu}\frac{\mathbf{p}_{\nu i}}{m_\nu}\cdot\mathbf{D}_{\nu i}\cdot\mathbf{F}_q-k_BT\sum_\nu^r\sum_i^{N_\nu}\frac{\partial}{\partial\mathbf{p}_{\nu i}}\cdot\mathbf{D}_{\nu i}\cdot\mathbf{F}_q\\
&=\sum_\nu^r\sum_i^{N_\nu}\frac{\mathbf{p}_{\nu i}}{m_\nu}\cdot\mathbf{S}_{\nu i}\cdot\mathbf{F}_q+\bar{\mathbf{S}}:\mathbf{F}_q\sum_\nu^r\sum_i^{N_\nu}(\chi_\nu-1)\frac{\mathbf{p}_{\nu i}}{m_\nu}\\
&\quad-k_BT\mathbf{1}:\mathbf{F}_q\sum_\nu^r\sum_i^{N_\nu}\left(1-\chi_\nu-\frac{1}{N}+\frac{\chi_\nu}{N}\right)\frac{\mathbf{p}_{\nu i}}{m_\nu}\\
&=\sum_\nu^r\sum_i^{N_\nu}\frac{\mathbf{p}_{\nu i}}{m_\nu}\cdot\mathbf{S}_{\nu i}\cdot\mathbf{F}_q\\
&=V\mathbf{J}_q\cdot\mathbf{F}_q\end{aligned} \tag{7.64}$$

as well as momentum conservation,

$$\sum_\nu^r\sum_i^{N_\nu}\mathbf{D}_{\nu i}=\sum_\nu^r\sum_i^{N_\nu}\left(\mathbf{S}_{\nu i}-\bar{\mathbf{S}}+\chi_\nu\bar{\mathbf{S}}+\chi_\nu k_BT\mathbf{1}\right)=0. \tag{7.65}$$

Due to the linear coupling between heat and mass transport, the heat field drives diffusive fluxes of both mass and heat. The thermal kinetic energy can be defined in terms of the velocities relative to the average velocities of the components as given by Equation (7.35), and the resulting expression for the thermostat multiplier α is again given by Equation (7.36c). The full equations of motion including the thermostat for the Evans-Cummings heat flow algorithm are

$$\dot{\mathbf{r}}_{\nu i}=\frac{\mathbf{p}_{\nu i}}{m_\nu} \tag{7.66a}$$

$$\dot{\mathbf{p}}_{\nu i}=\mathbf{F}_{\nu i}+\left(\mathbf{S}_{\nu i}-\bar{\mathbf{S}}+\chi_\nu\bar{\mathbf{S}}+\chi_\nu k_BT\mathbf{1}\right)\cdot\mathbf{F}_q-\alpha\left(\mathbf{p}_{\nu i}-\bar{\mathbf{p}}_\nu\right) \tag{7.66b}$$

$$\alpha=\frac{\sum_\nu^r\sum_i^{N_\nu}\frac{1}{m_\nu}\left(\mathbf{p}_{\nu i}-\bar{\mathbf{p}}_\nu\right)\cdot\left[\mathbf{F}_i+\left(\mathbf{S}_{\nu i}-\bar{\mathbf{S}}+\chi_\nu\bar{\mathbf{S}}+\chi_\nu k_BT\mathbf{1}\right)\cdot\mathbf{F}_q\right]}{\sum_\nu^r\sum_i^{N_\nu}\frac{1}{m_\nu}\left(\mathbf{p}_{\nu i}-\bar{\mathbf{p}}_\nu\right)^2}. \tag{7.66c}$$

Since we have already proven that the response theorem is satisfied, the linear response can readily be written down. We will only consider the case of a binary mixture, but the generalisation to multicomponent systems is straightforward.

If we now take the external field to be given by the x-component of the Evans-Cummings heat field F_q and the flux to be the x-component of the heat flux vector J_{qx} (which has zero mean in the equilibrium state), we obtain

$$\langle J_{qx}(t) \rangle = \frac{F_q V}{k_B T} \int_0^t \langle J_{qx}(s) J_{qx}(0) \rangle ds. \tag{7.67}$$

Just as we found for the colour diffusion algorithm, the heat field in the Evans-Cummings heat flow algorithm generates both a heat flux and a diffusive flux due to the linear coupling of these transport processes. The diffusive mass flux is then given by

$$\langle J_{1x}(t) \rangle = \frac{F_q V}{k_B T} \int_0^t \langle J_{1x}(s) J_{qx}(0) \rangle ds. \tag{7.68}$$

The relevant equilibrium auto- and cross-correlation functions of the heat and diffusive mass flux vectors are related to the phenomenological coefficients [236] by

$$L_{qq} = \frac{V}{k_B} \int_0^\infty \langle J_{qx}(t) J_{qx}(0) \rangle dt \tag{7.69}$$

$$L_{1q} = \frac{V}{k_B} \int_0^\infty \langle J_{1x}(t) J_{qx}(0) \rangle dt. \tag{7.70}$$

Comparing these equations with the linear response equations for the heat and diffusive fluxes, we can identify the quantities computed from the Evans-Cummings algorithm in the steady state and the linear response (zero field) limit as

$$L_{qq} = \lim_{F_q \to 0} \frac{\langle J_{qx} \rangle T}{F_q} \tag{7.71}$$

$$L_{1q} = \lim_{F_q \to 0} \frac{\langle J_{1x} \rangle T}{F_q}. \tag{7.72}$$

7.3.2 Results for Simple Fluids

The Evans-Cummings heat flow algorithm was used by Sarman and Evans [244, 245] to compute L_{qq} and L_{1q} for equimolar Argon-Krypton mixtures. Again, the values of these quantities obtained by nonequilibrium and equilibrium (Green–Kubo) simulations were in good agreement.

7.4 Evaluation of Thermodynamic Quantities

The colour diffusion and multicomponent heat flow algorithms described in the previous sections allow us to determine the phenomenological coefficients, but they do not directly give us the practical transport coefficients. For a binary mixture, the relevant linear phenomenological coefficients are L_{11}, L_{1q}, L_{qq} and L_{q1}, defined by Equation (2.83).

The practical transport coefficients defined by the flux–force relations expressed in terms of the gradients of measurable quantities such as temperature and concentration are defined by Equation (2.88). The practical transport coefficients are related to the primed phenomenological coefficients by Equation (2.89), and the primed phenomenological coefficients are related to the unprimed ones that we determine by either Green–Kubo or nonequilibrium methods by Equation (2.87). It is clear that additional thermodynamic quantities are required to compute the transport coefficients. To compute the mutual diffusion coefficient, we need the "thermodynamic factor" $(\partial \mu_1 / \partial c_1)_{T,p}$ and to relate the primed and unprimed phenomenological coefficients, we need the partial specific enthalpies h_ν.

The thermodynamic factor can be computed by using the Kirkwood-Buff theory of solutions [254], which gives the following relationship

$$\frac{1}{k_B T} \left(\frac{\partial \tilde{\mu}_1}{\partial n_1} \right)_{T,p} = \frac{1}{n_1} + \frac{G_{12} - G_{11}}{1 + n_1 \left(G_{11} - G_{12} \right)}, \tag{7.73}$$

where the $G_{\nu\mu}$ are given by

$$G_{\nu\mu} = 4\pi \int_0^\infty r^2 \left(g_{\nu\mu}(r) - 1 \right) dr \tag{7.74}$$

and the $g_{\nu\mu}(r)$ are the radial distribution functions. Note that the tilde indicates that the concentration n_ν in these equations is the number of molecules per unit volume, instead of the mass fractions c_ν used in our definitions of the transport coefficients. We will return to this issue later.

These integrals can be evaluated by directly computing the radial distribution functions and then numerically integrating them. But this procedure is highly problematic [255]. Since the radial distribution functions are multiplied by r^2 in the integrand, they need to be very accurate at large values of r. When simulations are carried out on systems with long ranged correlations, the system size may not be large enough to avoid truncation of the radial distribution functions. In such cases, it may be necessary to fit the tails of the radial distribution functions and evaluate the integral from the truncation point to infinite r from the fit to get accurate values of the total integrals. There is also the issue of normalisation of the radial distribution functions to consider. When they are computed in the canonical ensemble, they do not asymptote to 1 but rather to $(1 - 1/N)$ [256]. In light of these difficulties, several different methods for computing the thermodynamic factor and the $G_{\nu\mu}$ have been proposed [257, 258]. We have found that a method similar to that of Nichols *et al.* [258] gives excellent results [259].

The partial dynamic structure factors are defined as the zero-time values of the partial intermediate scattering functions

$$S_{\nu\mu}(\mathbf{k}) = \lim_{t \to 0} F_{\nu\mu}(\mathbf{k}, t), \tag{7.75}$$

where

$$F_{\alpha\mu}(\mathbf{k}, t) = \frac{1}{N} \left\langle \tilde{n}_\nu(\mathbf{k}, t) \tilde{n}_\mu^*(\mathbf{k}, 0) \right\rangle \tag{7.76}$$

and $\tilde{n}_\nu(\mathbf{k},t)$ is the Fourier transform of the number density of component ν,

$$\tilde{n}_\nu(\mathbf{k},t) = \int \sum_{i=1}^{N_\nu} \delta(\mathbf{r} - \mathbf{r}_i(t)) e^{i\mathbf{k}\cdot\mathbf{r}} d\mathbf{r} = \sum_{i=1}^{N_\nu} e^{i\mathbf{k}\cdot\mathbf{r}_i(t)}, \tag{7.77}$$

where N_ν is the number of molecules of component ν. The intermediate scattering function is then

$$F_{\nu\mu}(\mathbf{k},t) = \frac{1}{N}\left\langle \sum_{i=1}^{N_\mu} \sum_{j\neq i}^{N_\nu} e^{-i\mathbf{k}\cdot[\mathbf{r}_j(0)-\mathbf{r}_i(t)]} \right\rangle. \tag{7.78}$$

It is known [256] that the partial structure factors are related to the Fourier transforms of the radial distribution functions through

$$S_{\nu\mu}(\mathbf{k}) = x_\nu \delta_{\nu\mu} + x_\nu x_\mu n \int g_{\nu\mu}(r) e^{i\mathbf{k}\cdot\mathbf{r}} d\mathbf{r}, \tag{7.79}$$

where x_ν is the number fraction of component ν, n is the number density of component ν and n is the total number density. Assuming that the fluid is isotropic and writing the constant part of the radial distribution explicitly this can be rewritten as

$$S_{\nu\mu}(\mathbf{k}) = x_\nu \delta_{\nu\mu} + 4\pi x_\nu x_\mu n \int r^2 \left[g_{\nu\mu}(r) - 1\right] e^{i\mathbf{k}\cdot\mathbf{r}} dr + (2\pi)^3 x_\nu x_\mu n \delta(\mathbf{k}). \tag{7.80}$$

Comparing this with the expression given earlier for $G_{\nu\mu}$, we see that, if we ignore the contribution of the delta function at zero wavevector, we can write

$$\begin{aligned} G_{\nu\mu} &= 4\pi \int r^2 \left[g_{\nu\mu}(r) - 1\right] dr \\ &= \lim_{k\to 0} 4\pi \int r^2 \left[g_{\nu\mu}(r) - 1\right] e^{i\mathbf{k}\cdot\mathbf{r}} dr \\ &= \frac{1}{x_\nu x_\mu n}\left[\lim_{k\to 0} S_{\nu\mu}(\mathbf{k}) - x_\nu \delta_{\nu\mu}\right]. \end{aligned} \tag{7.81}$$

Using the structure factors evaluated at all available wavevectors in three dimensions, we can obtain a large amount of data from which a reliable extrapolation to zero wavevector can be obtained. Figure 7.2 shows data obtained in this way for the three partial structure factors for a mixture of WCA atoms (solvent, component 2) with atoms having a WCA potential on the outside of a hard core of diameter 3.03 (colloid, component 1), so that the colloid/solvent size ratio is 4.03 and the mass ratio is 50. This system is designed to model a colloidal suspension [259]. The reduced temperature is $T = 1.0$ and the reduced pressure is $p = 7.85$.

Once the $G_{\nu\mu}$ values have been obtained, it only remains to convert the thermodynamic factor from the number density concentration units to the mass fraction concentration units. This relationship can be derived using standard thermodynamics, giving

$$\left(\frac{\partial \mu_1}{\partial c_1}\right)_{T,p} = \left(\frac{\partial \tilde{\mu}_1}{\partial n_1}\right)_{T,p} \frac{\rho(1 - n_1 \tilde{v}_1)}{m_1^2(1 - c_1)}, \tag{7.82}$$

where the quantities with the tilde are in molecular number density concentration units, $\tilde{v}_1$ represents the partial molecular volume of species 1, c is the mass fraction and

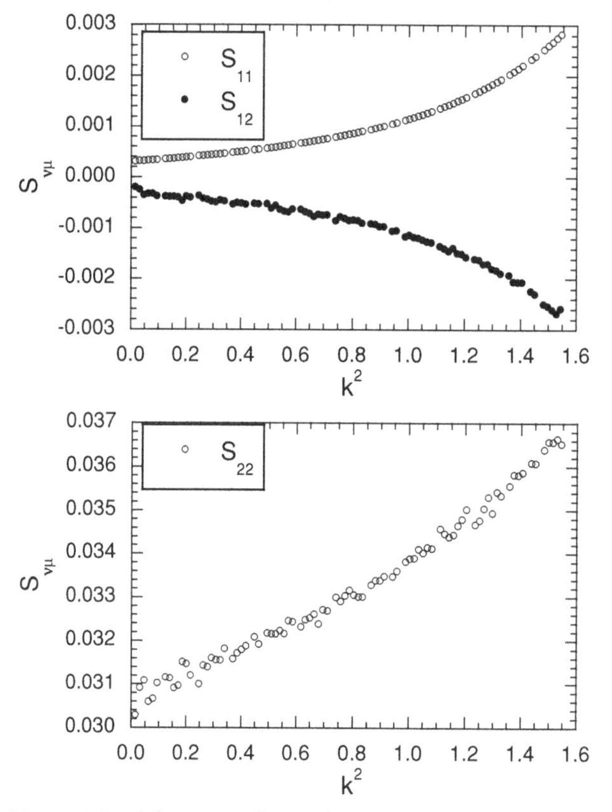

Figure 7.2 Partial structure factors for a model colloidal suspension at a packing fraction of 0.49 used to determine the thermodynamic factor for the mutual diffusion coefficient. The structure factors were plotted against k^2 and then fitted with a fifth-order polynomial to extract the zero wavevector intercept. Replotted from reference [259].

m_1 is the molecular mass of component 1. The partial molecular volume is given in Kirkwood-Buff theory by

$$\tilde{v}_1 = \frac{1 + (G_{22} - G_{12})\, n_2}{n_1 + n_2 + n_1 n_2 \left(G_{11} + G_{22} - 2G_{12}\right)}. \tag{7.83}$$

Zhou and Miller [260] have combined these to give the thermodynamic factor in mass fraction concentration units directly in terms of the $G_{\nu\mu}$ values as

$$\left(\frac{\partial \mu_1}{\partial c_1}\right)_{p,T} = \frac{V^2}{\rho \left\langle c_1^2 \right\rangle}, \tag{7.84}$$

where

$$\left\langle c_1^2 \right\rangle = \frac{m_1^2 m_2^2 x_1 x_2 n^2 N}{\rho^4}\left[1 + x_1 x_2 n \left(G_{11} + G_{22} - 2G_{12}\right)\right]. \tag{7.85}$$

The partial specific enthalpies h_ν are also required to convert the unprimed phenomenological coefficients into the primed ones. Several methods are available for the computation of partial enthalpies. The simplest is to just compute the enthalpy as a function of the mass of component ν keeping the temperature, pressure and total mass of each other component fixed and directly evaluate the partial derivative $h_\nu = \left(\frac{\partial H}{\partial M_\nu}\right)_{T,p,M_{\mu\neq\nu}}$ [255].

7.5 Heat Transport for Molecular Fluids

7.5.1 Heat Flow

The first NEMD algorithm for studying heat flow in molecular fluids was the Evans-Murad algorithm for rigid body molecules [261]. This algorithm has been widely applied to fluids of rigid body molecules such as hydrochloric acid, carbon dioxide and mixtures of methane and benzene [243]. However, it cannot be applied to flexible molecules, and a more general algorithm that can be used for both rigid and flexible molecules would be desirable. The heat flow algorithm originally presented by Daivis and Evans [262, 263] and generalised by Perronace *et al.* [251] is more generally useful because it can be applied to molecules of any degree of flexibility, from fully rigid to fully flexible. To derive the equations of motion, we will begin by deriving an expression for the heat flux vector in the molecular representation. Then, we deduce the form of the external field that is required to generate this flux in a homogeneous system.

The objective of an NEMD algorithm is to induce the appropriate flux with a synthetic force through the equations of motion so that when linear response theory is applied, we obtain a flux that is directly proportional to the field. In this case, the flux that we want to induce is the molecular heat flux. The expression for the internal energy density in the molecular localisation is

$$\rho u(\mathbf{r},t) = \sum_{i=1}^{N_m} u_i \delta(\mathbf{r}-\mathbf{r}_i), \tag{7.86}$$

where the internal energy of molecule i is given by

$$u_i = \sum_{\alpha=1}^{N_s}\left(\frac{\mathbf{p}_{i\alpha}^2}{2m_{i\alpha}} + \phi_{i\alpha}\right). \tag{7.87}$$

Note that the potential energy of interaction site α of molecule i consists of all types of inter- and intra-molecular potential energy, including Lennard-Jones, bond bending, bond stretching, and dihedral energy where appropriate. The Fourier transform of the internal energy density is

$$\widetilde{\rho u}(\mathbf{k},t) = \sum_{i=1}^{N_m} u_i e^{i\mathbf{k}\cdot\mathbf{r}_i}. \tag{7.88}$$

To derive the expression for the molecular heat flux vector from the balance equation for the molecular internal energy density, we begin by taking the time derivative of the internal energy density,

$$\frac{\partial}{\partial t}\widetilde{\rho u}(\mathbf{k},t)=\sum_{i=1}^{N_m}\dot{u}_i e^{i\mathbf{k}\cdot\mathbf{r}_i}+i\mathbf{k}\cdot\sum_{i=1}^{N_m}\dot{\mathbf{r}}_i u_i e^{i\mathbf{k}\cdot\mathbf{r}_i}. \tag{7.89}$$

The first term is given by

$$\begin{aligned}\sum_{i=1}^{N_m}\dot{u}_i e^{i\mathbf{k}\cdot\mathbf{r}_i}&=\sum_{i=1}^{N_m}\sum_{\alpha=1}^{N_s}\left(\frac{\mathbf{p}_{i\alpha}\cdot\dot{\mathbf{p}}_{i\alpha}}{m_{i\alpha}}+\dot{\phi}_{i\alpha}\right)e^{i\mathbf{k}\cdot\mathbf{r}_i}\\&=\sum_{i=1}^{N_m}\sum_{\alpha=1}^{N_s}\left(\frac{\mathbf{p}_{i\alpha}}{m_{i\alpha}}\cdot\mathbf{F}_{i\alpha}+\dot{\phi}_{i\alpha}\right)e^{i\mathbf{k}\cdot\mathbf{r}_i}.\end{aligned} \tag{7.90}$$

The exponential in the first term can be expanded and truncated at first order in the wavevector, giving

$$\sum_{i=1}^{N_m}\sum_{\alpha=1}^{N_s}\frac{\mathbf{p}_{i\alpha}}{m_{i\alpha}}\cdot\mathbf{F}_{i\alpha}e^{i\mathbf{k}\cdot\mathbf{r}_i}=\sum_{i=1}^{N_m}\sum_{\alpha=1}^{N_s}\mathbf{F}_{i\alpha}\cdot\mathbf{v}_{i\alpha}+i\mathbf{k}\cdot\sum_{i=1}^{N_m}\sum_{\alpha=1}^{N_s}\mathbf{r}_i\mathbf{F}_{i\alpha}\cdot\mathbf{v}_{i\alpha}+\cdots, \tag{7.91}$$

where the force on site α of molecule i is given by the sum of any internal bond constraint forces that may be applied to the sites, plus all two body, three body, four body and higher-order forces

$$\mathbf{F}_{i\alpha}=\mathbf{F}_{i\alpha}^{C}+\mathbf{F}_{i\alpha}^{(2)}+\mathbf{F}_{i\alpha}^{(3)}+\mathbf{F}_{i\alpha}^{(4)}+\cdots. \tag{7.92}$$

The contribution of intramolecular bond constraint forces to the molecular heat flux vector is zero, because

$$\begin{aligned}\sum_{i=1}^{N_m}\sum_{\alpha=1}^{N_s}\mathbf{v}_{i\alpha}\cdot\mathbf{F}_{i\alpha}^{C}e^{i\mathbf{k}\cdot\mathbf{r}_i}&=\sum_{i=1}^{N_m}\sum_{\alpha=1}^{N_s}\sum_{\beta=1}^{N_s}\mathbf{v}_{i\alpha}\cdot\mathbf{F}_{i\alpha i\beta}^{C}e^{i\mathbf{k}\cdot\mathbf{r}_i}\\&=\frac{1}{2}\sum_{i=1}^{N_m}\sum_{\alpha=1}^{N_s}\sum_{\beta=1}^{N_s}\left(\mathbf{v}_{i\alpha}-\mathbf{v}_{i\beta}\right)\cdot\mathbf{F}_{i\alpha i\beta}^{C}e^{i\mathbf{k}\cdot\mathbf{r}_i}\\&=0.\end{aligned} \tag{7.93}$$

The last equality follows from the fact that the constraint force must act in the direction of the constrained bond, and the component of the relative velocity of two constrained sites in the direction of the bond must be zero.

The time derivative of the potential energy can be written as a sum of two body, three body and higher terms. Let us consider only the two body terms first. We then have

$$
\begin{aligned}
\sum_{i=1}^{N_m}\sum_{\alpha=1}^{N_s}\dot{\phi}_{i\alpha}^{(2)}e^{i\mathbf{k}\cdot\mathbf{r}_{i\alpha}} &= \frac{1}{2}\sum_{i}^{N_m}\sum_{\alpha}^{N_s}\sum_{j}^{N_m}\sum_{\beta}^{N_s}\dot{\phi}_{i\alpha j\beta}\left(\mathbf{r}_{i\alpha},\mathbf{r}_{j\beta}\right)e^{i\mathbf{k}\cdot\mathbf{r}_i} \\
&= \frac{1}{2}\sum_{i}^{N_m}\sum_{\alpha}^{N_s}\sum_{j}^{N_m}\sum_{\beta}^{N_s}\left(\mathbf{v}_{i\alpha}\cdot\frac{\partial\phi_{i\alpha j\beta}}{\partial\mathbf{r}_{i\alpha}}+\mathbf{v}_{j\beta}\cdot\frac{\partial\phi_{i\alpha j\beta}}{\partial\mathbf{r}_{j\beta}}\right)e^{i\mathbf{k}\cdot\mathbf{r}_i} \\
&= -\frac{1}{2}\sum_{i}^{N_m}\sum_{\alpha}^{N_s}\sum_{j}^{N_m}\sum_{\beta}^{N_s}\left(\mathbf{v}_{i\alpha}\cdot\mathbf{F}_{i\alpha j\beta}^{(2)}+\mathbf{v}_{j\beta}\cdot\mathbf{F}_{j\beta i\alpha}^{(2)}\right)e^{i\mathbf{k}\cdot\mathbf{r}_i} \\
&= -\sum_{i}^{N_m}\sum_{\alpha}^{N_s}\mathbf{v}_{i\alpha}\cdot\mathbf{F}_{i\alpha}^{(2)} \\
&\quad -\frac{1}{2}i\mathbf{k}\cdot\sum_{i}^{N_m}\sum_{\alpha}^{N_s}\sum_{j}^{N_m}\sum_{\beta}^{N_s}\left(\mathbf{r}_i\mathbf{F}_{i\alpha j\beta}^{(2)}\cdot\mathbf{v}_{i\alpha}+\mathbf{r}_j\mathbf{F}_{i\alpha j\beta}^{(2)}\cdot\mathbf{v}_{i\alpha}\right)+O\left(k^2\right),
\end{aligned}
\tag{7.94}
$$

where we have again expanded the exponentials, because here we are only interested in deriving the zero wavevector component of the heat flux vector. Now if we combine terms we obtain

$$
\begin{aligned}
\sum_{i=1}^{N_m}\dot{u}_i e^{i\mathbf{k}\cdot\mathbf{r}_i} &= \sum_{i=1}^{N_m}\sum_{\alpha=1}^{N_s}\left(\mathbf{F}_{i\alpha}\cdot\mathbf{v}_{i\alpha}+\dot{\phi}_{i\alpha}\right)e^{i\mathbf{k}\cdot\mathbf{r}_i} \\
&= -\frac{1}{2}i\mathbf{k}\cdot\sum_{i}^{N_m}\sum_{\alpha}^{N_s}\sum_{j}^{N_m}\sum_{\beta}^{N_s}\mathbf{r}_{ij}\mathbf{F}_{i\alpha j\beta}^{(2)}\cdot\mathbf{v}_{i\alpha}+O\left(k^2\right).
\end{aligned}
\tag{7.95}
$$

Note that $\mathbf{r}_{ij}=\mathbf{r}_j-\mathbf{r}_i$ is the separation of the centres of mass of molecules i and j. For systems with only two body forces, we have

$$
\begin{aligned}
\frac{\partial}{\partial t}\widetilde{\rho u}\left(\mathbf{k},t\right) &= i\mathbf{k}\cdot\left(\sum_{i=1}^{N_m}\mathbf{v}_i u_i-\frac{1}{2}\sum_{i}^{N_m}\sum_{\alpha}^{N_s}\sum_{j}^{N_m}\sum_{\beta}^{N_s}\mathbf{r}_{ij}\mathbf{F}_{i\alpha j\beta}^{(2)}\cdot\mathbf{v}_{i\alpha}\right)+O\left(k^2\right) \\
&= i\mathbf{k}\cdot\widetilde{\mathbf{J}}_q\left(0,t\right)+O\left(k^2\right),
\end{aligned}
\tag{7.96}
$$

so that the heat flux vector in the molecular representation in this case is given by

$$
V\mathbf{J}_q\left(t\right)=\sum_{i=1}^{N_m}\mathbf{v}_i u_i-\frac{1}{2}\sum_{i}^{N_m}\sum_{\alpha}^{N_s}\sum_{j}^{N_m}\sum_{\beta}^{N_s}\mathbf{r}_{ij}\mathbf{F}_{i\alpha j\beta}^{(2)}\cdot\mathbf{v}_{i\alpha}. \tag{7.97}
$$

When three body forces are considered, we have

$$\begin{aligned}
&\sum_{i=1}^{N_m}\sum_{\alpha=1}^{N_s}\dot{\phi}_{i\alpha}^{(3)}e^{i\mathbf{k}\cdot\mathbf{r}_{i\alpha}}\\
&=\frac{1}{3}\sum_{i}^{N_m}\sum_{\alpha}^{N_s}\sum_{j}^{N_m}\sum_{\beta}^{N_s}\sum_{k}^{N_m}\sum_{\gamma}^{N_s}\dot{\phi}_{i\alpha j\beta k\gamma}\left(\mathbf{r}_{i\alpha},\mathbf{r}_{j\beta},\mathbf{r}_{k\gamma}\right)e^{i\mathbf{k}\cdot\mathbf{r}_i}\\
&=\frac{1}{3}\sum_{i}^{N_m}\sum_{\alpha}^{N_s}\sum_{j}^{N_m}\sum_{\beta}^{N_s}\sum_{k}^{N_m}\sum_{\gamma}^{N_s}\left(\mathbf{v}_{i\alpha}\cdot\frac{\partial\phi_{i\alpha j\beta k\gamma}}{\partial\mathbf{r}_{i\alpha}}+\mathbf{v}_{j\beta}\cdot\frac{\partial\phi_{i\alpha j\beta k\gamma}}{\partial\mathbf{r}_{j\beta}}+\mathbf{v}_{k\gamma}\cdot\frac{\partial\phi_{i\alpha j\beta k\gamma}}{\partial\mathbf{r}_{k\gamma}}\right)e^{i\mathbf{k}\cdot\mathbf{r}_i}\\
&=-\sum_{i}^{N_m}\sum_{\alpha}^{N_s}\mathbf{v}_{i\alpha}\cdot\mathbf{F}_{i\alpha}^{(3)}-i\mathbf{k}\cdot\frac{1}{3}\sum_{i}^{N_m}\sum_{\alpha}^{N_s}\sum_{j}^{N_m}\sum_{\beta}^{N_s}\sum_{k}^{N_m}\sum_{\gamma}^{N_s}\\
&\quad\times\left(\mathbf{r}_i\mathbf{F}_{i\alpha j\beta k\gamma}^{(3)}\cdot\mathbf{v}_{i\alpha}+\mathbf{r}_i\mathbf{F}_{j\beta k\gamma i\alpha}^{(3)}\cdot\mathbf{v}_{j\beta}+\mathbf{r}_i\mathbf{F}_{k\gamma i\alpha j\beta}^{(3)}\cdot\mathbf{v}_{k\gamma}\right)+O\left(k^2\right)\\
&=-\sum_{i}^{N_m}\sum_{\alpha}^{N_s}\mathbf{v}_{i\alpha}\cdot\mathbf{F}_{i\alpha}^{(3)}-i\mathbf{k}\cdot\frac{1}{3}\sum_{i}^{N_m}\sum_{\alpha}^{N_s}\sum_{j}^{N_m}\sum_{\beta}^{N_s}\sum_{k}^{N_m}\sum_{\gamma}^{N_s}\\
&\quad\times\left(\mathbf{r}_i\mathbf{F}_{i\alpha j\beta k\gamma}^{(3)}\cdot\mathbf{v}_{i\alpha}+\mathbf{r}_j\mathbf{F}_{i\alpha j\beta k\gamma}^{(3)}\cdot\mathbf{v}_{i\alpha}+\mathbf{r}_k\mathbf{F}_{i\alpha j\beta k\gamma}^{(3)}\cdot\mathbf{v}_{i\alpha}\right)+O\left(k^2\right).
\end{aligned}$$

Now we combine terms to find

$$\begin{aligned}
&\sum_{i=1}^{N_m}\sum_{\alpha=1}^{N_s}\left(\mathbf{F}_{i\alpha}^{(3)}\cdot\mathbf{v}_{i\alpha}+\dot{\phi}_{i\alpha}^{(3)}\right)e^{i\mathbf{k}\cdot\mathbf{r}_i}\\
&=-i\mathbf{k}\cdot\frac{1}{3}\sum_{i}^{N_m}\sum_{\alpha}^{N_s}\sum_{j}^{N_m}\sum_{\beta}^{N_s}\sum_{k}^{N_m}\sum_{\gamma}^{N_s}\left(\mathbf{r}_{ij}\mathbf{F}_{i\alpha j\beta k\gamma}^{(3)}\cdot\mathbf{v}_{i\alpha}+\mathbf{r}_{ik}\mathbf{F}_{i\alpha j\beta k\gamma}^{(3)}\cdot\mathbf{v}_{i\alpha}\right)+O\left(k^2\right),
\end{aligned}\tag{7.98}$$

so that the full expression for the molecular representation of the heat flux vector including three body force terms is

$$\begin{aligned}
V\mathbf{J}_q(t)=&\sum_{i=1}^{N_m}\mathbf{v}_i u_i-\frac{1}{2}\sum_{i}^{N_m}\sum_{\alpha}^{N_s}\sum_{j}^{N_m}\sum_{\beta}^{N_s}\mathbf{r}_{ij}\mathbf{F}_{i\alpha j\beta}^{(2)}\cdot\mathbf{v}_{i\alpha}\\
&-\frac{1}{3}\sum_{i}^{N_m}\sum_{\alpha}^{N_s}\sum_{j}^{N_m}\sum_{\beta}^{N_s}\sum_{k}^{N_m}\sum_{\gamma}^{N_s}\left(\mathbf{r}_{ij}\mathbf{F}_{i\alpha j\beta k\gamma}^{(3)}\cdot\mathbf{v}_{i\alpha}+\mathbf{r}_{ik}\mathbf{F}_{i\alpha j\beta k\gamma}^{(3)}\cdot\mathbf{v}_{i\alpha}\right)+O\left(k^2\right).
\end{aligned}\tag{7.99}$$

This is easily generalised to include terms due to four body forces. Three body forces are important in materials where directional interatomic interactions are present, e.g. liquid silicon. Three and four body force terms are also essential for calculations of the heat flux vector when bond angle bending potentials and dihedral potentials are present. In fact some of the earliest computations of the thermal conductivity for molecular fluids were for the Ryckaert-Bellemans united atom model of butane [264], which includes a dihedral potential. However, when the three and four body forces are purely intramolecular (as they are assumed to be in most models of molecular liquids), the expression for the heat flux vector simplifies significantly. In this case, we have $\mathbf{r}_{ij} = 0$ for all intramolecular ($i = j$) interactions, so that their contributions to the heat flux vector are equal to zero, and the only terms that remain are those due to intermolecular two body interactions,

$$V\mathbf{J}_q(t) = \sum_{i=1}^{N_m} \mathbf{v}_i u_i - \frac{1}{2} \sum_{i}^{N_m} \sum_{\alpha}^{N_s} \sum_{j \neq i}^{N_m} \sum_{\beta}^{N_s} \mathbf{r}_{ij} \mathbf{F}^{(2)}_{i\alpha j\beta} \cdot \mathbf{v}_{i\alpha}. \tag{7.100}$$

It is convenient to write this as

$$V\mathbf{J}_q = \sum_{i=1}^{N_m} \sum_{\alpha=1}^{N_s} \mathbf{v}_{i\alpha} \cdot \mathbf{S}_{i\alpha}. \tag{7.101}$$

We want to derive the equations of motion that generate heat flow in a homogeneous molecular fluid, consistent with the requirement that the rate of change of the internal energy satisfies

$$\frac{d}{dt} U(t) = V\mathbf{J}_q \cdot \mathbf{F}_q, \tag{7.102}$$

where $\mathbf{J}_q$ is the heat flux vector in the molecular representation. Let us assume that the equations of motion can be written in the form

$$\begin{aligned} \dot{\mathbf{r}}_{i\alpha} &= \frac{\mathbf{p}_{i\alpha}}{m_{i\alpha}} \\ \dot{\mathbf{p}}_{i\alpha} &= \mathbf{F}_{i\alpha} + \mathbf{D}_{i\alpha} \cdot \mathbf{F}_q. \end{aligned} \tag{7.103}$$

There is no term in the position equation of motion that couples to the external field because we do not expect the heat flow algorithm to generate a streaming velocity, in contrast to the SLLOD equations of motion. The rate of change of the internal energy is then

$$\begin{aligned} \frac{d}{dt} U(t) &= \frac{d}{dt} \sum_{i=1}^{N_m} \sum_{\alpha=1}^{N_s} \left(\frac{\mathbf{p}_{i\alpha}^2}{2m_{i\alpha}} + \phi_{i\alpha} \right) \\ &= \sum_{i=1}^{N_m} \sum_{\alpha=1}^{N_s} \left(\frac{\mathbf{p}_{i\alpha}}{m_{i\alpha}} \cdot \dot{\mathbf{p}}_{i\alpha} + \dot{\phi}_{i\alpha} \right). \end{aligned} \tag{7.104}$$

In the absence of the external field, the internal energy is conserved, so the only term that remains on the right-hand side is the one involving the external field. This enters when we substitute the momentum equation of motion for the derivative of the momentum. Then we find

$$\frac{d}{dt} U(t) = \sum_{i=1}^{N_m} \sum_{\alpha=1}^{N_s} \mathbf{v}_{i\alpha} \cdot \mathbf{D}_{i\alpha} \cdot \mathbf{F}_q. \tag{7.105}$$

This must be equal to the desired dissipation, $V\mathbf{J}_q \cdot \mathbf{F}_q$, giving

$$\sum_{i=1}^{N_m}\sum_{\alpha=1}^{N_s} \mathbf{v}_{i\alpha} \cdot \mathbf{D}_{i\alpha} \cdot \mathbf{F}_q = \sum_{i=1}^{N_m}\sum_{\alpha=1}^{N_s} \mathbf{v}_{i\alpha} \cdot \mathbf{S}_{i\alpha} \cdot \mathbf{F}_q. \tag{7.106}$$

Since we want homogeneous equations of motion, i.e. they must be the same for every particle in the system, we could conclude that the heat field coupling coefficient is given by $\mathbf{D}_{i\alpha} = \mathbf{S}_{i\alpha}$. However, with this coupling, the equations of motion do not conserve momentum. Momentum conservation is achieved by insisting that

$$\sum_{i=1}^{N_m}\sum_{\alpha=1}^{N_s} \dot{\mathbf{p}}_{i\alpha} = 0, \tag{7.107}$$

which requires us to redefine the coupling coefficient as

$$\mathbf{D}_{i\alpha} = \mathbf{S}_{i\alpha} - \frac{1}{N_m N_s}\sum_{i=1}^{N_m}\sum_{\alpha=1}^{N_s} \mathbf{S}_{i\alpha}. \tag{7.108}$$

This algorithm was originally derived by Daivis and Evans [262] for a united atom model of butane in which each atom in the system has the same reduced mass, explicitly taking $m_{i\alpha} = 1$.[3]

For more complicated molecular models, such as those having rigid body segments combined with flexible segments, or massless interaction sites (e.g. the TIP4P model of water and the anisotropic united atom models of alkanes [263, 265]), an approach based on the CFM formalism for molecules with constraints can be used to derive general expressions for the heat flux vector and the appropriate heat field coupling for NEMD algorithms [251].

[3] Note that the equations of motion in that paper are printed incorrectly, although the calculations were done using the correct equations.

8 Homogeneous Flows for Molecular Fluids

8.1 Explicit and Coarse-grained Molecular Models

The first NEMD simulations of molecular fluids were mostly performed with coarse-grained or simplified molecular models because simulations of more realistic, explicit models had high computational demands compared to the computing power then available. For simpler molecules, such as diatomic nitrogen and chlorine, nonequilibrium simulations of explicit molecular models were performed very early, but for more complicated molecules such as the n-alkanes, many of the earliest NEMD simulations were done with models that employed approximations such as the united atom approximation, where CH_2 and CH_3 groups were modelled as identical interaction sites with simplified intermolecular potentials. Furthermore, it was usually assumed that all bond lengths and bond angles were rigidly constrained and that dihedral angle potentials followed simplified functional forms. With the rapid growth in computational power, many of these assumptions have become unnecessary, but there are often good physical reasons for assuming that bond lengths should be constrained in classical molecular dynamics simulations. Therefore, constrained molecular models remain an important and useful part of the simulation toolkit, and we will devote some time to them in this chapter.

Many early attempts at coarse-graining were ad hoc and expedient, with little theoretical justification. This has recently changed quite dramatically, and the statistical-mechanical theory of coarse-graining and multiscale simulation is now an active field of research [266–268]. Here, we will only very briefly introduce a few simple coarse-grained models that are of interest mainly because they appear so prominently in the literature and they are so useful for simulating the general qualitative features of the dynamics of molecular fluids with minimal computational demands.

8.1.1 All-atom and United Atom Molecular Models

Explicit, all-atom molecular models and potential energy functions (force fields) are now available from many different sources, such as the enormous literature on molecular dynamics packages for biomolecular simulations and computational materials science. Generally speaking, most users of these potential energy functions encounter them in packages, which are already well-documented and do not need to be discussed here. There are also many discussions of the general features of interaction potentials for explicit all-atom modelling in standard texts [5, 7, 8, 62]. Here, our focus is on NEMD methodology rather than potential energy functions, so we will only cover a few of the essential features that are necessary for an understanding of the examples discussed later in this chapter.

Basic but fairly accurate explicit models of well-known simple molecules are very useful for investigations of less well understood properties and phenomena, because many of their properties can be benchmarked against extensive literature data. This is particularly important when new code is being developed. Typical examples in this class include the diatomic molecules nitrogen and chlorine, the linear triatomic molecule carbon disulphide and the planar molecules benzene and water. In all of these cases, simple molecular models based on Lennard-Jones site interactions, in some cases including partial charges, are available. In the case of water, many different models are available, including variants of the SPC, TIP3P and TIP4P families of models. The SPC/E model is probably the most commonly encountered one, giving reasonable values for many properties of interest. Obviously, if precise agreement with experimental data and quantitative prediction is the desired outcome, more sophisticated and accurate molecular models are required, but for testing new simulation methodology or studying less well-known features, it is important to begin with molecular models that are thoroughly understood, with well-documented and accepted values for commonly studied properties, which nevertheless may not be in precise agreement with experiment. For example, the viscosity of water at 25°C computed using the widely used SPC/E model is now well-known to be around 20% lower than the experimental value [269], but the commonly computed quantities of the SPC/E model are so comprehensively and accurately known, that it remains a favoured model for testing new methodology or computing less commonly studied observables.

Rigid-body molecules can be simulated with an algorithm based on the quaternion formulation of the rigid body equations of motion [5, 270]. This is a very efficient method, but it is not easily applied to molecules with partial or total flexibility. A mixture of rigid body and flexible molecules could not be simulated with one simple set of equations of motion using this technique, because different equations of motion would apply to the rigid molecules and the flexible molecules. We will concentrate our attention on constrained dynamics methods that explicitly account for constraints whenever they are present and can accommodate fully rigid or partially flexible molecules with the same equations of motion.

8.1.2 Freely Jointed Chain and FENE Chain Models for Polymeric Fluids

NEMD simulations have been used to compute the rheological properties of model polymer solutions and melts using a variety of molecular models, including bead-rod, bead-spring and all-atom explicit models. One of the important discoveries of polymer physics is that many features of polymer melts and solutions obey a principle of universality. This principle states that universal properties such as the rms end to end distance, the radius of gyration, the self-diffusion coefficient and the intrinsic viscosity of polymeric fluids can be scaled in such a way that all polymers behave in the same way, provided that suitable prefactors that account for individual differences are chosen. The essential features that determine polymer behaviour are therefore rather generic. They include the noncrossability of the chains, the connectivity and the general aspects of their interactions with solvent and other polymer beads. All of these generic features are represented in the molecular dynamics versions of the bead-rod and bead-spring models, so they should exhibit all of the characteristics that polymeric liquids are assumed to possess. As we shall see later, this has been amply demonstrated. This means that it is unnecessary to use all-atom explicit models to study the universal features of polymeric liquids. However, if detailed agreement with experimental data is required, or if the values of the prefactors appearing in scaling relationships are needed, explicit models must be used.

One of the earliest NEMD simulations of model polymeric fluids was reported by Hess [271] who modelled the bonds between beads by simply increasing the LJ attractions between bonded beads. This method suffered from uncontrolled chain scission under shear, which although realistic in certain respects, is inconvenient in other ways.

Bead-spring models represent a polymer chain as a set of interacting spheres joined by bond stretching potential energy functions. There are no bond bending or dihedral potentials, so the small scale chain structure is not explicitly represented. The bond stretching could be represented by Hookean springs, similar to those used for simulations of small molecules and also used in the Rouse model of polymer dynamics. However, it is well-known that the Hookean spring model fails to correctly describe the elongational rheology of polymeric fluids because of the ability of the Hookean spring to extend infinitely as the elongation strain rate is increased [29]. The finitely extensible nonlinear elastic (FENE) chain model, attributed to Warner [272] and used extensively for molecular dynamics simulations since the seminal work of Kremer and Grest [273, 274], does not possess this disadvantage because it includes a spring force that rapidly hardens as the spring stretches, thus introducing finite extensibility. In the version of the FENE model introduced by Grest and Kremer for molecular dynamics simulations, all beads interact with the truncated and shifted form of the Lennard-Jones interaction known as the WCA potential,

$$\phi_{ij}^{WCA} = \begin{cases} 4\epsilon\left[\left(\frac{\sigma}{r_{ij}}\right)^{12} - \left(\frac{\sigma}{r_{ij}}\right)^{6}\right] + \epsilon & r_{ij} < 2^{1/6}\sigma \\ 0 & r_{ij} \geqslant 2^{1/6}\sigma \end{cases} \tag{8.1}$$

and bonded beads interact with the nonlinear spring potential

$$\phi_{ij}^{FENE} = \begin{cases} -0.5kR_0^2 \ln\left[1 - \left(\frac{r_{ij}}{R_0}\right)^2\right] & r_{ij} < R_0 \\ \infty & r_{ij} \geqslant R_0 \end{cases}. \tag{8.2}$$

The combination of these two interactions results in strong repulsive interactions between any pairs of beads that get close to overlapping, preventing chain crossings, a strong restoring force when bonds are stretched, and most importantly, finite extensibility. The nonlinear spring force here is analogous to the harmonic spring force in the Rouse model that represents the entropic resistance of a polymer subchain to stretching. In the original implementation of this model, a Langevin thermostat was proposed for temperature control. The thermal noise of the Langevin thermostat can be thought of as a consequence of the coarse-graining of the explicit polymer model in the sense of the generalised Langevin equation [275], but this is not necessarily required and in fact identical results are found when conventional molecular dynamics thermostats such as the velocity rescaling thermostat are used with this model at equilibrium [274].

The bead-rod molecular model, also known as the freely-jointed chain model, represents a polymer chain by a set of interacting spheres joined by bonds of fixed length with no explicit bond bending or dihedral interactions. This model obviously satisfies the finite extensibility requirement, and is very simple to implement with the same constraint algorithm that is used for constraints in smaller rigid body molecules. The equilibrium properties of a version of this model in which the bond length is exactly the same as the Lennard-Jones diameter σ (the Lennard-Jones tangent chain model) have been studied by Johnson and Gubbins [276]. In this model, interaction sites separated by one (fixed) bond do not interact at all because they are already constrained, but all others interact with a standard Lennard-Jones interaction. The implementation of this model requires a method for calculating the constraint forces, which we will discuss next.

8.1.3 Constraint Algorithms

Molecular dynamics algorithms for molecules with internal constraints have been formulated in many different ways. One approach is to eliminate the constrained degrees of freedom and use generalised coordinates and Lagrange's equations of motion [277]. In this approach, the generalised coordinates would be different for each different molecular architecture, requiring considerable theoretical work and recoding before simulating each different type of molecule. Another approach is to use Cartesian coordinates, explicitly calculate the constraint forces required to satisfy the constraints, and solve the equations of motion with the constraint forces explicitly computed at each time step (this is sometimes called the matrix method). This method is in principle exact, but the discretisation error of converting the differential equations to finite difference algorithms, combined with finite numerical precision of the calculations, means that the equations cannot in practice be solved with infinite precision. This leads to errors in the constraint forces which can gradually accumulate, ending in catastrophic failure of the algorithm. This problem can be avoided by solving the equations of motion iteratively in such a way that the constraints are guaranteed to be satisfied to a predetermined numerical precision at each time step. This is the main idea behind the most popular family of constraint solvers, based on the SHAKE algorithm, originally proposed by Ryckaert, Ciccotti and Berendsen [278] for use with Verlet type integrators and modified for use with the velocity Verlet type integrators by Andersen [279]. One disadvantage of this type of constraint algorithm is that it must be modified or rederived whenever the equations of motion change. Algorithms for constant pressure simulations [280], and NEMD simulations of shear flow with the SLLOD equations of motion for constrained molecules require additional (ROLL) steps in the implementation of the

SHAKE/RATTLE algorithm that make it even more complex. This algorithm has been successfully used in simulations of shear flow for the SPC/E model of water [281].

Here, we will describe an alternative approach that retains the advantages of explicit computation of the constraint forces, and deals with the constraint drift problem in a simple way. This approach, originally proposed by Edberg Evans and Morriss [282], is implemented directly in the equations of motion, rather than a modified form of the finite difference algorithm, so it is therefore independent of the integrator chosen. It can be used equally easily with predictor-corrector, velocity Verlet or Runge-Kutta integrators, and it can be used in the same form for both equilibrium and nonequilibrium simulations. The procedure for correcting the errors in the constraints due to numerical drift proposed by Edberg was somewhat cumbersome to apply but was later improved by Baranyai and Evans, who applied it to NEMD simulations of the shear flow of benzene and naphthalene [173, 283].

We begin with an example, deriving the equations of motion for a simple homonuclear diatomic molecule using Gauss's principle of least constraint. Gauss's principle states that difference between the actual motion of a constrained system and the motion of the corresponding unconstrained system is minimised in a least squares sense. Mathematically, this is expressed by the condition that we minimise the curvature subject to the fixed bond length constraint, i.e.

$$\frac{\partial}{\partial \ddot{\mathbf{r}}_i}(C - \lambda G) = 0, \tag{8.3}$$

where C is the curvature, defined as

$$C = \frac{1}{2}\sum_{i=1}^{2} m_i \left(\ddot{\mathbf{r}}_i - \frac{\mathbf{F}_i}{m_i}\right)^2 \tag{8.4}$$

and λ is a Lagrangian (or in this case, Gaussian) multiplier which will be determined later. The most direct way of expressing the bond length constraint is to equate the instantaneous distance between atoms 1 and 2 to a constant

$$g(\mathbf{r}_i, \dot{\mathbf{r}}_i, t) = \mathbf{r}_{12}^2 - d^2 = 0. \tag{8.5}$$

However, Gauss's principle is an equation in acceleration space, so the constraint equation must be expressed in terms of the accelerations. This is easily achieved by differentiating twice with respect to time

$$G(\ddot{\mathbf{r}}_i) = \mathbf{r}_{12} \cdot \ddot{\mathbf{r}}_{12} + \dot{\mathbf{r}}_{12}^2 = 0. \tag{8.6}$$

Note that the constant value of the bond length has now disappeared from the constraint equation. This is consistent with the classification of the Gaussian constraint mechanism as a differential control mechanism, in which the rate of change of the constrained distance is set to zero, and the set value of the constrained distance only appears as an initial condition. Substituting for the constraint equation and minimising the curvature subject to the constraint, we find the equation of motion for each of the particles from

$$\frac{\partial}{\partial \ddot{\mathbf{r}}_j}\left(\frac{1}{2}\sum_{i=1}^{2} m_i\left(\ddot{\mathbf{r}}_i - \frac{\mathbf{F}_i}{m_i}\right)^2 - \lambda\left(\mathbf{r}_{12}\cdot\ddot{\mathbf{r}}_{12} + \dot{\mathbf{r}}_{12}^2\right)\right) = 0. \tag{8.7}$$

Using the fact that the interatomic forces, the positions and the velocities are independent of the accelerations, we find the equations of motion

$$\begin{aligned} m_1\ddot{\mathbf{r}}_1 &= \mathbf{F}_1 - \lambda\mathbf{r}_{12} \\ m_2\ddot{\mathbf{r}}_2 &= \mathbf{F}_2 + \lambda\mathbf{r}_{12}. \end{aligned} \tag{8.8}$$

The sign difference between the two equations of motion arises from the fact that

$$\frac{\partial \ddot{\mathbf{r}}_{12}}{\partial \ddot{\mathbf{r}}_1} = \frac{\partial}{\partial \ddot{\mathbf{r}}_1}\left(\ddot{\mathbf{r}}_2 - \ddot{\mathbf{r}}_1\right) = -\mathbf{1} = -\frac{\partial}{\partial \ddot{\mathbf{r}}_2}\left(\ddot{\mathbf{r}}_2 - \ddot{\mathbf{r}}_1\right). \tag{8.9}$$

Physically, this means that the constraint force on particle 1 has the same magnitude but opposite direction to the constraint force on particle 2. Therefore, as expected, the sum of the constraint forces on the whole molecule is zero and the motion of the centre of mass of the molecule is unaffected by the constraint forces. By subtracting the equation of motion of atom one from that for atom two, we obtain an equation of motion for the bond vector,

$$\ddot{\mathbf{r}}_{12} = \frac{\mathbf{F}_2 + \lambda\mathbf{r}_{12}}{m_2} - \frac{\mathbf{F}_1 - \lambda\mathbf{r}_{12}}{m_1}. \tag{8.10}$$

The constraint multiplier is determined by substituting this into the acceleration-dependent constraint equation, giving

$$\lambda = -\frac{\mathbf{r}_{12}\cdot\left(\frac{\mathbf{F}_2}{m_2} - \frac{\mathbf{F}_1}{m_1}\right) + \dot{\mathbf{r}}_{12}^2}{\mathbf{r}_{12}^2\left(\frac{1}{m_2} + \frac{1}{m_1}\right)}, \tag{8.11}$$

which can be evaluated from the positions and momenta at each timestep. This method of evaluating the constraint multiplier is particularly simple because, even for a larger set of constraints, it results in a set of linear simultaneous equations. It differs from the matrix method of Ryckaert, Ciccotti and Berendsen [278], in which the position-dependent constraint equation is substituted into a finite difference approximation to the equations of motion, yielding a quadratic equation, which must be solved iteratively.

This method can easily be generalised to larger molecules with arbitrarily complex combinations of bond length and bond angle constraints, both of which can be expressed as distance constraints. For a system with many constraints we write the constraint equations as

$$\left(\mathbf{r}_\beta - \mathbf{r}_\alpha\right)^2 - d_{\alpha\beta}^2 = 0, \tag{8.12}$$

where the $d_{\alpha\beta}$ are the constrained distances. Introducing a new index that runs over all constraints in a molecule, this can be written as

$$\mathbf{R}_\delta^2 - d_\delta^2 = 0, \tag{8.13}$$

where $\mathbf{R}_\delta$ is the bond vector for constraint δ. The acceleration-dependent constraint equation obtained from this is

$$G_\delta = \mathbf{R}_\delta \cdot \ddot{\mathbf{R}}_\delta + \dot{\mathbf{R}}_\delta^2 = 0. \tag{8.14}$$

Application of Gauss's principle results in the equations of motion

$$m_\alpha \ddot{\mathbf{r}}_\alpha = \mathbf{F}_\alpha^\phi + \sum_{\delta=1}^{n_c} M_{\alpha\delta}(\lambda \mathbf{R})_\delta, \tag{8.15}$$

where n_c is the number of intramolecular constraints to be applied, and M is a selector matrix consisting only of ones, minus ones and zeroes. For a rigid, 3 particle model of water such as the SPC/E model which has two bond length constraints and one bond angle constraint, the constraint matrix takes the form

$$M = \begin{pmatrix} -1 & 0 & -1 \\ 1 & -1 & 0 \\ 0 & 1 & 1 \end{pmatrix}, \tag{8.16}$$

where each column corresponds to one constraint and each row to one atom. The first two columns represent the bond length constraints and we see that the first bond length constraint involves atoms 1 and 2 (with opposite signs) and the second bond length constraint involves atoms 2 and 3. There is only one bond angle constraint (column 3), which is represented as a distance constraint between atoms 1 and 3. The ordering of the constraints and atoms is, of course completely arbitrary but it is convenient to put them into some kind of logical sequence.

We now need to construct an equation that will allow the constraint multipliers to be evaluated. This is done by taking the difference between the equations of motion for the two atoms involved in each constraint, so that we obtain equations for each of the bond acceleration vectors,

$$\ddot{\mathbf{r}}_\beta - \ddot{\mathbf{r}}_\alpha = \frac{\mathbf{F}_\beta^\phi}{m_\beta} - \frac{\mathbf{F}_\alpha^\phi}{m_\alpha} + \sum_{\delta=1}^{n_c} \left(\frac{M_{\beta\delta}}{m_\beta} - \frac{M_{\alpha\delta}}{m_\alpha} \right) (\lambda \mathbf{R})_\delta. \tag{8.17}$$

There is one of these equations for each constraint, so they can be indexed with a single index that runs over all constraints, and we can write

$$\ddot{\mathbf{R}}_\gamma = \mathbf{H}_\gamma + \sum_{\delta=1}^{n_c} N_{\gamma\delta}(\lambda \mathbf{R})_\delta, \tag{8.18}$$

where

$$\mathbf{H}_\gamma = \frac{\mathbf{F}_\beta^\phi}{m_\beta} - \frac{\mathbf{F}_\alpha^\phi}{m_\alpha} \tag{8.19}$$

and

$$N_{\gamma\delta} = \frac{M_{\beta\delta}}{m_\beta} - \frac{M_{\alpha\delta}}{m_\alpha}. \tag{8.20}$$

Inserting this into the constraint equation, we have

$$\sum_{\delta=1}^{n_c} \left(\mathbf{R}_\gamma \cdot \mathbf{R}_\delta N_{\gamma\delta}\right) \lambda_\delta = -\left(\mathbf{R}_\gamma \cdot \mathbf{H}_\gamma + \dot{\mathbf{R}}_\gamma^2\right), \tag{8.21}$$

which is a set of linear algebraic equations that can be solved by standard numerical methods. For large constraint matrices, the computational cost of solving the set of linear equations by direct methods can become prohibitively expensive. However, in such cases, the N matrix is always sparse and sparse matrix solvers substantially reduce the computational cost, making this method competitive with any other constraint algorithm.

The only remaining drawback of this direct method of applying the constraint forces is that although it is in principle exact, the constraints drift due to the imprecise numerical solution of the differential equations. This problem was solved in a convenient computational form that is again independent of the precise algorithm chosen for the solution of the differential equations, by Baranyai and Evans [173, 283]. The small numerical error that occurs at each timestep can be prevented from accumulating by applying proportional feedback to oppose the numerical drift. The final form of the equations of motion for atom α of molecule i including the constraint forces and proportional feedback is

$$\begin{aligned}
\dot{\mathbf{r}}_{i\alpha} &= \frac{\mathbf{p}_{i\alpha}}{m_{i\alpha}} + C\sum_{\delta=1}^{n_c} M_{\alpha\delta}\left(\mathbf{R}_\delta^2 - d_\delta^2\right)\hat{\mathbf{R}}_\delta \\
\dot{\mathbf{p}}_{i\alpha} &= \mathbf{F}_{i\alpha}^{\phi} + \sum_{\delta=1}^{n_c} M_{\alpha\delta}\lambda_\delta \mathbf{R}_\delta + D\sum_{\delta=1}^{n_c} M_{\alpha\delta}\left(\mathbf{R}_\delta \cdot \dot{\mathbf{R}}_\delta\right)\hat{\mathbf{R}}_\delta .
\end{aligned} \tag{8.22}$$

In the first equation, the proportional feedback is proportional to the bond length error and it is applied in the direction of a unit vector along the bond, $\hat{\mathbf{R}}_\delta$. The constraint selector matrix $M_{\alpha\delta}$, which is already available, is used to select the atoms to which the proportional feedback must be applied for a given constraint. In the second equation, we see the constraint force itself, followed by the proportional feedback applied to correct for bond velocity errors. The component in the direction of the bond of the relative velocity of two atoms involved in a constraint should be zero, but if it is not, the proportional feedback acts to change the velocity appropriately. Note that the direction of each feedback term is given by a unit vector in the direction of the bond. This makes the magnitude of the feedback term dependent only on the coefficients C and D and the error in the bond length or velocity, but not on the magnitude of the bond length itself, so that the numerical errors of all constrained bonds in the system, regardless of their length, can be controlled with equal values of the feedback strengths.

This method cannot be used to model rigid body molecules for which the number of degrees of freedom of the molecule is not equal to $3n - n_c$. For a rigid diatomic molecule, there are $6 - 1 = 5$ degrees of freedom, 3 translational and 2 rotational. There are only 2 rotational degrees of freedom because the moment of inertia about the axis through the centre of mass and in the direction of the bond is zero. A rigid linear triatomic molecule, such as CS_2 also has 5 degrees of freedom, but in this case, the number of degrees of freedom (incorrectly) calculated from the number of atoms and the number of distance constraints is $9 - 3 = 6$ if we count two bond length and one bond angle constraint. Here, the number of degrees of freedom should instead be calculated from $9 - 4 = 5$ because atoms 1 and 2 have one constraint, but all 3 degrees of freedom of atom 3 are lost because its position is completely determined by the positions of the other two. This is not the case for a planar triatomic molecule, because atom 3 has an additional rotational degree of freedom not specified by the positions of atoms 1 and 2. The problem arises again for a planar 4-atom molecule. For example, a rigid version of the united atom model for butane fixed in the planar all-trans conformation, should have 6 degrees of freedom, but from the number of unconstrained degrees of freedom minus the number of bond length and bond angle constraints, we would find $12 - 3 - 2 = 7$. The correct number of degrees of freedom for the molecule is obtained from the first three atoms, and the fourth should not add any. In this sense, we should calculate the number of degrees of freedom from $12 - 3 - 3 = 6$, because between the first three atoms there are three constraints (two bond length and one bond angle) and the fourth atom's three degrees of freedom should be subtracted completely.

These considerations led Ciccotti, Ferrario and Ryckaert (CFR) [284] to propose a method in which the molecule is divided into primary (or basic) particles and secondary particles. The position and orientation of the molecule is completely specified by the degrees of freedom of the submolecule consisting of the primary atoms. The equations of motion are solved for the primary atoms, using the forces on all of the atoms, and the positions of the secondary atoms are found by using the distance constraints and the positions of the primary atoms. This method is usually applied with the SHAKE or RATTLE constraint algorithms [5] but it is also possible to use the CFR method with matrix methods that exactly solve the constraint equations.

In the CFR method, there are three possible types of rigid body polyatomic molecule in three dimensions; those with two basic atoms (linear molecules), those with three noncolinear basic atoms (planar molecules) and those with four noncolinear and noncoplanar basic atoms. In this formulation, if there are n_b basic particles and n_s secondary particles, the basic particles are held in a rigid structure by $l_b = n_b(n_b - 1)/2$ constraints and the secondary particles' positions are defined by $l_s = 3n_s$ constraints.

The positions of the secondary particles (denoted by primed position vectors, $\mathbf{r}'_\gamma$) can be defined in terms of the position vectors of the basic particles as

$$\sum_{\alpha=1}^{n_b} C_{\gamma\alpha}\mathbf{r}_\alpha - \mathbf{r}'_\gamma = 0, \tag{8.23}$$

where $\gamma = 1, \ldots, n_s$.

For each value of γ, the sum of the $C_{\gamma\alpha}$ factors over all values of the basic particle index can be written as

$$\sum_{\alpha=1}^{n_b} C_{\gamma\alpha} = 1. \tag{8.24}$$

The equilibrium equations of motion for the momenta, $\dot{\mathbf{p}}_\alpha = m_\alpha \ddot{\mathbf{r}}$ of the basic particles can be written as

$$m_\alpha \ddot{\mathbf{r}}_\alpha = \mathbf{F}_\alpha^\phi + \sum_{\delta=1}^{n_c} M_{\alpha\delta}(\lambda \mathbf{R})_\delta - \sum_{\gamma=1}^{n_s} C_{\gamma\alpha}\boldsymbol{\mu}_\gamma \tag{8.25}$$

and the equations of motion for the momenta of the secondary particles (which do not need to be solved but are given for completeness) are

$$m_\gamma \ddot{\mathbf{r}}'_\gamma = \mathbf{F}'^\phi_\gamma + \boldsymbol{\mu}_\gamma \qquad \gamma = 1, \ldots, n_s. \tag{8.26}$$

Using Equations (8.23) and (8.26), we can obtain an equation for the constraint force on a secondary particle

$$\boldsymbol{\mu}_\gamma = m_\gamma \sum_{\alpha=1}^{n_b} C_{\gamma\alpha}\ddot{\mathbf{r}}_\alpha - \mathbf{F}'^\phi_\gamma. \tag{8.27}$$

Substituting the equation of motion for the primary particle and simplifying the resulting expression, we find

$$\boldsymbol{\mu}_\gamma = \sum_{\epsilon=1}^{n_s} \left(A^{-1}\right)_{\gamma\epsilon} \left[\mathbf{T}_\epsilon + \sum_{\alpha=1}^{n_b} \frac{C_{\epsilon\alpha}}{m_\alpha} \sum_{\gamma=1}^{n_c} M_{\alpha\gamma}(\lambda \mathbf{R})_\gamma \right], \tag{8.28}$$

where

$$\mathbf{T}_\epsilon = -\frac{\mathbf{F}'^\phi_\epsilon}{m_\epsilon} + \sum_{\alpha=1}^{n_b} \frac{C_{\epsilon\alpha}}{m_\alpha}\mathbf{F}_\alpha^\phi \tag{8.29}$$

and

$$A_{\gamma\epsilon} = \frac{\delta_{\gamma\epsilon}}{m_\gamma} + \sum_{\alpha=1}^{n_b} \frac{C_{\gamma\alpha}}{M_\alpha} C_{\epsilon\alpha} = A_{\epsilon\gamma}. \tag{8.30}$$

Note that this expression differs from the one appearing in the paper by Baranyai and Evans, due to the omission of a summation in Equation (6) of that paper [173].

The constraint multipliers, λ are found by solving a set of linear equations that are formulated in a similar way to the original Edberg-Evans-Morriss treatment [173, 282, 285],

$$-\left(\mathbf{H}_\zeta \cdot \mathbf{R}_\zeta + \ddot{\mathbf{R}}_\zeta^2\right) = \sum_{\delta=1}^{n_c} \left(\mathbf{R}_\zeta \cdot N_{\zeta\delta}\mathbf{R}_\delta\right) \lambda_\delta, \tag{8.31}$$

where $\mathbf{H}_\zeta = \mathbf{H}_\beta - \mathbf{H}_\alpha$ is the difference in $\mathbf{H}$ between two primary particles in a constraint, defined as

$$\mathbf{H}_\alpha = \frac{\mathbf{F}_\alpha^\phi - \sum_{\delta=1}^{n_s}\sum_{\epsilon=1}^{n_s} C_{\epsilon\alpha}\left(A^{-1}\right)_{\delta\epsilon}\mathbf{T}_\delta}{m_\alpha}. \tag{8.32}$$

$N_{\zeta\delta}$ is a generalisation of the matrix formed from the difference between rows of the constraint matrix that appears in the original Edberg-Evans-Morriss formulation, and it is given by

$$N_{\zeta\delta} = \frac{M_{\zeta\delta} - \sum_{\gamma=1}^{n_s} \sum_{\epsilon=1}^{n_s} C_{\epsilon\zeta} \left(A^{-1}\right)_{\gamma\epsilon} \sum_{\alpha=1}^{n_b} \frac{C_{\gamma\alpha}}{m_\alpha} M_{\alpha\delta}}{m_\zeta}. \tag{8.33}$$

8.2 Molecular Representation of the Pressure Tensor

Although Equation (4.22) is a valid representation for the momentum density, it may not provide as much physical insight as a representation that explicitly accounts for the existence of molecules. In this section, we show how the molecular representation of the densities and fluxes can be obtained from the atomic representation, using the momentum density and the pressure tensor as examples.

The atomic representations of the mass and momentum densities for a molecular fluid can be written as

$$\rho(\mathbf{r}, t) = \sum_{i=1}^{N_m} \sum_{\alpha=1}^{N_s} m_{i\alpha} \delta(\mathbf{r} - \mathbf{r}_{i\alpha}) \tag{8.34}$$

and

$$\mathbf{J}(\mathbf{r}, t) = \rho\mathbf{v}(\mathbf{r}, t) = \sum_{i=1}^{N_m} \sum_{\alpha=1}^{N_s} m_{i\alpha} \mathbf{v}_{i\alpha} \delta(\mathbf{r} - \mathbf{r}_{i\alpha}), \tag{8.35}$$

where the inner summation extends over the number of atoms (or interaction sites) N_s in a molecule and the outer summation extends over the number of molecules N_m in the system. In general, N_s could depend on the molecule index i for a multicomponent system, but for simplicity, we will assume that this is not the case. The Fourier transforms of the mass and momentum densities are

$$\tilde{\rho}(\mathbf{k}, t) = \sum_{i=1}^{N_m} \sum_{\alpha=1}^{N_s} m_{i\alpha} e^{i\mathbf{k}\cdot\mathbf{r}_{i\alpha}} \tag{8.36}$$

and

$$\tilde{\mathbf{J}}(\mathbf{k}, t) = \sum_{i=1}^{N_m} \sum_{\alpha=1}^{N_s} m_{i\alpha} \mathbf{v}_{i\alpha} e^{i\mathbf{k}\cdot\mathbf{r}_{i\alpha}}. \tag{8.37}$$

Now it is convenient to introduce the position of site α of molecule i relative to the centre of mass of molecule i by the definition $\mathbf{R}_{i\alpha} = \mathbf{r}_{i\alpha} - \mathbf{r}_i$, where the centre of mass of molecule i is defined by

$$\mathbf{r}_i = \frac{\sum_{\alpha=1}^{N_s} m_{i\alpha} \mathbf{r}_{i\alpha}}{m_i} \tag{8.38}$$

and $m_i = \sum_{\alpha=1}^{N_s} m_{i\alpha}$ is the mass of molecule i. This means that the Fourier transform of the atomic mass density can be written as

$$\begin{aligned}\tilde{\rho}(\mathbf{k},t) &= \sum_{i=1}^{N_m}\sum_{\alpha=1}^{N_s} m_{i\alpha} e^{i\mathbf{k}\cdot(\mathbf{r}_i+\mathbf{R}_{i\alpha})} \\ &= \sum_{i=1}^{N_m}\sum_{\alpha=1}^{N_s} m_{i\alpha}\left(1 + i\mathbf{k}\cdot\mathbf{R}_{i\alpha} + \frac{1}{2}[i\mathbf{k}\cdot\mathbf{R}_{i\alpha}]^2 + \cdots\right) e^{i\mathbf{k}\cdot\mathbf{r}_i} \\ &= \tilde{\rho}^{(M)}(\mathbf{k},t) + i\mathbf{k}\cdot\sum_{i=1}^{N_m}\sum_{\alpha=1}^{N_s} m_{i\alpha}\mathbf{R}_{i\alpha}e^{i\mathbf{k}\cdot\mathbf{r}_i} + \frac{1}{2}i\mathbf{k}i\mathbf{k} : \sum_{i=1}^{N_m}\sum_{\alpha=1}^{N_s} m_{i\alpha}\mathbf{R}_{i\alpha}\mathbf{R}_{i\alpha}e^{i\mathbf{k}\cdot\mathbf{r}_i} + \cdots\end{aligned} \tag{8.39}$$

in which we define the Fourier transformed mass density in the molecular representation as $\tilde{\rho}^{(M)}(\mathbf{k},t) = \sum_{i=1}^{N_m} m_i e^{i\mathbf{k}\cdot\mathbf{r}_i}$. The local molecular mass density in real space is clearly just $\rho^{(M)}(\mathbf{r},t) = \sum_{i=1}^{N_m} m_i \delta(\mathbf{r} - \mathbf{r}_i)$, which localises the entire mass of each molecule at the position of the molecular centre of mass. The second term on the right-hand side of Equation (8.39) is equal to zero because $\sum_{\alpha=1}^{N_s} m_{i\alpha}\mathbf{R}_{i\alpha} = 0$. The third term, sometimes called the mass dispersion tensor [286], is related to the local moment of inertia density, which can be shown by writing

$$\begin{aligned}\sum_{\alpha=1}^{N_s} m_{i\alpha}\mathbf{R}_{i\alpha}\mathbf{R}_{i\alpha} &= \sum_{\alpha=1}^{N_s} m_{i\alpha}R_{i\alpha}^2\mathbf{1} - \left[\sum_{\alpha=1}^{N_s}\left(m_{i\alpha}R_{i\alpha}^2\mathbf{1} - m_{i\alpha}\mathbf{R}_{i\alpha}\mathbf{R}_{i\alpha}\right)\right] \\ &= \sum_{\alpha=1}^{N_s} m_{i\alpha}R_{i\alpha}^2\mathbf{1} - \mathbf{I}_i \\ &= m_i R_{g,i}^2\mathbf{1} - \mathbf{I}_i,\end{aligned} \tag{8.40}$$

where $\mathbf{1}$ is the unit tensor and $\mathbf{I}_i$ is the instantaneous moment of inertia tensor of molecule i, and $R_{g,i}^2$ is the instantaneous squared radius of gyration about the centre of mass of molecule i, as defined in polymer science [287].

Following a similar procedure to expand the atomic momentum density about the molecular centre of mass, we find

$$\begin{aligned}\tilde{\mathbf{J}}(\mathbf{k},t) &= \sum_{i=1}^{N_m}\sum_{\alpha=1}^{N_s} m_{i\alpha}\mathbf{v}_{i\alpha}\left(1 + i\mathbf{k}\cdot\mathbf{R}_{i\alpha} + \cdots\right)e^{i\mathbf{k}\cdot\mathbf{r}_i} \\ &= \tilde{\mathbf{J}}^{(M)}(\mathbf{k},t) + i\mathbf{k}\cdot\sum_{i=1}^{N_m}\sum_{\alpha=1}^{N_s} m_{i\alpha}\mathbf{R}_{i\alpha}\mathbf{v}_{i\alpha}e^{i\mathbf{k}\cdot\mathbf{r}_i} + \cdots\end{aligned} \tag{8.41}$$

where $\sum_{\alpha=1}^{N_s} m_{i\alpha}\mathbf{v}_{i\alpha} = m_i\mathbf{v}_i$ is the momentum of the centre of mass of molecule i and $\tilde{\mathbf{J}}^{(M)}(\mathbf{k},t)$ is the Fourier transform of the momentum density in the molecular representation. The second term of Equation (8.41) can be simplified by defining the velocity of site α of molecule i relative to the velocity of the molecular centre of mass as $\mathbf{V}_{i\alpha} = \mathbf{v}_{i\alpha} - \mathbf{v}_i$ and recalling that $\sum_{\alpha=1}^{N_s} m_{i\alpha}\mathbf{R}_{i\alpha} = 0$, giving

$$\tilde{\mathbf{J}}(\mathbf{k}, t) = \tilde{\mathbf{J}}^{(M)}(\mathbf{k}, t) + i\mathbf{k} \cdot \sum_{i=1}^{N_m}\sum_{\alpha=1}^{N_s} m_{i\alpha}\mathbf{R}_{i\alpha}\mathbf{V}_{i\alpha}e^{i\mathbf{k}\cdot\mathbf{r}_i} + \cdots \tag{8.42}$$

to first order in $i\mathbf{k} \cdot \mathbf{R}_{i\alpha}$. This can be simplified further by separating the last term into symmetric and antisymmetric parts, i.e.

$$\begin{aligned}\mathbf{R}_{i\alpha}\mathbf{V}_{i\alpha} &= \frac{1}{2}\left(\mathbf{R}_{i\alpha}\mathbf{V}_{i\alpha} + \mathbf{V}_{i\alpha}\mathbf{R}_{i\alpha}\right) + \frac{1}{2}\left(\mathbf{R}_{i\alpha}\mathbf{V}_{i\alpha} - \mathbf{V}_{i\alpha}\mathbf{R}_{i\alpha}\right) \\ &= \left(\mathbf{R}_{i\alpha}\mathbf{V}_{i\alpha}\right)^s + \left(\mathbf{R}_{i\alpha}\mathbf{V}_{i\alpha}\right)^a.\end{aligned} \tag{8.43}$$

The symmetric part can be written in terms of the time derivative of the mass dispersion tensor [286] (also known as the mass-weighted orientational order parameter [288]). Adopting the terminology of Olmsted and Snider (with minor differences), we define the mass dispersion tensor in such a way that its Fourier transform is

$$\widetilde{\mathsf{M}}(\mathbf{k}, t) = \sum_{i=1}^{N_m}\sum_{\alpha=1}^{N_s} m_{i\alpha}\mathbf{R}_{i\alpha}\mathbf{R}_{i\alpha}e^{i\mathbf{k}\cdot\mathbf{r}_i}. \tag{8.44}$$

As mentioned earlier, this is closely related to the local moment of inertia density. The time derivative of the mass dispersion tensor is

$$\dot{\widetilde{\mathsf{M}}}(\mathbf{k}, t) = \sum_{i=1}^{N_m}\sum_{\alpha=1}^{N_s} m_{i\alpha}\left(\mathbf{R}_{i\alpha}\mathbf{V}_{i\alpha} + \mathbf{V}_{i\alpha}\mathbf{R}_{i\alpha}\right)e^{i\mathbf{k}\cdot\mathbf{r}_i} + i\mathbf{k} \cdot \sum_{i=1}^{N_m}\sum_{\alpha=1}^{N_s} m_{i\alpha}\mathbf{v}_i\mathbf{R}_{i\alpha}\mathbf{R}_{i\alpha}e^{i\mathbf{k}\cdot\mathbf{r}_i}. \tag{8.45}$$

The second term, representing the flux of the mass dispersion tensor, is of higher order in wavevector and we will assume that it can be neglected.

The antisymmetric part of a second-rank tensor $\mathsf{A}^a = \frac{1}{2}(\mathsf{A} - \mathsf{A}^T)$ is related to its pseudovector dual $\mathbf{A}^d$, by $\mathsf{A}^a = \boldsymbol{\epsilon} \cdot \mathbf{A}^d$, where $\boldsymbol{\epsilon}$ is the completely antisymmetric, isotropic third-rank tensor ($\epsilon_{ijk} = +1$ for indices ijk equal to cyclic permutations of 123, -1 for all other permutations of these indices, and 0 when two or more of the indices are equal). The cross product of two vectors can also be written in terms of the isotropic third-rank tensor as $\mathbf{a} \times \mathbf{b} = -\boldsymbol{\epsilon} : \mathbf{ab}$. If the tensor A can be represented as a dyadic, $\mathsf{A} = \mathbf{ab}$, $\mathbf{A}^d$ can also be written as $\mathbf{A}^d = \frac{1}{2}\left(\mathbf{a} \times \mathbf{b}\right)$. Using these results, we obtain

$$\begin{aligned}i\mathbf{k} \cdot \left(\mathbf{R}_{i\alpha}\mathbf{V}_{i\alpha}\right)^a &= i\mathbf{k} \cdot \left[\boldsymbol{\epsilon} \cdot \left(\mathbf{R}_{i\alpha}\mathbf{V}_{i\alpha}\right)^d\right] \\ &= -i\mathbf{k} \times \left(\mathbf{R}_{i\alpha}\mathbf{V}_{i\alpha}\right)^d \\ &= -\frac{1}{2}i\mathbf{k} \times \left(\mathbf{R}_{i\alpha} \times \mathbf{V}_{i\alpha}\right).\end{aligned} \tag{8.46}$$

Note that this can also be obtained by using the vector identity $\mathbf{a} \times (\mathbf{b} \times \mathbf{c}) = (\mathbf{a} \cdot \mathbf{c})\, \mathbf{b} - (\mathbf{a} \cdot \mathbf{b})\, \mathbf{c} = \mathbf{a} \cdot (\mathbf{cb} - \mathbf{bc})$. These can then be combined to obtain the result

$$\tilde{\mathbf{J}}(\mathbf{k}, t) = \tilde{\mathbf{J}}^{(M)}(\mathbf{k}, t) + \frac{1}{2} i\mathbf{k} \cdot \tilde{\dot{\mathbf{M}}}(\mathbf{k}, t) - \frac{1}{2} i\mathbf{k} \times \tilde{\mathbf{S}}(\mathbf{k}, t), \tag{8.47}$$

where

$$\tilde{\mathbf{S}}(\mathbf{k}, t) = \sum_{i=1}^{N_m} \sum_{\alpha=1}^{N_s} m_{i\alpha} \mathbf{R}_{i\alpha} \times \mathbf{V}_{i\alpha} e^{i\mathbf{k} \cdot \mathbf{r}_i} \tag{8.48}$$

is the Fourier transform of the local density of spin angular momentum. This shows that the total momentum density can be broken into three parts with definite physical interpretations in the molecular picture. The first part is the flux of molecular centre of mass momentum, the second is a flux of momentum due to the change in the mass dispersion tensor, and the third is the momentum flux due to molecular rotation about the centre of mass. The molecular picture is seen to be useful mainly due to the way that it allows us to construct a clear physical interpretation of molecular transport processes. This is particularly beneficial in situations where molecular rotation needs to be considered explicitly, for example in the construction of thermostats for molecular fluids, which we will discuss later.

The spatially averaged, instantaneous pressure tensor for a system of particles with two-body central forces is given by the zero wavevector limit of Equation (4.48). Rewriting this in terms of molecular centre of mass positions and velocities, we have

$$\begin{aligned} V\mathbf{P}(t) &= \sum_{i=1}^{N_m} \sum_{\alpha=1}^{N_s} m_{i\alpha} \mathbf{c}_{i\alpha} \mathbf{c}_{i\alpha} - \frac{1}{2} \sum_{i=1}^{N_m} \sum_{\alpha=1}^{N_s} \sum_{j=1}^{N_m} \sum_{\beta=1}^{N_s} \mathbf{r}_{i\alpha j\beta} \mathbf{F}_{i\alpha j\beta} \\ &= \sum_{i=1}^{N_m} \sum_{\alpha=1}^{N_s} m_{i\alpha} \left(\mathbf{c}_i + \mathbf{C}_{i\alpha}\right) \left(\mathbf{c}_i + \mathbf{C}_{i\alpha}\right) - \frac{1}{2} \sum_{i=1}^{N_m} \sum_{\alpha=1}^{N_s} \sum_{j=1}^{N_m} \sum_{\beta=1}^{N_s} \left(\mathbf{r}_{ij} + \mathbf{R}_{j\beta} - \mathbf{R}_{i\alpha}\right) \mathbf{F}_{i\alpha j\beta} \\ &= V\mathbf{P}^{(M)}(t) + \sum_{i=1}^{N_m} \sum_{\alpha=1}^{N_s} m_{i\alpha} \mathbf{C}_{i\alpha} \mathbf{C}_{i\alpha} + \sum_{i=1}^{N_m} \sum_{\alpha=1}^{N_s} \mathbf{R}_{i\alpha} \mathbf{F}_{i\alpha}, \end{aligned} \tag{8.49}$$

where we have defined $\mathbf{C}_{i\alpha} = \mathbf{c}_{i\alpha} - \mathbf{c}_i$, with $\mathbf{c}_{i\alpha} = \mathbf{v}_{i\alpha} - \mathbf{v}\left(\mathbf{r}_{i\alpha}\right)$ and $\mathbf{c}_i = \mathbf{v}_i - \mathbf{v}\left(\mathbf{r}_i\right)$. The zero-wavevector molecular pressure tensor for molecules consisting of atoms with only two-body central forces is given by

$$V\mathbf{P}^{(M)}(t) = \sum_{i=1}^{N_m} m_i \mathbf{c}_i \mathbf{c}_i - \frac{1}{2} \sum_{i=1}^{N_m} \sum_{j=1}^{N_m} \mathbf{r}_{ij} \mathbf{F}_{ij}. \tag{8.50}$$

Note that this need not be instantaneously symmetric, even though the atomic pressure tensor is. The difference between the atomic and molecular pressure tensors can be expressed as

$$
\begin{aligned}
V\mathbf{P}(t) - V\mathbf{P}^{(M)}(t) &= \sum_{i=1}^{N_m}\sum_{\alpha=1}^{N_s} m_{i\alpha}\mathbf{C}_{i\alpha}\mathbf{C}_{i\alpha} + \sum_{i=1}^{N_m}\sum_{\alpha=1}^{N_s} \mathbf{R}_{i\alpha}\mathbf{F}_{i\alpha} \\
&= \sum_{i=1}^{N_m}\sum_{\alpha=1}^{N_s} m_{i\alpha}\mathbf{C}_{i\alpha}\mathbf{C}_{i\alpha} + \sum_{i=1}^{N_m}\sum_{\alpha=1}^{N_s} (\mathbf{R}_{i\alpha}\mathbf{F}_{i\alpha})^s + \sum_{i=1}^{N_m}\sum_{\alpha=1}^{N_s} (\mathbf{R}_{i\alpha}\mathbf{F}_{i\alpha})^a \\
&= \frac{1}{2}V\ddot{\mathbf{M}}(t) - V\mathbf{P}^{(M)a}(t),
\end{aligned}
\tag{8.51}
$$

where

$$
\frac{1}{2}V\ddot{\mathbf{M}}(t) = \sum_{i=1}^{N_m}\sum_{\alpha=1}^{N_s} m_{i\alpha}\mathbf{C}_{i\alpha}\mathbf{C}_{i\alpha} + \sum_{i=1}^{N_m}\sum_{\alpha=1}^{N_s} (\mathbf{R}_{i\alpha}\mathbf{F}_{i\alpha})^s \tag{8.52}
$$

and

$$
V\mathbf{P}^{(M)a}(t) = \sum_{i=1}^{N_m}\sum_{\alpha=1}^{N_s} (\mathbf{R}_{i\alpha}\mathbf{F}_{i\alpha})^a = \frac{1}{2}\sum_{i=1}^{N_m} (\mathbf{r}_i\mathbf{F}_i - \mathbf{F}_i\mathbf{r}_i). \tag{8.53}
$$

Observe that the antisymmetric part of the molecular pressure tensor arises solely from the configurational component of the pressure tensor, because the kinetic component is inherently symmetric.

Several studies have verified that the time averaged value of the antisymmetric part of the molecular pressure tensor in a shearing steady state is zero [264, 288–290]. This means that the steady state viscometric functions can equally well be evaluated from either the molecular or atomic pressure tensor. However, the molecular formalism has some advantages over the atomic formalism. One important advantage is that the molecular stress tensor varies more slowly with time than the atomic stress tensor, particularly when high frequency bond length vibrations are present. This means that evaluation of the viscosity from the integral of the stress autocorrelation function is numerically easier to perform when the molecular stress tensor is used [291]. Another important advantage of using the molecular stress tensor is that its antisymmetric part indicates whether any torques are being applied to the system, either intentionally, through a synthetic spin force [292], or unintentionally through an incorrectly formulated thermostat [293].

8.3 Molecular SLLOD

The SLLOD algorithm for atomic fluids can be directly applied to molecular fluids without modification, provided that all of the relevant forces are included. In fact, this is the most commonly implemented form of the equations of motion for NEMD simulations of molecular fluids, including polymeric fluids. In the *atomic* SLLOD algorithm, the streaming term in the equations of motion is applied at the position of each atom in a molecule, i.e.

$$\dot{\mathbf{r}}_{i\alpha} = \frac{\mathbf{p}_{i\alpha}}{m_{i\alpha}} + \mathbf{r}_{i\alpha} \cdot \nabla \mathbf{v}$$
$$\dot{\mathbf{p}}_{i\alpha} = \mathbf{F}_{i\alpha} - \mathbf{p}_{i\alpha} \cdot \nabla \mathbf{v}. \tag{8.54}$$

The *molecular* SLLOD algorithm differs in the point at which the streaming velocity is evaluated:

$$\dot{\mathbf{r}}_{i\alpha} = \frac{\mathbf{p}_{i\alpha}}{m_{i\alpha}} + \mathbf{r}_{i} \cdot \nabla \mathbf{v}$$
$$\dot{\mathbf{p}}_{i\alpha} = \mathbf{F}_{i\alpha} - \left(\frac{m_{i\alpha}}{m_i}\right) \mathbf{p}_{i} \cdot \nabla \mathbf{v}, \tag{8.55}$$

where $\mathbf{F}_{i\alpha}$ includes all intermolecular forces due to potential energy functions, constraints etc., $m_{i\alpha}$ is the mass of site α of molecule i and m_i is the mass of molecule i. The molecular version of the SLLOD equations of motion was originally proposed by Ladd [127]. Edberg, Morriss and Evans [282] showed that, although the transient responses of a system to the molecular and atomic SLLOD equations of motion are different, they produce identical averages in the steady state. In particular, both forms of the equations of motion generate a steady state in which the antisymmetric part of the time averaged molecular pressure tensor is zero when a correctly formulated thermostat is applied [294, 295].

When we consider molecular thermostats later, we will see that an additional complication arises from the difference between the atomic and molecular forms of the SLLOD equations of motion. The first equation of Equation (8.55) acts as a *definition* of the peculiar momentum of site α of molecule i. This peculiar momentum is used to calculate the pressure, temperature and other properties. The molecular pressure tensor and temperature will be correctly calculated, but, as Equation (8.47) shows, the local atomic momentum density (and therefore, the atomic streaming velocity) differs from the molecular momentum density due to terms that account for molecular rotation and deformation. In other words, the assumption that the streaming velocity at site α of molecule i is given by the perfectly linear expression $\mathbf{v}(\mathbf{r}_{i\alpha}) = \mathbf{r}_{i\alpha} \cdot \nabla \mathbf{v}$ cannot possibly be correct when it is applied to atoms within a molecule, because nearby bonded sites of a molecule cannot be separated from each other by an indefinite amount by the velocity gradient in the same way that independent atoms can. Instead, molecular rotation and limited deformation occur. On the other hand, the molecular centre of mass streaming velocity does not suffer from this problem, making it a more attractive quantity for the calculation of the molecular peculiar velocity, which can then be used to compute and control the molecular pressure, temperature and other properties. Thus, the main advantage of the molecular SLLOD equations of motion is that they provide us with a definition of the molecular centre of mass peculiar momentum that does not rely on assumptions about the systematic internal molecular motions associated with flow, such as molecular rotation and deformation. It is worth noting here that irrotational flows such as elongation should be less strongly affected by these problems due to the absence of shear-induced molecular rotation. Of course, a molecular thermostat introduces its own problems. In a system consisting of a very small number of extremely long molecules, only a few degrees of freedom will be thermostatted. This may lead to severe non-equipartitioning of energy at high deformation rates. In such cases, the configurational thermostat introduced by Braga and Travis [162] may have distinct advantages. We will present the algorithms for the implementation of the molecular SLLOD equations of motion in Sections 8.6 and 8.7, but first we need to consider the issues of energy and momentum balance, and thermostats for molecular fluids.

8.4 Momentum and Internal Energy Balance in the Presence of a Homogeneous Thermostat

The thermostat term in the equations of motion for homogeneous flow effectively removes heat from the system, making it possible to generate a steady state with constant time averaged values of the static thermodynamic properties such as the pressure and internal energy and the relevant fluxes. The equations of motion for an atomic system with a deterministic thermostat are given by Equation (5.82). The details of some of the more widely used thermostats that have been proposed were discussed in Section 5.3 and have been thoroughly reviewed in previous publications [162, 163, 192, 293, 294]. However, two subjects that have received far less attention are the detailed analysis of the place of the thermostat in the balance equations for momentum and internal energy, and the subtle issues surrounding the formulation of thermostats for molecular fluids. These will be the topics of our discussion in this section. We will only consider the simplest deterministic thermostat, the Gaussian isokinetic thermostat, in our discussion. This choice is not intended to imply that this thermostatting mechanism is superior to others. Rather, we have chosen it because it enables us to study the main issues of interest, energy balance and the thermostatting of molecular fluids, using a simple and familiar model.

The effect of the thermostat on the momentum balance equation will be considered first. Physically, it seems best to regard the thermostat term in the atomic equations of motion as an internal force because, as we will soon see, it cannot contribute to the acceleration of the centre of mass of the system. However, it remains distinct from interatomic constraint forces such as the bond length or bond angle constraints, which must be included in the atomic pressure tensor. Therefore, we add a term to account for the thermostat in addition to the external body force term, $\rho \mathbf{F}^e$ in Equation (4.26). Macroscopically, we then have

$$\frac{\partial (\rho \mathbf{v})}{\partial t} = -\nabla \cdot \mathbf{P} - \nabla \cdot (\rho \mathbf{v}\mathbf{v}) + \rho \mathbf{F}^e + \rho \mathbf{T}, \tag{8.56}$$

where $\rho \mathbf{T} = \rho(\mathbf{r}, t)\, \mathbf{T}(\mathbf{r}, t)$ is the local force density due to the thermostat term in the equations of motion. The microscopic expression for this force density is easily obtained by considering the effect of the additional term in the equations of motion on the microscopic expression for the rate of change of the momentum density. Once again, this is most easily calculated after Fourier transformation:

$$
\begin{aligned}
\frac{\partial \left(\widetilde{\rho \mathbf{v}}\right)}{\partial t} &= \frac{\partial}{\partial t} \sum_i m_i \mathbf{v}_i e^{i\mathbf{k}\cdot\mathbf{r}_i} \\
&= \sum_i m_i \dot{\mathbf{v}}_i e^{i\mathbf{k}\cdot\mathbf{r}} + i\mathbf{k} \cdot \sum_i m_i \mathbf{v}_i \mathbf{v}_i e^{i\mathbf{k}\cdot\mathbf{r}_i} \\
&= i\mathbf{k} \cdot \tilde{\mathbf{P}}(\mathbf{k}, t) + i\mathbf{k} \cdot \left[\widetilde{\rho \mathbf{v}\mathbf{v}}(\mathbf{k}, t)\right] + \widetilde{\mathbf{G}}(\mathbf{k}, t) - \alpha \sum_i m_i \mathbf{c}_i e^{i\mathbf{k}\cdot\mathbf{r}_i}.
\end{aligned} \tag{8.57}
$$

This shows that the microscopic expression for the density of the force due to the thermostat is

$$\rho(\mathbf{r}, t)\, \mathbf{T}(\mathbf{r}, t) = -\alpha \sum_i m_i \mathbf{c}_i \delta(\mathbf{r} - \mathbf{r}_i). \tag{8.58}$$

The integral of this quantity over the whole volume is equal to zero, because the peculiar velocity is defined in such a way that the sum of the peculiar momenta is zero at all times. Equivalently, in a periodic system, the zeroth Fourier component of the force exerted on the system by the thermostat is zero. This assumes the streaming velocity used to define the peculiar velocity has been evaluated correctly, e.g. using a PUT (see Section 5.3.3).

The effect of the thermostat on the internal energy is taken into account by adding a term to the internal energy balance equation

$$\frac{\partial(\rho u)}{\partial t} = -\nabla \cdot \mathbf{J}_q - \nabla \cdot (\rho u \mathbf{v}) - \mathbf{P}^T : \nabla \mathbf{v} + R, \tag{8.59}$$

where R represents the density of the rate at which internal energy is added to the system by the homogeneous thermostat in the equations of motion. By substituting the equations of motion into the Fourier transform of the left-hand side of Equation (8.59), we find that the Fourier transform of R is

$$\tilde{R} = -\alpha \sum_i m_i \mathbf{c}_i^2 e^{i\mathbf{k}\cdot\mathbf{r}_i}. \tag{8.60}$$

The zeroth Fourier component of this quantity for a periodic system gives us the total rate of change of the internal energy of the particles in the primary simulation box due to the thermostat. In a spatially homogeneous system in which work is done by stresses, we can keep the internal energy constant by applying the condition that

$$\frac{dU}{dt} = \int \frac{\partial(\rho u)}{\partial t} d\mathbf{r} = -V\mathbf{P}^T : \nabla \mathbf{v} + VR = 0, \tag{8.61}$$

which, when combined with Equation (8.60), provides us with the expression for α that is required to keep the internal energy constant (see also Section 6.1.1):

$$\alpha = -\frac{V\mathbf{P}^T : \nabla \mathbf{v}}{\sum_i m_i c_i^2}. \tag{8.62}$$

8.5 Molecular Thermostats

The equipartition theorem ensures that the temperatures of all degrees of freedom are equal at equilibrium. It is well known that this is not the case for systems that are far from equilibrium, even for simple atomic fluids. For example, the shear rate dependence of the differences between the temperatures calculated from the x, y and z components of the peculiar momentum in a shearing atomic fluid have been shown to be significant at extremely high shear rates by Baranyai [84]. This is linked to a greater question regarding the validity of our usual thermodynamic definition of temperature, $T = (\partial U/\partial S)_V$ for systems far from equilibrium. In fact, the existence of a nonequilibrium temperature has itself been questioned and the issue is far from being resolved [80]. The zeroth law of thermodynamics, which deals with the definition of thermal equilibrium and the existence of a unique temperature, has been shown to be invalid for an atomic system in a strong, spatially varying strain rate field [296]. From a purely pragmatic point of view, it is convenient to assume that temperature does exist in far from equilibrium systems, although any chosen means of calculating it may no longer be unique, and we must accept that the thermodynamic meaning of such a quantity is questionable when we consider systems far from equilibrium.

The kinetic temperature, which has its basis in the equipartition theorem, the ideal gas equation of state and kinetic theory, is defined for an atomic system through the relationship

$$\langle K \rangle = \left\langle \sum_i \frac{1}{2} m_i \mathbf{c}_i^2 \right\rangle = \frac{N_f k_B T}{2}, \tag{8.63}$$

where N_f is the number of independent degrees of freedom contributing to the peculiar kinetic energy. For a system in which all of the momenta are independent, this would be equal to $3N$ where N is the number of atoms. In the presence of constraints, this must be modified, because the number of degrees of freedom is reduced by one for every constraint on the momenta. This means that in a molecular dynamics simulation in which the equations of motion include no explicit constraints, only the three components of the total momentum are fixed and $N_f = 3N - 3$. If a Gaussian thermostat is used to additionally fix the total peculiar kinetic energy, we have $N_f = 3N - 4$. For an equilibrium constant energy simulation of a system of molecules with N_c internal constraints per molecule, the number of degrees of freedom that should be used to calculate the atomic temperature is $N_f = N_m (3N_s - N_c) - 3$, where N_m is the number of molecules and N_s is the number of sites per molecule. The kinetic temperature has an attractive interpretation as the residual (or random) part of the kinetic energy that results from a least squares estimation of the streaming velocity from the velocities of the individual atoms (discussed in the first edition of reference [2]).

Another measure of temperature that can be used in molecular systems is the kinetic temperature of the molecular centre of mass degrees of freedom. In this case, the temperature is obtained from the centre of mass peculiar kinetic energy of each molecule using Equation (8.63) again, but with $\mathbf{c}_i$ now representing the molecular centre of mass peculiar velocity instead of the atomic peculiar velocity, m_i representing the mass of molecule i and the sum now extending over all molecules. The number of degrees of freedom for the calculation of the molecular centre of mass kinetic temperature in a system with fixed total momentum and a constrained value of the total molecular peculiar kinetic energy is $N_f = 3N_m - 4$, where N_m is the number of molecules in the system. The molecular and atomic temperatures have identical average values at equilibrium, but far from equilibrium these two temperatures differ. In steady planar shear flow, the difference is proportional to the strain rate squared and the atomic temperature is higher than the molecular temperature [297] when the molecular centre of mass degrees of freedom are thermostatted. This occurs because the relaxation time for the heat generated by viscous flow processes to flow from the internal molecular degrees of freedom to the thermostatted centre of mass translational degrees of freedom is significant.

All calculations of the kinetic temperature assume that the relevant (atomic or molecular) peculiar velocity has been correctly computed. The SLLOD equations of motion provide us with a definition of the peculiar velocity, but is there a difference between the peculiar velocities that we obtain from the atomic (Equation (8.54)) and molecular (Equation (8.55)) forms of the equations of motion and is this difference significant? This question has been examined by Travis *et al.* [293–295], and the brief answer is that a significant difference does exist, and that an incorrectly formulated thermostat can produce misleading results. In particular, enhanced orientational ordering and nonzero values of the antisymmetric part of the molecular stress tensor in a shearing steady state can result from the use of an incorrectly formulated atomic thermostat. Travis *et al.* [293–295] have shown that when the antisymmetric part of the molecular stress tensor is nonzero in a steady state, it indicates that a torque is being applied to the molecules. If this is done unintentionally by applying an incorrectly formulated thermostat, errors result.

It is easily shown that if a thermostat is formulated using the molecular centre of mass kinetic temperature calculated from either the atomic or molecular SLLOD equations of motion, correct results are obtained. From the atomic SLLOD equations of motion, the peculiar momentum of the molecular centre of mass is given by summing over all sites on a given molecule

$$\begin{aligned} m_i \mathbf{c}_i &= \sum_{\alpha} m_{i\alpha} \dot{\mathbf{r}}_{i\alpha} - \sum_{\alpha} m_{i\alpha} \mathbf{r}_{i\alpha} \cdot \nabla \mathbf{v} \\ &= m_i \dot{\mathbf{r}}_i - m_i \mathbf{r}_i \cdot \nabla \mathbf{v} \end{aligned} \tag{8.64}$$

and similarly, from the molecular SLLOD equations of motion, we find

$$\begin{aligned} m_i \mathbf{c}_i &= \sum_{\alpha} m_{i\alpha} \dot{\mathbf{r}}_{i\alpha} - \sum_{\alpha} m_{i\alpha} \mathbf{r}_i \cdot \nabla \mathbf{v} \\ &= m_i \dot{\mathbf{r}}_i - m_i \mathbf{r}_i \cdot \nabla \mathbf{v}. \end{aligned} \tag{8.65}$$

Thus, atomic SLLOD with a molecular thermostat (ASMT) is equivalent to molecular SLLOD with a molecular thermostat (MSMT). However, in simulations where an atomic thermostat is formulated using the peculiar velocity defined by the atomic SLLOD equations of motion (ASAT), artifacts are generated.

The correct formulation of a thermostat that keeps the atomic kinetic temperature constant in a far from equilibrium molecular system requires that the streaming velocity at an atomic site of any molecule be computed accurately. Equation (8.47) shows that the difference between the momentum density at the centre of mass of a molecule and one of the atoms on that molecule can be used to calculate the difference in their streaming velocities. When this difference is properly taken into account, artifacts such as the enhanced orientational ordering and the nonzero antisymmetric part of the steady state molecular pressure tensor are eliminated [293–295].

The analysis of Travis *et al.* [293–295] showed that two issues must be addressed when formulating a homogeneous thermostat for a molecular liquid. The first issue, which is common to simple monatomic and polyatomic molecular fluids, is the determination of the correct form of the instantaneous velocity profile. At sufficiently high strain rates, the assumption of a stable, homogeneous velocity profile may not be valid, and the streaming velocity must be computed during the simulation. The calculated streaming velocity is then used in the thermostat term of the equations of motion. As previously discussed in Section 5.3.3 this type of thermostat is called a profile unbiased thermostat (PUT). The second issue is unique to molecular fluids. A calculation of the streaming velocity of an atom that resides on a molecule with a nonzero average angular velocity superimposed on its centre of mass translational streaming motion *must* also take the rotational component into consideration when a thermostat is applied. Similarly, if there are any other systematic internal motions, such as an oscillatory molecular stretching, these should also be taken into consideration. This is clearly an extremely complicated task. The formulation of a rotationally unbiased atomic thermostat has only ever been successfully carried out for the relatively simple case of a fluid of rigid diatomic molecules [293–295].

The approach taken by Travis *et al.* [293–295] was as follows. First, it was assumed that the streaming velocity at site α of molecule i could be represented as

$$\begin{aligned}\mathbf{v}(\mathbf{r}_{i\alpha}) &= \mathbf{v}_T(\mathbf{r}_{i\alpha}) + \mathbf{v}_R(\theta_i, \phi_i)\\ &= \mathbf{v}_T(\mathbf{r}_i) + \boldsymbol{\omega}(\theta_i, \phi_i) \times \mathbf{R}_{i\alpha},\end{aligned} \tag{8.66}$$

where $\mathbf{v}_T(\mathbf{r}_i)$ represents the translational centre of mass streaming velocity of molecule i, and $\mathbf{v}_R(\theta_i, \phi_i)$ represents the angularly resolved average rotational streaming velocity of the fluid, assumed here to be independent of position. The introduction of an orientational dependence of the average angular velocity was found to be crucial. The average streaming velocity must be allowed to vary with the orientation of the molecule, specified by the angles (θ_i, ϕ_i), in order to eliminate spurious enhanced orientational ordering. Deviations of $\mathbf{v}_T(\mathbf{r}_i)$ from a strictly linear velocity profile were accounted for by expanding deviations from the linear velocity profile as a Fourier series, following the technique of Evans *et al.* [188].

The average streaming component of the angular velocity is conventionally defined by the equation

$$\mathbf{S} = \boldsymbol{\Theta} \cdot \omega, \tag{8.67}$$

where $\mathbf{S}$ is the total spin angular momentum of the system and $\boldsymbol{\Theta}$ is the total moment of inertia tensor for all molecules in the system [4]. It should be remarked that the numerical average of the angular velocity vectors of all molecules in the system, $\bar{\boldsymbol{\omega}}$, is different from the average angular velocity obtained by solving Equation (8.67) and should not be regarded as a valid macroscopic definition of $\boldsymbol{\omega}$. In fact, for diatomic molecules, Travis *et al.* [293] have shown that in the limit of zero shear rate, the numerically averaged angular velocity is 2/3 of the streaming angular velocity. A simple physical interpretation of the difference between these two quantities is that while $\bar{\boldsymbol{\omega}}$ represents an unweighted average of molecular angular velocities, $\boldsymbol{\omega}$ represents a weighted average angular velocity that takes the molecular inertia tensor as the weighting factor. This is similar to the conventional definition of the translational streaming velocity as the mass-weighted average velocity (see Equation (5.98)).

Taking the orientation dependence of the streaming angular velocity into account, the angular velocity was expressed as a spherical harmonic expansion in which the expansion coefficients were determined at each timestep by the method of least squares. The value of the atomic streaming velocity was then determined for each atom using Equation (8.66), and the equations of motion, given by

$$\begin{aligned}
\dot{\mathbf{r}}_{i\alpha} &= \frac{\mathbf{p}_{i\alpha}}{m_{i\alpha}} + \mathbf{r}_i \cdot \nabla\mathbf{v} \\
\dot{\mathbf{p}}_{i\alpha} &= \mathbf{F}_{i\alpha} - \left(\frac{m_{i\alpha}}{m_i}\right)\mathbf{p}_i \cdot \nabla\mathbf{v} - \alpha^{(T)}\left[\mathbf{v}_i - \mathbf{v}\left(\mathbf{r}_i, t\right)\right] - \alpha^{(R)}\left[\boldsymbol{\omega}_i - \boldsymbol{\omega}\left(\theta_i, \phi_i\right)\right] \\
\dot{\alpha}^{(T)} &= \xi\left[T_T\left(t\right) - T\right] \\
\dot{\alpha}^{(R)} &= \xi\left[T_R\left(t\right) - T\right]
\end{aligned} \tag{8.68}$$

were integrated. Even with the simplest rotationally unbiased atomic thermostat (RUAT), which assumes that the translational streaming velocity is strictly linear but takes account of the molecular streaming velocity, the results showed a dramatic reduction in the antisymmetric part of the molecular pressure tensor, indicating that spurious thermostat-induced molecular torques were eliminated.

The extension of this approach to flexible molecules is expected to be very difficult, because no technique for the determination of the streaming components of internal motions has yet been devised. The recent development of configurational thermostats offers much better prospects for the development of thermostatting techniques that are free of artifacts. Configurational thermostats operate purely on the atomic positions and do not require an estimate of the atomic streaming velocity. The first applications of configurational thermostat to molecular fluids were published by Lue *et al.* [160] and Delhommelle and Evans [190]. The equations of motion for a molecular fluid with a configurational atomic thermostat in the presence of bond constraints are written as

$$\begin{aligned}
\dot{\mathbf{r}}_{i\alpha} &= \frac{\mathbf{p}_{i\alpha}}{m_{i\alpha}} + \mathbf{r}_i \cdot \nabla\mathbf{v} + \frac{s}{T}\frac{\partial T_c}{\partial \mathbf{r}_{i\alpha}} + \left(\mathbf{L}\cdot\boldsymbol{\zeta}\right)_{i\alpha} \\
\dot{\mathbf{p}}_{i\alpha} &= \mathbf{F}_{i\alpha} - \left(\frac{m_{i\alpha}}{m_i}\right)\mathbf{p}_i \cdot \nabla\mathbf{v} \\
\dot{s} &= -Q\frac{\left(T_c - T\right)}{T},
\end{aligned} \tag{8.69}$$

where $\mathbf{F}_{i\alpha}$ is the total force including constraint forces, $\mathbf{L}$ is obtained from the bond constraint matrix and $\boldsymbol{\zeta}$ is a modified constraint multiplier [160]. The last term in the first equation is added to ensure that bond constraints are not violated by the thermostat force, which is different for each atom in a molecule. For more detail, refer to Lue *et al.* and Delhommelle and Evans [160, 190]. The results obtained with this thermostat generally show that at reduced shear rates above 0.5, the intramolecular properties of the configurational thermostatted system are far less perturbed from their equilibrium values than when a centre of mass kinetic thermostat is used. Travis and Braga [164] have recently extended their version of the molecular thermostat to molecular systems at equilibrium, but as yet have not considered systems under flow conditions.

8.6 Molecular SLLOD Algorithms for Shear Flow

8.6.1 (*NVT*) Ensemble

Implementation of the SLLOD algorithm for homogeneous planar shear flow in an (*NVT*) ensemble is straightforward after the basic elements are in place. From the discussion in previous sections, we can conclude that the simplest implementation that suffers least from distortions associated with incorrect assumptions about the streaming velocity, is the molecular SLLOD algorithm with a thermostat applied to either the molecular centre of mass kinetic temperature, or the configurational temperature. Here, we will focus on the version that uses the molecular centre of mass kinetic thermostat, because that is the one that has been most commonly adopted. The results for linear transport properties such as the zero shear rate viscosity, are independent of the details of the thermostat, but nonlinear properties such as the normal stress coefficients and the shear rate dependent contributions to the viscosity are sensitive to the details of the thermostat [170].

The velocity gradient tensor for planar shear flow where the velocity is in the x-direction and the gradient is in the y-direction, is given by

$$\nabla\mathbf{v} = \begin{pmatrix} 0 & 0 & 0 \\ \dot{\gamma} & 0 & 0 \\ 0 & 0 & 0 \end{pmatrix}. \tag{8.70}$$

The equations of motion for planar homogeneous shear flow with the molecular SLLOD algorithm and molecular centre of mass kinetic temperature are

$$\begin{aligned} \dot{\mathbf{r}}_{i\alpha} &= \frac{\mathbf{p}_{i\alpha}}{m_{i\alpha}} + \dot{\gamma} y_i \mathbf{i} \\ \dot{\mathbf{p}}_{i\alpha} &= \mathbf{F}_{i\alpha} + \mathbf{F}_{i\alpha}^{C} - \frac{m_{i\alpha}}{M_i} p_{yi}\dot{\gamma}\mathbf{i} - \frac{m_{i\alpha}}{M_i}\zeta\mathbf{p}_i, \end{aligned} \tag{8.71}$$

where $\mathbf{F}_{i\alpha}$ represents the sum of all forces on atom α of molecule i due to interaction potentials, which includes all intermolecular and intramolecular Lennard-Jones type forces and may also include intramolecular forces due to bond stretching, bending and dihedral angles. $\mathbf{F}_{i\alpha}^{C}$ represents all forces on atom α of molecule i due to any constraint forces that may be present. M_i and $\mathbf{p}_i$ are the molecular centre of mass and momentum of the centre of mass of molecule i, respectively. The thermostat multiplier ζ for the Gaussian isokinetic molecular thermostat is given by

$$\zeta = \frac{\sum_{i=1}^{N_m} \left(\mathbf{F}_i \cdot \mathbf{p}_i - \dot{\gamma} p_{xi} p_{yi}\right) / M_i}{\sum_{i=1}^{N_m} \mathbf{p}_i^2 / M_i}. \quad (8.72)$$

This implementation of the SLLOD algorithm requires calculation of the molecular centre of mass positions and peculiar momenta at each time step. However, this calculation is computationally inexpensive and these quantities are also required for application of the periodic boundary conditions and for computation of properties such as the molecular temperature and pressure tensor. It is very useful to compute both the atomic and molecular kinetic temperatures, and similarly the atomic and molecular pressure tensors, because they enable important consistency checks to be carried out. At equilibrium, the averages of the molecular and atomic temperatures should be equal. Similarly, the atomic and molecular pressure tensors should also be equal. In steady shear flow, the atomic and molecular temperatures for molecular fluids will differ, but the atomic and molecular pressure tensors are equal. This is closely related to the fact that for central force interactions, the atomic pressure tensor is rigorously symmetric, but the antisymmetric part of the molecular pressure tensor decays to zero in the approach to the steady state. In fact the symmetry of the (average) molecular pressure tensor in the shearing steady state is an important tool that can be used to detect unintentionally applied external torques resulting from incorrectly formulated thermostats.

These equations of motion have been successfully applied to molecular fluids of many different types including small, rigid body molecules such as nitrogen and chlorine (diatomic), carbon disulphide (linear triatomic), simplified united atom models of alkanes, larger rigid body molecules including benzene and naphthalene, and coarse-grained models of polymers.

Results for these types of systems are summarised in the reviews by Cummings and Evans [298], Sarman, Evans and Cummings [243] and Todd and Daivis [1].

8.6.2 (NpT) Ensemble

The algorithm described in the previous section simulates planar shear flow at constant volume, but when rheological measurements are carried out in the laboratory, they are usually done in such a way that the material that is being measured has at least one surface in equilibrium with the atmosphere. When this surface reaches mechanical equilibrium with its surroundings (in the shearing steady state), the component of the pressure normal to the surface must be equal to the pressure of the atmosphere. This allows the volume of the system to change as the shear rate is changed. This phenomenon is known as shear dilatancy (from the term originally coined by Reynolds [299]). A similar condition is achieved by setting the isotropic part of the pressure tensor (one third of the trace) to be constant. The equations of motion below use a Gaussian thermostat to keep the centre of mass kinetic temperature constant and a Nosé-Hoover type algorithm to keep the pressure constant:

$$\begin{aligned}\dot{\mathbf{r}}_{i\alpha} &= \frac{\mathbf{p}_{i\alpha}}{m_{i\alpha}} + \dot{\gamma} y_i \mathbf{i} + \dot{\xi} \mathbf{r}_i \\ \dot{\mathbf{p}}_{i\alpha} &= \mathbf{F}_{i\alpha} + \mathbf{F}_{i\alpha}^C - \frac{m_{i\alpha}}{M_i} p_{yi} \dot{\gamma} \mathbf{i} - \frac{m_{i\alpha}}{M_i} \zeta \mathbf{p}_i - \frac{m_{i\alpha}}{M_i} \dot{\xi} \mathbf{p}_i.\end{aligned} \tag{8.73}$$

Both the thermostat and barostat act on the molecular centres of mass. Similarly to the atomic case, the volume and the barostat multiplier are found by solving the additional differential equations

$$\begin{aligned}\dot{V} &= 3\dot{\xi} V \\ \ddot{\xi} &= \frac{(p - p_0) V}{Q N_m k_B T}\end{aligned} \tag{8.74}$$

where p is the instantaneous isotropic pressure, given by $p = \frac{1}{3}\mathrm{Tr}(\mathbf{P})$, p_0 is the target pressure and Q is the barostat relaxation time (since it has dimensions of time). The thermostat multiplier given by Gauss's principle of least constraint when applied to the centre of mass kinetic temperature is given by

$$\zeta = \frac{\sum_{i=1}^{N_m} \left(\mathbf{F}_i \cdot \mathbf{p}_i - \dot{\gamma} p_{xi} p_{yi}\right)/M_i}{\sum_{i=1}^{N_m} \mathbf{p}_i^2/M_i} - \dot{\xi}. \tag{8.75}$$

8.7 Molecular SLLOD Algorithms for Elongational Flow

8.7.1 (*NVT*) Ensemble

Planar elongational flow is defined by the velocity gradient tensor

$$\nabla \mathbf{v} = \begin{pmatrix} \dot{\epsilon} & 0 & 0 \\ 0 & -\dot{\epsilon} & 0 \\ 0 & 0 & 0 \end{pmatrix}, \tag{8.76}$$

where the extension occurs in the x-direction and contraction occurs in the y-direction, while the z-direction is neutral. Using the Kraynik-Reinelt boundary conditions described in Section 5.1.5, it is possible to simulate planar elongational flow of molecular fluids in periodic boundary conditions without violating the minimum image convention, despite the continual contraction of the simulation box in the y-direction. The equations of motion to implement the molecular SLLOD algorithm with a Gaussian isokinetic molecular centre of mass thermostat under constant volume conditions are given by

$$\dot{\mathbf{r}}_{i\alpha} = \frac{\mathbf{p}_{i\alpha}}{m_{i\alpha}} + \dot{\epsilon}\left(x_i \mathbf{i} - y_i \mathbf{j}\right) \tag{8.77}$$

$$\dot{\mathbf{p}}_{i\alpha} = \mathbf{F}_{i\alpha} + \mathbf{F}_{i\alpha}^C - \dot{\epsilon} \frac{m_{i\alpha}}{M_i}\left(p_{xi}\mathbf{i} - p_{yi}\mathbf{j}\right) - \frac{m_{i\alpha}}{M_i} \zeta \mathbf{p}_i \tag{8.78}$$

with the thermostat multiplier

$$\zeta = \frac{\sum_{i=1}^{N_m} \left[\mathbf{F}_i \cdot \mathbf{p}_i - \dot{\epsilon} \left(p_{xi}^2 - p_{yi}^2 \right) \right] / M_i}{\sum_{i=1}^{N_m} \mathbf{p}_i^2 / M_i}. \tag{8.79}$$

8.7.2 (NpT) Ensemble

Planar elongation flow of a molecular fluid can also be simulated at constant isotropic pressure, with the equations of motion given by

$$\dot{\mathbf{r}}_{i\alpha} = \frac{\mathbf{p}_{i\alpha}}{m_{i\alpha}} + \dot{\epsilon}\left(x_i \mathbf{i} - y_i \mathbf{j}\right) + \dot{\xi}\mathbf{r}_i \tag{8.80}$$

$$\dot{\mathbf{p}}_{i\alpha} = \mathbf{F}_{i\alpha} + \mathbf{F}_{i\alpha}^{C} - \dot{\epsilon}\frac{m_{i\alpha}}{M_i}\left(p_{xi}\mathbf{i} - p_{yi}\mathbf{j}\right) - \frac{m_{i\alpha}}{M_i}\zeta\mathbf{p}_i - \frac{m_{i\alpha}}{M_i}\dot{\xi}\mathbf{p}_i. \tag{8.81}$$

The volume and the barostat multiplier are again found by solving their equations of motion

$$\dot{V} = 3\dot{\xi}V \tag{8.82}$$

$$\ddot{\xi} = \frac{(p - p_0)V}{QN_m k_B T} \tag{8.83}$$

and the centre of mass isokinetic thermostat multiplier is given by

$$\zeta = \frac{\sum_{i=1}^{N_m} \left[\mathbf{F}_i \cdot \mathbf{p}_i - \dot{\epsilon} \left(p_{xi}^2 - p_{yi}^2 \right) \right] / M_i}{\sum_{i=1}^{N_m} \mathbf{p}_i^2 / M_i} - \dot{\xi}. \tag{8.84}$$

8.8 Results for Molecular Fluids

In this section, we will give examples of results for the rheological properties of molecular fluids computed using the SLLOD equations of motion discussed in the previous sections.

8.8.1 Carbon Disulphide

Carbon disulphide is a linear triatomic molecule, which can be modelled using the three-centre model of Tildesley and Madden [300]. It can be treated as a rigid body and the equations of motion for the translational and rotational degrees of freedom solved directly, using the quaternion method. Alternatively, a CFR type constraint algorithm can be used to solve the equations of motion for the individual interaction sites. In fact, CS_2 is a useful test system for the development of a CFR type constraint code. Prathiraja, Daivis and Snook [301] applied the Baranyai and Evans [173, 283] version of the

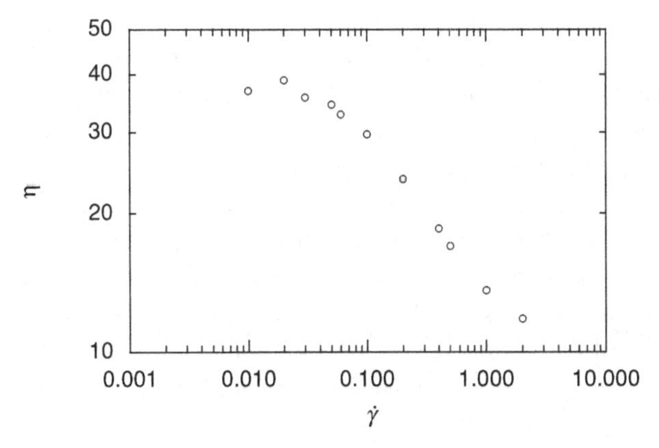

Figure 8.1 Shear viscosity versus shear rate for CS_2 at T =193 K and ρ = 1420 kg/m^3. Error bars are only slightly larger than the plot symbols and have been omitted for clarity. Replotted from reference [301].

CFR algorithm to simulations of the rheology of CS_2 at T =193 K and ρ = 1420 kg/m^3 using both equilibrium and nonequilibrium methods. The simulations were conducted with a system size of N = 256 molecules under constant (N, V, T) conditions.

Figure 8.1 shows the shear viscosity of CS_2 computed using the molecular SLLOD algorithm with a molecular centre of mass kinetic thermostat, plotted against strain rate on a logarithmic scale to display the strong strain rate dependence of the viscosity over the range of strain rates investigated. All properties are expressed in reduced units using the Lennard-Jones parameters ϵ_{CC}, σ_{CC}, and the mass m_C as the base units. Similar to the behaviour observed for atomic fluids, we see a Newtonian region at low strain rates where the viscosity becomes almost independent of strain rate, followed by a region where the fluid is strongly shear thinning. The high shear rates at which shear thinning occurs for this molecular fluid are not experimentally accessible. Nevertheless, the strain

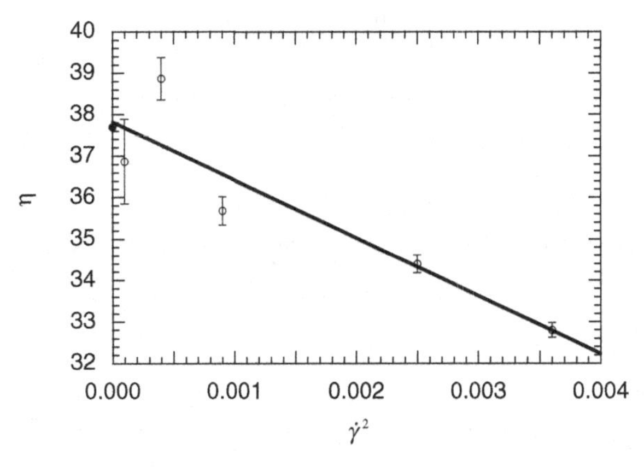

Figure 8.2 Shear viscosity versus squared strain rate for CS_2. The linear fit shown has an intercept giving $\eta_0 = 37.8 \pm 0.8$ in reduced units, while the Green–Kubo result at zero shear rate is 37.7 ± 0.1. Replotted from reference [301].

rate dependence is of theoretical interest, and the viscosity can be extrapolated to zero shear rate to obtain the experimentally measurable zero strain rate viscosity. As for atomic liquids, there is no consensus on the most appropriate functional form to fit to the strain rate dependence of the viscosity for this purpose. From a purely practical point of view, the agreement between the equilibrium (Green–Kubo) and extrapolated NEMD values of the zero strain rate viscosities is usually found to be excellent if the viscosity is computed at very low strain rates and then fitted with a polynomial in the squared strain rate. Figure 8.2 shows the results of this procedure for CS_2. The range of strain rates used in the fit has been restricted such that the plot is linear and the viscosity is a quadratic function of the strain rate. The zero strain rate viscosity obtained from the linear fit is in excellent agreement with the Green–Kubo result.

8.8.2 Results for Polymeric Liquids

The rheology of polymeric liquids, including melts, blends and solutions has been extensively studied using NEMD with the SLLOD algorithm. Fully explicit molecular dynamics studies have so far generally been limited to relatively short chains due to the enormous computational requirements of molecular dynamics simulations of long chain systems. This means that the effects of entanglements are rather difficult to study by standard molecular dynamics methods. From this point of view, simulation techniques that make use of coarse-graining to eliminate fast degrees of freedom are very attractive. However, standard NEMD simulations have several advantages that ensure that they will remain useful for studies of the rheology of polymeric liquids. Firstly they can be regarded as benchmarks for comparisons with other simulation methods that rely on strong assumptions to achieve coarse-graining. They also enable consistent computations of widely differing properties with a single molecular model. For example, both thermodynamic and transport properties can be studied simultaneously. The assumption of incompressibility, which is often made in coarse-grained simulations, is not required in molecular dynamics, so shear dilatancy – the change in the density of a liquid when it is sheared strongly – emerges naturally in molecular dynamics simulations. Hydrodynamic interactions also emerge naturally in molecular dynamics simulations of colloids and polymer solutions, and do not need to be included by computing complicated many-body hydrodynamic mobilities. First and second normal stress differences can easily be computed for melts and solutions without neglecting solvent or hydrodynamic interaction effects. As computer power continues to grow and simulation techniques continue to improve, many of the disadvantages of molecular dynamics simulations will gradually become less significant and studies of entangled polymeric liquids with huge disparities in timescales will become possible.

We will now discuss some examples to illustrate the type of detailed information that can be extracted from NEMD simulations of polymeric liquids.

As we have seen in previous chapters, it is possible to study both shear flow and planar elongational flow by combining appropriate periodic boundary conditions with the SLLOD equations of motion. This is a substantial advantage for studies of nonlinear rheological constitutive equations. It is assumed that the very general constitutive equations derived from the retarded motion expansion should self-consistently describe the nonlinear response of a viscoelastic fluid to different types of deformation for low to moderate values of the deformation rate. By this, we mean deformation rates that are large enough to generate a measurable nonlinear response, but not so large that the viscosity and normal stresses depart from a polynomial dependence on the rate of deformation.

Matin, Daivis and Todd have described the development of algorithms and simulation methodology for studying planar shear flow and planar elongational flow of molecular fluids in a series of papers. They computed the rheological properties of molecular fluids with a wide range of chain lengths ranging from simple diatomic molecules to short chain (unentangled) oligomers and longer chain (entangled) polymers [197, 302–304]. Extending these studies to very low strain rates and chain lengths that are sufficiently high to show the effects of entanglement, Daivis Matin and Todd [217] reported results for steady state shear and planar elongational flow of freely jointed chain polymer melts with chain lengths ranging from $N = 4$ up to $N = 100$. Fixed bond length constraints were implemented by the Gaussian constraint method described earlier, and the molecular SLLOD algorithm with a molecular centre of mass kinetic thermostat was used to generate the flow and control the temperature of the system. In these simulations, the reduced temperature was $T = 1.0$ and the reduced atomic (or interaction site) number density was fixed at $n = 0.84$. The number of molecules in the system was $N_m = 2500$ for $N = 4$, 10 and $N_m = 500$ for $N = 20, 50, 100$.

By calculating the Kuhn length b_K of these model polymer chains in the melt, Daivis, Matin and Todd were able to compare the simulated system with typical polymer melts. Using the computed value of the mean squared radius of gyration for their 100-site chains, they obtained $b_K = 6\langle R_g^2\rangle/L = 1.74$, where L is the contour length of the chain ($L = 99$). This gave the number of Kuhn steps per molecule as $N_K = 99/b_K = 57$. This means that a 100-site model polymer is equivalent to polyethylene with a molar mass of approximately 8,300 g/mol if the characteristic ratio of polyethylene is taken as 7.2 and the molar mass per bond is 14 g/mol. This procedure calibrates the characteristic length of the LJ potential representing one chain segment as $\sigma = 7.7$ Å and the segment molar mass as $m = 83$ g/mol. Choosing the temperature for comparison with experiment as 443 K, the energy parameter for the LJ potential is given by $\epsilon/k_B = 443$ K. With this mapping, we find that a reduced shear rate of 1.0 corresponds to a real shear rate for polyethylene of $2.7 \times 10^{11} s^{-1}$.

This correspondence should be regarded as approximate at best. The actual density of polyethylene at this temperature is 768 kg/m^3, but the density of the model polymer melt is only 254 kg/m^3 when expressed in real units based on polyethylene. Furthermore, the critical chain length for the onset of entanglement coupling N_c, which depends mainly on the geometry of the molecule expressed in terms of the packing length, occurs at a far smaller number of Kuhn lengths for the model polymer melt than it does for real polyethylene. More sophisticated and realistic coarse-graining schemes for polymer melts have been discussed in the literature [305, 306].

When we compare results for shear and elongational flows, it is useful to define a measure of deformation rate that is comparable for both. The second scalar invariant of the rate of strain tensor, defined by $II = \mathrm{Tr}(\mathbf{D}\cdot\mathbf{D})$ where $\mathbf{D} = \nabla\mathbf{v} + (\nabla\mathbf{v})^T$ is very useful for this purpose. II is directly proportional to the rate at which work is done by the deformation. In the steady state, when transients due to viscoelasticity have decayed, it is proportional to the rate of energy dissipation. The second scalar invariant is $II = 2\dot{\gamma}^2$ for shear flow and $II = 8\dot{\epsilon}^2$ for planar elongational flow.

The retarded motion expansion (RME) is a systematic power series expansion of the stress (or pressure) tensor of a *rheologically simple* [31, 32] viscoelastic fluid that is based on continuum mechanics and is therefore purely phenomenological. The RME does not depend on the validity of any particular microscopic molecular model, so it can be equally well applied to colloids, polymer solutions, melts and other complex fluids. The details of the retarded motion expansion are given in standard references [28, 31]. The third-order retarded motion expansion of the stress tensor for an incompressible fluid gives the following expressions for the viscometric functions in planar shear flow [28],

$$\eta = b_1 - 2\left(b_{12} - b_{1:11}\right)\dot{\gamma}^2 \tag{8.85}$$

$$\Psi_1 = -2b_2 \tag{8.86}$$

$$\Psi_2 = b_{11} \tag{8.87}$$

and the following expressions for the viscometric functions in planar elongational flow:

$$\eta_1 = 4b_1 + 16\left(b_3 - 2b_{12} + 2b_{1:11}\right)\dot{\epsilon}^2 \tag{8.88}$$

$$\eta_2 = 2b_1 - 4\left(b_{11} - b_2\right)\dot{\epsilon} + 8\left(b_3 - 2b_{12} + 2b_{1:11}\right)\dot{\epsilon}^2 \tag{8.89}$$

where the b_{ij} coefficients are constants in the RME. The relationship between these coefficients and the Rivlin-Eriksen tensors defined in Section (2.2.2) is discussed in reference [28]. These equations show that the initial strain rate dependence of the shear viscosity is expected to be quadratic and the first and second normal stress coefficients should be constant at this level of approximation. Likewise, the two elongational viscosities are expected to be quadratic functions of the strain rate in the third-order RME approximation.

The coefficients in these equations were evaluated by fitting the low strain rate data with the appropriate polynomial functions. Examples for $N = 4$ and $N = 50$ are shown in Figure 8.3 and Figure 8.4. Although the data analysis was performed in terms of the shear and elongational strain rates $\dot{\gamma}$ and $\dot{\epsilon}$, the results for the two different flows are presented in terms of the second scalar invariant of the strain rate tensor II where $II = 2\dot{\gamma}^2$ for shear flow and $II = 8\dot{\epsilon}^2$ for planar elongational flow. According to the third-order RME, the shear viscosity and the first planar elongational viscosity should be proportional to the square of the shear and elongational strain rates at low strain rates. We should also observe that the initial strain rate dependence of the second planar elongational viscosity is given by the sum of a term proportional to $\dot{\epsilon}$ and a term proportional to $\dot{\epsilon}^2$. These expectations were fulfilled, but the ranges of validity of the assumed functional forms for each of the viscosity coefficients differed greatly. For example, for $N = 4$, the third-order RME expression was obeyed by the shear viscosity up to $II = 0.008$, whereas the first planar elongational viscosity obeyed the third-order RME

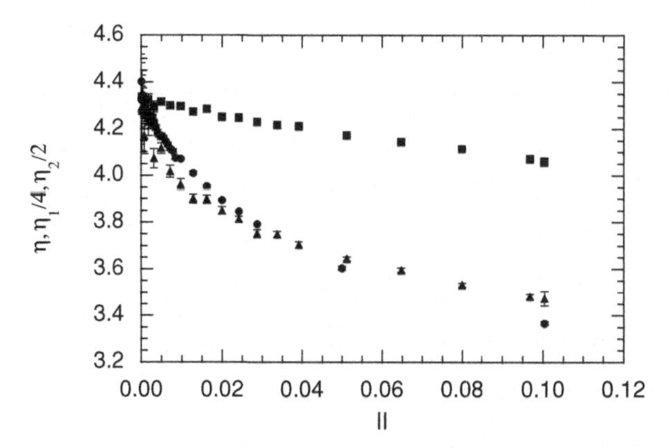

Figure 8.3 Shear viscosity (circles), first planar elongational viscosity divided by four (squares), and second planar elongational viscosity divided by two (triangles), for molecules with chain length N = 4, plotted against the second scalar invariant of the strain rate tensor, II, at low strain rates. Reprinted from reference [217] with permission from Elsevier.

form up to $II = 0.04$ and the second planar elongational viscosity up to $II = 0.2$. A clear difference between the results for $N = 4$ and $N = 50$ shown in Figures 8.3 and 8.4 is that the initial slope of the first planar elongational viscosity is negative for the short chain molecules and positive for the long chain molecules. This is an indication of a change from molecular to polymeric behaviour as the molecular orientation and stretch become significant in the elongational stresses for chain lengths above $N = 4$.

The results of the data analysis using the third-order fluid model to describe the steady shear and steady planar elongational flow results are shown in Tables 8.1 and 8.2 respectively. Results for $N = 2$ [197] have also been included.

Results for the zero strain rate viscosities, η_0, obtained from the shear and first and second elongational viscosities are all in good agreement for each N. The values of

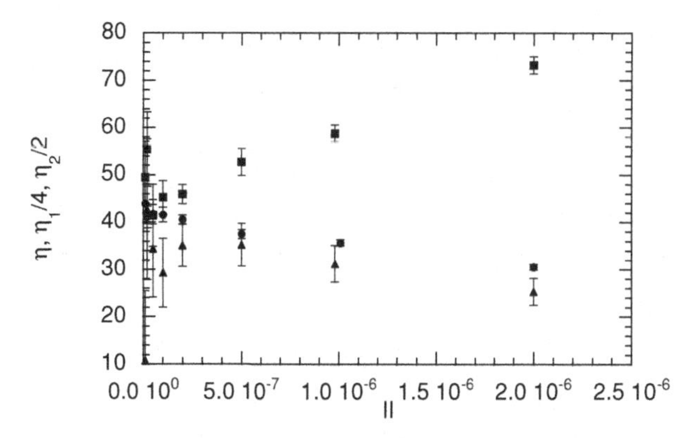

Figure 8.4 Shear viscosity (circles), first planar elongational viscosity divided by four (squares), and second planar elongational viscosity divided by two (triangles), for molecules with chain length N = 50, plotted against the second scalar invariant of the strain rate tensor, II, at low strain rates. Reprinted from reference [217] with permission from Elsevier.

Table 8.1 Third-order fluid model parameters calculated from planar shear flow data. Numbers in brackets represent the standard errors in the means.

N	$b_1 = \eta_0$	$2\,(b_{12} - b_{1:11})$	$-2b_2 = \Psi_{1,0}$	$-b_{11} = -\Psi_{2,0}$	$(b_{11} - b_2)$
2	2.957(2)	4.9(1)	3.44(7)	1.47(6)	0.25(9)
4	4.307(6)	53(2)	20.5(3)	5.2(2)	5.1(4)
10	8.42(4)	3900(340)	320(10)	48(5)	110(10)
20	15.6(1)	105000(9000)	3090(240)	490(170)	1100(300)
50	41.3(6)	$1.09(9) \times 10^7$	56000(3000)	6200(1600)	22000(3000)
100	108(3)	$7(1) \times 10^8$	$1.10(4) \times 10^6$	$4.3(2.5) \times 10^5$	120000(260000)

$(b_{11} - b_2) = (\Psi_{1,0}/2 + \Psi_{2,0})$ evaluated directly from the normal pressure differences in shear flow (column 6 of Table 8.1) and from the strain rate dependence of the second planar elongational viscosity (column 5 of Table 8.2) agree within uncertainties of one standard error for all values of N except $N = 2$. The discrepancy for $N = 2$ is almost certainly due to an overestimation of $(b_{11} - b_2)$ in the fit to η_2, because it also coincides with a value of $(b_3 - 2b_{12} + 2b_{1:11})$ evaluated from η_2 that is small and positive, compared to the *negative* value obtained from η_1.

The values of $-\Psi_{2,0}/\Psi_{1,0}$ for the simulated systems range from 0.43 at $N = 2$ down to 0.11 at $N = 50$. These values agree with those found in previous work on similar systems [197, 307]. For $N = 100$, the value of $-\Psi_{2,0}/\Psi_{1,0}$ calculated from the results in Table 8.1 is 0.39 ± 0.24, which is larger than expected. This is almost certainly due to the poorer statistical accuracy of the data for $N = 100$. When the large uncertainty is taken into account, the lower limit is closer to the expected range of values.

Comparing the results for $(b_3 - 2b_{12} + 2b_{1:11})$ obtained from the strain rate dependence of the first planar elongational viscosity η_1 (column 3 of Table 8.2) and from the second planar elongational viscosity η_2 (column 6 of Table 8.2), we see significant discrepancies. The third-order RME fits to η_2 require the evaluation of three coefficients while the fits to η_1 only require two, resulting in lower uncertainties, but the discrepancy appears to be statistically significant. This discrepancy may be due to the sensitivity of the nonlinear transport properties to the specific details of the thermostat, which we

Table 8.2 Third-order fluid model parameters calculated from planar elongational flow data. Numbers in brackets represent the standard errors.

	From η_1		From η_2		
N	$b_1 = \eta_0$	$(b_3 - 2b_{12} + 2b_{1:11})$	$b_1 = \eta_0$	$(b_{11} - b_2)$	$(b_3 - 2b_{12} + 2b_{1:11})$
2	2.958(3)	−2.6(1)	2.99(1)	0.83(5)	0.30(9)
4	4.32(1)	−5.9(6)	4.31(2)	5.2(2)	6.6(8)
10	8.55(5)	1100(100)	8.4(2)	100(20)	880(250)
20	16.1(2)	94000(13000)	14.8(5)	670(90)	14000(4000)
50	44(2)	$29(3) \times 10^6$	39(3)	16000(3500)	$2.5(9) \times 10^6$
100	120(10)	$1.1(9) \times 10^9$	105(35)	$2(2.5) \times 10^5$	–

Table 8.3 Relaxation times and values of the shear rate and elongation rate at which $De = 1$ for different chain lengths.

N	$\tau = -b_2/b_1$	$\dot{\gamma}^*$	$\dot{\epsilon}^*$
2	0.58	1.72	0.86
4	2.4	0.42	0.21
10	19	0.053	0.026
20	99	0.010	0.005
50	670	0.0015	0.0007
100	5100	0.0002	0.0001

already know must first appear at second order in the deformation rate. A similar effect observed for simple liquids has already been discussed in Section 6.4.2.

The RME is only valid for small values of the Deborah number, $De < 1$. Note that in this section, we follow Bird *et al.* [28] in using De as the dimensionless number to describe steady flow, where only one characteristic time can be identified, although this is not a universal convention. A reasonable estimate of De that is valid for both planar shear flow and planar elongational flow is given by

$$De = \frac{\Psi_{1,0}}{2\eta_0}\sqrt{II/2}. \tag{8.90}$$

which reduces to $De = \dot{\gamma}\tau$ for shear flow, where $\tau = \Psi_{1,0}/2\eta_0$ is the viscous relaxation time, defined by $\tau = \int_0^\infty tG(t)dt/\int_0^\infty G(t)dt$ and $G(t)$ is the stress relaxation modulus from the theory of linear viscoelasticity [308]. Relaxation times and values of the shear rate and elongation rate at which $De = 1$ are given in Table 8.3. The condition that $De < 1$ was easily satisfied in all cases where we have applied the RME to evaluate the shear viscosity, first and second normal stress coefficients and the first planar elongational viscosity. The fits to the second planar elongational viscosity η_2 vs. $\dot{\epsilon}$ extended to the greatest values of De ($De \approx 1$) because it was observed that the expected function form was obeyed to much higher values of De for this property than for the others.

These considerations imply that we cannot attribute the discrepancy between the two values of $(b_3 - 2b_{12} + 2b_{1:11})$ to the application of the RME beyond an appropriate range of strain rates. Therefore, it must be due to the failure of one of the underlying assumptions of the RME, most likely the neglect of thermal effects, which may be significant in NEMD simulations unlike most experimental measurements.

We will now examine the chain length dependence of the coefficients of the RME.

The zero shear rate viscosity is plotted against the number of bonds in the chain, $N - 1$ in Figure 8.5. For short chains ($N < 10$), the molecules are too small for Rouse theory to apply. The slope of 1 predicted by the Rouse model passes through the $N = 10, 20, 50$ points, but the viscosity of the $N = 100$ system is clearly greater than the value predicted by the Rouse model, indicating the onset of entanglement effects. A value of N_c of around 100 for this density and temperature agrees very well with the value found by Kröger and Hess [309] and also by Bosko *et al.* [310] from computations of the shear viscosity of the FENE chain model under the same thermodynamic

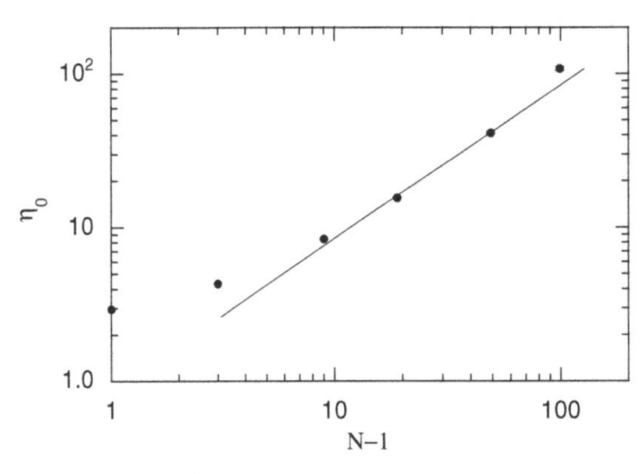

Figure 8.5 Zero shear rate shear viscosity plotted against the number of bonds in the chain. The line has a slope of 1, predicted by the Rouse model. Reprinted from reference [217] with permission from Elsevier.

conditions. From the zero shear rate viscosity in the range of chain lengths that satisfy Rouse scaling, we obtain the value of the monomeric friction coefficient defined by $\zeta = 36\eta/nNb^2$ for $N = 20, 50$ of $\zeta = 22 \pm 1$.

Although the FENE chain model superficially seems to be a more coarse-grained model due to the inclusion of a spring potential between beads, the conformational and rheological data for the FENE chain and the freely jointed bead-rod chain used here are almost identical. Considering the work by Padding and Briels [306], this is not surprising – a truly coarse grained polymer model should use an appropriate potential of mean force to determine the bond stretching force and the direct forces between the "blobs". Thus, the FENE and bead-rod model chains, which both use the same hard potential to represent direct bead interactions, and stiff or totally rigid bonds between monomers, operate at essentially the same level of coarse-graining. The same conclusion was drawn for comparative studies of the rheological and structural properties of polymer melts with FENE and freely jointed chain models [311].

The chain length scaling of the first and second normal stress coefficients is shown in Figure 8.6. Our improved low strain rate data and the additional data for $N = 100$ clarify the chain length dependence which we previously reported. The chain length dependence of both normal stress coefficients is consistent with a power law exponent of 3. This agrees with the value predicted by the Rouse model for the first normal stress coefficient. Rouse theory predicts that the second normal stress coefficient is zero, contrary to our results. Figure 8.7 shows the chain length dependence of the RME coefficient $2\,(b_{12} - b_{1:11})$ defined in Equation (8.85). We see that over the range of chain lengths where other properties are well described by the Rouse model, its chain length dependence is consistent with a power law exponent of 6. At the largest chain length, $N = 100$, the data deviates from the low N prediction due to the onset of entanglement coupling, similar to the data for the viscosity and normal stress coefficients. For simulations performed at constant volume and number of particles, the reduced site number density remains constant as the deformation rate is increased. At the high strain rates accessible

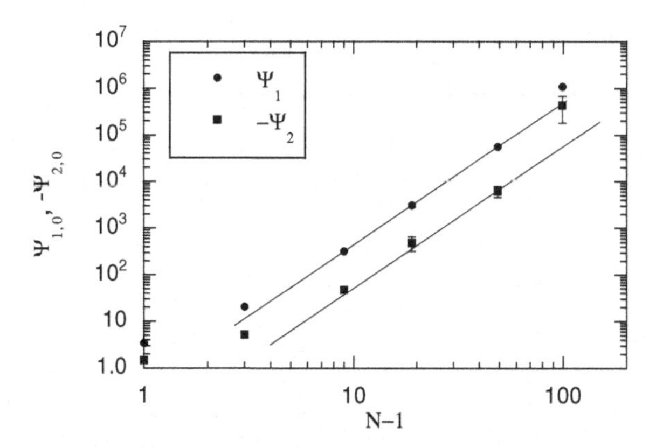

Figure 8.6 Normal stress coefficients at zero shear rate and plotted as $\Psi_{1,0}$ and $-\Psi_{2,0}$ against the number of bonds in the chain. The lines indicate power laws with an exponent of 3. Reprinted from reference [217] with permission from Elsevier.

in NEMD simulations, the pressure and internal energy both vary as the deformation rate is increased. This occurs because the strain rate is sufficiently high to cause distortion of the microscopic structure of the fluid (observable in the pair correlation function $g(\mathbf{r})$), despite the fact that the Deborah number remains small due to the relatively short viscoelastic relaxation times of these fluids. It should be recalled that our largest chain length is approximately equivalent to a polyethylene of molar mass of approximately 8,300 g/mol and a strain rate of 1.0 in reduced units is equivalent to $2.7 \times 10^{11} s^{-1}$ in real units for the equivalent polyethylene. The variation of pressure with deformation rate indicates the presence of compressibility, although here it is the pressure rather than the volume that changes with deformation rate.

Matin, Daivis and Todd computed the strain rate dependence of the isotropic pressure, $p = \frac{1}{3}\mathrm{Tr}(\mathbf{P})$ in both shear and elongational flow. At sufficiently low reduced shear rates,

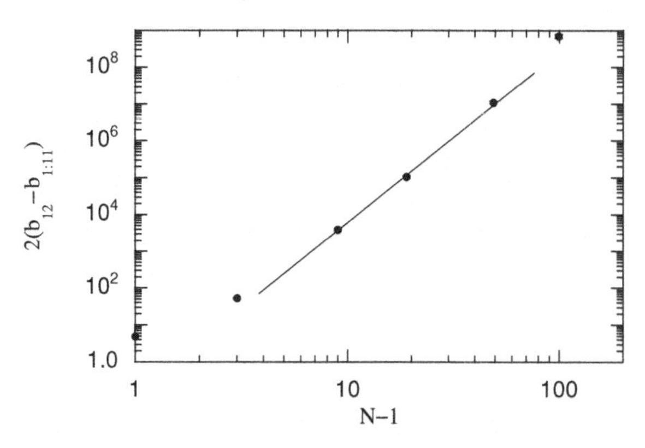

Figure 8.7 Chain length dependence of the coefficient of the quadratic term in the RME expression for the strain rate dependence of the shear viscosity. The line indicates a power law exponent of 6. Reprinted from reference [217] with permission from Elsevier.

Table 8.4 Results of fits to pressure versus deformation rate data for shear and elongational flow. The numbers in brackets represent the standard error in the mean.

N	p_0 (shear)	A	p_0 (elong)	B	B/A
2	6.5511(1)	0.990(6)	6.5510(1)	3.83(2)	3.87(4)
4	5.8335(3)	−0.77(40)	5.8350(4)	−2.14(70)	3(2)
10	5.3950(2)	−35(4)	5.3940(2)	−149(9)	4.3(7)
20	5.2500(2)	−311(50)	5.2500(2)	−1361(240)	4.4(15)
50	5.1660(4)	−7600(2400)	5.1670(3)	−51580(5800)	6.8(30)

the pressure was quadratic in the strain rate for shear and elongational flows. Therefore, at low shear rates, the pressure could always be represented in the form:

$$p = p_0 + A\dot{\gamma}^2 \tag{8.91}$$

$$p = p_0 + B\dot{\epsilon}^2 \tag{8.92}$$

for shear and elongational flows respectively, where p_0 represents the equilibrium pressure. The values of p_0, A and B are given in Table 8.4. The results for $N = 100$ had very large uncertainties, due to the small range of strain rates studied at this chain length, so they are not shown. In continuum mechanics, it is usually assumed that the density of a fluid does not vary with deformation rate. This incompressibility assumption is often mathematically expressed as an internal constraint which opposes any tendency towards isotropic expansion or compression [31]. In a real fluid, an idealised constraint of this type does not exist and the small but nonzero change in the pressure of the fluid due to shear or elongational deformation at constant volume must be described by an equation of state. Apart from a few detailed computational investigations for atomic fluids [209, 214], this equation of state has received little attention, but it is possible to obtain one from the Retarded Motion Expansion.

Most derivations of the RME introduce the incompressibility assumption at an early stage [28, 31, 32], but the derivation can be carried out without making this assumption. Prud'homme and Bird [312] studied dilational flows using a second-order fluid model derived from a truncated RME in the rate of strain tensor and its Jaumann derivatives. However, they did not study shear or elongational flows. Daivis, Matin and Todd [217] have obtained a second-order fluid model from the RME following the same conventions as Bird *et al.* [28], using the strain rate tensor, its lower convected derivative and their invariants, without making the compressibility approximation. The result is:

$$\begin{aligned}\mathbf{P} = p_0\mathbf{1} - \big[&b_1\boldsymbol{\gamma}_{(1)} + b_{0:1}\left(\mathbf{1} : \boldsymbol{\gamma}_{(1)}\right)\mathbf{1} + b_2\boldsymbol{\gamma}_{(2)} + b_{11}\left(\boldsymbol{\gamma}_{(1)} \cdot \boldsymbol{\gamma}_{(1)}\right) + b_{0:1,1}\left(\mathbf{1} : \boldsymbol{\gamma}_{(1)}\right)\boldsymbol{\gamma}_{(1)} \\ &+ b_{0:2}\left(\mathbf{1} : \boldsymbol{\gamma}_{(2)}\right)\mathbf{1} + b_{0:1,0:1}\left(\mathbf{1} : \boldsymbol{\gamma}_{(1)}\right)^2\mathbf{1} + b_{1:1}\left(\boldsymbol{\gamma}_{(1)} : \boldsymbol{\gamma}_{(1)}\right)\mathbf{1}\big],\end{aligned} \tag{8.93}$$

where $\mathbf{1}$ is the unit tensor, $\boldsymbol{\gamma}_{(1)}$ is the first rate of strain tensor, $\boldsymbol{\gamma}_{(1)} = \boldsymbol{\gamma}$ and $\boldsymbol{\gamma}_{(2)}$ is the second rate of strain tensor defined in terms of the lower convected derivative [28]. Explicit expressions for the viscometric functions for planar shear and elongational flow are obtained by substituting the appropriate rate of strain tensors into Equation (8.93)

[217]. For planar shear flow, we find

$$\eta_0 = b_1 \tag{8.94}$$

$$\Psi_{1,0} = -2b_2 \tag{8.95}$$

$$\Psi_{2,0} = b_{11} \tag{8.96}$$

$$p = p_0 + \frac{2}{3}\left[3\left(b_{0:2} - b_{1:1}\right) - \left(b_{11} - b_2\right)\right]\dot{\gamma}^2 \tag{8.97}$$

and for planar elongational flow the desired relationships are

$$\eta_{1,0} = 4b_1 \tag{8.98}$$

$$\eta_2 = 2b_1 + 4\left(b_2 - b_{11}\right)\dot{\epsilon} \tag{8.99}$$

$$p = p_0 + \frac{8}{3}\left[3\left(b_{0:2} - b_{1:1}\right) - \left(b_{11} - b_2\right)\right]\dot{\epsilon}^2. \tag{8.100}$$

The shear and elongational viscosities and the normal stress coefficients to second order in the strain rate are identical to the predictions of the usual incompressible second-order fluid model, but we also now obtain predictions of the strain rate dependence of the pressure. The predicted value of the ratio B/A of the coefficients in Equation (8.91) and Equation (8.92) in the zero deformation rate limit is 4.0. The results in Table 8.4 show that this prediction is reasonably well obeyed. It is interesting to note that if the second scalar invariant of the strain rate tensor is used as a measure of deformation rate, a universal curve for the pressure is obtained for both types of flow, because, as noted earlier, there is a factor of 4 difference between the second scalar invariants of the rate of strain tensors in shear and elongational flows.

When this expansion is carried through to third order, the relationship between the coefficients of the quadratic terms in the two elongational viscosities is the same as in the case where an incompressible fluid is considered, i.e. the coefficient of the quadratic term in the first elongational viscosity is twice the value of the coefficient of the quadratic term in the second elongational viscosity. This means that the discrepancy observed in the values of $(b_3 - 2b_{12} + 2b_{1:11})$ is not resolved by taking compressibility into account – more evidence that thermal effects are responsible for the discrepancy.

Before closing, we remark that there are now many NEMD simulation studies of molecular fluids under shear or elongational flow using the homogeneous techniques we have outlined in this chapter, including studies of branched and hyperbranched molecules and dendrimers. We refer interested readers to the literature review presented in Table 2 of our 2007 review paper for further details [1], though since that time many more simulation studies have been published. While not discussed specifically in this chapter, diffusion of molecules under either shear or elongational flow is another useful property to study and we refer readers to the papers by Sarman *et al.* [313] and Hunt and Todd [314] for further details.

9 Inhomogeneous Flows for Atomic Fluids

So far we have largely considered NEMD simulation methods for fluids undergoing homogeneous flows. The exceptions to this were the “method of planes” and volume averaging techniques, in which expressions were derived that allow us to compute densities and fluxes for inhomogeneous fluids for various geometries (see Chapter 4). In this chapter we will use some of these expressions to compute relevant properties of highly confined fluids under several different flow conditions. In addition, we will show how to implement appropriate equations of motion to faithfully model fluids subject to spatially periodic fields and fluids under extreme confinement. It is with the latter application in mind that NEMD techniques are particularly important in the field of nanofluidics.

For over 150 years classical Navier-Stokes hydrodynamics [315] has been a wonderfully successful theoretical tool for predicting the properties of fluids and gases under a large variety of conditions. Its success extends from describing the dynamics of galactic motion [316], the aerodynamics of flight [317], the hydrodynamics of substances from liquid water to dense polymer melts [28], and right down to the flows of fluids on the microscale [318, 319]. It has even been shown to be accurate down to nanoscale dimensions [96], as long as certain conditions are maintained. In a numerical study on the Lennard-Jones fluid, it was first clearly demonstrated that, for an atomic fluid confined by atomistic walls, the Navier-Stokes equations were valid down to confinement spaces as low as around 5 to 10 atomic diameters [96, 320]. Below this spacing the fluid becomes highly inhomogeneous in space so the assumptions of constant density, constant viscosity, etc. break down, as do the Navier-Stokes equations.

At such small length scales another significant problem arises: the transport properties of fluids become nonlocal in nature. Although this effect has implicitly been built into the theory of generalised hydrodynamics [321] (see also Chapter 11) it has only recently been validated when the spatial extent of variations in the velocity gradient of the fluid are of the order of the width of the viscosity kernel [322]. The kernel itself is a nonlocal material property of the system with both wavevector and frequency dependence [321, 323] and has been accurately computed and parameterised for the Lennard-Jones fluid [324]. It has also been computed for hard sphere fluids at several state points [321], liquid water [325, 326] and polymeric fluids [327, 328].

The most significant conclusion to follow from reference [322] is that accurate predictions of the shear stress profile for a fluid under time-independent flow (or equivalently, the velocity profile of a shearing fluid) require the full nonlocal viscosity kernel to be invoked, rather than the local Navier-Stokes (infinite wavelength) viscosity. This also raises an interesting question: *to what extent does curvature in the velocity gradient of fluids affect the Navier-Stokes predictions of the shear stress?* We will answer this question in Section 11.1. First however, we consider the case of an unbounded fluid under the influence of a spatially periodic external field. We will then consider highly confined simple atomic fluids under planar Couette and Poiseuille flow, and then molecular fluids and binary mixtures in the following chapter.

9.1 Sinusoidal Transverse Field (STF) Method

One of the first NEMD techniques devised was in fact the sinusoidal transverse field method (STF), developed by Gosling *et al.* [16]. In this method, an external field is applied to fluid atoms such that it is spatially modulated in the direction transverse to the direction of the field, as depicted in Figure 9.1. In what follows we will assume we have a three-dimensional system of atoms interacting via pairwise forces. The system is fully periodic in all three dimensions and the external field is directed in the x-direction but modulated in y. The STF method is a very useful and powerful device to extract a number of transport properties of fluids, including thermal conductivities and wavevector dependent viscosities (and hence the nonlocal transport properties of fluids, as described in Chapter 11 [16, 296, 322, 324, 329]). It has also been successfully used to discover the coupling of the heat flux to the gradient of the square of the strain rate tensor [296, 330, 331]. We will consider these methods and applications in what follows.

The external force can be expressed as

$$\mathbf{F}^e(y) = \mathbf{i}F_{x1}\sin(k_1 y), \tag{9.1}$$

where $\mathbf{i}$ is the unit vector in the x-direction and F_{x1} is the magnitude of the external force. The wavevector k_1 is given as $k_1 = (2\pi/L_y)$ where L_y is the length of the simulation box in the y direction. At low Reynolds number, such a field can be expected to generate

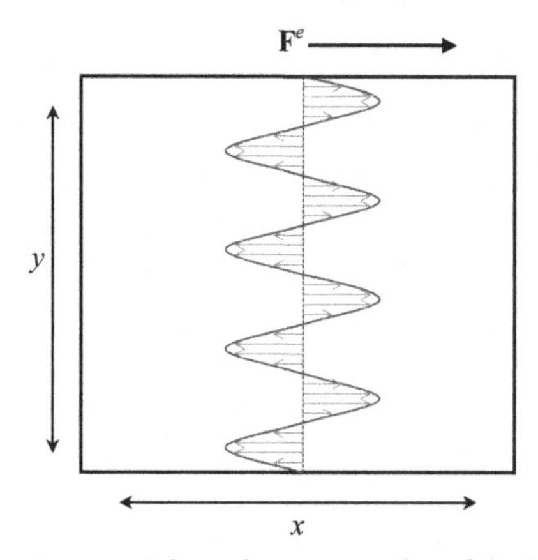

Figure 9.1 Schematic representation of the STF system.

a streaming velocity in the x-direction only, which will also be a function only of y [16, 296]. We can expand the streaming velocity as

$$v_x(y) = \sum_n v_x(k_n)\sin(k_n y)\,;\;\; n = 1, 3, 5... \tag{9.2}$$

where $k_n = (2\pi n/L_y)$. Similarly, we find that the other thermodynamic quantities of interest can all be written as Fourier expansions involving purely odd or even sine or cosine terms [16, 296]. In particular, we note the forms of the xy-element of the pressure tensor, temperature, heat flux vector and number density to be:

$$P_{xy}(y) = \sum_n P_{xy}(k_n)\cos(k_n y)\,;\;\; n = 1, 3, 5... \tag{9.3}$$

$$T(y) = T_0 + \sum_n T(2k_n)\cos(2k_n y)\,;\;\; n = 1, 2, 3... \tag{9.4}$$

$$J_q(y) = \sum_n J_q(2k_n)\sin(2k_n y)\,;\;\; n = 1, 2, 3... \tag{9.5}$$

$$n(y) = n_0 + \sum_n n(2k_n)\cos(2k_n y)\,;\;\; n = 1, 2, 3... \tag{9.6}$$

Here T_0 and n_0 are the zero-field values of temperature and number density, respectively. The coefficients of these expansions can be calculated by using the instantaneous microscopic representations of the relevant quantities in Fourier space, or by performing least-squares fits for the expansions [2, 296]. We will include a method that employs the IK1 approximation for the O_{ij} terms (see Section 4.3) for the purposes of comparison, and to demonstrate convincingly the errors associated with using the IK1 methods for inhomogeneous fluids.

In the following we use two exact ways to compute the pressure tensor and heat flux vector, one of course being the method of planes technique, which is ideally suited for this type of geometry, where the relevant expressions are given by Equations (4.97), (4.102), (4.111) and (4.114). However, a further technique that requires no statistical mechanical derivation is to simply integrate the governing momentum and energy continuity equations either analytically (if possible) or numerically [86, 88]. We previously showed that the addition of a thermostatting term in the equations of motion does not affect the momentum continuity equation *as long as the correct functional form of the streaming velocity is assumed* (see Section 5.3.3). In the case of the pressure tensor, Equation (4.26) this reduces to

$$0 = -\frac{dP_{xy}(y)}{dy} + n(y)F^e(y), \tag{9.7}$$

which leads to the solution

$$P_{xy}(y) = \int n(y)F^e(y)\,dy. \tag{9.8}$$

Substituting Equations (9.1) and (9.6) into Equation (9.8) gives us an analytic solution for the first two harmonics of P_{xy}:

$$P_{xy}(k_1) = \frac{F_{x1}}{k_1}\left(\frac{n(2k_1)}{2} - n_0\right) \tag{9.9}$$

$$P_{xy}(k_3) = -\frac{F_{x1}n(2k_1)}{6k_1} \tag{9.10}$$

and we will assume that the field strength is sufficient only to excite the first two harmonics. As these expressions involve an integration of the momentum continuity equation, they are termed the "IMC method", as first proposed in reference [86].

The evaluation of the heat flux vector by integrating the energy continuity equation (Equation (4.27)) is more involved. This is because the thermostat term must now be included in the energy continuity expression to account for the heat extracted by the thermostat. If we assume equations of motion of a general form

$$\begin{aligned} \dot{\mathbf{r}}_i &= \frac{\mathbf{p}_i}{m_i} \\ \dot{\mathbf{p}}_i &= \mathbf{F}_i - \alpha(\mathbf{r}_i)\left(\frac{\mathbf{p}_i}{m_i} - \mathbf{v}(\mathbf{r}_i)\right), \end{aligned} \tag{9.11}$$

where the $\mathbf{p}_i$ are laboratory momenta and $\mathbf{F}_i$ is the total force (including both interatomic and external forces) acting on atom i, then by Fourier transforming the energy continuity equation, substituting the microscopic definition of the energy density and the equations of motion into this equation, and then inverse transforming, one obtains the resultant expression for the energy continuity equation in the presence of a thermostat:

$$\begin{aligned} \rho(\mathbf{r},t)\frac{du(\mathbf{r},t)}{dt} &= -\nabla\cdot\mathbf{J}_q(\mathbf{r},t) - \mathbf{P}^T : \nabla\mathbf{v}(\mathbf{r},t) \\ &\quad -\alpha(\mathbf{r})\sum_i m_i(\mathbf{v}_i(t) - \mathbf{v}(\mathbf{r},t))^2\delta(\mathbf{r}-\mathbf{r}_i(t)). \end{aligned} \tag{9.12}$$

In the steady-state the left-hand side of this expression is zero, leaving us with an alternative equation for the heat flux vector, involving only integration of the energy continuity expression:

$$J_q(y) = -\int \mathbf{P}^T : \nabla\mathbf{v}dy - \int \alpha(\mathbf{r})\sum_i m_i(\mathbf{v}_i(t) - \mathbf{v}(\mathbf{r},t))^2\,\delta(\mathbf{r}-\mathbf{r}_i(t))\,dy. \tag{9.13}$$

Following the procedure developed by Baranyai *et al.* [296], the equation of motion for particle momenta (i.e. the second expression in Equation (9.12)) is given as

$$\dot{\mathbf{p}}_i = \mathbf{F}_i^{\phi} + \mathbf{i}F_x(y_i) - \left[\alpha_0 + \sum_n \frac{\alpha_n}{\rho(y_i)}\cos(2k_n y_i)\right](\mathbf{p}_i - m_i\mathbf{i}v_x(y_i)). \tag{9.14}$$

The thermostatting mechanism is wavevector dependent, enabling temperature harmonics to be nulled out at will. This will be instructive when we examine the heat flux in the presence of a velocity gradient shortly. It is simpler to use a Nosé-Hoover feedback scheme to compute the thermostat multipliers, and these are given by solving the differential equations:

$$\dot{\alpha}_0 = \frac{T - T_0}{Q_0}; \quad \dot{\alpha}_n = \frac{T_n(2k_n) - T_{n0}(2k_n)}{Q_n}, \tag{9.15}$$

where T is the instantaneous zero-wavevector kinetic temperature of the *entire* system, T_0 is the target temperature, Q_0 is the "mass" of the temperature feedback control, and $T(2k_n)$ and Q_n are the instantaneous Fourier coefficients of the temperature and relevant feedback control constant.

If we now use the microscopic definition of the kinetic temperature (setting $k_B = 1$),

$$3n(\mathbf{r},t)\,T(\mathbf{r},t) = \sum_i m_i\,(\mathbf{v}_i(t) - \mathbf{v}(\mathbf{r},t))^2\,\delta(\mathbf{r} - \mathbf{r}_i(t)) \tag{9.16}$$

as well as the specific form of the thermostatting mechanism defined in Equation (9.14), the resulting energy continuity equation can be expressed as

$$\begin{aligned} \rho(\mathbf{r},t)\frac{du(\mathbf{r},t)}{dt} &= -\nabla\cdot\mathbf{J}_q(\mathbf{r},t) - \mathbf{P}^T : \nabla\mathbf{v}(\mathbf{r},t) - 3n(\mathbf{r},t)\,T(\mathbf{r},t) \\ &\quad\times\left[\alpha_0 + \frac{1}{n(y,t)}\sum_n \alpha_n\cos(2k_n y, t)\right]. \end{aligned} \tag{9.17}$$

Bearing in mind that all simulations are performed in the steady-state and that all thermodynamic and structural properties are functions of y alone, Equation (9.17) may now be integrated to give the resulting alternative to the method of planes expression for the heat flux vector, the so-called "integrated energy continuity" expression, proposed in reference [88]:

$$J_q(y) = -\int P_{xy}(y)\,\dot{\gamma}(y)\,dy - 3\int T(y)\left[\alpha_0 n(y) + \sum_n \alpha_n\cos(2k_n y)\right]dy. \tag{9.18}$$

We may now substitute the known functional forms for $P_{xy}(y)$, $\dot{\gamma}(y)$, $T(y)$ and $n(y)$ into Equation (9.18) to obtain the Fourier coefficients of $J_q(y)$. However, for the purposes of demonstration we assume that only the zero wavevector (for $T(y)$ and $n(y)$) and first two allowed Fourier components of the above quantities are nonzero, except in the case of $n(y)$, where $n(2k_2)$ is insignificant in comparison to $n(2k_1)$. This assumption is certainly true in the low field regime. We furthermore only calculate the first Fourier coefficient for $J_q(y)$, bearing in mind that the higher-order terms are negligible in comparison. With these approximations in mind we find that $J_q(y)$ is composed of several terms proportional to y and to $\sin(2k_1 y)$. The coefficient of the term proportional to y is

$$\begin{aligned} J_q^{lin} &= -\frac{k_1 v_x(k_1)\,P_{xy}(k_1)}{2} \\ &\quad - 3\left[\alpha_0 T_0 n_0 + \frac{\alpha_0}{2}T(2k_1)\,n(2k_1) + \frac{1}{2}\left(T(2k_1)\,\alpha_1 + T(2k_2)\,\alpha_2\right)\right]. \end{aligned} \tag{9.19}$$

This term, were it to be nonzero, would obviously break the symmetry of the system. However, as we will demonstrate shortly, the first term in Equation (9.19), due to viscous heating, *exactly* cancels the second term, due to the extraction of heat by the thermostat. Thus the linear term is zero. The term involving $\sin(2k_1y)$ is

$$\begin{aligned}J_q(2k_1) = &-\frac{1}{4}\left[(v_x(k_1)+3v_x(k_3))P_{xy}(k_1)+v_x(k_1)P_{xy}(k_3)\right]\\&-\frac{3}{2k_1}\left\{\alpha_0\left[(T_0n(2k_1)+T(2k_1)n_0)+\frac{1}{2}T(2k_2)n(2k_1)\right]\right.\\&\left.+\alpha_1T_0+\frac{1}{2}(\alpha_1T(2k_2)+\alpha_2T(2k_1))\right\}.\end{aligned} \tag{9.20}$$

Baranyai *et al.* [296] postulated an additional coupling of the gradient of the square of the strain rate tensor to the heat flux vector, even in the weak-field limit. They performed STF simulations and confirmed this effect. Several other authors have also postulated such couplings – including heat flux components parallel to the velocity streamlines – and verified these by numerical simulations [80, 331–335]. Baranyai *et al.* proposed the constitutive equation for the heat flux as

$$\mathbf{J}_q(\mathbf{r}) = -\lambda\nabla T(\mathbf{r}) - \xi\nabla\left[\nabla\mathbf{v}(\mathbf{r}):(\nabla\mathbf{v}(\mathbf{r}))^T\right], \tag{9.21}$$

where λ is the thermal conductivity and ξ is the strain rate coupling coefficient. Note that if there is no gradient in the square of the strain rate tensor, then Equation (9.21) just reduces to the well-known Fourier law of heat conduction. For our geometry, this expression simplifies to

$$J_q(y) = -\lambda\frac{dT(y)}{dy} - \xi\frac{d(\dot{\gamma}^2)}{dy}. \tag{9.22}$$

In order to show this effect, as well as to demonstrate the versatility of the various thermostatting options available to us, we discuss the results of some simulations on a WCA fluid of 540 particles. The zero wavevector state point is the Lennard-Jones triple point $(T_0, n_0) = (0.722, 0.8442)$. The simulation box has dimensions $L_x = L_z = 5.0388$, $L_y = 25.194$. All quantities were computed from least squares fitting of the known functional forms of the various quantities of interest. In addition, all quantities except the pressure tensor and heat flux vector were computed in bins of width $\Delta = 0.50388$. The pressure tensor and heat flux vector were instead additionally computed by the MoP technique with the same spatial resolution of planes. In this way we are able to confirm that both the binned/MoP values agree with the least-squares functional fits.

The simulations performed are of three types, as with the simulations performed in Baranyai *et al.* [296]: Type I, in which only the zero wavevector component of the temperature is fixed ($T_0 = 0.722$); Type II, in which additionally the next four allowable temperature harmonics are constrained to zero; and Type III, where the Fourier coefficients of the temperature field were adjusted to that value required to null out the heat flux, whist still maintaining a zero wavevector temperature of 0.722. The heat flux we null for this Type III simulation is that calculated instantaneously by the MoP technique, as this is exact and relies upon no approximations. Furthermore, in contrast to

Table 9.1 The first two terms of the linear component of the heat flux vector given by Equation (9.19), as well as the sum of both terms.

F_{x1}	$J_q^{lin}(1)$	$J_q^{lin}(2)$	ΣJ_q^{lin}
0.2	0.1165	−0.1173	−0.0008
0.3	0.2964	−0.2972	−0.0008
0.4	0.6062	−0.6062	0.0000
0.5	1.0952	−1.0923	0.0029
0.6	2.0364	−2.0332	0.0032

Baranyai *et al.*, we now constrain the first four allowable harmonics of the temperature for type II simulations to ensure we have nulled out the temperature gradient.

Type II simulations, in which the temperature gradient is nulled, would allow us to determine the value of ξ for a number of field strengths:

$$\xi = -\frac{J_q(2k_1)}{(d\dot{\gamma}^2/dy)_1}, \tag{9.23}$$

where $(d\dot{\gamma}^2/dy)_1$ means the first Fourier coefficient of the gradient of the square of the strain rate. In principle, from the known values of ξ we can then calculate the thermal conductivity for either type I or type III simulations.

In Table 9.1 we show the first two terms of Equation (9.19), involving the coefficient of the linear term in the integrated energy continuity (IEC) equation. Also shown is the sum of both terms, which is clearly zero to within estimated statistical uncertainties. This allows us to be confident that there is no net resultant linear term arising in the heat flux, which is required by the symmetry of the system. It also confirms that the heat generated by viscous flow is exactly removed by the thermostat.

We plot the first allowable harmonics of the velocity, density and temperature profiles as a function of field strength for all three types of simulations in Figure 9.2. Also plotted is the second allowable harmonic of the temperature field for Type I simulations. Clearly this is nonzero, justifying the need to apply temperature constraints for the higher-order terms for the Type I simulations at least.

In Figure 9.3 we display $P_{xy}(k_1)$ and $P_{xy}(k_3)$ for Type I and II simulations calculated by the IK1, MoP and IMC methods. For both types of simulation, $P_{xy}(k_1)$ is in good agreement with all three methods. The MoP and IEC values do agree with each other better, though this is difficult to see on the scale of the plot. However, there is a clear disparity in the results for $P_{xy}(k_3)$. In this case we see excellent agreement between the IEC and MoP results for both types of simulations, but the IK1 results are significantly higher. Evidently, at wavevector k_3 the distances involved are too short for the IK1 approximation to be reasonable. As discussed in Section 4.3 and detailed in references [86, 88], this anomaly in the IK1 calculation is a consequence of neglecting the higher-order O_{ij} contributions in the operator expansion. Clearly, ignoring these higher-order terms will lead to significantly incorrect values in the shear stress.

Similar behaviour is also observed for the heat flux. Figure 9.4 shows $J_q(2k_1)$ for Type I and II simulations, once again calculated for all three methods. In Type II

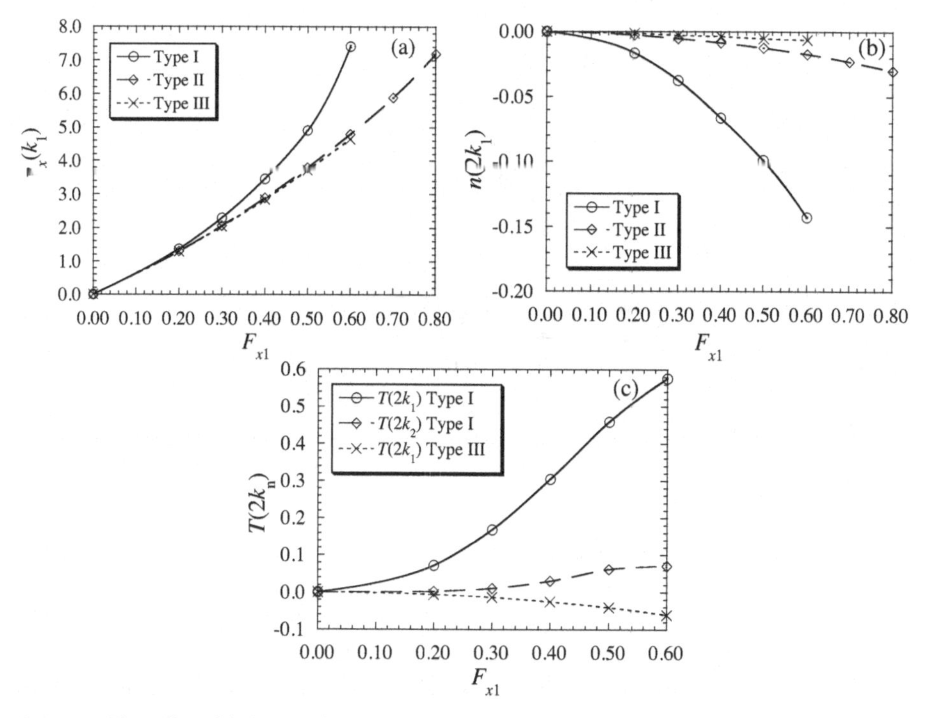

Figure 9.2 First allowable harmonics of (a) streaming velocity, (b) number density, and (c) temperature as a function of external force field strength for different simulation types. Error bars are smaller than symbol sizes.

simulations all methods agree reasonably well, though MoP and IEC values are in better agreement, again difficult to see on the scale of the plot. However for Type I simulations the IK1 values are significantly lower than those for the MoP and IEC methods, which are in excellent agreement with each other. The disparity between the IK1 and IEC/MoP results is again due to the neglect of the higher-order terms in the O_{ij} expansion. Similar

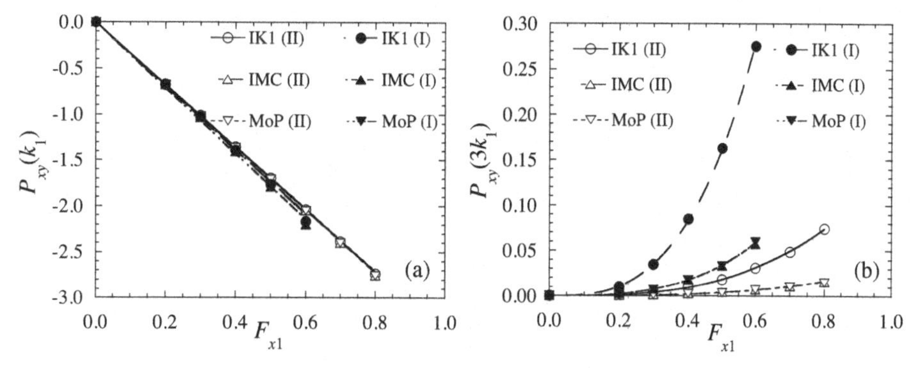

Figure 9.3 First (a) and second (b) allowable harmonics of the negative shear stress for Type I and II simulations. Displayed are Fourier coefficients computed via IK1, IMC and MoP methods.

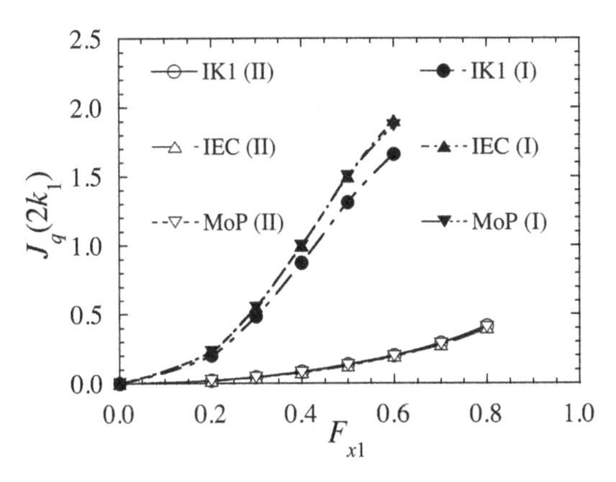

Figure 9.4 First allowable harmonic of the heat flux vector for Type I and II simulations. Displayed are Fourier coefficients computed via IK1, IEC and MoP methods.

results were also obtained for both the heat flux vector and pressure tensor for Type III simulations (not shown).

It is significant that Type II simulations, in which the temperature gradient is nulled, still produce a significant heat flux, confirming the predictions and observations made by Baranyai *et al.* and other authors [80, 296, 331–334]. The sinusoidal form of the heat flux is compatible with a coupling to $\nabla\dot{\gamma}^2$. The only other candidate to have the correct functional form for the heat flux would be ∇P_{xy}^2. However, P_{xy} is related to $\dot{\gamma}$ by the shear viscosity, so $\nabla\dot{\gamma}^2$ and ∇P_{xy}^2 are formally equivalent representations.

In Figure 9.5 we plot the cross coupling coefficient ξ against the first Fourier coefficient of its postulated thermodynamic driving force, $\nabla\dot{\gamma}^2$. We fit the data to the functional form $\xi = a + b(d\dot{\gamma}^2/dy)_1^{1/4}$, though we note that other irrational exponents apart from the $1/4$ we have used can give at least as good a fit. The parameters a and b, along

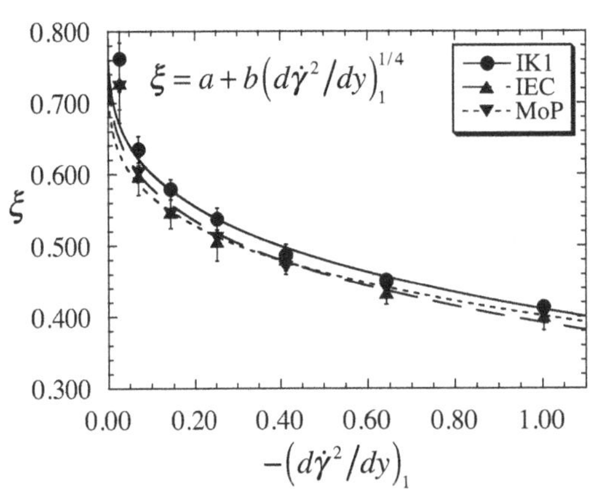

Figure 9.5 Cross-coupling coefficient ξ computed for Type II simulations plotted against the first harmonic of the gradient of the square of the strain rate. ξ is computed via IK1, IEC and MoP methods.

Table 9.2 Coefficients a and b, along with the residual (R), for the fits to the cross-coupling coefficient displayed in Figure 9.5.

Parameter	IK1	IEC	MoP
a	0.85 ± 0.02	0.82 ± 0.04	0.78 ± 0.01
b	-0.43 ± 0.02	-0.43 ± 0.05	-0.38 ± 0.01
R	0.998	0.986	0.998

with the residual error of the plot, are displayed in Table 9.2. Here ξ was calculated by Type II simulations for various field strengths. Use of the IK1 value of $J_q(2k_1)$ overestimates ξ by around 10%. The IEC and MoP estimates for ξ are in good agreement with each other, though errors for the MoP results are significantly lower than corresponding IEC errors. While the errors in the MoP determinations of ξ result from statistical fluctuations of the exact microscopic heat flux, the errors in the IEC values result from errors in the Fourier coefficients of the velocity, density, temperature and shear stress, as well as errors in the thermostat multipliers. It is therefore not surprising that they compound to produce an overall larger error in the IEC heat flux. It should be noted that Ayton *et al.* [336] have also computed the cross-coupling coefficient, except that they used the definition of the normal temperature, rather than the kinetic temperature. Their values of ξ are comparable to our values based on the kinetic temperature.

The thermal conductivity can be computed from Type I and Type III simulations via Equation (9.22). Baranyai *et al.* [296] found that the thermal conductivity was around 25% overestimated if the effects of strain-rate coupling were not included in the constitutive equation.

The strain rate coupling effect is significant since it implies that even in the weak flow limit the temperature profile induced by viscous heating in planar Poiseuille flow is *not* quartic, as predicted by the classical Navier-Stokes equations, but instead contains a contribution which is quadratic. This has in fact been verified by NEMD simulations [330, 336]. The deviation from quartic behaviour is however only noticeable over short distances $y \ll 12\xi/\eta$ [296, 330], near the centre of the flow of the channel, where η is the shear viscosity. It is expected to be important in flows of highly confined fluids and has relevance in nanofluidic engineering as well as capillary flows and lubrication of contacts. We will demonstrate this effect in Section 9.2.

There is one final comment to make on the matter of heat flux for flowing liquids. It has been demonstrated that for a fluid undergoing planar Poiseuille flow, the heat flux remains unchanged from its classical cubic profile by the inclusion of the cross-coupling term in the linear constitutive equation for the heat flux [337]. This result is in fact general for fluids undergoing planar flow [330]. To demonstrate this, for simplicity assume that η and λ are constant. Consider an arbitrary weak flow planar strain rate profile $\dot{\gamma}$. Because the flow is weak we may ignore the effects that viscous heating has on the transport coefficients. The energy conservation equation in the steady-state can be written as

$$\frac{dJ_{qy}(y)}{dy} + P_{xy}(y)\,\dot{\gamma}(y) = 0. \tag{9.24}$$

Subsituting the relevant constitutive relations for shear stress and heat flux we find

$$\frac{dT(y)}{dy} = -\frac{\eta}{\lambda}\int_0^y \dot{\gamma}^2(y')dy' - \frac{\xi}{\lambda}\frac{d\dot{\gamma}^2(y)}{dy}, \tag{9.25}$$

where we have used the symmetry about the centre of the flow $y = 0$ to eliminate the integration constant. This equation may be integrated one more time to give the temperature profile $T(y)$. This profile will clearly depend on the value of the strain rate coupling coefficient ξ and is thus sensitive to it. We can also calculate the heat flux vector assuming the generalised constitutive relation given by Equation (9.22) and substituting the temperature gradient from Equation (9.25) into it. This gives

$$J_{qy}(y) = -\eta\int_0^y \dot{\gamma}^2(y')dy'. \tag{9.26}$$

This is clearly the same expression we would have derived had we not known about the phenomenon of strain rate coupling. Thus we see that strain rate coupling affects the temperature profile, but not the heat flux vector [337]. As far as measurable thermodynamic quantities are concerned, we could have derived exactly the same temperature and heat flux profiles by defining a new nonequilibrium temperature

$$T_{ne} = T_{eq} + \frac{\xi}{\lambda}\dot{\gamma}^2 \tag{9.27}$$

and not invoking a generalised constitutive relation for heat flow [337]. This has also been noticed by Jou and co-workers [80, 334].

In addition to investigating the transport of heat, the STF method can be used to determine the wavevector dependent shear viscosity of fluids. This has been performed by several authors [16, 296, 324, 329] and we refer readers to these papers for greater detail. We will however revisit the wavevector dependent viscosities when discussing nonlocal transport in Section 11.1.

Finally, the sinusoidal field method can also be applied in a longitudinal manner (SLF method), rather than a transverse shearing (STF) manner. Thus, if the field is applied in the y rather than x direction, the net result is to induce a steady-state stationary oscillatory density distribution in the fluid [338, 339]. Furthermore, one can use a combination of both SLF and STF to generate a fully periodic three-dimensional fluid with oscillatory density distribution and shear that can mimic the structure and rheology of a highly confined fluid under shear, yet without the complications of interfacial solid-liquid transport processes. Using such a combination of SLF and STF fields, Dalton *et al.* [339–342] were able to precisely quantify the wavevector dependence of an inhomogeneous fluid under shear and gain considerable insight into the nature of the coupling of the density and velocity gradients, including a theoretical study of the linear and nonlinear density response functions for inhomogeneous fluids. Hoang and Galliero used the technique to investigate the local average density model (LADM) approach to viscosity inhomogeneity [338, 343, 344]. The ultimate goal of such a study is to quantify for the first time the nonlocal viscous behavior of highly inhomogeneous (confined) fluids, analogous to what has already been accomplished for unconfined homogeneous fluids [322, 345]. We refer interested readers to these papers for further details, as well as Section 11.1.

9.2 Poiseuille Flow

Planar Poiseuille, or pressure driven flow, strictly speaking occurs when a pressure gradient exists to drive a fluid through walls separated by some distance, L_y, as depicted in Figure 9.6. In our geometry we consider flow in the x-direction, with all inhomogeneity existing only in the y-direction. The system is fully periodic in all three spatial dimensions. We can express the momentum continuity equation as

$$\rho(\mathbf{r},t)\frac{d\mathbf{v}(\mathbf{r},t)}{dt} = -\nabla\cdot\mathbf{\Pi} - \nabla p + n(\mathbf{r},t)\mathbf{F}^e, \tag{9.28}$$

where $\mathbf{\Pi}$ is the viscous part of the pressure tensor, p is the isotropic pressure (one third the trace of the full pressure tensor), $n(\mathbf{r},t)$ is the number density and $\mathbf{F}^e$ is an external force per particle. Note that it is the *gradient* of the pressure that drives the flow. Because of this we can choose an external field such that

$$\nabla p = -n(\mathbf{r},t)\mathbf{F}^e \tag{9.29}$$

and perform simulations in the absence of an actual pressure gradient. In our geometry an actual pressure gradient to drive the flow would have to exist in the x-direction. However, in computer simulations one would necessarily require very large spatial gradients in this direction to generate the flow. This in turn could generate significant density gradients in the x-direction which then makes the analysis of the system very complicated. For these reasons it is far simpler to use Equation (9.29) to generate the driving field and ensure that the system is entirely homogeneous in both the x and z directions.

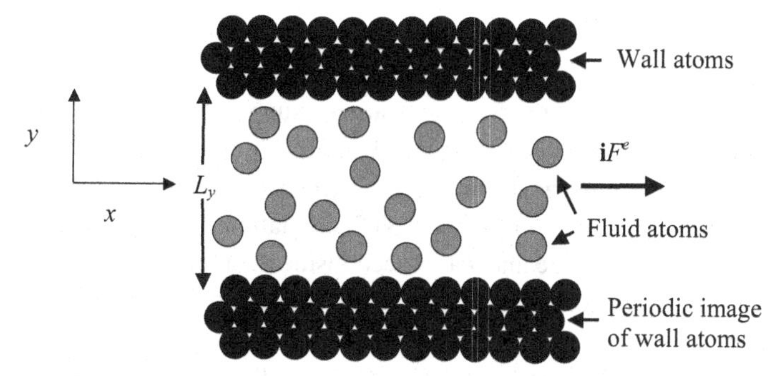

Figure 9.6 Schematic diagram for Poiseuille flow. The system is fully periodic in all three directions, with flow in the x direction and the channel separated by L_y in the y direction. z is normal to the page. Simulations are such that the second wall is just the periodic image of the first wall.

As pointed out previously [86], one might object to this method by observing that in Equation (9.29) the pressure gradient is not constant (as it depends on the number density of the fluid) and therefore this is not equivalent to the standard interpretation of classical Poiseuille flow. This is particularly so near the walls where there are significant density gradients, even though for a particular value of the coordinate y the density is constant. However, as we are interested in material properties of fluids, these are independent of whether the field is constant or inversely proportional to the density. Simulations using either method would yield identical results for material properties as long as these are in the weak field limit. Furthermore, as our goal is to study flows inside very narrow channels the standard classical interpretation of Poiseuille flow may no longer be valid in any case. In the situation where large density gradients of the order of molecular diameters exist near walls, it is no longer necessarily true that the pressure gradient will be independent of position throughout the channel.

With the above clarification in mind, we consider now flow governed by the following hydrodynamic equation:

$$\rho(\mathbf{r},t)\frac{d\mathbf{v}(\mathbf{r},t)}{dt} = -\nabla\cdot\mathbf{\Pi} + n(\mathbf{r},t)\mathbf{F}^e. \tag{9.30}$$

This would be equivalent to gravity-driven flow, where the external field could be equated with gravity via $\mathbf{F}^e = m\mathbf{g}$ and m is the mass of fluid atoms and we assume here that we have a single-component system. Binary systems will be considered in Section 10.3. We also assume that the system is sufficiently small such that differences in atmospheric hydrostatic pressure can be ignored. In the steady-state and for our geometry, Equation (9.30) reduces to

$$0 = -\frac{dP_{xy}(y)}{dy} + n(y)F^e, \tag{9.31}$$

where $\mathbf{F}^e = \mathbf{i}F^e$. If we assume that the flow is symmetric about the mid-point $y = 0$ (leading to zero stress in the middle of the channel), then Equation (9.31) can be integrated to obtain the expression:

$$P_{xy}(y) = F^e\int_0^y n(y')dy'. \tag{9.32}$$

As we have already seen for the STF simulations, such expressions give an alternative to the method of planes statistical mechanical approach. In the case of planar Poiseuille flow it is particularly simple as it involves a numerical integration of the density profile. This is very similar to Equation (9.8) for STF simulations, except that in that case the field was not constant in space and was placed inside the integral. Both of these expressions are obtained by directly integrating the governing momentum continuity equation (the IMC method) [86].

We display the number density, pressure normal to the walls (P_{yy}) and negative shear stress (P_{xy}) for a fluid of WCA atoms in Figure 9.7 for a high density fluid with relatively wide channel width of 24.2σ. Note that "channel width" here defines the accessible width available to fluid atoms, and we label it as h. There will always be some degree of arbitrariness in how this is defined, and in our case we define it to be the width between

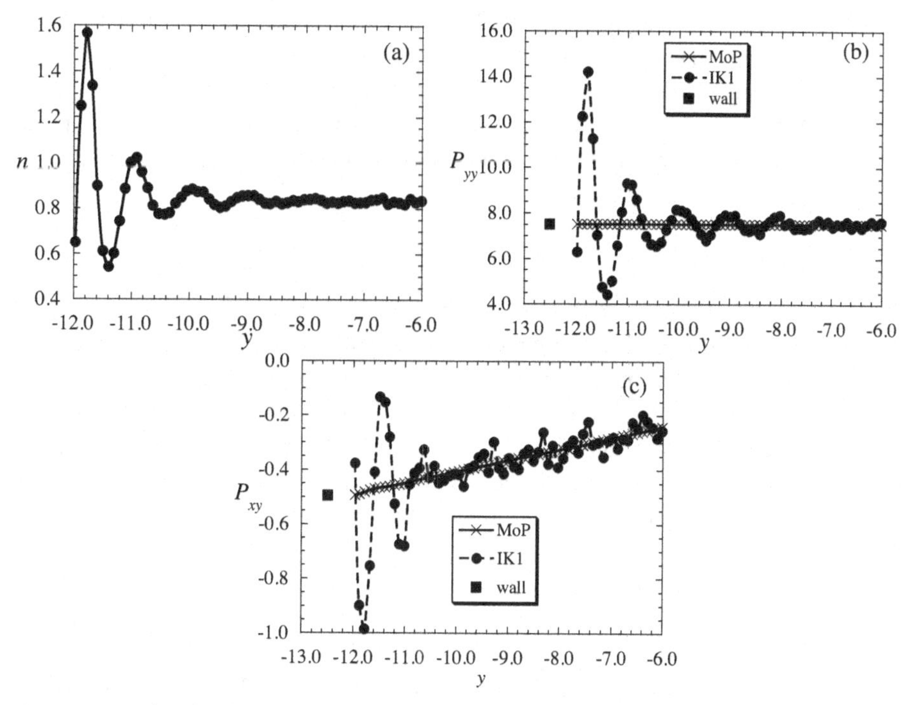

Figure 9.7 Number density (a), normal pressure (b) and negative shear stress (c) for a WCA atomic fluid of 558 atoms confined by walls separated by length $L_y = 25.2$. There are three layers of real wall atoms consisting of 18 atoms per layer. The second wall is just the periodic image of the first wall. Liquid density is 0.836; wall density is 1.255. Walls are thermostatted at $T = 0.722$, the external field strength is 0.05 and wall atoms are held together by a Hookean force with spring constant $k = 57.15$. Average fluid temperature = 0.97. Only the first few layers away from the first wall are shown. The fluid is symmetric about $y = 0$. Along with MoP and IK1 values, the wall pressure and stress are also shown. Replotted from reference [86].

centres of mass of the first layers of opposite walls minus the width of one atomic layer, i.e. $h = L_y - 1.0\sigma$. In addition to plotting pressure values computed via the MoP technique, we also display the values computed via the IK1 method. Clearly IK1 pressures and stresses contain artificial oscillations that are strongly correlated with oscillations in the fluid density that result from layering near the wall-fluid boundary [86]. This again highlights the importance of *not* using IK1 pressures or stresses for systems that are inhomogeneous in space. The wall pressure and stress are also displayed, where these are computed by dividing the y and x components respectively of interaction forces between wall atoms and fluid atoms on one side of the wall by the surface area of the wall. Because one wall is just the periodic image of the other, the total force exerted on the "real" wall by fluid atoms is zero. Because of mechanical stability in the steady-state P_{yy} at the wall must equal P_{yy} in the fluid, and must be constant, which is indeed what is seen from the MoP calculations. The shear stress is almost linear, but has some weak oscillations near the walls, as seen from the MoP data.

One can of course solve the governing Navier-Stokes equation for this type of flow and obtain a predicted streaming velocity profile. For sufficiently wide channels, it has been shown that the Navier-Stokes prediction is in very good agreement with actual NEMD velocity profiles [96]. This agreement breaks down for channel widths less than about 5–10 atomic diameters. Substitution of Newton's law of viscosity $P_{xy}(y) = -\eta\dot{\gamma}(y)$ into Equation (9.31), assuming a constant density profile, and integrating twice over y with stick boundary conditions (i.e. $v_x(y = \pm L_y/2) = 0$) gives us the Navier-Stokes prediction:

$$v_x(y) = -\frac{\bar{n}F^e}{2\eta}\left(y^2 - \frac{L_y^2}{4}\right), \tag{9.33}$$

where $\bar{n}$ is an effective number density (constant). This expression gives the classical Navier-Stokes quadratic solution to the velocity profile and can be fit to actual NEMD streaming velocity data to extract the effective viscosity η. Alternatively η can be estimated by either homogeneous NEMD (e.g. SLLOD) or equilibrium Green–Kubo calculations for the weak-field limit at a state point that is the average of the system density and temperature, and then used in Equation (9.33) to predict the streaming velocity profile.

The streaming velocity profile is presented in Figure 9.8 for the system described in Figure 9.7. For such a wide channel excellent agreement is found between the Navier-Stokes prediction and the NEMD simulation data. The effective viscosity, determined from the fit to Equation (9.33), is found to be $\bar{\eta} = 2.32$. As mentioned, the Navier-Stokes prediction will fail for very narrow channel widths. This can be seen in Figure 9.9 in which the actual and Navier-Stokes streaming velocities are plotted for a channel width of 4.1σ. Clearly the Navier-Stokes prediction fails dismally for such a system. Instead, a reasonable fit to the streaming velocity is found by including an additional cosine series to the classical quadratic form [96]:

$$v_x(y) = v_1 + v_2 y^2 + \sum_n v_n^c \cos\left(\frac{2\pi}{L_y} y\right). \tag{9.34}$$

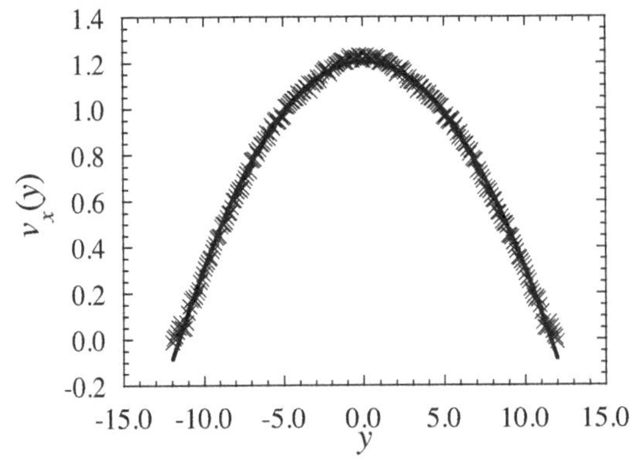

Figure 9.8 Streaming velocity profile (crosses) for the system described in Figure 9.7. The solid curve represents the quadratic fit to the data, compatible with the Navier-Stokes prediction. Replotted from reference [86].

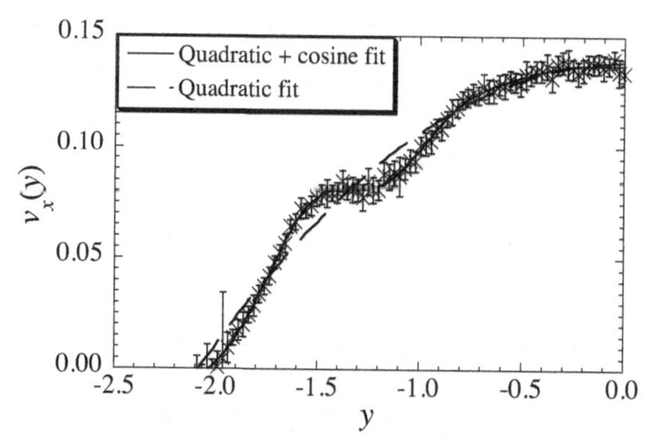

Figure 9.9 Left half of the streaming velocity profile (crosses) for the $L_y = 5.1\sigma$ system. The solid curve represents the fit to the data given by applying Equation (9.34) using a 10-term cosine function [96]. The dashed line is the standard Navier-Stokes quadratic fit. The system consisted of 360 fluid atoms and three "real" wall layers of 72 atoms per layer. $\bar{n}_f = 0.715$; $n_w = 0.85$; $\bar{T}_f = 0.9$; $T_w = 0.722$; $F^e = 0.1$; $k = 150.15$. Replotted from reference [96].

In Figure 9.10 we show the negative shear stress for this system computed via the MoP and IMC methods, showing excellent agreement. Note the strong oscillations due to high fluid layering near the walls.

Another important quantity to measure is the flux of heat through the system. We want to mimic nature as closely as possible and the fluid to dissipate heat as faithfully as possible compared to a real physical experiment. The way we do this is to thermostat the walls only and allow the heat generated by the viscous flow to dissipate through the walls. In this case the flux of heat should be as close to a natural system as possible, since the details of the thermostatting mechanism are neither known nor needed by fluid atoms (see following section for details of thermostatting mechanism). This in turn should generate the expected profile for the heat flux. The heat flux vector can be

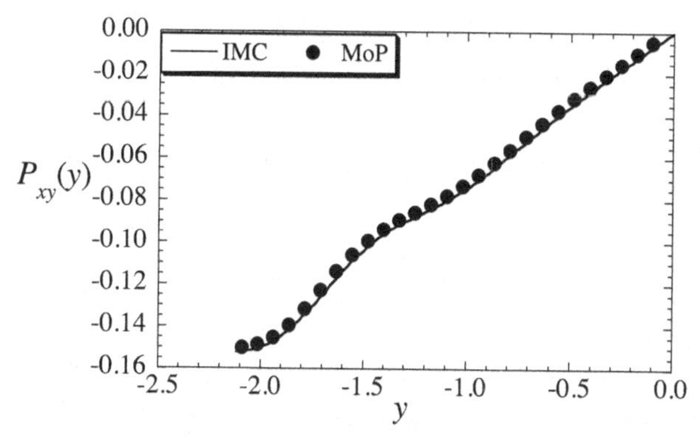

Figure 9.10 Left half of the negative shear stress profile for the system described in Figure 9.9 computed by the MoP and IMC methods. Error bars are smaller than symbol sizes. Replotted from reference [96].

computed either by the method of planes technique detailed in Section 4.3 or, as with the STF system, by directly integrating the energy continuity equation. We recall Equation (2.38c)

$$\rho(\mathbf{r},t)\frac{du(\mathbf{r},t)}{dt} = -\nabla\cdot\mathbf{J}_q(\mathbf{r},t) - \mathbf{P}^T(\mathbf{r},t) : \nabla\mathbf{v}(\mathbf{r},t). \tag{9.35}$$

In the steady-state the left-hand side is zero, so for our geometry this expression reduces to

$$0 = -\frac{dJ_{qy}(y)}{dy} - P_{xy}(y)\dot{\gamma}(y). \tag{9.36}$$

This is the same expression that we found for the STF simulation (see Equation (9.24)). Its solution for the heat flux vector is simply

$$J_{qy}(y) = -\int_0^y P_{xy}(y')\dot{\gamma}(y')\,dy', \tag{9.37}$$

where we note again that the flow is symmetric about $y = 0$, hence $J_{qy}(y) = 0$ in the middle of the channel.

It is also useful to compute the heat flux *at the walls*, i.e. the heat removed by the thermostat. This is useful because it can be used as a check to ensure that the heat flux computed either by the MoP or IEC techniques are correct at the fluid-wall interface. Gaussian thermostats remove heat from a system at the rate given by [2]

$$\dot{Q}(t) = \alpha(t)\sum_{i=1}^{N}\frac{\mathbf{p}_i^2}{m_i} = 2\alpha(t)K(t), \tag{9.38}$$

where here N refers to the number of particles in the system being thermostatted and K is the kinetic energy of the particles. For our system, N would be the number of wall atoms. If our simulation models two separate walls, then the heat flux at each wall is obtained by dividing this expression by the surface area of the walls, A, and averaging over time, i.e.

$$\langle J_{qy}(y = \pm L_y/2)\rangle = \frac{2}{A}\langle K_w\alpha\rangle, \tag{9.39}$$

where K_w and α are the kinetic energy and thermostat multiplier of the particular wall considered. If, however, we have only one "real" wall, and model the second wall as the periodic image of the first wall, then the heat extracted by this wall (and its image) is halved, i.e.

$$\langle J_{qy}(y = \pm L_y/2)\rangle = \frac{1}{A}\langle K_w\alpha\rangle. \tag{9.40}$$

In Figure 9.11 we plot the MoP and IK1 values for the heat flux vector for a system of WCA fluid atoms in a channel of width 49σ [88]. Not shown are IEC values because they are identical to MoP to within very small statistical uncertainty. Also shown is the value of the heat flux at the walls, computed by Equation (9.40). While the MoP and IEC values are in excellent agreement, the IK1 values again display spurious oscillations, particularly near the walls. The cubic profile is in agreement with the cubic profile one

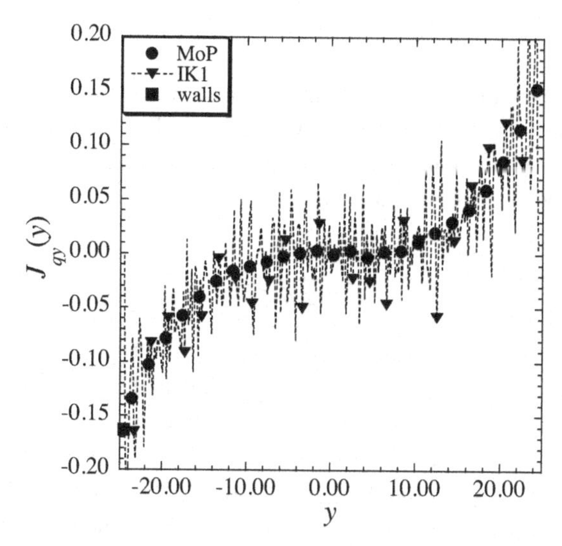

Figure 9.11 Heat flux vector for a system of WCA fluid atoms confined to a channel where $L_y = 50.0$. The system consisted of 738 fluid atoms and three "real" layers of wall atoms with 18 atoms per layer. $\bar{n}_f = n_w = 0.844$; $T_w = 0.722$; $F^e = 0.01$; $k = 57.15$. IEC values are identical to MoP data to within statistical uncertainty but are not shown for clarity of display. Replotted from reference [88].

expects from classical Navier-Stokes theory [88, 337]. The situation of course changes for very narrow channels, as displayed in Figure 9.12 for a channel of width 4.1σ [96]. Now we find that the classical cubic profile is incorrect and instead real oscillations are superimposed on a cubic background. Note that it is essential to use the correct form of the streaming velocity profile in either the MoP or IEC methods. Use of an incorrect

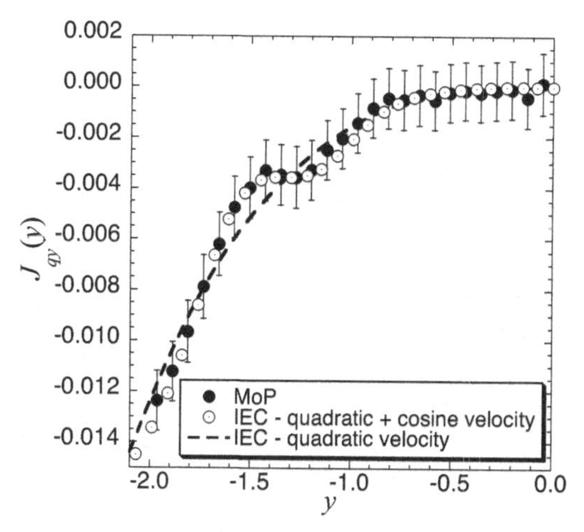

Figure 9.12 Heat flux vector for the system described in Figure 9.9, with $L_y = 5.1\sigma$. Along with MoP results, IEC values are displayed based on (i) assumption of a classical quadratic streaming velocity, and (ii) use of the quadratic plus 10-term cosine function defined in Equation (9.34). Replotted from reference [96].

form (perhaps by assuming a classical quadratic velocity profile) would result in an incorrect heat flux calculation [96].

For Poiseuille flow the solution of the governing Navier-Stokes equation for the heat flux is obtained by substituting Newton's law of viscosity and Fourier's law of heat conduction, $J_{qy}(y) = -\lambda dT(y)/dy$, into the energy conservation equation, Equation (9.36) and solving for the temperature profile to obtain

$$T(y) = T_0 - \frac{(nF^e)^2}{12\eta\lambda} y^4, \tag{9.41}$$

where T_0 is the temperature at the middle of the channel ($y = 0$) and again n, η and λ are the effective number density, viscosity and thermal conductivity, respectively. This expression gives us the classical Navier-Stokes quartic temperature profile for Poiseuille (or gravity-driven) flow. As with the case of the Navier-Stokes prediction for the streaming velocity profile, we can fit Equation (9.41) to actual NEMD simulation data to estimate λ. Or, if our aim is prediction of the temperature profile, then Green–Kubo or NEMD techniques such as the Evans heat flow algorithm [235] (see Chapter 7) can be used to estimate λ at an effective state point representative of the fluid and substituted into Equation (9.41) to predict the profile.

However, as we have already seen in Section 9.1 above, the assumption of Fourier's law for systems where the gradient of the square of the strain rate tensor is not zero is actually incorrect. For flows such as planar Couette flow, where the strain rate is constant, this poses no problem and so classical Navier-Stokes theory should accurately predict the temperature profile for a sufficiently wide channel. However, for Poiseuille flow the strain rate is linear, which implies a gradient in the square of the strain rate which is also linear. The additional cross-coupling term in the heat flux constitutive equation given by Equation (9.22) is therefore nonzero and *must* be included in a modified Navier-Stokes energy equation to accurately predict the temperature profile [330]. As was already noted by Baranyai *et al.* [296], this is only ever noticeable when $y \ll \sqrt{12\xi/\eta}$, i.e. in regions very close to the middle of the channel [330]. For macroscopic flows, this will never be noticed, but for microscopic and nanoscale flows, it could play a subtle role that determines the temperature profile. Note that we have already demonstrated in Section 9.1 that while the temperature profile will change, the heat flux itself will remain unaffected by the cross-coupling term [330, 337].

In order to accurately predict the temperature profile we also need to account for spatial variation in the transport coefficients. Subsitution of Equation (9.22) into the energy continuity equation gives us

$$\frac{d}{dy}\left[\lambda(y)\frac{dT(y)}{dy}\right] + \frac{d}{dy}\left[\xi(y)\frac{d\dot{\gamma}^2(y)}{dy}\right] + \eta(y)\dot{\gamma}^2(y) = 0. \tag{9.42}$$

This is a difficult equation to solve, requiring four independent boundary conditions. To avoid this complication, a perturbation expansion can be applied [330]. If for the moment, we ignore the spatial variations and treat the viscosity, thermal conductivity and cross-coupling coefficient as constants, the solution of Equation (9.42) gives [330, 337]

$$T(y) = T_0 - \frac{(nF^e)^2 \xi}{\lambda \eta^2} y^2 - \frac{(nF^e)^2}{12\lambda\eta} y^4. \tag{9.43}$$

Notice that the coefficients of the quadratic and quartic terms both have the *same* dependence on the external field strength. Therefore, even in the weak-field limit, cross-coupling can not be ignored.

As the pressure is a function of density and temperature, we can express its variation in y as

$$\frac{dp}{dy} = \frac{\partial p}{\partial \rho}\frac{d\rho}{dy} + \frac{\partial p}{\partial T}\frac{dT}{dy} = 0. \tag{9.44}$$

This is zero due to the requirement of mechanical stability for the system at steady-state as seen from NEMD simulation results, such as displayed in Figure 9.7(b). Since $\partial p/\partial \rho$ and $\partial p/\partial T$ are constant coefficients, the variations in density and temperature are proportional, i.e. $\Delta\rho \propto \Delta T$. In the linear regime the leading-order terms in the temperature are quartic and quadratic in y. This then implies that the density variations must also be quartic and quadratic, so the number density must vary as $n(y) = n_0 + n_2 y^2 + n_4 y^4$. If we now assume that the transport coefficients are *local* in space (this will be true for channels of width greater than around 5 to 10 atomic diameters in general [96]; nonlocality is discussed in Section 11.1), then they must vary as the density, i.e.

$$\begin{aligned} \eta(y) &= \eta_0 + \eta_2 y^2 + \eta_4 y^4 \\ \lambda(y) &= \lambda_0 + \lambda_2 y^2 + \lambda_4 y^4 \\ \xi(y) &= \xi_0 + \xi_2 y^2 + \xi_4 y^4. \end{aligned} \tag{9.45}$$

Assuming a quadratic streaming velocity profile in the weak-field limit (again valid for channel widths greater than 5–10 atomic diameters in general [96]) $v_x(y) = v_0 + v_2 y^2 + O(y^4)$, substitution of these spatial varying transport coefficients into Equation (9.42) leads to a modified Navier-Stokes temperature profile prediction:

$$T(y) = T_0 + T_2 y^2 + T_4 y^4 + T_6 y^6 + O(y^8) \tag{9.46}$$

with coefficients given by [330]

$$\begin{aligned} T_2 &= -\frac{4v_2^2\xi_0}{\lambda_0} \\ T_4 &= -\frac{v_2^2(\lambda_0\eta_0 + 6\lambda_0\xi_2 - 6\lambda_2\xi_0)}{3\lambda_0^2} \\ T_6 &= -\frac{2v_2^2(30\lambda_0^2\xi_4 + 3\lambda_0^2\eta_2 + 30\lambda_2^2\xi_0 - 30\lambda_0\lambda_4\xi_0 - 5\lambda_2\lambda_0\eta_0 - 30\lambda_2\lambda_0\xi_2)}{45\lambda_0^3}. \end{aligned} \tag{9.47}$$

We display the temperature data and predictions based on the extended Navier-Stokes (i.e. including spatial variation in transport coefficients) and the extended Navier-Stokes plus cross-coupling theories in Figure 9.13. Here we have simulated a WCA fluid confined to a relatively wide channel of approximately 70σ [330]. The temperature is

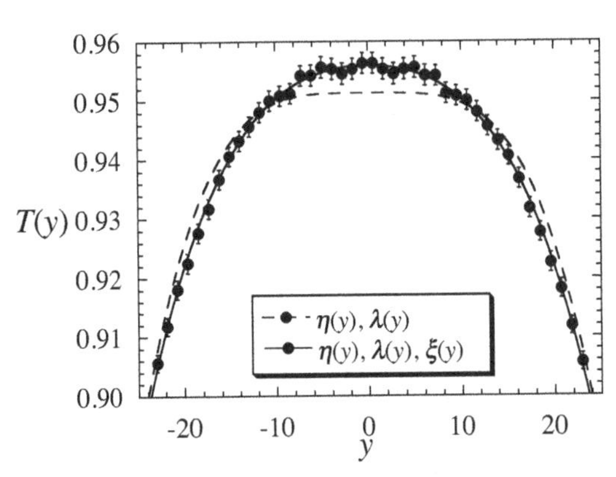

Figure 9.13 Symmetrised temperature profile for a system confined to $L_y = 70\sigma$. Only the central part of the profile is displayed and fits to the extended Navier-Stokes theory (including spatial variation in transport coefficients) with and without the inclusion of the cross-coupling term are included. The system consisted of 1278 WCA fluid atoms confined by 54 wall atoms (3 layers of 18 atoms each). $\bar{n}_f = 0.84$; $n_w = 0.844$; $T_w = 0.722$; $F^e = 0.005$. Replotted from reference [330].

computed in bins as

$$\langle T(y_{bin}) \rangle = \frac{\left\langle \sum_{i \in bin}^{N_{bin}} m_i \left[\mathbf{v}_i - \mathbf{v}(y,t) \right]^2 \right\rangle}{\left\langle 3N_{bin} - d\left(N_{bin}/N\right) \right\rangle}, \tag{9.48}$$

where d is the number of degrees of freedom lost to the system by fitting the streaming velocity profile, and we set $k_B = 1$. The streaming velocity is computed at each time step by least-squares fitting to the instantaneous velocity profile, and since $d = k/2 + 1$ degrees of freedom are lost in fitting the coefficients of the streaming velocity to a symmetric polynomial of order k, a factor of $d\,(N_{bin}/N)$ degrees of freedom are lost in each bin. An equivalent method would be to compute the laboratory kinetic energy density in each bin, given by $\langle K(y_{bin}) \rangle = \left\langle \frac{1}{2} \sum_{i \in bin} m_i \mathbf{v}_i^2 \delta(y_{bin} - y_i) \right\rangle$ and subtract out the streaming kinetic energy density $\frac{1}{2} \langle \rho(y_{bin}) \rangle \langle \mathbf{v}(y_{bin}) \rangle^2$ at the end of the simulation. This gives the peculiar kinetic energy density, which when normalised by the density of each bin, gives the bin temperature. One could even do this at planes using the MoP methodology [89, 96]. This is an example of post-processing of data and can be applied to all properties of interest, not just the binned temperature. The advantage of post-processing of data such as this is that one does not need to make any assumption about the functional form of the fluid streaming velocity. This is particularly important for very narrowly confined fluids, in which the streaming velocity may not be strictly quadratic (in the case of Poiseuille flow) but could have a complicated functional form [96]. The same is true for any other type of flow geometry. To use this post-processing methodology one must insist that the thermal (peculiar) momentum of *fluid* atoms is conserved (i.e. $\sum_i^{N_f} m_i \mathbf{c}_i = 0$, where i indexes only fluid atoms, not walls, $\mathbf{c}_i$ is the thermal velocity of atom i, and N_f is the total number of fluid atoms), otherwise an unwanted residual net nonzero thermal component is introduced.

The effects of cross-coupling become apparent near the mid-point of the flow, as predicted. Note that merely including the spatial variation in the transport coefficients will not alone explain the change in shape of the temperature profile, or accurately predict the temperature at the mid-point. Only the inclusion of the cross-couping coefficient accounts for this. A reasonable approximation can be obtained by assuming constant transport coefficients, but again one must include the cross-coupling term to obtain the correct temperature profile prediction [330]. However, note that no evidence for the cross-coupling term in the temperature profile was observed by Travis and Gubbins [320], who computed the temperature by a post-processing method without assuming a functional form for the velocity field.

9.2.1 Microscopic Equations of Motion

Consider a simple atomic fluid in which particles interact via some smooth, continuous interaction potential. Furthermore, the interaction between wall atoms and fluid atoms is also governed by the same interaction potential. While it is obvious that different interaction potentials will affect the quantitative features of fluid structure and transport properties, for the purposes of instruction the details are irrelevant. What is essential is the underlying dynamics, dictated by the equations of motion. Once these are established it is relatively straightforward to modify the various interaction terms. In addition to the WCA interaction, the wall atoms must also be tethered to their equilibrium lattice sites. We do this by a simple Hookean-spring model, in which the tethering potential ϕ^t is defined as

$$\phi^t(\mathbf{r}_i) = \frac{1}{2}k\left(\mathbf{r}_i - \mathbf{r}_i^{eq}\right)^2, \tag{9.49}$$

where $\mathbf{r}_i^{eq}$ is the "frozen" lattice site position of atom i and k is the "spring" constant. Wall atoms are thus allowed to vibrate around their lattice sites.

However, if one simulates the walls as thus, with fluid atoms under the influence of some driving field, one finds that the fluid exerts a pressure on the walls such that the walls actually move slightly apart. This was seen in simulations performed by Liem *et al.* [168], for example. This has the undesirable effect of increasing the system volume and furthermore depends on the strength of the driving field (or, as would be the case for boundary-driven flow like Couette flow, the strain rate of the fluid, which in turn depends on the relative speed of the moving boundaries and the separation between them). The increase in system volume can be quite significant when one is interested in studying small-scale confinement, such as would be the case for nanofluidic systems. In order to ensure the walls remain fixed at their centre of mass sites we can impose further constraints on the equations of motion. In particular, if we wish to constrain the centre of mass of each wall layer such that its average y position is invariant, we can construct constraints based either on Gauss's principle of least constraint, or the Nosé-Hoover integral feedback procedure. The former ensures that the centre of mass y coordinate of each layer is fixed instantaneously, whereas the latter would be constrained only in the time-averaged sense. It is a matter of preference which method to employ, but our method of choice is the Gaussian constraint since it is more efficient and guarantees no fluctuation of the system volume at all times.

Let us assume we have a system of N_W wall layers, each of which consists of n_w wall atoms. Each wall layer is labelled as L_i, where $i = 1, 2, 3 \ldots N_W$, so we have a total of $N_W n_w$ atoms in a wall. There are a total of N combined wall and fluid atoms, i.e. $N = N_W n_w + N_f$, where N_f is the number of fluid atoms. Furthermore, in order to faithfully mimic the processes that occur in a real fluid, we thermostat the walls such that viscous heat produced in the fluid is extracted through the walls via the wall thermostat. Note that we *do not* thermostat the fluid itself. Such a process can be dangerously misleading and can cause many unphysical side-effects that affect the fluid structure and its mechanical, thermodynamic and even dynamical properties [346, 347].[1] Noting that the constraints are themselves coupled, the equations of motion for wall atoms are [86, 349]:

$$\dot{\mathbf{r}}_i = \frac{\mathbf{p}_i}{m_i} \tag{9.50}$$

$$\dot{\mathbf{p}}_i = \mathbf{F}_i - \alpha \mathbf{p}_i - \mathbf{j}\lambda_{L_j}, \; i \in L, \tag{9.51}$$

where $\mathbf{F}_i$ is defined as

$$\mathbf{F}_i \equiv -k\left(\mathbf{r}_i - \mathbf{r}_i^{eq}\right) + \sum_{j=1}^{N} \mathbf{F}_{ij}^{\phi}. \tag{9.52}$$

α is the thermostat multiplier for the walls and λ_{L_j} is the wall layer multiplier, given by

$$\lambda_{L_j} = \frac{\mathbf{j}}{n_w} \cdot \sum_{i \in L_j}^{n_w} \left[\mathbf{F}_i - \alpha \mathbf{p}_i\right] \tag{9.53}$$

and

$$\alpha = \frac{\sum\limits_{i \in L}^{N_W n_w} \left\{\mathbf{F}'_i \cdot \mathbf{p}'_i\right\}}{\sum\limits_{i \in L}^{N_W n_w} \mathbf{p}'^2_i}, \tag{9.54}$$

where

$$\mathbf{F}'_i = \mathbf{F}_i - \frac{\mathbf{j}}{n_w} \sum_{i \in L_j}^{n_w} F_{yi}$$

$$\mathbf{p}'_i = \mathbf{p}_i - \frac{\mathbf{j}}{n_w} \sum_{i \in L_j}^{n_w} p_{yi} \tag{9.55}$$

and we note $\sum_{L_j=1}^{N_W} \sum_{i \in L_j}^{n_w} 1 = n_w N_W$. Equations (9.53) and (9.54) are obtained by setting the time derivative of the centre of mass y component of each layer and total wall

[1] For complex walls that are difficult to thermostat, a new "virtual particle" thermostat was designed to ensure structural stability while avoiding the need to thermostat the liquid directly. Details can be found in reference [348].

temperature to zero, respectively, and substituting the equations of motion in the resulting expressions. Note that this necessarily implies that momentum must be conserved by each wall layer independently. It is not sufficient to just set the initial total momentum of $n_w N_W$ wall atoms to zero. Note also that as the walls do not translate, there is no distinction between an atom's thermal (peculiar) and laboratory momentum: they are equivalent.

As we have already seen, Gaussian constraints will drift over very long simulation times. Therefore, one needs to add the following temperature feedback term to the equation of motion for the momenta, Equation (9.51):

$$-B\left[T_w(t) - T_w(0)\right]\mathbf{p}_i, \tag{9.56}$$

where $T_w(t)$ is the instantaneous wall kinetic temperature and T_w is the desired temperature. Typically, a value of $B = 10$ or thereabouts suffices, though trial and error is the best judge. Similarly, we must account for drift in the wall constraints. Failure to account for this drift results in the walls moving apart gradually over very long simulation times. This can be accomplished by adding the following feedback terms to the position and momentum equations of motion, Equations (9.50) and (9.51) respectively:

$$-\frac{C}{n_w}\left[\sum_{i \in L_j}^{n_w} y_i - \sum_{i \in L_j}^{n_w} y_i(t=0)\right] - \frac{D}{n_w}\sum_{i \in L_j}^{n_w} \dot{y}_i. \tag{9.57}$$

For a system of WCA atoms, Travis and Evans [349] note that D can be set to zero, whereas $C \sim 10$ is appropriate.

The equations of motion for fluid atoms are just given by Newton's equations, since there are no constraints at all acting on these:

$$\begin{aligned}\dot{\mathbf{r}}_i &= \frac{\mathbf{p}_i}{m_i} \\ \dot{\mathbf{p}}_i &= \sum_{j=1}^{N} \mathbf{F}_{ij}^{\phi} + \mathbf{F}_i^{e}\end{aligned} \tag{9.58}$$

where the $\mathbf{p}_i$ are laboratory momenta (i.e. the sum of thermal and streaming velocities).

9.3 Couette Flow

The procedure for performing NEMD simulations on fluids undergoing planar Couette flow is very similar to what has been described above for Poiseuille flow. The only significant difference is that now the fluid does not come under the influence of an external force field directly. Instead, the walls move with respect to each other in equal and opposite directions. If the flow is symmetric about $y = 0$ then the top and bottom walls move at velocities $\mp\frac{1}{2}\dot{\gamma}L_y$ in the x-direction, respectively. The geometry for this flow is depicted in Figure 9.14.

The equations of motion for Couette flow are similar but simpler than those for Poiseuille flow. For the wall atoms, one can generate flow by displacing the harmonic

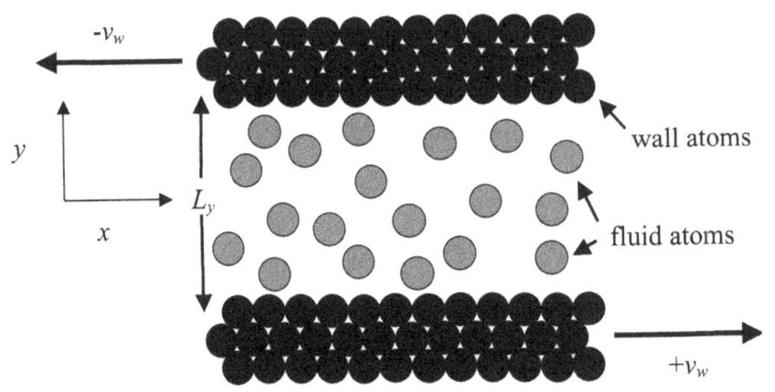

Figure 9.14 Schematic diagram for planar Couette flow. The system is periodic in the $x-z$ directions, with flow in the x direction and the channel separated by L_y in the y direction. The two sets of walls are independent and not periodically imaged, unlike the case for Poiseuille flow. z is normal to the page. The top wall moves at constant velocity $-v_w$, whereas the bottom wall moves at $+v_w$.

tethering lattice sites by an amount $\mp\frac{1}{2}\dot{\gamma}L_y\Delta t$ at each time step, where $\dot{\gamma}$ is the desired strain rate input into the program. In the case of fluid atoms, there is no external force term appearing in the second line of Equation (9.58). The flow is entirely generated by the displacement of atoms in the walls (i.e. it is boundary driven). Note that while it is simple to use pbc images of one wall to generate the second wall for Poiseuille flow, this is not the case for Couette flow, since now walls move in equal but opposite directions. As such, two sets of unique walls are required.

The governing Navier-Stokes equation for planar Couette flow is particularly simple. Equation (9.30) can be readily solved in the steady-state by setting both the left-hand side of the equation and the external force term to zero. The solution is

$$P_{xy} = C \tag{9.59}$$

where C is a constant. Thus, unlike the situation in Poiseuille flow, the shear stress will *always* be constant no matter how highly confined the fluid is. This is simply a consequence of momentum balance. Newton's law of viscosity then implies

$$\dot{\gamma} = -\frac{C}{\eta} \Rightarrow v_x(y) = -\frac{C}{\eta}y + D, \tag{9.60}$$

where $D = 0$ for symmetric flow about $y = 0$. The constant C is readily determined from the flow boundary conditions at the walls. If the velocities of the top and bottom walls are defined to be $\mp v_w$, respectively, then Equation (9.60) implies

$$\begin{aligned} v_x\left(\pm\frac{L_y}{2}\right) &= \mp\frac{C}{\eta}\frac{L_y}{2} = \mp v_w \\ &\Rightarrow C = \frac{2v_w\eta}{L_y} = P_{xy}. \end{aligned} \tag{9.61}$$

Substitution of Equation (9.61) into Equation (9.60) in turn leads to the velocity (hence strain rate) expressed independently of the viscosity:

$$v_x(y) = -\frac{2v_w}{L_y}y \tag{9.62}$$

$$\dot{\gamma}(y) = -\frac{2v_w}{L_y}. \tag{9.63}$$

In a similar manner we can solve the energy balance equation. By substitution of the appropriate quantities derived above, Equation (9.36) reduces to

$$\begin{aligned} \frac{dJ_{qy}(y)}{dy} &= -P_{xy}(y)\,\dot{\gamma}(y) \\ &= \eta\dot{\gamma}^2(y) \\ &= 4\eta\left(\frac{v_w}{L_y}\right)^2. \end{aligned} \tag{9.64}$$

Integrating over y and noting by symmetry that the heat flux at $y = 0$ is zero, we obtain the linear expression

$$J_{qy}(y) = 4\eta\left(\frac{v_w}{L_y}\right)^2 y. \tag{9.65}$$

Substitution of Fourier's law of heat conduction into Equation (9.65) and solving for the temperature gives the standard classical quadratic temperature profile

$$T(y) = -\frac{2\eta}{\lambda}\left(\frac{v_w}{L_y}\right)^2 y^2 + D, \tag{9.66}$$

where D is a constant. D can be determined by the boundary condition $T\left(\pm\frac{L_y}{2}\right) = T_w$, where T_w is the wall temperature. Doing this yields the final expression:

$$T(y) = \frac{2\eta}{\lambda}\left(\frac{v_w}{L_y}\right)^2\left[\frac{L_y^2}{4} - y^2\right] + T_w. \tag{9.67}$$

This also implies that the temperature in the middle of the channel is always $\geq T_w$ by the amount $(\eta/2\lambda)v_w^2$. This derivation does not consider the strain rate coupling to the heat flux described in previous sections. For Couette flow the strain rate is constant in all but very highly confined geometries, as we will see below. This in turn implies that the strain rate coupling term would typically be zero, since coupling requires the gradient of the strain rate squared to be non-zero. However, for very highly confined geometries in which significant departure from linearity in the streaming velocity occurs, then the same procedures used in Sections 9.1 and 9.2 could be used.

We first display the results of an NEMD simulation of a relatively large system with $L_y = 11$, i.e. confinement dimension of $2h = 10$. Results for density, streaming velocity, normal and off-diagonal pressures (negative shear stress) and temperature are presented in Figure 9.15. We find layering of fluid immediately adjacent to the walls, becoming less prominent towards the centre of the channel. The streaming velocity is well

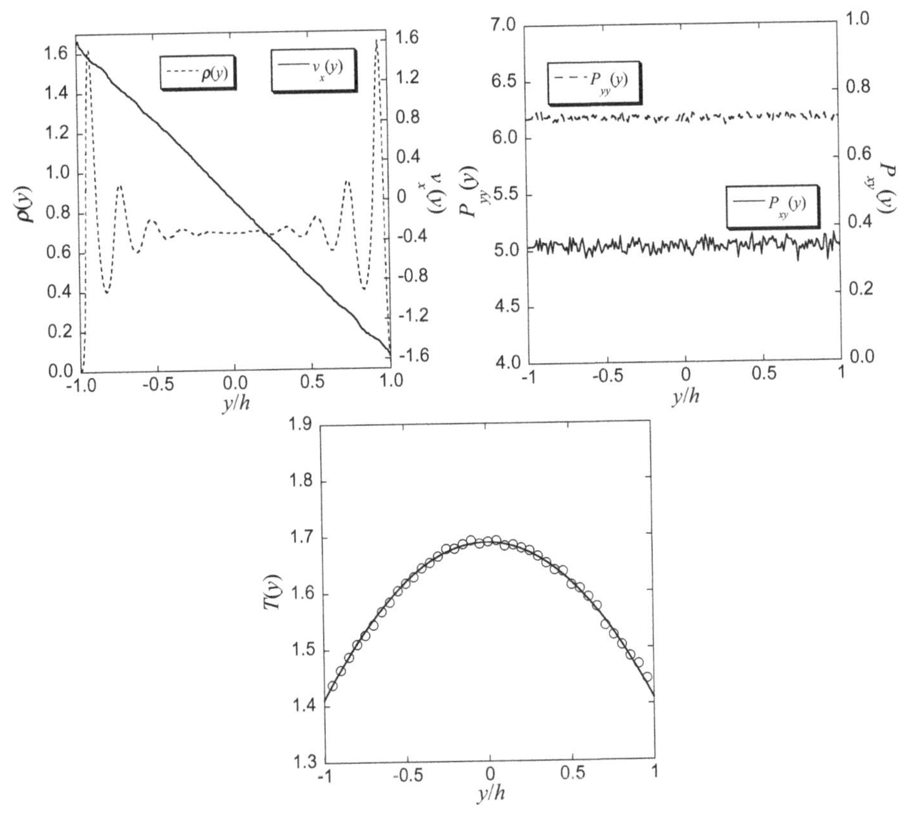

Figure 9.15 System profiles for a system of 1444 WCA fluid atoms undergoing planar Couette flow. The fluid is confined by atomistic walls separated in the y-direction by an effective width of ~ 10 (a) density and streaming velocity, (b) normal and off-diagonal pressures, and (c) temperature (open circles are NEMD data, solid curve is the classical quadratic fit to the data). Lower and upper walls are each two layers thick and move at constant velocities of ± 2.547, respectively. Total number of wall atoms is 784. $\rho_f = 0.7$, $\rho_w = 0.95$. The system was equilibrated for 1 million timesteps of $\Delta t = 0.001$ before accumulating steady-state averages for a further 1 million timesteps [350].

described by the classical linear profile, except immediately near the walls, where deviations from linearity result from the high degree of fluid layering. The lower and upper walls move at velocities ± 2.547, respectively, which indicates a slip velocity magnitude of approximately 0.95 at each wall. Both the normal and off-diagonal pressures are constant to within statistical fluctuations, as required by conservation of momentum. The temperature also displays the expected classical quadratic profile.

In the case of extreme confinement we expect classical-like behaviour to be destroyed, as we have observed for Poiseuille flow. In Figure 9.16 we display the same properties as the system above, except that now $L_y = 3.85$. This gives a confinement dimension of $2h = 2.85$. Clearly, the fluid forms three distinct layers and this high degree of layering causes significant deviations in the streaming velocity, which has oscillations of the order of σ superposed upon a linear profile. Note that significant slip occurs at the

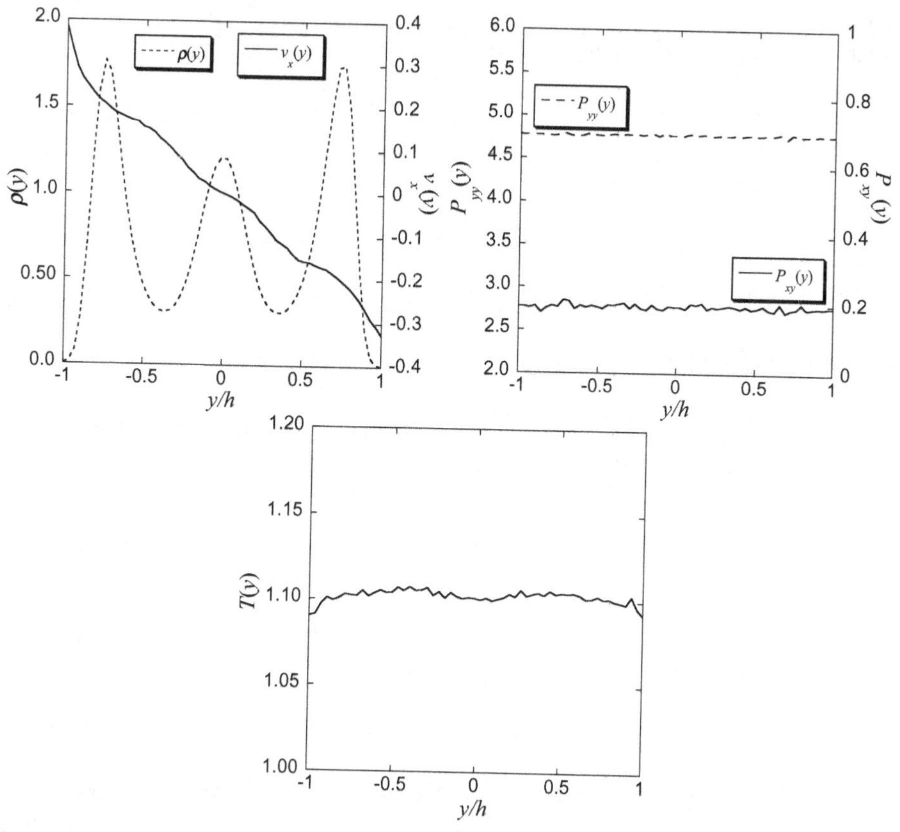

Figure 9.16 System profiles for a system of 108 WCA fluid atoms undergoing planar Coeutte flow. The fluid is confined by atomistic walls separated in the y-direction by an effective width of ~ 2.85 (a) density and streaming velocity, (b) normal and off-diagonal pressures, and (c) temperature. Lower and upper walls are each two layers thick and move at constant velocities of ± 0.767, respectively. Total number of wall atoms is 196. $\rho_f = 0.7$, $\rho_w = 0.95$. The system was equilibrated for 1 million timesteps of $\Delta t = 0.001$ before accumulating steady-state averages for a further 19 million timesteps [350].

walls, which move at lower and upper velocities of ± 0.767, respectively. Slip will be discussed in greater detail in Chapter 11. As expected, both the shear stress and normal pressure are constant across the channel. The high degree of confinement significantly affects the temperature, which does not develop the classical quadratic profile, but rather tends to be almost flat.

10 Confined Molecular Fluids

10.1 Molecular Fluids

In Chapter 2 we discussed the modifications required to formulate the Navier-Stokes equations to take into account the intrinsic spin of molecularly structured fluids. For uniaxial molecules in the linear regime, these can be expressed as [4, 36]

$$\rho\frac{d\mathbf{v}}{dt} = -\nabla p + \left(\eta_v + \eta/3 - \eta_r\right)\nabla\left(\nabla\cdot\mathbf{v}\right) + \left(\eta + \eta_r\right)\nabla^2\mathbf{v} + 2\eta_r\nabla\times\boldsymbol{\omega} + \rho\mathbf{F}^e \tag{10.1}$$

$$\rho\frac{d\mathbf{S}}{dt} = 2\eta_r\left(\nabla\times\mathbf{v} - 2\boldsymbol{\omega}\right) + \left(\zeta_v + \zeta/3 - \zeta_r\right)\nabla\left(\nabla\cdot\boldsymbol{\omega}\right) + \left(\zeta + \zeta_r\right)\nabla^2\boldsymbol{\omega} + \rho\boldsymbol{\Gamma}^e, \tag{10.2}$$

where $\mathbf{S}$ is the spin angular momentum, η_r is the vortex viscosity, ζ, ζ_v and ζ_r are the vortex spin viscosities, $\boldsymbol{\Gamma}^e$ is an external torque, and the other quantities are as previously defined. In the case of steady-state low Reynolds number incompressible flows, for a fluid under the influence of an external field (such as that already discussed above) but no external torque, these equations reduce to

$$\rho\frac{d\mathbf{v}}{dt} = \left(\eta + \eta_r\right)\nabla^2\mathbf{v} + 2\eta_r\nabla\times\boldsymbol{\omega} + \rho\mathbf{F}^e \tag{10.3}$$

$$\rho\frac{d\mathbf{S}}{dt} = 2\eta_r\left(\nabla\times\mathbf{v} - 2\boldsymbol{\omega}\right) + \left(\zeta + \zeta_r\right)\nabla^2\boldsymbol{\omega}. \tag{10.4}$$

Note that for uniaxial molecules the streaming angular momentum is given as $\mathbf{S} = \boldsymbol{\Theta}\cdot\boldsymbol{\omega} = \Theta\boldsymbol{\omega}$, where $\Theta = 1/3\mathrm{Tr}\left(\boldsymbol{\Theta}\right)$. $\boldsymbol{\Theta}$ is the average moment of inertia tensor per unit mass, given by

$$\boldsymbol{\Theta} = \langle\boldsymbol{\Theta}_i\rangle = \left\langle\frac{\sum\limits_{\alpha\in i} m_{i\alpha}\left(\mathbf{r}'^2_{i\alpha}\mathbf{1} - \mathbf{r}'_{i\alpha}\mathbf{r}'_{i\alpha}\right)}{\sum\limits_{\alpha\in i} m_{i\alpha}}\right\rangle, \tag{10.5}$$

where the vector $\mathbf{r}'_{i\alpha}$ is the vector from the centre of mass of molecule i to site α of the same molecule and $\mathbf{1}$ is the unit tensor. Note also that the coupling of these two equations is through the vortex viscosity and that there is a continuous exchange between spin angular velocity and vorticity. The solution of these two coupled equations leads to analytic expressions for the linear and rotational streaming velocities [349, 351, 352]

$$v_x(y) = v_c\left[1 - \bar{y}^2 + \frac{2\eta_r}{(\eta + \eta_r) Kh}\coth(Kh)\left(\frac{\cosh(Kh\bar{y})}{\cosh(Kh)} - 1\right)\right] \tag{10.6}$$

$$\omega_z = \frac{v_c}{h}\left(\bar{y} - \frac{\sinh(Kh\bar{y})}{\sinh(Kh)}\right), \tag{10.7}$$

where v_c is a parameter which can be obtained by fitting to NEMD streaming velocity data, and h is half the effective width of the channel (which could, for example, be $(L_y - 1.0\sigma)/2$ as previously defined). $\bar{y}$ is defined as $\bar{y} \equiv y/h$ and $K = [4\eta_0\eta_r/(\eta_0 + \eta_r)(\zeta_0 + \zeta_r)]^{1/2}$.

10.1.1 Rigid Diatomic Molecules under Poiseuille Flow

For rigid diatomic liquid molecules the equations of motion are

$$\dot{\mathbf{r}}_{i\alpha} = \frac{\mathbf{p}_{i\alpha}}{m_i} \tag{10.8}$$

$$\dot{\mathbf{p}}_{i\alpha} = \mathbf{F}^{\phi}_{i\alpha} + (-1)^{\alpha}\chi_i \mathbf{r}_{i12} + \frac{1}{2}\mathbf{i}F^e, \tag{10.9}$$

where here we consider that χ_i is a Gaussian bond constraint multiplier and $\mathbf{r}_{i12}$ is defined as $\mathbf{r}_{i12} = \mathbf{r}_{i2} - \mathbf{r}_{i1}$. Application of Gauss's principle of least constraint gives χ_i as

$$\chi_i = -\frac{\left[\mathbf{r}_{i12}\cdot\mathbf{F}^{\phi}_{i\alpha} + m_i(\dot{\mathbf{r}}_{i12})^2\right]}{2\mathbf{r}^2_{i12}}. \tag{10.10}$$

Once again we must apply proportional feedback to ensure there is no drift in the Gaussian bond constraints. This is accomplished by adding the following terms to Equations (10.8) and (10.9), respectively:

$$-F(r_{i12} - d_{12})\hat{\mathbf{r}}_{i12} \tag{10.11}$$

$$-G(\dot{\mathbf{r}}_{i12}\cdot\hat{\mathbf{r}}_{i12})\hat{\mathbf{r}}_{i12}, \tag{10.12}$$

where r_{i12} is the magnitude of the bond vector in molecule i, d_{12} is the desired bond length, $\hat{\mathbf{r}}_{i12}$ is the unit vector in the bond direction and $\dot{\mathbf{r}}_{i12}$ is the bond velocity. Again typical values of the constants F and G are of the order of 10.

In Figure 10.1 we plot the streaming velocity profile for a rigid diatomic fluid interacting via the WCA potential. This is for a system with effective channel half width $h = 4.5\sigma$. The expected quadratic profile is clearly seen.

To compare simulation results with Navier-Stokes theory for the angular streaming velocity, we rearrange Equation (10.7) such that the left-hand side is dimensionless and the right-hand side is a function only of the dimensionless quantity Kh [351, 352]:

$$\frac{h\omega_z}{v_c} = \bar{y} - \frac{\sinh(Kh\bar{y})}{\sinh(Kh)}. \tag{10.13}$$

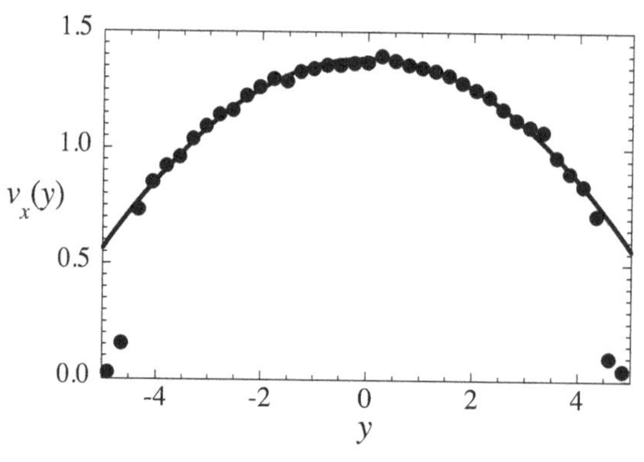

Figure 10.1 Translational streaming velocity profile for a diatomic fluid confined to a channel of effective width 9.0σ. The system consists of 360 diatomic fluid molecules confined by 3 layers of wall atoms, each of 72 atoms per layer. Interaction potential between all atoms is the WCA potential. $n_w = 0.90$; $n_f = 0.544$; $T_w = 0.97$; $F^e = 0.5$. Quadratic fit to the data gives $v_x(y) = 1.389 - 0.033y^2$. Replotted from reference [352].

This gives us an expression which has both linear and hyperbolic terms for the streaming angular velocity. Without the hyperbolic term, a purely linear term would imply that the streaming angular velocity is half the vorticity, which would be the case for a homogeneous fluid [353]. However, the presence of boundaries effectively adds a "switch" function, in the form of the hyperbolic term. It controls the rate at which the streaming angular velocity changes from its boundary value (zero) to its preferred value of half the vorticity [349, 352]. This is seen in Figure 10.2, in which the theoretical values of the streaming angular velocity are plotted against normalised channel position. At low values of Kh the switching function is broad, implying that the streaming angular velocity is never able to attain its preferred value of half the vorticity. On the other hand, large values of Kh show that over most of the channel the spin angular velocity is able to approach and equal this value. Near the walls it naturally fails, since the boundaries

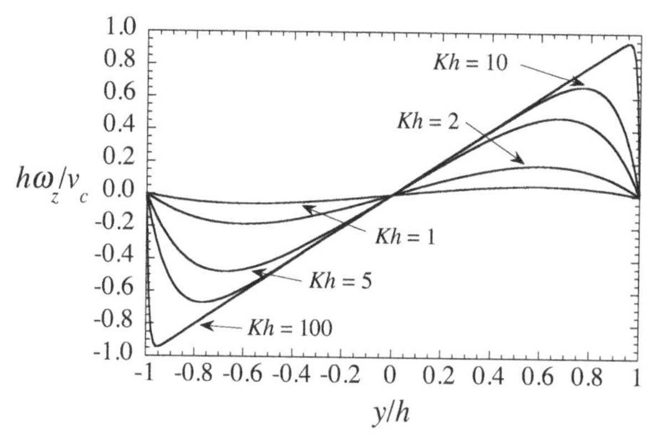

Figure 10.2 Predicted spin angular velocity profiles for various values of Kh for the diatomic system. Replotted from reference [352].

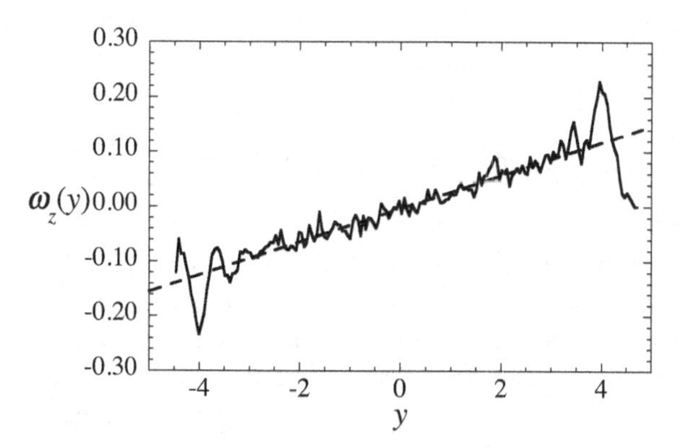

Figure 10.3 NEMD spin angular velocity profile for the system described in Figure 10.1. Linear fit to the streaming angular velocity gives $\omega_z(y) \sim 0.030y$. Replotted from reference [352].

necessarily force the streaming angular velocity to zero under the assumption of no-slip angular velocity boundary conditions.

In Figure 10.3 the actual streaming angular velocity profile obtained from NEMD simulation is plotted. The molecular spin angular momentum was first computed in bins of finite width. If $\mathbf{S}_{bin}$ is the total intrinsic angular momentum in a bin and $\boldsymbol{\omega}$ is the streaming angular velocity in that bin, then

$$\mathbf{S}_{bin} = \boldsymbol{\Theta}_{bin} \cdot \boldsymbol{\omega}, \tag{10.14}$$

where

$$\mathbf{S}_{bin} = \sum_{i \in bin} \sum_{\alpha} \mathbf{r}'_{i\alpha} \times \mathbf{p}_{i\alpha} \tag{10.15}$$

$$\boldsymbol{\Theta}_{bin} = \sum_{i \in bin} \boldsymbol{\Theta}_i \tag{10.16}$$

and $\boldsymbol{\Theta}_i$ and $\mathbf{r}'_{i\alpha}$ are defined in Equation (10.5). The streaming angular velocity is then computed by solving Equation (10.14). An alternative method is to compute the angular velocity at a plane, using the method of planes approach [89, 349]. In this method one computes the spin angular momentum density at a plane:

$$\rho(y) S_z(y) = \lim_{\tau \to \infty} \frac{1}{\tau A} \sum_i \sum_{\alpha} \frac{S_{zi}(t_\alpha)}{|\dot{y}_i(t_\alpha)|}. \tag{10.17}$$

Using this expression along with $\mathbf{S} = \boldsymbol{\Theta} \cdot \boldsymbol{\omega}$ gives the angular velocity $\omega_z(y)$. To obtain the spin at a plane, one computes the relative positions and velocities at the time the molecular centre of mass crosses a plane in the y direction. Full details can be found in reference [349].

The profile is similar to the theoretical profile for $10 \le Kh \le 100$. The oscillations near the walls are due to layering of the fluid and are not accounted for by the classical theory. Also shown is a linear fit to the angular velocity profile, from which we can deduce that the slope is $\partial\omega_z/\partial y \sim 0.030 \pm 0.001$. This agrees well with half the vorticity, $\frac{1}{2}\nabla \times \mathbf{v}$. For our system, this implies $\omega_z(y) = -\dot{\gamma}/2 = 0.033y$.

10.1.2 Flexible Polymeric Molecules under Poiseuille Flow

Flexible coarse-grained polymers are often modeled by combining a typical Lennard-Jones interaction potential with some tethering harmonic potential. Of the latter, the FENE potential is commonly used [274] (see discussion in Chapter 8). It has the advantage of preventing chains from breaking or stretching indefinitely, as well as ensuring excluded volume. There are a number of other potentials available in the literature for more sophisticated modeling of polymers (e.g. including semi-flexibility, etc.) and the reader is referred to the book by Kröger and references therein for further details [354]. The FENE potential is as given by Equation (8.2) [274] where R_0 is the finite extensibility allowable for the model and k is a spring constant. Typical values in the literature are $R_0 = 1.5\sigma$ and $k = 30\epsilon/\sigma^2$. These values have been chosen to prevent bond crossing at temperatures even higher than those used in our simulations [274].

The equations of motion for fluid molecules are

$$\dot{\mathbf{r}}_{i\alpha} = \frac{\mathbf{p}_{i\alpha}}{m_{i\alpha}} \tag{10.18}$$

$$\dot{\mathbf{p}}_{i\alpha} = \mathbf{F}^{\phi}_{i\alpha} + \mathbf{F}^{FENE}_{i\alpha} + \mathbf{i}F^e, \tag{10.19}$$

where $\mathbf{F}^{FENE}_{i\alpha}$ is the FENE force acting on atom α of molecule i due to intramolecular interactions. We will again set all interaction parameters equal to unity, as well as all site masses.

One of the interesting static properties in confined systems is the shape and size of the chains near surfaces and how they deviate from the respective bulk shape and sizes. The most appropriate quantities to describe these are the atomic and molecular density profiles across the channel and radius of gyration of the chains. In Figure 10.4 we plot the atomic and molecular densities for linear polymer melts of chain lengths varying from $N = 2$ to $N = 50$ for a channel width of $L_y = 7.0\sigma$. The densities are respectively defined as

$$\rho_a(\mathbf{r}, t) = \sum_{i=1}^{N_m}\sum_{\alpha=1}^{N_s} m_{i\alpha}\,\delta(\mathbf{r} - \mathbf{r}_{i\alpha}(t)) \tag{10.20}$$

$$\rho_m(\mathbf{r}, t) = \sum_{i=1}^{N_m} M_i\,\delta(\mathbf{r} - \mathbf{r}^c_i(t)), \tag{10.21}$$

where N_m is the number of fluid molecules, N_s is the number of atomic sites per molecule, $\mathbf{r}_{i\alpha}(t)$ is the position vector of site α of molecule i at time t and $\mathbf{r}^c_i(t)$ is the centre of mass position of molecule i. Here we assume all sites have identical mass, $m_{i\alpha} = m$, and the molecular mass, M, is given as $M = mN_s$.

To compare it with the atomic number density profile, the molecular density is multiplied by the number of atoms N_s per chain. The influence of the chain length on the atomic number density is small as shown in Figure 10.4(a), unlike the case for the

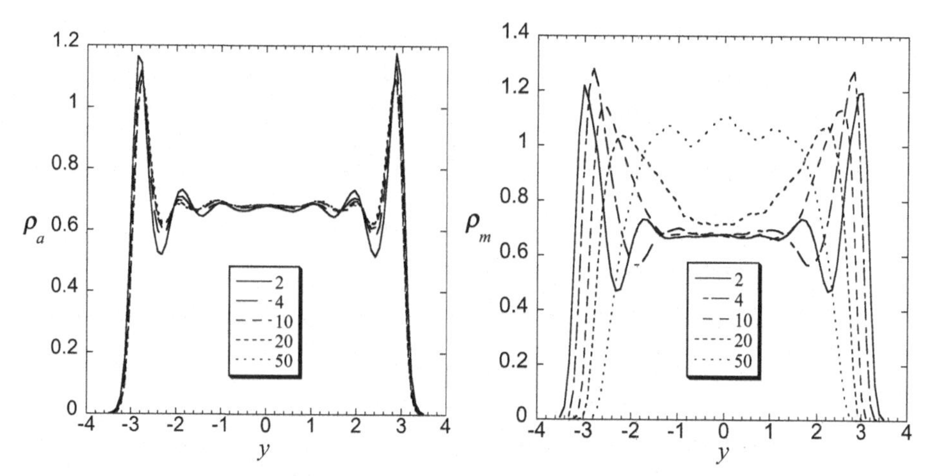

Figure 10.4 (a) Atomic density ρ_a and (b) molecular density ρ_m multiplied by N_s sites per chain. Average fluid density $= 0.65$; $F^e = 0.05$; $L_y = 7.0$. Replotted from reference [355].

molecular number density shown in Figure 10.4(b), which strongly depends on the chain length. Layering effects are clearly seen and the layered structure is more obvious if the chain length is small compared to the width of the channel. Because atoms in the first atomic layer have their centre of mass of the molecular chain in the region of the depletion zone between the first and the second atomic layers, the molecular number density does not follow the atomic layering [355].

Due to a large number of internal degrees of freedom, a polymer molecule can exist in many different configurations. Two commonly used parameters to quantify the configuration of polymers are the average end-to-end distance, R_e and the radius of gyration, R_g, which are measures of the distribution of beads in polymer chains [356]. For ideal (Gaussian) chains we have $R_e^2 = N_s b^2$ and $R_g^2 = \frac{1}{6} N_s b^2$, where b is the bond length. For an ideal chain $R_e^2/R_g^2 = 6$.

In Figure 10.5 we plot the mean squared radius of gyration as a function of y for several fluid densities and chain lengths. The radius of gyration is computed by dividing the fluid channel between the walls into bins, where for each bin the mean squared radius of gyration is computed as an average evaluated at the midpoint of each bin. The mean squared radius of gyration tensor is defined by

$$\mathbf{R}^2_{g,bin} \equiv \left\langle \frac{\sum\limits_{\alpha=1}^{N_s} m_\alpha \left(\mathbf{r}_\alpha - \mathbf{r}^c\right)\left(\mathbf{r}_\alpha - \mathbf{r}^c\right)}{\sum\limits_{\alpha=1}^{N_s} m_\alpha} \right\rangle_{bin} \tag{10.22}$$

from which we obtain $R_g^2 = \mathrm{Tr}(\mathbf{R}_g^2)$.

Once again layering effects are clearly observed and a full discussion can be found in reference [355], including a discussion of the end-to-end distance R_e. Furthermore, the ratio R_e^2/R_g^2 oscillates about 6 for 10-site chains, but deviates significantly for higher N_s since the chains unfold parallel to the walls. Note that the theoretical prediction is based

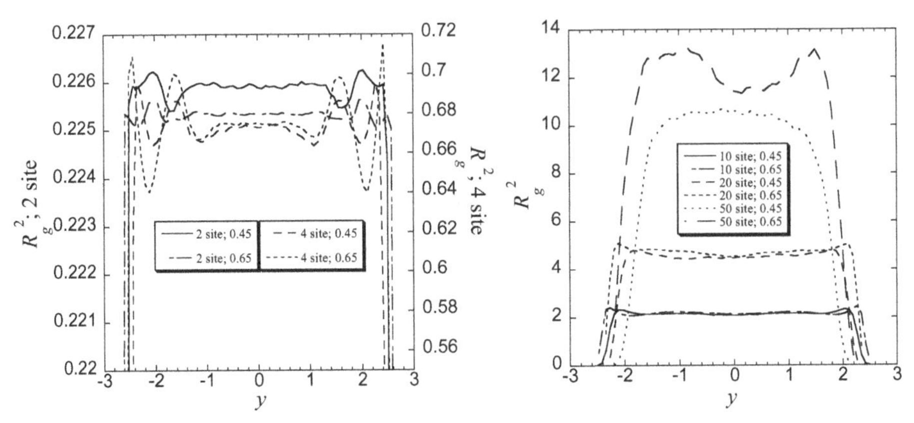

Figure 10.5 Mean squared radius of gyration for average fluid densities of 0.45 and 0.65 (a) $N_s = 2, 4$ and (b) $N_s = 10, 20, 50$. $F^e = 0.05$; $L_y = 7.0$. Replotted from reference [355].

only on systems that are homogeneous in space, so it is not expected to hold for highly confined systems [355].

The atomic and molecular streaming velocities can be defined as

$$\mathbf{v}_a(\mathbf{r}, t) = \frac{\sum_{i=1}^{N_m} \sum_{\alpha=1}^{N_s} m \mathbf{v}_{i\alpha}(t) \delta(\mathbf{r} - \mathbf{r}_{i\alpha}(t))}{\sum_{i=1}^{N_m} \sum_{\alpha=1}^{N_s} m \delta(\mathbf{r} - \mathbf{r}_{i\alpha}(t))} \tag{10.23}$$

$$\mathbf{v}_m(\mathbf{r}, t) = \frac{\sum_{i=1}^{N_m} M \mathbf{v}_i^c(t) \delta(\mathbf{r} - \mathbf{r}_i^c(t))}{\sum_{i=1}^{N_m} M \delta(\mathbf{r} - \mathbf{r}_i^c(t))}, \tag{10.24}$$

where $\mathbf{v}_{i\alpha}(t)$ is the velocity of site α of molecule i at time t and $\mathbf{v}_i^c(t)$ is the centre of mass velocity. Once again, we divide the fluid into bins of width Δy, compute the number and momentum densities in these bins, and then divide the quantities according to Equation (10.23) and Equation (10.24) to obtain the streaming velocities.

We plot the molecular streaming velocity for 2, 20 and 50 site chains in Figure 10.6 for several densities. For the 2-site system, a typical quadratic velocity profile is seen, as was found for the rigid diatomic fluid. However, significant deviations occur for the 20 and 50 site systems, including the appearance of plug-flow-like behaviour associated with significant slip. This plug-like flow behaviour is qualitatively similar to experimental observations of high density polyethylene melts in a slit die [357, 358]. The degree of slip is largely determined by the wall roughness, chain length relative to channel width and wetting effects that depend upon the nature of the wall–fluid and fluid–fluid interactions [355] (see Chapter 11 for more detailed discussion on modeling slip). Clearly, the magnitude of the streaming velocity decreases with increasing density for a given field strength, chain length and channel width, due to increased resistance from surrounding molecules as the average fluid density increases.

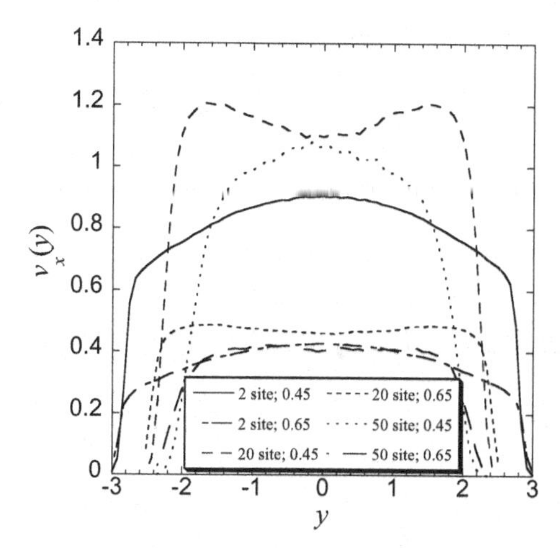

Figure 10.6 Streaming velocity profiles for 2, 20 and 50-site systems at average fluid densities of 0.45 and 0.65. $F^e = 0.05$; $L_y = 7.0$. Replotted from reference [355].

In Figure 10.7 we plot the streaming angular velocity, determined via Equation (10.14) for 2, 10 and 50 site chains. The magnitude of ω_z decreases as the chain length increases for the given values of the overall system density and external field. Note that for the 50-site chain there is no freedom for the molecule to rotate at all anywhere in the channel and ω_z is close to zero everywhere. The rotation of molecules depends on the value of the velocity gradient at the centre of mass location and its radius of gyration compared to the pore width. Rotation requires both sufficiently large velocity gradients and adequate available space to rotate. For molecules with relatively small R_g, rotation is strongest near the wall boundaries since this is where the velocity gradients are strongest. Near the centre of the channel all chains have small ω_z (due to the weakest velocity gradients here) which become identically zero at the channel centre ($y = 0$)

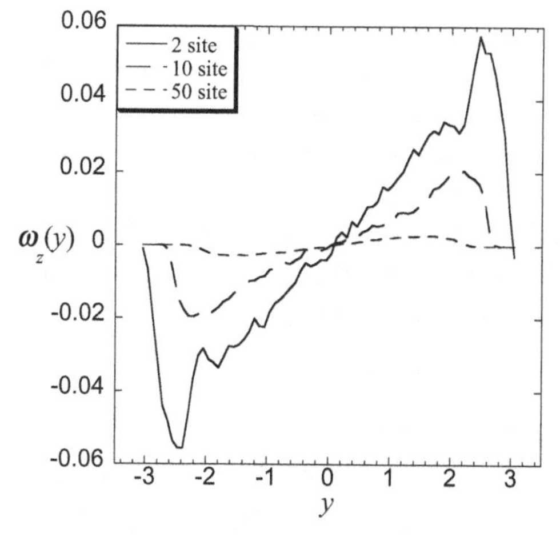

Figure 10.7 Streaming angular velocity profiles for 2, 10 and 50-site systems at an average fluid density of 0.65. $F^e = 0.05$; $L_y = 7.0$. Replotted from reference [355].

where the velocity gradient is zero. 50-site molecules are sufficiently large such that $R_g > L_y$, thus inhibiting molecular rotation everywhere.

Not shown is the comparison between the streaming angular velocity and the vorticity. While it was shown to be half the vorticity for the rigid diatomic system, this is not expected to hold for nonrigid molecules, and this is indeed what is observed as N_s increases [355].

10.2 Spin Coupling, Flow Reduction and Manipulation on the Nanoscale

One of the important consequences of the coupling of intrinsic molecular spin to the vorticity of a fluid is that it leads to a reduction in the flow rate, compared to what would be obtained if no spin-coupling existed. In the case of Poiseuille flow, one can see this clearly if one plots the ratio of the extended Navier-Stokes solution for the linear streaming velocity (Equation (10.6)) to the classical Navier-Stokes quadratic solution that assumes no molecular spin at all, as done by Hansen *et al.* [359]:

$$v^{rel}(\bar{y}) \equiv v_x(\bar{y})/v_x^C(\bar{y}) = 1 + \frac{1}{1-\bar{y}^2}\frac{2\eta_r \coth(Kh)}{(\eta_0+\eta_r)Kh}\left[\frac{\cosh(Kh\bar{y})}{\cosh(Kh)} - 1\right], \quad (10.25)$$

where $v_x^C(\bar{y})$ is the classical Navier-Stokes quadratic prediction. Hansen *et al.* showed that as the length of rigid linear molecules increased, v^{rel} decreased, and was particularly reduced near the walls. Further, by integrating over the velocity profiles of both the classical and extended Navier-Stokes predictions, they found a flow rate reduction that went as $3\eta_r[\coth(Kh) - 1]/(\eta_0 + \eta_r)(Kh)^2$. In the case of buta-triene, this implies a flow rate reduction of $\sim 10\%$ for a channel width of 1 nm, rapidly falling to $\sim 2\%$ at 7 nm. Thus it was seen that flow reduction due to spin-coupling only becomes significant on the confinement scale of nanometers, but it is insignificant on larger scales. This is why it is not invoked for classical macro-scale Navier-Stokes hydrodynamics, but *must* be included for nanofluidic applications that require accurate prediction of fluid flow rates. Similar predictions were made for other alkenes and even nano-confined water [359, 360], as well as for oscillatory flows [361, 362]. In the case of water, these predictions were compared to actual NEMD simulation results, and represents the first convincing demonstration that spin-coupling significantly affects the flow profiles of molecularly structured fluids under nanoscale confinement [363].

We now mention a recent innovation in applying this spin-coupling process to a nano-confined system to affect fluid actuation. As we have seen, a molecularly structured fluid undergoing shear flow will result in the molecules themselves spinning with angular velocity ω. We could very well invert the situation, and ask: *can an induced molecular spin cause a molecular fluid to shear?* In other words, can we generate a flow by somehow manipulating the molecules in a fluid to spin? Bonthuis *et al.* [364, 365] demonstrated theoretically that indeed this can occur. Their ingenious scheme involved a polar molecular liquid, such as water. Application of an external rotating electric field to water causes the permanent dipole moment of water to follow the rotating field with a phase lag. If the fluid is confined to the nanoscale by planar walls, and both walls are identical (symmetric boundary conditions), they showed that the fluid rotational angular

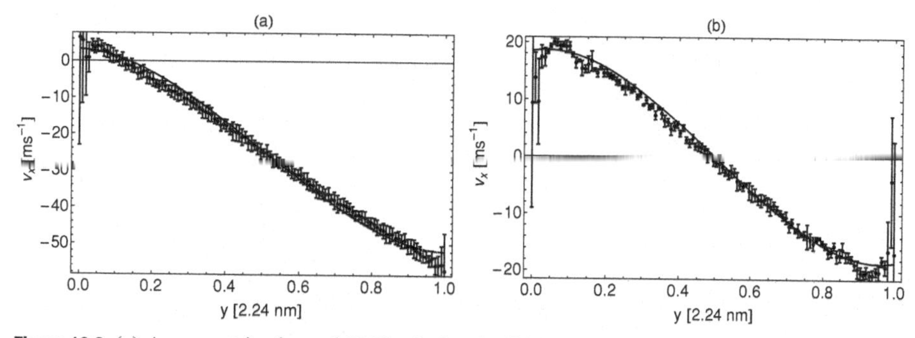

Figure 10.8 (a) Asymmetric channel ENS solution (solid line), compared with NEMD simulation results (solid circles) for electric field strength of $0.184\ \mathrm{VÅ}^{1}$ and frequency 23.9 GHz. (b) Symmetric channel. Reprinted from [366] with the permission of AIP Publishing.

momentum couples to its linear streaming momentum resulting in equal and opposite flow on each side of the axis of symmetry. However, if one wall is hydrophobic while the other hydrophilic (asymmetric boundary conditions), the asymmetry of boundary conditions means that there is little or no fluid flow near the hydrophilic wall, whereas significant flow occurs at the hydrophobic wall. The net result is a velocity profile with a characteristic "S" shape and a nonzero net flow, in other words, uni-directional pumping.

Hansen *et al.* [360] shortly thereafter used more accurate values of the spin viscosities and provided a simplified solution, concluding that flow rates can be significant. De Luca *et al.* [366] then performed the first NEMD simulations of such a system and additionally solved the coupled extended Navier-Stokes (ENS) equations with a torque source term (Equations (10.1) and (10.2)) numerically. Their numerical predictions agreed very well with the NEMD streaming velocity profiles. While they were able to reproduce the magnitude of the angular velocity profile, they found the ENS model *did not* accurately predict its shape, which could have been due to a number of factors, including the nonuniformity of the position dependent torque and the fact that all the transport coefficients used were assumed to be constant, even though strong density inhomogeneity in a highly confined fluid means that there can be a significant spatial inhomogeneity and even nonlocality in all transport coefficients. Figure 10.8, taken from De Luca *et al.* [366], shows the NEMD and ENS results for an asymmetric and symmetric system of water confined to planar atomistic walls. Later work applied the method to an asymmetric system of water confined to hydrophobic graphene and hydrophilic β-cristobalite surfaces [367]. The significance of this work, apart from demonstrating the viability of the technology, is that it leads to the conclusion that flows can be achieved under experimental microwave frequencies and amplitudes, and without excessive heating of water. It suggests a novel method of pumping polar liquids at the nanoscale without the complications of large pressure gradients or the attachment of electrodes.

10.3 Binary Mixtures

The emergence of nanofluidic technology, along with the importance of highly confined flows to separation science, are strong motivation for studies of highly confined binary

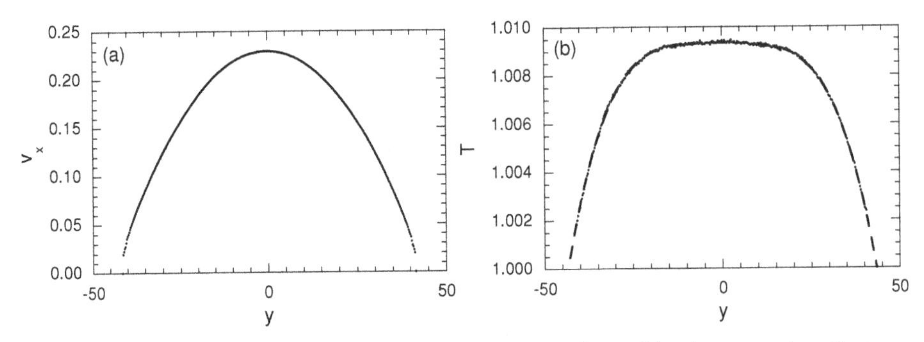

Figure 10.9 (a) Streaming velocity and (b) molecular centre of mass kinetic temperature for a highly confined model polymer solution. The quartic fit to the temperature profile predicted by Fourier's law is also shown by a dashed line which is almost indistinguishable from the data.

liquid mixtures. In particular, studies of model polymer solutions can tell us whether the Navier-Stokes equations are likely to be satisfied for macromolecular solutions flowing in channels of width as little as a few times the radius of gyration of the polymer chain. Using a freely jointed chain of WCA type atoms in a simple WCA atomic solvent and equations of motion similar to those used for confined polymer melts in Equation (10.19), we can obtain results for the velocity, temperature and concentration profiles of solutions in planar Poiseuille flow. Menzel [368] has carried out simulations of a polymer solution consisting of 20-site polymer chains dissolved in solution at a concentration of 20% by mass. The temperature of the equilibrium system (and the wall temperature in the nonequilibrium simulations) was 1.0 and the input total site number density was 0.84.

The results shown in Figure 10.9 are for an external field of 1×10^{-3}. There are small but measurable velocity and temperature jumps at the fluid-solid interface, indicating the presence of velocity and temperature "slip". Apart from small undulations close to the wall, the velocity profile appears to follow the classical quadratic Poiseuille flow profile. Closer examination of the residuals shows that even at this small value of the external field, higher order terms are present in the velocity. The value of the viscosity (3.66) extracted from a quadratic plus quartic fit is in excellent agreement with the Green–Kubo value (3.69) and the nonequilibrium MD value (3.66). The temperature profile shown is purely quartic as predicted by Fourier's law. For this system, the strain rate coupling is apparently absent under these conditions. An important consideration in the determination of the temperature profile is that the streaming kinetic energy component was in this case removed from the kinetic energy profile by post-processing using the extremely accurate binned velocity profile instead of subtracting the streaming kinetic energy using a quadratic fit to the (noisy) instantaneous velocity profile at each accumulation of data. The small (almost indiscernible) difference between the purely quadratic and quadratic plus quartic fits to the velocity profile at this value of the external field could result in a significant error in the temperature profile. This issue is the subject of current investigation.

Figure 10.10 shows the concentration and normal stress difference for the same system. Due to depletion effects, the polymer concentration begins to smoothly decrease at

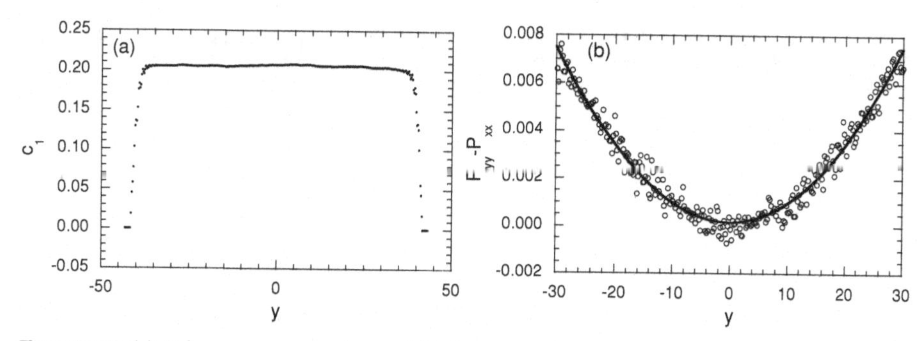

Figure 10.10 (a) Polymer concentration and (b) first normal stress difference for a highly confined model polymer solution. The full line shows that the first normal stress difference is well described by a quadratic fit, as expected.

around ± 38 when the walls, which are located at ± 41.5, are nearer than approximately one radius of gyration of the polymer. The concentration does not show the strong oscillations due to atomic packing effects seen in the density.

The first normal stress differences $N_1 = P_{yy} - P_{xx}$ are also shown in Figure 10.10. Data outside the range $-30 < y < 30$ have been omitted because they showed strong oscillations due to interfacial effects. At low strain rates, N_1 is expected to be proportional to the strain rate squared. We have seen in Figure 10.9 that the velocity profile is very nearly quadratic, so the strain rate varies linearly with y. The quadratic fit shown in the plot agrees very well with the data, indicating that the normal stress coefficient is constant over the range of strain rates present in the plot. The value of the first normal stress coefficient for the fluid obtained from the fit is $\Psi_1 = 180 \pm 10$. Because the fluid is naturally thermostatted by heat transfer to the walls, this value of the first normal stress coefficient is expected to be free of artefacts that can affect nonlinear rheological properties of synthetically thermostatted, homogeneous nonequilibrium molecular dynamics simulations using the SLLOD algorithm as discussed in section 6.4.2.

11 Generalised Hydrodynamics and Slip

11.1 Generalised Hydrodynamics

The realisation that classical Navier-Stokes hydrodynamics would fail at molecular length-scales is not new. It has long been expected that generalised hydrodynamics could play an important part in the prediction of transport properties of inhomogeneous fluids. In 1983 Alley and Alder [321] noted that generalised hydrodynamic models could be very useful to the development of molecular-level predictive tools. In generalised hydrodynamics, the transport coefficients are no longer regarded as constants, nor are they simply material properties of a fluid at the local thermodynamic state point. Transport now becomes a fully nonlocal property of fluids, and likewise all the transport "coefficients" now become nonlocal in both time and space. In other words, the transport "coefficients" are now replaced by kernels and the governing constitutive equations are integral functions (convolutions) over both space and time. The kernels themselves now have both a wavelength and frequency dependence. In fact, significant progress in the development of generalised hydrodynamics was made in the 1970s by Akcasu and Daniels [369] and Ailawadi *et al.* [370]. The books by Boon and Yip [371], Eu [372] and Hansen and McDonald [61] also treat this subject in considerable detail, and we refer readers to these references for a much more thorough theoretical foundation.

Our purpose is to show how generalised hydrodyamics can be used as a predictive tool for inhomogeneous fluids. In particular, we will show how classical Navier-Stokes hydrodynamics breaks down at molecular length scales when the spatial variation in the driving thermodynamic force (e.g. the strain rate) is significant over the range of molecular interactions. Such conditions can occur not only for highly confined fluids, but also for fluids under shock conditions [373–375], where moment and gradient expansions originally derived from gas kinetic theory are commonly employed [372, 376]. Nonlocal effects are also observed in shear-banding phenomena [377], micellar solutions [378], Brownian suspensions of rigid fibres [379] and jammed glassy systems [380] and may well play a significant role in turbulence.

We will first demonstrate where breakdown in classical theory occurs in both confined fluids and systems in which the curvature in the strain rate exceeds the width of the relevant transport kernel. We then express the governing constitutive equation in its generalised form and demonstrate several techniques to compute the transport kernels. Once the kernels are computed, it is a straightforward matter to use them to predict properties of interest. Finally, we will provide a mathematical tool for determining whether or not nonlocal effects are significant in the system being studied. We will limit our discussion strictly to momentum transport. However, the generalisation to the transport of mass and heat follows analogously.

11.1.1 Breakdown of Navier-Stokes Predictions

The breakdown of Navier-Stokes hydrodynamics was clearly demonstrated by NEMD simulations in the 1990s. The first of these papers by Travis *et al.* [96] considered the failure of the Navier-Stokes and Fourier formulations of momentum and heat transport, respectively, to correctly predict the velocity and heat flux profiles for highly confined fluids. This was motivated by the (then) recent development of the MoP methods, as has been described in previous sections. It was noted in this, and a previous study by Akhmatskaya *et al.* [381], that a local viscosity coefficient was inconsistent with the type of velocity profiles that were actually measured in NEMD simulations. As had already been noted by Alley and Alder [321] and by Evans and Morriss [2], at least for the case of homogeneous fluids, Newton's local law of viscosity could well be the source of the discrepancy between prediction and observation when the strain rate varies appreciably over the range of molecular interactions. Numerically convincing evidence of this breakdown of Navier-Stokes theory was given by Travis and Gubbins [320] in which the flow profiles for very highly confined fluids (effective channel half-width, $h = 2.0$) indicated incontrovertible proof of the failure of the theory.

In Figures 11.1 and 11.2 we show the actual density and streaming velocity profiles for the confined atomic fluid studied by Travis and Gubbins [320]. The system undergoes gravity driven (Poiseuille) flow and fluid and wall atoms interact via either (a) WCA (fluid)-WCA (fluid-wall)-WCA (wall) interactions, (b) LJ (fluid)-LJ (fluid-wall)-LJ (wall) interactions, or (c) a combination of WCA (fluid)-LJ (fluid-wall)-WCA (wall) interactions. The different types of interaction potentials encourage different layering

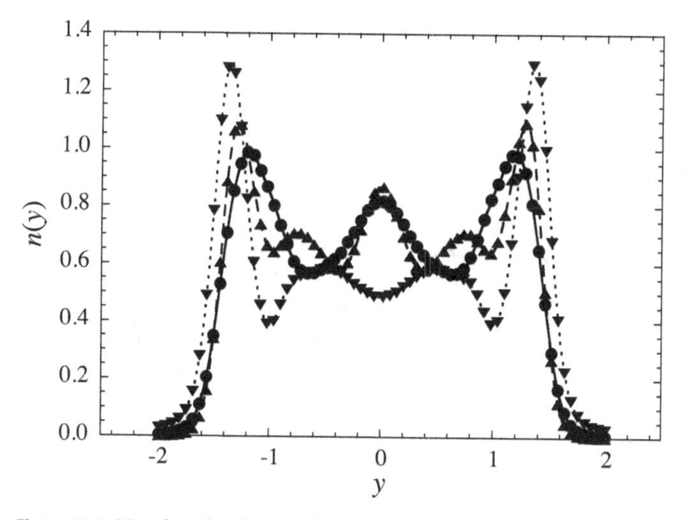

Figure 11.1 Number density profiles for the three distinct systems (a) WCA system (circles), (b) LJ system (upright triangles), and (c) fluid-fluid and solid-solid WCA, fluid-solid LJ (upside down triangles) for a channel of half-width $h = 2.0$ [320]. Error bars are smaller than symbol sizes. Data provided by K.P. Travis [382].

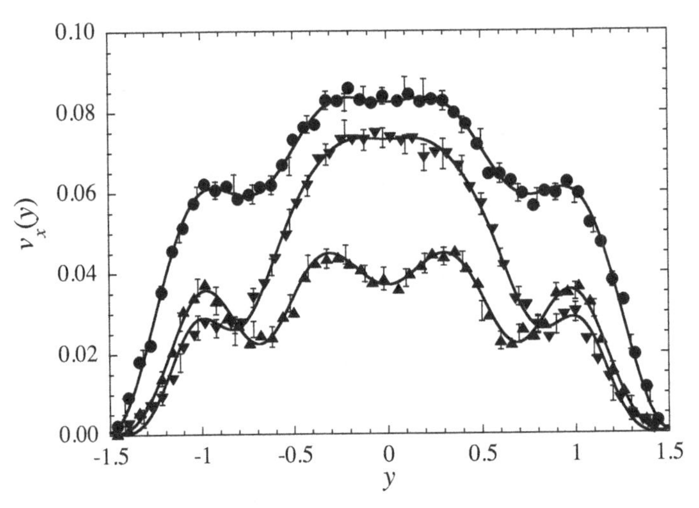

Figure 11.2 Streaming velocity profiles for the three distinct systems depicted in Figure 11.1 [320] (symbols are as defined in that figure). Data provided by K.P. Travis [382].

arrangements for such a tightly confined fluid, as indicated by the density profiles in Figure 11.1. Clearly, in Figure 11.2 the streaming velocity profiles for all three systems are completely different to the quadratic profile predicted by Navier-Stokes theory (see Chapter 9).

More revealing than this, however, is the strain rate profile, obtained by differentiating the streaming velocity profile, and presented in Figure 11.3. Of particular significance here is the presence of zeros at a number of locations, including the channel mid-point. The classical Navier-Stokes equation relies upon the use of Newton's local consititutive

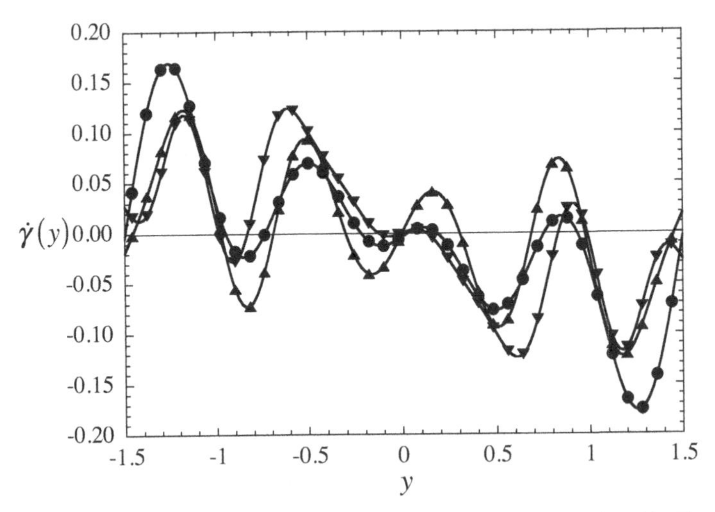

Figure 11.3 Strain rate profiles for the three distinct systems depicted in Figure 11.1 [320] (symbols are as defined in that figure). Data provided by K.P. Travis [382].

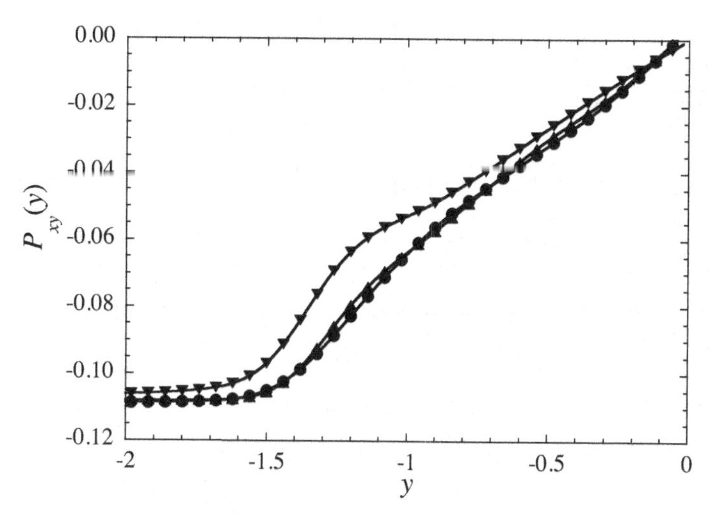

Figure 11.4 Negative shear stress profiles for the three distinct systems depicted in Figure 11.1 [320] (symbols are as defined in that figure). Data provided by K.P. Travis [382].

equation, $P_{xy}(y) = -\eta(y)\dot{\gamma}(y)$. In other words, the strain rate $\dot{\gamma}$ is proportional to the shear stress, $-P_{xy}$. If this is true, then P_{xy} *must* have zeros in *exactly* the same locations as the zeros in $\dot{\gamma}$. In Figure 11.4 we show P_{xy} as a function of y for half the channel width (the system is symmetric about $y = 0$). In this case, P_{xy} is computed by the IMC method. Clearly, P_{xy} is only zero at the channel mid-point, but is finite and nonzero elsewhere. Therefore, the simple proportionality of Newton's law simply can not be true for this system. Furthermore, the absurdity of a local viscosity is demonstrated by trying to compute it, i.e. using the local definition $\eta(y) = -P_{xy}(y)/\dot{\gamma}(y)$. Plotting this (again only for half the channel due to symmetry) leads to Figure 11.5, in which the

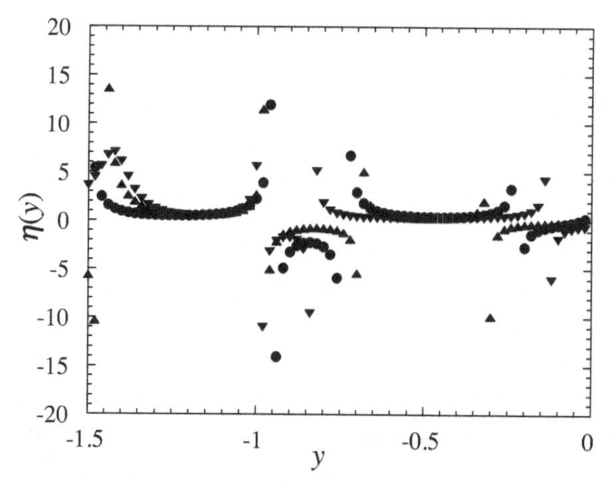

Figure 11.5 Viscosity profiles for the three distinct systems depicted in Figure 11.1 [320] (symbols are as defined in that figure). Data provided by K.P. Travis [382].

shear viscosity displays a number of physically meaningless singularities. Travis and Gubbins also show similar breakdown of the local Fourier law for heat flow, $\lambda(y) = -J_q(y)/\nabla T(y)$, leading to similarly absurd results [320].

11.1.2 Computation of Nonlocal Viscosity Kernel

The generalised form for the constitutive equation relating shear stress to strain rate for a *homogeneous* fluid is given by [2, 321]

$$\sigma_{xy}(\mathbf{r},t) = -P_{xy}(\mathbf{r},t) = \int_0^t \int_{-\infty}^{\infty} \eta\left(\mathbf{r}-\mathbf{r}', t-t'\right)\dot{\gamma}\left(\mathbf{r}',t'\right) d\mathbf{r}'dt'. \tag{11.1}$$

For an inhomogeneous fluid, the following form has been postulated [96, 381]

$$\sigma_{xy}(\mathbf{r},t) = \int_0^t \int_{-\infty}^{\infty} \eta\left(\mathbf{r}, \mathbf{r}-\mathbf{r}', t-t'\right)\dot{\gamma}\left(\mathbf{r}',t'\right) d\mathbf{r}'dt'. \tag{11.2}$$

In Equation (11.1) the kernel is homogeneous in space and time, i.e. at any point $\mathbf{r}$ in the fluid and time t, the kernel has the same form in *any* direction in space and moment in time. Furthermore, the shear stress is now nonlocal in space and time: it is formed by the convolution of the viscosity kernel with the strain rate over all space and time. However, Equation (11.2) is different: while the kernel is still homogeneous in time, it is no longer homogeneous in space. A point $\mathbf{r}_1$ in space has a viscosity kernel different to a point located at some other position $\mathbf{r}_2$. Furthermore, at any point $\mathbf{r}$, the kernel itself has a different form depending on which direction in space one views it from. This will be made clear further on.

If we now consider only the steady-state, we can drop the explicit time dependence of the viscosity kernel, and again if we use the same flow geometry as before (i.e. flow in the x-direction with confinement/spatial inhomogeneity in the y-direction) then Equations (11.1) and (11.2) become

$$\sigma_{xy}(y) = \int_{-\infty}^{\infty} \eta\left(y-y'\right)\dot{\gamma}\left(y'\right) dy' \tag{11.3}$$

$$\sigma_{xy}(y) = \int_{-\infty}^{\infty} \eta\left(y, y-y'\right)\dot{\gamma}\left(y'\right) dy' \tag{11.4}$$

for spatially homogeneous and inhomogeneous fluid systems, respectively.

11.1.2.1 Homogeneous Kernel

There are two straightforward methods to extract the homogeneous viscosity kernel for atomic or molecular fluids. One method relies upon the well-established methods of generalised hydrodynamics, in which the wavevector and frequency dependent viscosity is computed from equilibrium time correlation functions of either the transverse momentum current or shear stress. This method has been used successfully in the past to compute kernels for hard spheres [321], water [325, 326, 383] and most recently for the Lennard-Jones fluid [324] and mixtures [384]. The second method makes use of the NEMD STF algorithm, described in Chapter 9. We briefly describe both methods in what follows, starting with the derivation of the transverse momentum current and shear stress time correlation functions. Derivations of the former can be found in the books by Hansen and MacDonald [61] and Evans and Morriss [2], with Evans and Morriss also including a derivation of the latter. However, these derivations are simplified to scalar components. Our derivation is more general.

We begin with the linearised wavevector dependent momentum conservation equation given by Equation (4.29), describing fluctuations in the momentum density for an equilibrium fluid

$$\frac{\partial \mathbf{J}(\mathbf{k},t)}{\partial t} = i\mathbf{k} \cdot \mathsf{P}(\mathbf{k},t). \tag{11.5}$$

For an atomic fluid the pressure tensor can be decomposed into a sum of isotropic ($p\mathbf{1}$) and traceless symmetric components (P^{ts}), since the antisymmetric part is zero. Both these terms can be expressed as generalised constitutive equations for the compressive and viscous momentum fluxes (see Equation (2.56)):

$$\Pi(\mathbf{k},t) = \int_0^t ds\, \eta_v(\mathbf{k},t-s)\, i\mathbf{k} \cdot \mathbf{v}(\mathbf{k},s) \tag{11.6}$$

$$\mathsf{P}^{ts}(\mathbf{k},t) = 2\int_0^t ds\, \eta(\mathbf{k},t-s)\, i(\mathbf{kv})^{ts}(\mathbf{k},s). \tag{11.7}$$

Mori-Zwanzig theory shows that each of these microscopic flux equations should include a random flux term to correctly describe the fluctuations [2], but we have omitted them because they are uncorrelated with their respective momentum density components and do not appear in the final expressions. Substituting these two constitutive equations into Equation (11.5) gives

$$\frac{\partial \mathbf{J}(\mathbf{k},t)}{\partial t} = i\mathbf{k} \cdot \int_0^t ds \left\{\mathbf{1}\,\eta_v(\mathbf{k},t-s)\, i\mathbf{k} \cdot \mathbf{v}(\mathbf{k},s) + 2\eta(\mathbf{k},t-s)\, i(\mathbf{kv})^{ts}(\mathbf{k},s)\right\}. \tag{11.8}$$

Unfortunately Equation (11.8) alone is insufficient to obtain an expression involving only $\eta(\mathbf{k})$ since it also contains the bulk viscosity. If, for example, we define $\mathbf{k}$ to be in the y-direction, we will find that the momentum density has contributions due to both η and η_v in this direction and we can not decouple them to find an expression for η alone. However, the decoupling *can* be accomplished by looking instead at the direction transverse to $\mathbf{k}$, which would be in the x-z plane. We can obtain this transverse momentum density by multiplying Equation (11.8) by $\mathbf{1} - \hat{\mathbf{k}}\hat{\mathbf{k}}$, where $\hat{\mathbf{k}}$ is the unit vector in the direction of $\mathbf{k}$. Using the same notation as Evans and Morriss we define this transverse density as $\mathbf{J}_\perp(\mathbf{k},t)$. Thus we obtain the transverse momentum density equation as

$$\frac{\partial \mathbf{J}_\perp(\mathbf{k},t)}{\partial t} = \left(\mathbf{1} - \hat{\mathbf{k}}\hat{\mathbf{k}}\right) i\mathbf{k} : [\mathsf{I}_1 + \mathsf{I}_2], \tag{11.9}$$

where I_1 and I_2 are the two (separated) integrals on the right-hand side of Equation (11.8), respectively. It is straightforward to show that $\left(\mathbf{1} - \hat{\mathbf{k}}\hat{\mathbf{k}}\right)\mathbf{k} : \mathsf{I}_1 = \mathbf{0}$, so it does not contribute to the transverse momentum density. Noting that $\mathbf{J} \equiv \rho\mathbf{v}$, Equation (11.9) can be expressed as

$$\frac{\partial \mathbf{J}_{\perp}(\mathbf{k},t)}{\partial t} = -\frac{k^2}{\rho}\int_0^t ds\,\eta(\mathbf{k},t-s)\mathbf{J}_{\perp}(\mathbf{k},s). \tag{11.10}$$

This applies no matter which direction is chosen for $\mathbf{k}$, because the material is isotropic at equilibrium.

We now have an expression involving only the wavevector and frequency dependent shear viscosity, which is precisely what we want. However, to usefully extract the viscosity we can form time-correlation functions with a measurable property of the system. This can conveniently (though not uniquely) be achieved by taking the time-correlation function of both sides of Equation (11.10) with $\mathbf{J}_{\perp}(-\mathbf{k},0)$, ensemble averaging, and taking the time derivative outside the ensemble average, giving

$$\frac{\partial C(\mathbf{k},t)}{\partial t} = -\frac{k^2}{\rho}\int_0^t ds\,\eta(\mathbf{k},t-s)C(\mathbf{k},s), \tag{11.11}$$

where $C(\mathbf{k},t) \equiv \frac{1}{2V}\langle \mathbf{J}_{\perp}(-\mathbf{k},0)\cdot\mathbf{J}_{\perp}(\mathbf{k},t)\rangle$. Finally, taking the Fourier-Laplace transform of Equation (11.11) and re-arranging terms gives us the required expression for the wavevector and frequency dependent shear viscosity:

$$\eta(\mathbf{k},\omega) = \rho\frac{C(\mathbf{k},t=0) - i\omega C(\mathbf{k},\omega)}{k^2 C(\mathbf{k},\omega)}, \tag{11.12}$$

where $C(\mathbf{k},\omega)$ is the Fourier-Laplace transform of $C(\mathbf{k},t)$. If, for example, $\mathbf{k}$ is chosen to be in thc y direction, the transverse momentum density will be explicitly given by

$$\mathbf{J}_{\perp}(\mathbf{k},t) = \sum_{i=1}^{N}\mathbf{p}_{\perp i}e^{ik_y y_i} = \sum_{i=1}^{N}(p_{ix}\hat{\mathbf{x}} + p_{iz}\hat{\mathbf{z}})e^{ik_y y_i}. \tag{11.13}$$

Clearly, $C(\mathbf{k},t=0) = \rho k_B T$, and in practice this is obtained from the actual correlation function simulation data, rather than by this expression due to finite system size effects of order $O(1/N)$ [324].

Equation (11.12) diverges at $\mathbf{k} = 0$ and can not be used to obtain the zero wavevector value of the viscosity kernel. However, an alternative method is to compute the wavevector and frequency dependent viscosity kernel from the autocorrelation function of the shear stress [2], for which we provide a more generalised derivation as follows.

If $\mathbf{A}(\mathbf{k},t)$ is some general flux variable and we define

$$C_A(\mathbf{k},t) \equiv \langle \mathbf{A}(\mathbf{k},t)\cdot\mathbf{A}(-\mathbf{k},0)\rangle \tag{11.14}$$

and

$$\phi_A(\mathbf{k},t) \equiv \left\langle \dot{\mathbf{A}}(\mathbf{k},t)\cdot\dot{\mathbf{A}}(-\mathbf{k},0)\right\rangle, \tag{11.15}$$

then it can be shown [2] that

$$\frac{d^2}{dt^2}C_A(\mathbf{k},t) = -\phi(\mathbf{k},t). \tag{11.16}$$

The Fourier-Laplace transform of this equation is

$$-\phi(\mathbf{k},\omega) = (i\omega)^2 C_A(\mathbf{k},\omega) - i\omega C_A(\mathbf{k},t=0)\,. \tag{11.17}$$

If we let $\mathbf{A}(\mathbf{k},t) \equiv \mathbf{J}(\mathbf{k},t)$, then we know from Equation (11.5) that

$$\frac{d\mathbf{J}(\mathbf{k},t)}{dt} = i\mathbf{k}\cdot\mathsf{P}(\mathbf{k},t) \equiv \dot{\mathbf{A}}(\mathbf{k},t) \tag{11.18}$$

and so Equation (11.16) takes the form

$$\frac{d^2}{dt^2}\langle \mathbf{J}(\mathbf{k},t)\cdot\mathbf{J}(-\mathbf{k},0)\rangle = -\mathbf{kk} : \langle \mathsf{P}(\mathbf{k},t)\mathsf{P}(-\mathbf{k},0)\rangle\,. \tag{11.19}$$

Noting that only the transverse momentum density current contributes to the shear stress, Equation (11.19) can be expressed as

$$\frac{d^2}{dt^2}C(\mathbf{k},t) = -\frac{k^2}{10V}\left\langle \mathsf{P}^{ts}(\mathbf{k},t) : \mathsf{P}^{ts}(-\mathbf{k},0)\right\rangle \tag{11.20}$$

where the factor of 10 in the denominator arises from averaging over all independent terms of the tensor contraction (true for a homogeneous fluid), as shown previously in the discussion of the Green–Kubo shear viscosity given in Chapter 3. If we denote the Fourier-Laplace transform of some quantity B as $L\{B\}$ and define

$$N(\mathbf{k},\omega) \equiv \frac{1}{10Vk_BT}L\left\{\left\langle \mathsf{P}^{ts}(\mathbf{k},t) : \mathsf{P}^{ts}(-\mathbf{k},0)\right\rangle\right\}, \tag{11.21}$$

then taking the Fourier-Laplace transform of Equation (11.20) (noting the general form in Equation (11.17)) gives us

$$(i\omega)^2 C(\mathbf{k},\omega) - i\omega C(\mathbf{k},t=0) = -k^2 k_BTN(\mathbf{k},\omega)\,, \tag{11.22}$$

where C is as defined in Equations (11.11) and (11.12). Finally, substituting Equation (11.22) into Equation (11.12) leads to the expression for the wavevector and frequency dependent shear viscosity in terms of the shear-stress time autocorrelation function:

$$\eta(\mathbf{k},\omega) = \rho\frac{N(\mathbf{k},\omega)}{\frac{C(\mathbf{k},t=0)}{k_BT} - \frac{k_y^2N(\mathbf{k},\omega)}{i\omega}}\,. \tag{11.23}$$

Equation (11.23) does not suffer from singularities at the origin and can be used to compute the zero wavevector viscosity. Note also that the Green–Kubo viscosity given by Equation (3.71) is trivially obtained by taking the zero wavevector and frequency limits of Equation (11.23).

The advantage of computing the wavevector dependent viscosity via Equation (11.23) is the statistical superiority that results from averaging over all possible off-diagonal components of the shear stress autocorrelation functions.

Once the wavevector dependent viscosity is computed it can be readily inverse-transformed back into real space to obtain the position-dependent viscosity kernel, which can then be used to predict the stress profile via Equation (11.3). Alternatively, if the stress profile is already known (as we have shown to be the case by using either MoP or IMC techniques), one can invert the expression and predict the strain rate profile, and hence (by one further integration) the streaming velocity flow profile.

The alternative method to computing equilibrium time-correlation functions is to use the STF method described in Chapter 9. In this method we make use of the harmonic excitations implicit in the external driving field $F_x(y)$. As we have already seen, this induces harmonics in all the useful fields we are interested in, including velocity, temperature, stress and heat flux. If the field is sufficiently large, then harmonics also appear in the density field. However, for the purposes of computing a homogeneous visosity kernel, we apply weak fields such that the density remains constant throughout the simulation box.

The simplicity of the STF method stems from the deconvolution of the integral constitutive expression given in Equation (11.3) in reciprocal space. Taking the Fourier transform of this expression and rearranging in terms of the wavevector dependent viscosity gives

$$\eta(k_m) = -\frac{P_{xy}(k_m)}{\dot{\gamma}(k_m)}, \tag{11.24}$$

where k_m is the wavevector and P_{xy} and $\dot{\gamma}$ are the Fourier coefficients of the stress and strain rate fields, as described in Chapter 9. Explicitly, for an external force given by

$$F_x(y) = F_0 \sin(k_n y), \tag{11.25}$$

where $k_n = 2\pi n/L_y$ and $n = 1, 2, 3, \ldots$ is varied to control the wavevector of the external force, we have

$$P_{xy}(y) = \sum_{m=n}^{N} P_{xy}(k_m)\cos(k_m y) \tag{11.26}$$

$$\dot{\gamma}(y) = \sum_{m=n}^{N} \dot{\gamma}(k_m)\cos(k_m y), \tag{11.27}$$

where $\dot{\gamma}(k_m) = k_m v_x(k_m)$. As we deal only with small field strengths with no higher density harmonics, this simplifies to

$$\eta(k_m) = -\lim_{\dot{\gamma}\to 0} \frac{\rho_0 F_0}{k_n^2 v_x(k_m)}, \tag{11.28}$$

where ρ_0 is the zero wavevector density (i.e. the fluid density for a homogeneous fluid).

In Figure 11.6 we plot the wavevector dependent viscosity for a WCA fluid of 1700 atoms, with $L_y = 17.61$ ($\sigma = 1.0$) and density of $\rho_0 = 0.685$. Data are presented for both the equilibrium time correlation function and nonequilibrium STF methods. We only present the positive half of the curve, as it is symmetric about $k_n = 0$. Clearly we can see that both methods agree to within statistical error. The STF data are extrapolated to zero field strength, and full details can be found in reference [324]. We are able to fit the wavevector dependent viscosity to two functional forms. The first is a sum of two

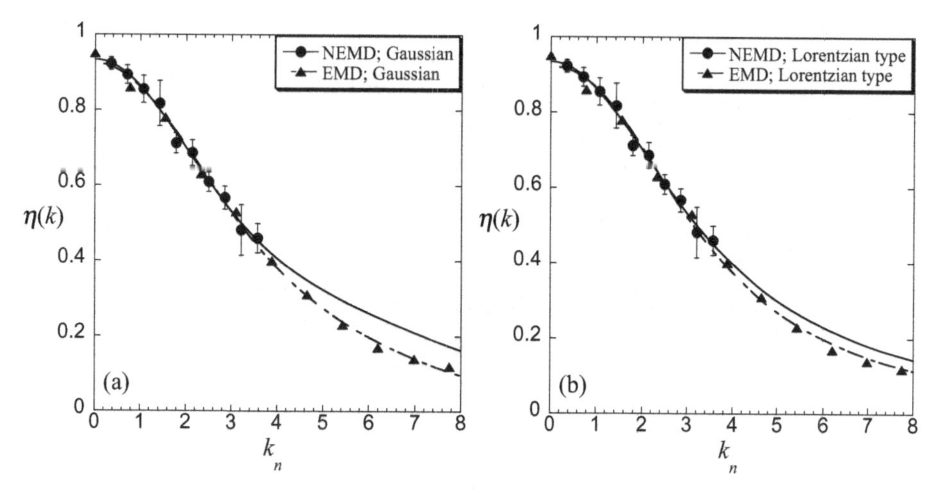

Figure 11.6 k-space viscosity kernels for a WCA fluid, $T = 0.765$, $\rho_0 = 0.685$. Equilibrium time-correlation function data represented by triangles; NEMD STF data depicted by circles. (a) Two-term Gaussian function fits given by Equation (11.30). (b) Same data as in (a), but fit to a Lorentzian-type function given by Equation (11.31). Replotted from reference [324].

Gaussians, given by

$$\eta_G(k_n) = \eta_0 \left(A e^{-k_n^2/2\sigma_1^2} + [1 - A] e^{-k_n^2/2\sigma_2^2} \right); \; A, \, \sigma_1, \, \sigma_2 \in R_+. \tag{11.29}$$

As can be seen from Table 11.1, since $A \sim 0.5$, Equation (11.29) can be approximated by a simpler two-parameter function

$$\eta_G(k_n) \approx \frac{\eta_0}{2} \left(e^{-k_n^2/2\sigma_1^2} + e^{-k_n^2/2\sigma_2^2} \right). \tag{11.30}$$

It is clear from Figure 11.6 that the extrapolated fits to the NEMD data for high wavevectors are not very good. An alternative fit, which does extrapolate well at high wavevectors, is given by a Lorentizian-type function of the form

$$\eta_L(k_n) = \frac{\eta_0}{1 + \alpha |k_n|^\beta}; \; \alpha, \, \beta \in R_+. \tag{11.31}$$

Table 11.1 Zero frequency, zero wavevector shear viscosity and fitted parameter values for Equation (11.29) and Equation (11.31) at different state points. Data sourced from reference [324].

	$T_0 = 0.726$				$T_0 = 0.765$			
ρ_0	0.374	0.438	0.485	0.698	0.375	0.450	0.480	0.685
η_0	0.274	0.340	0.397	0.971	0.273	0.354	0.390	0.929
A	0.417	0.498	0.406	0.628	0.440	0.507	0.399	0.492
σ_1	1.166	1.445	1.616	2.180	1.376	1.615	1.604	1.929
σ_2	4.587	4.674	4.892	4.442	4.750	4.762	4.879	4.497
α	0.234	0.160	0.131	0.048	0.180	0.126	0.131	0.080
β	1.547	1.753	1.805	2.438	1.662	1.886	1.808	2.142

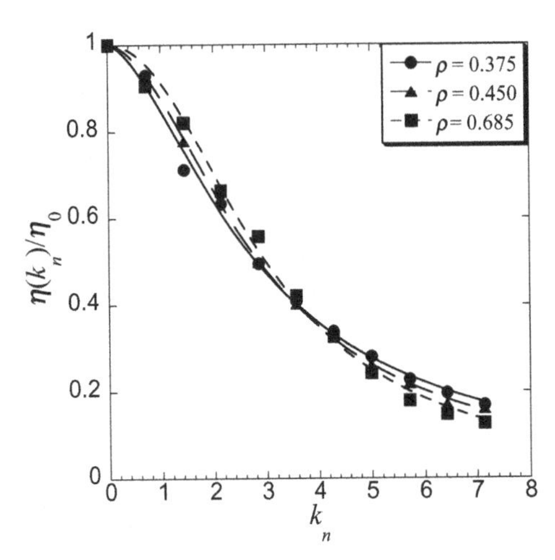

Figure 11.7 Normalised k-space viscosities for the WCA fluid at different densities at reduced temperature $T = 0.765$. The fits are those based on Equation (11.31). Replotted from reference [324].

All parameters are given in Table 11.1. In Figure 11.7 we display the normalised wavevector viscosity kernel for three different fluid densities. As the density increases, the shape of the viscosity curve is broader, though they all decay to zero at approximately the same rate at high k_n.

The advantage of using a Gaussian form for the wavevector viscosity is that its inverse Fourier transform can be obtained analytically. While the Lorentzian-type form does have an inverse transform (it is absolutely integrable, square integrable, and the function and its derivative are piecewise continuous), it can not be computed analytically. Instead, it needs to be computed numerically. Note that for both types of functional forms (and indeed, a necessary property of the Fourier transform), we have that the zero wavevector viscosity is given as

$$\eta_0 = \int_{-\infty}^{\infty} \eta\,(k_n) dk_n. \tag{11.32}$$

The real-space viscosity kernel, obtained by inverse-transforming Equation (11.30) is thus

$$\eta_G\,(y) \approx \frac{\eta_0}{2\sqrt{2\pi}} \left(\sigma_1 e^{-(\sigma_1 y)^2/2} + \sigma_2 e^{-(\sigma_2 y)^2/2}\right). \tag{11.33}$$

As with the wavevector dependent viscosity, the following relationship also holds true:

$$\eta_0 = \int_{-\infty}^{\infty} \eta\,(y) dy\,, \tag{11.34}$$

where $\eta\,(y)$ is either the analytical Gaussian form, or that obtained by numerical transform of Equation (11.31). We note here that numerical transforms were performed by Simpson, trapezoid or Lado methods [385], and all were of similar numerical accuracy [324].

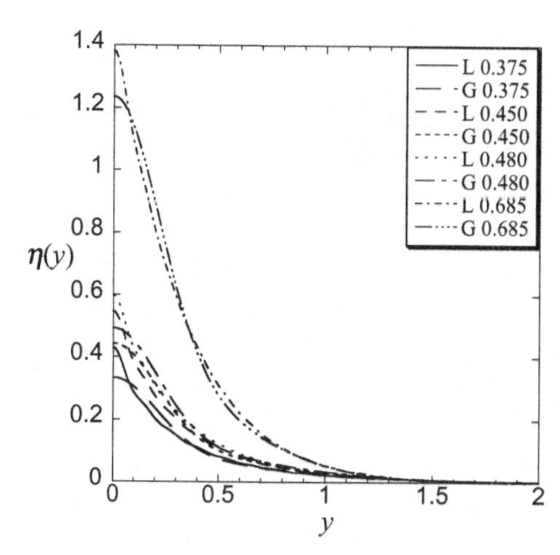

Figure 11.8 Real space viscosity kernels for the WCA fluid at different densities at reduced temperature $T = 0.765$. The fits are those based on the symmetric Gaussian form Equation (11.30) (G) and the Lorentzian type form Equation (11.31) (L). The number in the identifying labels indicates the reduced fluid density. Replotted from reference [324].

The real-space kernels are presented in Figure 11.8 for a number of fluid densities, for both the Gaussian and Lorentzian-type functional forms. Once again, only half the kernel is presented, with $y < 0$ data being symmetrical about the origin. Higher density fluids have higher-peaked $\eta\,(0)$ values and the curves are broader. However, for all kernels we see that the base-width is of the order of 2–3 atomic diameters, and all kernels decay to zero rapidly. As we will see in what follows in Sections 11.1.3 and 11.1.4, this implies that the Navier-Stokes relations will break down when the strain rate varies significantly over distances < 2–3 atomic diameters for such a fluid.

While a 2-term Gaussian is a good analytical choice for the viscosity kernel, it has been pointed out that a 4-term Gaussian can give more accurate fits to the k-space kernel simulation data [386]. Both k- and y-space kernels have been computed for simple Lennard-Jones fluids, as well as liquid chlorine [386], butane and polymer melts [327]. By computing the viscosity kernel of a polymer melt, considerable insight was given into the approach to the glass transition by Puscasu *el al.* [328]. In this study it was found that the width of the k-space kernel dramatically decreased on close approach to the glass transition, thus implying a broad real-space kernel, reflective of the strong degree of nonlocality for such liquids. The implication of this is that constitutive modeling of polymers close to their glass transition states may need to incorporate nonlocality into the theoretical formalism. It also implies that classical Navier-Stokes hydrodynamics will fail to correctly predict flow or stress profiles near the glass transition for glassy polymers under flow conditions.

11.1.2.2 Inhomogeneous Kernel

In principle one might imagine that the extraction of an inhomogeneous kernel might be possible by directly Fourier transforming Equation (11.4) and then solving for the

k-space kernel. However, the fact that the kernel now has arguments $y - y'$ *and* y makes such a transform nontrivial. For a low density fluid one might try to approximate the kernel by $\eta\,(y, y - y') \approx \eta\,(y - y')$ and then perform the operation. However, this also has serious problems for confined fluids in particular, where the presence of walls implies that the integrals are no longer infinite in extent, but must be bounded at the walls. This in turn implies that the Fourier transform of Equation (11.4) is truncated, and this has serious ramifications because the truncation of the strain rate at $y = \pm h$ means that it must be convoluted with a top-hat function. The Fourier transform of such a convolution leads to superposition of the desired signal with the undesired sinc function, which leads to unphysical singularities, and makes the extraction of the nonlocal viscosity kernel very problematic. This method was attempted and analysed in significant detail by Cadusch *et al.* [387], and readers should refer to this paper for details.

The solution to this problem is highly nontrivial and involves additional higher-order couplings to density gradients. While a complete solution for confined fluids has not yet been produced, significant progress has been made by artificially inducing density gradients in an unconfined fluid through use of the sinusoidal longitudinal force (SLF) method that was briefly mentioned in Chapter 9. By avoiding the added complication of a solid-fluid interface, in which slip and surface stresses need to be considered, one can study an unconfined system that behaves periodically in space with density wavelength variations of the order of atomic diameters. By further applying an external sinusoidal transverse field (STF), shear can be induced. The coupling between density and strain rate can then be studied extremely precisely and a full nonlocal treatment for such an inhomogeneous system can be explored. This has been done by Dalton, Glavatskiy and colleagues very recently [339–342], with further results to be published shortly [388–390]. They furthermore convincingly show that the local average density model (LADM) of Bitsanis and colleagues [391, 392] is insufficient to explain or predict the velocity profile for highly inhomogeneous fluids, and that only a complete treatment accounting for nonlocality and higher-order couplings can give accurate predictions of flow profiles that agree with NEMD simulation results.

11.1.3 Prediction via Generalised Hydrodynamics

To demonstrate the significance and accuracy of the generalised hydrodynamics approach, we consider a homogeneous fluid under the influence of a sinusoidal field of form $F_x\,(y) = F_0 \cos\,(k_n y)$. This demonstration was first given in reference [322], which we summarise here. If the field is sufficiently weak so that only the first harmonics are excited, then the shear stress, streaming velocity and strain rate are given by

$$\sigma_{xy}\,(y) = -\frac{F_0 \rho_0}{k_n} \sin\,(k_n y) \tag{11.35}$$

$$v_x\,(y) = v_x\,(k_n) \cos\,(k_n y) \tag{11.36}$$

$$\dot{\gamma}\,(y) = \partial v_x\,(y) / \partial y = -\,k_n v_x\,(k_n) \sin\,(k_n y)\,, \tag{11.37}$$

Table 11.2 Parameters used for the prediction of shear stress [322].

	T	ρ_0	F_0	η_0	σ_1	σ_2	k_n	$v_x(k_n)$
$n = 1$	0.765	0.685	0.15	0.929	1.929	4.497	0.357	0.886
$n = 10$	0.765	0.685	0.225	0.929	1.929	4.497	3.57	0.027

where we note that the density ρ_0 is constant. These are the known values of these quantities that can be compared directly with NEMD data and be confirmed, as we did in Chapter 9 when we first discussed the STF method. However, we can also predict what the stress or velocity profile should be according to Navier-Stokes hydrodynamics or generalised hydrodynamics. Considering only the predictions of shear stress, in the case of the former we substitute Equation (11.37) into Newton's law of viscosity to obtain an expression for the *local* shear stress:

$$\sigma_{xy}^{L}(y) = \eta_0 \dot{\gamma}(y) = -\eta_0 k_n v_x(k_n) \sin(k_n y). \tag{11.38}$$

We should expect this expression to compare well to Equation (11.35) in those cases where k_n is small, i.e. the wavelength of the excitations in strain rate is large compared to the width of the viscosity kernel. Alternatively, in those cases where the opposite is true, one should expect the generalised nonlocal constitutive expression given by Equation (11.3) to be valid, i.e.

$$\begin{aligned} \sigma_{xy}^{NL}(y) &= \int_{-\infty}^{\infty} \eta(y - y')\dot{\gamma}(y')\, dy' \\ &= -\frac{\eta_0 k_n v_x(k_n)}{2}\left[e^{-k_n^2/2\sigma_1^2} + e^{-k_n^2/2\sigma_2^2}\right]\sin(k_n y), \end{aligned} \tag{11.39}$$

where we have used the form of the nonlocal viscosity kernel given by Equation (11.33) for analytical simplicity.

In order to compare both these predictive expressions for the shear stress we use the kernel parameters obtained from the *equilibrium* MD method described in Section 11.1.2 for two different wavevectors, one relatively small and one relatively large, compared to the kernel width of 2–3 atomic diameters. The parameters are given in Table 11.2. We note that use of the STF method to compute the kernel parameters would lead to a trivial circular argument. Rather, our aim here is to demonstrate by a simple example the predictive power of generalised hydrodynamics.

In Figure 11.9 (a) and (b) we display the stress, computed (1) exactly by the IMC method (Equation (11.35)), (2) by the local Newtonian constitutive equation (Equation (11.38)) and (3) by the nonlocal constitutive equation (Equation (11.39)). In (a), $n = 1$ which implies a wavelength of excitation field to be $\lambda = L_y = 17.61$, where we note that the number of WCA atoms is $N = 1700$. As the base width of the kernel is ~ 3, the ratio of this width to the wavelength is ~ 0.17, i.e. very little variation in strain rate over this length scale. As such, we would expect both the local and nonlocal stress predictions to agree with the exact stress, which indeed is seen to be the case. Comparing all these quantities for a system in which now $n = 10$, i.e. $\lambda = L_y/10 = 1.76$ in Figure 11.9(b), we find considerable difference between the local and nonlocal

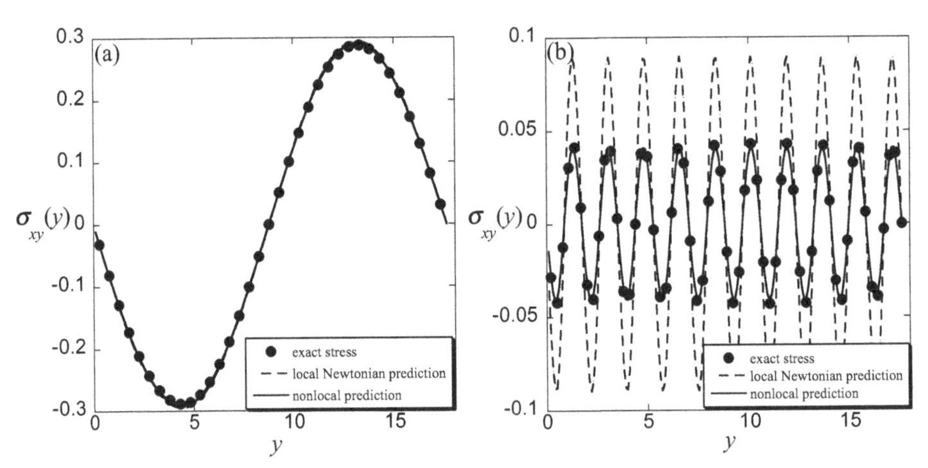

Figure 11.9 Exact and predicted stress profiles for (a) $n = 1$ and (b) $n = 10$. Replotted from reference [322].

predictions. Now the ratio between kernel width and wavelength is ~ 1.7, which means that the variation in strain rate is significant over this length scale. Indeed, Figure 11.9(b) clearly shows that the local stress prediction is significantly in error by almost 100% in places (notably where the greatest curvature occurs near the peaks and troughs of the stress/strain rate). In stark contrast, the nonlocal stress prediction is in *exact* agreement with the actual stress measurements, confirming the validity of the generalised hydrodynamic predictive approach. While we do not show further demonstration here, the results are confirmed by many variations in input parameters, such as n, F_0, L_y, etc. as long as the field strength is sufficiently weak to ensure the system is homogeneous in space.

To conclude this section we point out that in many applications of generalised hydrodynamics the exact form of the kernel is never actually used, as we have done. Rather, Equation (11.3) can be expanded in a Taylor series about the strain rate at $y = y'$. In an equivalent method, one can Fourier transform Equation (11.3) and expand the kernel in k-space about $k = 0$, finally inverse-tranforming into real space to give the identical expansion. Either method leads to the following expansion up to second order:

$$\sigma_{xy}^{NL}(y) = \eta_0 \dot{\gamma}(y) + \eta_1 \left.\frac{d\dot{\gamma}}{dy'}\right|_{y'=y} + \eta_2 \left.\frac{d^2\dot{\gamma}}{dy'^2}\right|_{y'=y} + \cdots \tag{11.40}$$

where $\eta_1 = -\int_{-\infty}^{\infty} y\eta(y) = 0$ and $\eta_2 = \frac{1}{2}\int_{-\infty}^{\infty} y^2\eta(y)$. For our system, with the kernel defined by Equation (11.33), we obtain

$$\sigma_{xy}^{NL}(y) \approx -\eta_0 k_n v_x(k_n)\left[1 + \frac{1}{4}\left(\frac{1}{\sigma_1^2} + \frac{1}{\sigma_2^2}\right)k_n^2\right]\sin(k_n y). \tag{11.41}$$

The accuracy of such an expansion can be readily checked, as displayed in Table 11.3, in which the predicted stress coefficient via Equation (11.41) is compared to the exact stress coefficient (Equation (11.35)). A 10% deviation in stress calculated by the gradient expansion (Equation (11.41)) compared to the exact nonlocal stress (Equation (11.39)) is found when the wavelength of the external field is approximately

Table 11.3 Comparison of exact nonlocal stress coefficient, σ_{xy}, with the gradient expansion approximation, σ_{xy}^{NL}.

F_0	n	k_n	σ_{xy}	σ_{xy}^{NL}
0.15	1	0.357	−0.288	−0.296
0.225	2	0.714	−0.216	−0.236

8 in reduced units. The stress calculated by the Newtonian local expression (Equation (11.38)) gives a similar deviation when the wavelength is 2π in reduced units [322]. Such a comparison shows that the gradient expansion can be a useful approximation to the stress for small wavevectors.

11.1.4 Nonlocality and the Degree of Departure from Classical Predictions

The most significant conclusion to follow from reference [322] is that accurate predictions of the shear stress profile for a fluid subject to rapid changes in strain rate over several molecular diameters, is that the full nonlocal viscosity kernel must be used, rather than the local Navier-Stokes (infinite wavelength) viscosity. This then raises an interesting question: *to what extent does curvature in the velocity gradient of fluids affect the Navier-Stokes predictions of the shear stress?* This question was explored in reference [345] and is discussed in what follows.

11.1.4.1 Local and Nonlocal Kernels with Constant Strain Rate

Consider first a homogeneous fluid with a linear time-independent velocity profile (constant strain rate). The shear stress is given by Equation (11.1), which for time-independent flow is

$$\sigma_{xy}(\mathbf{r}) = \int_{-\infty}^{\infty} \eta\left(\mathbf{r}-\mathbf{r}'\right)\dot{\gamma}\left(\mathbf{r}'\right)d\mathbf{r}'. \tag{11.42}$$

In statistical mechanics we traditionally express the infinite wavelength transport coefficients as delta functions in space, which gives us a delta-function kernel for the viscosity [2]

$$\eta\left(\mathbf{r}-\mathbf{r}'\right) = \eta_0\delta\left(\mathbf{r}-\mathbf{r}'\right), \tag{11.43}$$

where $\eta(\mathbf{r}-\mathbf{r}')$ is the kernel, localised in space at $\mathbf{r}'=\mathbf{r}$. Substitution of Equation (11.43) into Equation (11.42) gives

$$\begin{aligned}\sigma_{xy}(\mathbf{r}) &= \eta_0\int_{-\infty}^{\infty}\delta\left(\mathbf{r}-\mathbf{r}'\right)\dot{\gamma}\left(\mathbf{r}'\right)d\mathbf{r}' \\ &= \eta_0\dot{\gamma}(\mathbf{r}).\end{aligned} \tag{11.44}$$

Equation (11.44) is the standard local Newtonian expression relating the shear stress to the strain rate at point $\mathbf{r}$ with constant of proportionality η_0. For constant strain rate we have $\sigma_{xy}=\eta_0\dot{\gamma}$, the standard planar Couette flow result.

If we are interested in fluids where the actual width of the kernel is significant, such as is the case in nanofluidic systems where the strain rate varies significantly over this length scale, then we can no longer use the delta function approximation. We have already seen that the actual kernels for a simple atomic fluid are several atomic diameters in width and can be approximated by a sum of Gaussian and Lorentzian-type functions. In an overall sense then, Equation (11.43) is not correct and should be replaced by the more general form:

$$\eta\left(\mathbf{r}-\mathbf{r}'\right)=\eta_0 f\left(\mathbf{r}-\mathbf{r}'\right), \tag{11.45}$$

where f is some general normalised, even function with nonzero finite width satisfying

$$\int_{-\infty}^{\infty} f(\mathbf{r})\,d\mathbf{r}=1. \tag{11.46}$$

Substitution of Equation (11.45) into Equation (11.42) gives

$$\sigma_{xy}(\mathbf{r})=\eta_0\int_{-\infty}^{\infty} f\left(\mathbf{r}-\mathbf{r}'\right)\dot{\gamma}\left(\mathbf{r}'\right)d\mathbf{r}'. \tag{11.47}$$

For constant $\dot{\gamma}$ this results in $\sigma_{xy}=\eta_0\dot{\gamma}$, which is *exactly* the same expression as that obtained when assuming a delta function form of the kernel. Thus, for a homogeneous fluid with constant strain rate, any fluid behaves as if it had a local viscosity kernel, η_0. This is entirely due to there being no variation in strain rate over the width of the kernel, enabling us to take the strain rate outside of the integral in Equation (11.47). This is why we can represent the true, finite-width viscosity kernel as a delta-function, or equivalently, as a constant transport coefficient in such circumstances.

11.1.5 Nonlocal Viscosity Kernel with Linear Strain Rate

Now consider the case of a nonlocal viscosity kernel and a linear strain rate. As we have done consistently, we consider a three-dimensional fluid and assume that flow is in the x-direction, with linear velocity gradient αy in the y-direction, giving

$$\nabla\mathbf{v}(\mathbf{r})=\begin{pmatrix}0&0&0\\ \alpha y&0&0\\ 0&0&0\end{pmatrix}. \tag{11.48}$$

This is the flow geometry for planar Poiseuille flow (or equivalently, gravity driven flow) with quadratic velocity profile, as described in Chapter 9. Equation (11.47) gives

$$\sigma_{xy}(y)=\eta_0\alpha\int_{-\infty}^{\infty} f\left(y-y'\right)y'dy' \tag{11.49}$$

which can be expressed via suitable variable substitution as

$$\int_{-\infty}^{\infty} f\left(y-y'\right)y'dy'=y\int_{-\infty}^{\infty} f(u)\,du-\int_{-\infty}^{\infty} f(u)\,u\,du. \tag{11.50}$$

The first integral on the RHS of Equation (11.50) is unity from the normalisation condition, Equation (11.46), while the second integral is zero since the integrand is an odd function of u. This gives

$$\int_{-\infty}^{\infty} f(y - y')y'dy' = y \tag{11.51}$$

and so from Equation (11.49) we have the result

$$\sigma_{xy}(y) = \eta_0 \alpha y \tag{11.52}$$

for any choice of $f(y)$. This is a noteworthy result because it is equivalent to Newton's law of viscosity, $\sigma_{xy}(y) = \eta_0 \dot{\gamma}(y)$. It says that even in the case of a quadratic streaming velocity profile we can always expect a linear shear stress profile, no matter what the precise mathematical form of the true viscosity kernel is.

The implication of this is that in the case of the two most common simple flow types – planar Couette and Poiseuille flow – the inclusion of a nonlocal viscosity kernel gives *exactly* the same predictions as classical hydrodynamics that assumes a constant local viscosity. This is independent of the width of the true kernel. Mathematically, this is equivalent to saying that Newton's law of viscosity is always exact as long as $d\dot{\gamma}(y)/dy = c$, where c is a constant or zero. This is a suitable condition for determining the validity of the Navier-Stokes treatment for viscous transport no matter how nanoscopic the flow is. In more general cases where this condition is not met, the use of a nonlocal constitutive equation, such as Equation (11.47), will be required when variations in the strain rate occur over molecular length-scales.

11.1.6 Nonlocal Viscosity Kernel with Nonlinear Strain Rate

Consider first the simple case of a nonlinear strain rate with form

$$\dot{\gamma}(y) = \alpha y^n; \qquad n = 0, 1, 2... \tag{11.53}$$

The result obtained for this single nonlinear polynomial term can be generalised to that of any functional form for the strain rate by taking the infinite series sum of the result, which then represents the Taylor series expansion of some general analytic function $\dot{\gamma}(y)$ around $y = 0$, as will be shown later. We consider the general case where n can be either odd (symmetric velocity profile) or even (asymmetric velocity profile) about $y = 0$. From Equation (11.47) we have

$$\sigma_{xy}(y) = \eta_0 \alpha \int_{-\infty}^{\infty} dy' f(y - y') y'^n. \tag{11.54}$$

By making suitable mathematical manipulations (details found in reference [345]), and taking note of symmetry arguments, we find that for n even

$$\sigma_{xy}(y) = \eta_0 \alpha \left[b_n + b_{n-2} y^2 + b_{n-4} y^4 + \cdots + y^n \right], \tag{11.55}$$

where the coefficients b_k are given by the even moments of the kernel,

$$b_k \equiv a_k \int_{-\infty}^{\infty} du\, f(u)\, u^k. \tag{11.56}$$

Similarly, for n odd we have

$$\sigma_{xy}(y) = \eta_0 \alpha \left[b_{n-1} y + b_{n-3} y^3 + b_{n-5} y^5 + \cdots + y^n \right]. \tag{11.57}$$

The equivalent local (Newtonian) shear stress is given as

$$\sigma_{xy}^L(y) = \eta_0 \alpha y^n. \tag{11.58}$$

This tells us that the local and nonlocal shear stresses are not equivalent in general for flows where the curvature of the strain rate is not constant, and that the degree of non-locality depends on the relative contribution of the different moments to the total stress given by Equations (11.55) and (11.57). This in turn depends on how much variation in strain rate takes place over the width of the kernel (i.e. the degree of curvature of the strain rate profile with respect to the kernel width). The general case for some arbitrary function can be obtained by summing over an infinite series of terms given in Equations (11.55) and (11.57), as long as the strain rate function is analytic. Therefore, a full Taylor series expansion of a more complicated strain rate function still results in a polynomial series such as given by Equations (11.55) and (11.57), with the same consequences.

To demonstrate this, we represent $\dot{\gamma}(y)$ as a Taylor expansion about the flow mid-point (assumed to be at $y = 0$)

$$\dot{\gamma}(y) = \sum_{n=0}^{\infty} \frac{\dot{\gamma}'^n(0)}{n!} y^n, \tag{11.59}$$

where $\dot{\gamma}'^n$ is the n-th derivative of $\dot{\gamma}$ with respect to y. Therefore, the nonlocal shear stress can be expressed as

$$\begin{aligned} \sigma_{xy}(y) &= \eta_0 \int_{-\infty}^{\infty} dy' f(y-y') \sum_{n=0}^{\infty} \frac{\dot{\gamma}'^n(0)}{n!} y'^n \\ &= \eta_0 \sum_{n=0}^{\infty} \frac{\dot{\gamma}'^n(0)}{n!} \int_{-\infty}^{\infty} dy' f(y-y') y'^n. \end{aligned} \tag{11.60}$$

The integrand in Equation (11.60) can be expanded further to give [345]

$$\begin{aligned} \sigma_{xy}(y) &= \eta_0 \sum_{n=0}^{\infty} \frac{\dot{\gamma}'^n(0)}{n!} \left[b_n^n + b_{n-1}^n y + b_{n-2}^n y^2 + b_{n-3}^n y^3 + \cdots + y^n \right] \\ &= \eta_0 \sum_{n=0}^{\infty} \frac{\dot{\gamma}'^n(0)}{n!} \left[\sum_{i=0}^{n} b_{n-i}^n y^i \right], \end{aligned} \tag{11.61}$$

where $b_0^n = a_0 \int_{-\infty}^{\infty} du\, f(u) = a_0 \equiv 1$. The coefficients b_k^n are analogous to those defined in Equation (11.56) except that now an entire set of coefficients b_k exists for each value of n in the Taylor expansion (hence the notation b_k^n). Equation (11.61) could be simplified further by collecting all coefficients of terms y^n together (in effect removing the term in square brackets and simply having a sum over n with new coefficients c_n). However, we keep it in this form to compare directly with the expression for the local shear stress, which is

$$\sigma_{xy}^{L}(y) = \eta_0 \sum_{n=0}^{\infty} \frac{\dot{\gamma}'^{n}(0)}{n!} y^{n}. \tag{11.62}$$

Equations (11.61) and (11.62) are not the same. This means that for any general strain rate in which the curvature is nonzero (i.e. the gradient of the strain rate is not zero or constant), nonlocality is significant when the contributions of the even moments of the kernel become non-negligible in Equation (11.61). This approach gives greater insight into the effect of nonzero strain rate curvature on the fluid stress than traditional gradient expansion treatments (see Section 11.1.3). This condition can also be used to check whether nonlocality should be included into any appropriate constitutive equation used for modelling nanofluidic flows.

Before we close this section, we note that recent work by Hansen and colleagues has further generalised hydrodynamics to include the wavevector-dependence of the various spin viscosities, in effect generalizing the extended Navier-Stokes equations. They termed this the Generalised Extended Navier-Stokes equations (GENS) and further details may be found in references [413] and [414]. A discussion on extended and generalised hydrodynamics for nanofluidic systems can also be found in reference [415].

11.2 Predicting Slip

Slip, i.e. the "sliding" of the layer of fluid immediately adjacent to the interface between a solid wall and the fluid with nonzero velocity, has long been known to be a feature of fluid flow. In many instances, particularly on the meso and macro length scales, the effect of slip is negligible and the governing Navier-Stokes equations with no-slip (or "stick") boundary conditions may be used. However, as more work has been done in recent years at the micro and nano length scales, and on surfaces in which slip is actually very high (e.g. flow in carbon nanotubes), slip becomes highly nontrivial and very important to consider if one desires accurate predictions of volumetric flow rates, efficiencies of heat transfer between fluid and confining walls, etc. Failure to include slip in the governing Navier-Stokes equations can lead to significant errors in estimating such properties of interest. Therefore, more than ever before, a sound knowledge of slip and an accurate ability to predict it via new theoretical techniques, is essential in the development of the field of nanofluidics.

In 1823 Navier [393] derived an expression for the so-called slip length of a fluid. The slip length is defined as that length extrapolated into the solid wall at which the fluid velocity would be zero, assuming a simple linear extrapolation. The larger the slip length, the greater the degree of fluid slip at the wall and hence a larger slip velocity. This is indicated in Figure 11.10. Navier defined the slip length by

$$|L_s| = \eta/\zeta, \tag{11.63}$$

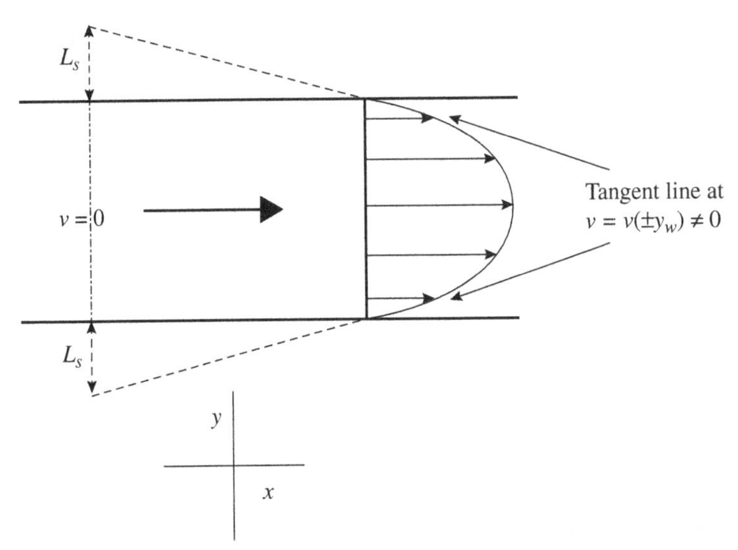

Figure 11.10 Schematic representation of the slip length. In this example, the fluid flows in the x-direction by application of some external driving force or pressure gradient, while the direction of confinement is the y-direction. The fluid velocity at the wall is some nonzero value, $v(\pm y_w)$, where the channel is symmetric about the mid-point, $y = 0$. The slip length, L_s, is defined as the extrapolated distance into the wall where the fluid velocity would be zero.

where η and ζ are the fluid shear viscosity and wall-fluid friction coefficients, respectively. However, at the time Navier did not have any microscopic (i.e. statistical mechanical) means of computing the friction coefficient.

The first statistical mechanical approach towards calculating the friction coefficient was the method proposed by Bocquet and Barrat in 1994 [394]. They presented two methods, but in particular they related the friction coefficient to the integral of the autocorrelation function of the wall-fluid force. Their resulting expression for the friction coefficient resembles a Green–Kubo relation, but they noted it did not produce results in quantitative agreement with a method based on the transverse momentum density autocorrelation function (an alternative method proposed in their paper). The former method was revised by Petravic and Harrowell [395, 396], who suggested that the friction coefficient computed in the force autocorrelation function method is actually the total fluid friction for shear flow in the confined fluid, including the slip friction at both interfaces as well as the viscous friction in the fluid. Consider planar shear (Coeutte) flow between two walls separated by L_y in the y-direction, with flow in the x-direction. If slip occurs at either or both walls, the total velocity difference between them is

$$\Delta u = u_{s1} + \Delta u_L + u_{s2} \tag{11.64}$$

where u_{s1} and u_{s2} are slip velocities at the two confining wall surfaces and Δu_L is the velocity difference across the shearing fluid. For mechanical equilibrium to be maintained the shear stress (σ_{xy}) is constant across the entire system. The stress at each wall is given by $\sigma_{xy} = \zeta_1 u_{s1} = \zeta_2 u_{s2}$, where ζ_1 and ζ_2 are the friction coefficients at walls

1 and 2, respectively. In the fluid Newton's law applies, i.e. $\sigma_{xy} = \eta \Delta u_L / L_y$. Introducing an effective friction coefficient μ that represents the combined effect of all sources of friction, we can write $\sigma_{xy} = \mu \Delta u$. Substituting all these constitutive relations into Equation (11.64), we have

$$1/\mu = 1/\zeta_1 + 1/\zeta_2 + L_y/\eta. \tag{11.65}$$

This relation, which is similar to what Petravic and Harrowell derived [395, 396], prompted them to question the validity of the Green–Kubo relation obtained by Bocquet and Barrat [394]. Bhatia and Nicholson [397] draw a similar conclusion in relation to the friction coefficient calculated from the fluid centre-of-mass velocity autocorrelation function. When the channel width L_y approaches zero or when the slip friction coefficients approach zero, leading to plug flow, as is often found for the flow of water near a hydrophobic wall, the slip friction should make a dominant contribution to the effective friction. In these situations the effective friction should give a good estimate of the slip friction, but this may not be the case in general and it is therefore necessary to remove the viscous contribution to the total friction in order to compute the slip friction coefficient. Note also that the effective friction is system size dependent, since the viscous friction term includes L_y. It is therefore not a true material (intensive) property of the system. Other authors have attempted analogous calculations (see references [394–396, 398–407]). A direct method of computing the slip friction coefficient was proposed by Hansen *et al.* [408], and in what follows we describe this method.

11.2.1 Calculation of the Friction Coefficient and Slip Length

We consider the system described in Figure 11.11, in which a fluid confined by two solid walls is partitioned into two regions: one a small slab of width Δ ($0 \leq y \leq \Delta$) immediately adjacent to wall 1 (or equivalently, wall 2), and the second region being the remainder of the fluid ($\Delta < y \leq L_y$), where once again y is the direction of confinement and flow occurs in the x-direction. The slab of width Δ is of the order of one molecular diameter and has volume $V = L_x \Delta L_z$, where L_x and L_z are the simulation box dimensions in the x and z dimensions, respectively. If we now subject the fluid to a constant body force in the x-direction, the slab dynamics is dictated by Newton's second law of motion:

$$m\frac{du_s}{dt} = F_x'(t) + F_x''(t) + mF^e, \tag{11.66}$$

where m is the slab mass and u_s is the x-component of the centre-of-mass velocity of the slab. Note that Δ is actually an average value in any simulation, since is must fluctuate slightly as fluid molecules enter or exit the slab, in order to keep the slab mass a constant of the motion. F_x' is the x-component of the force due to all wall-slab atomic interactions, while F_x'' is the x-component of force between all fluid atoms in the slab and all other fluid atoms in the system. F_x'' also contains an impulsive (kinetic) force to account for fluid atoms crossing the boundary at $y = \Delta$. F^e is the external force per unit mass acting

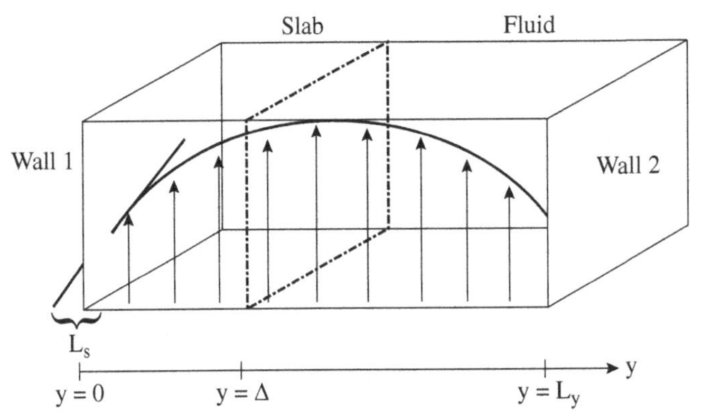

Figure 11.11 Schematic representation of the simulation system. Reprinted with permission from reference [408]. Copyright 2011 by the American Physical Society.

in the x-direction that drives the flow. Thus,

$$u_s(t) = \frac{1}{m} \sum_{i \in slab} m_i v_{i,x}(t) \tag{11.67}$$

$$F'_x(t) = \sum_{\substack{i \in slab \\ j \in wall}} F_{ij,x}(t) \tag{11.68}$$

$$F''_x(t) = \sum_{\substack{i \in slab \\ j \notin wall,\ slab}} F_{ij,x}(t) + \sum_{i \in slab} F^K_{i,x}(t) \tag{11.69}$$

where $F^K_{i,x}$ only acts if a fluid particle crosses into or out of the slab from or to the rest of the fluid, and we note that Newton's third law dictates that all fluid-fluid interaction forces within a slab sum to zero.

F'_x is in effect the frictional shear force that depends on the relative velocity difference between the fluid and the wall. We can therefore postulate a *linear* constitutive equation valid for sufficiently small velocity difference $\Delta u = u_s - u_w$:

$$F'_x(t) = -A \int_0^t \xi(t-\tau)\Delta u(\tau)\, d\tau + F'_{r,x}(t), \tag{11.70}$$

where ξ is the friction kernel, A is the wall surface area in the $x-z$ plane, and the expression is written in a generalised form to account for memory effects. $F'_{r,x}$ is a zero-mean random force, assumed to be uncorrelated with u_s, i.e.

$$\langle F'_{r,x}(t)\rangle = 0; \quad \langle u_s(0) F'_{r,x}(t)\rangle = 0. \tag{11.71}$$

Equation (11.71) is valid in the weak field regime but would not hold under stronger flows. For steady-state flows we can write the time averaged form of the constitutive equation as

$$\langle F'_x(t)\rangle = -\xi_0 \langle \Delta u(t)\rangle. \tag{11.72}$$

It is important to note that Equations (11.70) and (11.72) now define a slip friction coefficient excluding viscous friction contributions. The friction kernel ξ only depends on the velocity difference between the wall and an adjacent fluid layer, overcoming the system size dependence of other methods, and is thus true in spirit to that proposed by Navier.

The viscous forces between the slab and the rest of the fluid can also be recast into the well-known Newtonian constitutive equation

$$\langle F''(t)\rangle = A\int_0^t \eta(t-\tau)\dot{\gamma}(\tau)d\tau = A\eta_0\langle\dot{\gamma}\rangle = A\eta_0 \left.\frac{\partial u}{\partial y}\right|_{y=\Delta}, \tag{11.73}$$

where the final two equalities are valid in the steady-state.

The slab width, Δ, is an important parameter. If it is chosen to be significantly smaller than the width of the first fluid layer adjacent to the wall, the centre-of-mass velocity of the slab will be a poor approximation to the velocity at a certain slip plane [405] due to the extremely small number of molecules inside the slab. For large values of Δ, the slab centre-of-mass velocity will be an average that includes a large contribution from fluid far away from the wall. In this case, the difference between the centre-of-mass slab velocity and the wall velocity will not equal the slip plane velocity, thus invalidating Equation (11.70). In the end, the optimal slab width will be a compromise between a value large enough to completely include the layer of fluid that slips over the surface and one small enough to exclude fluid experiencing purely viscous flow. It will depend on the details of the intermolecular interactions and the thermodynamic state of the system.

To compute the friction coefficient, we set the wall velocity, u_w, to zero, multiply both sides of Equation (11.70) by $u_s(0)$ and take the ensemble average. In doing this we form a simple relation between the slab velocity-force time correlation function $C_{uF'_x}$ and the slab velocity autocorrelation function, C_{uu}, strictly true at equilibrium but which should remain valid in the linear regime:

$$\begin{aligned} C_{uF'_x}(t) \equiv \langle u_s(0)F'(t)\rangle &= -\left\langle \int_0^t \zeta(t-\tau)C_{u\Delta u}(\tau)d\tau \right\rangle \\ &= -\int_0^t \zeta(t-\tau)C_{uu}(\tau)d\tau \end{aligned} \tag{11.74}$$

where we have defined the friction kernel to be $\zeta(t) \equiv \xi(t)/A$ and $C_{uu}(t) \equiv \langle u_s(0)u_s(t)\rangle$. It is more convenient to take the Laplace transform of Equation (11.74) to convert it into an algebraic form:

$$\tilde{C}_{uF'_x}(s) = -\tilde{\xi}(s)\tilde{C}_{uu}(s). \tag{11.75}$$

In principle, one now performs *equilibrium* molecular dynamics simulations on the confined fluid, computes the two time-correlation functions $C_{uF'_x}$ and C_{uu}, Laplace transforms them, and then fits the data to the friction kernel ζ by Equation (11.75). However, in the steady-state further simplifications can take place. Firstly, it was noted [408] that the friction kernel is a simple n-term Maxwellian function

$$\zeta(t) = \sum_{i=1}^{n} B_i e^{-\lambda_i t}. \tag{11.76}$$

Laplace transforming this expression, integrating out the time dependence for steady-state flows and substituting into Equation (11.75) results in

$$\tilde{C}_{uF'_x}(s) = -\sum_{i=1}^{n} \frac{B_i}{(\lambda_i + s)} \tilde{C}_{uu}(s), \tag{11.77}$$

where the zero-frequency (steady-state) friction coefficient is

$$\zeta_0 = \sum_{i=1}^{n} \frac{B_i}{\lambda_i}. \tag{11.78}$$

Typically, for simple fluids a one term Maxwellian function is sufficient, but this may increase as fluid complexity increases. Typical Laplace transformed correlation functions are shown in Figure 11.12 [409]. The predicted model slip length, in the limit as $\Delta \to 0$, is shown to reproduce the Navier slip length, given by Equation (11.63).

Slip lengths and slip velocities were computed using the model and directly from NEMD simulation by Kannam *et al.* [409] and excellent agreement between model prediction and NEMD simulation was found, displayed in Figures 11.13–11.14. The slip

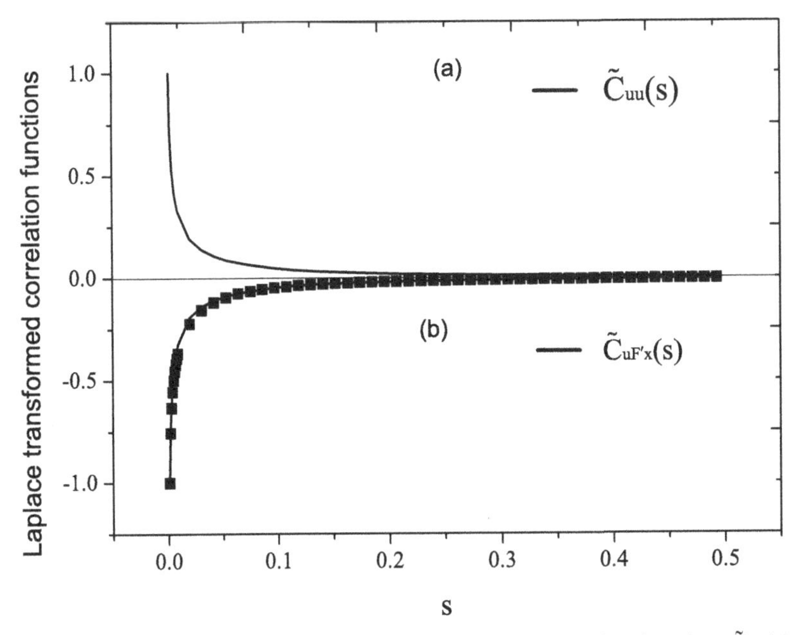

Figure 11.12 Normalised and Laplace transformed time correlation functions $\tilde{C}_{uu}$ (a) and $\tilde{C}_{uF'_x}$ (b). The system modelled was methane confined to graphene walls (see [409] for details). Note that (b) also contains the Maxwellian fit (square symbols) given by Equation (11.77). Reprinted from reference [409] with the permission of AIP Publishing.

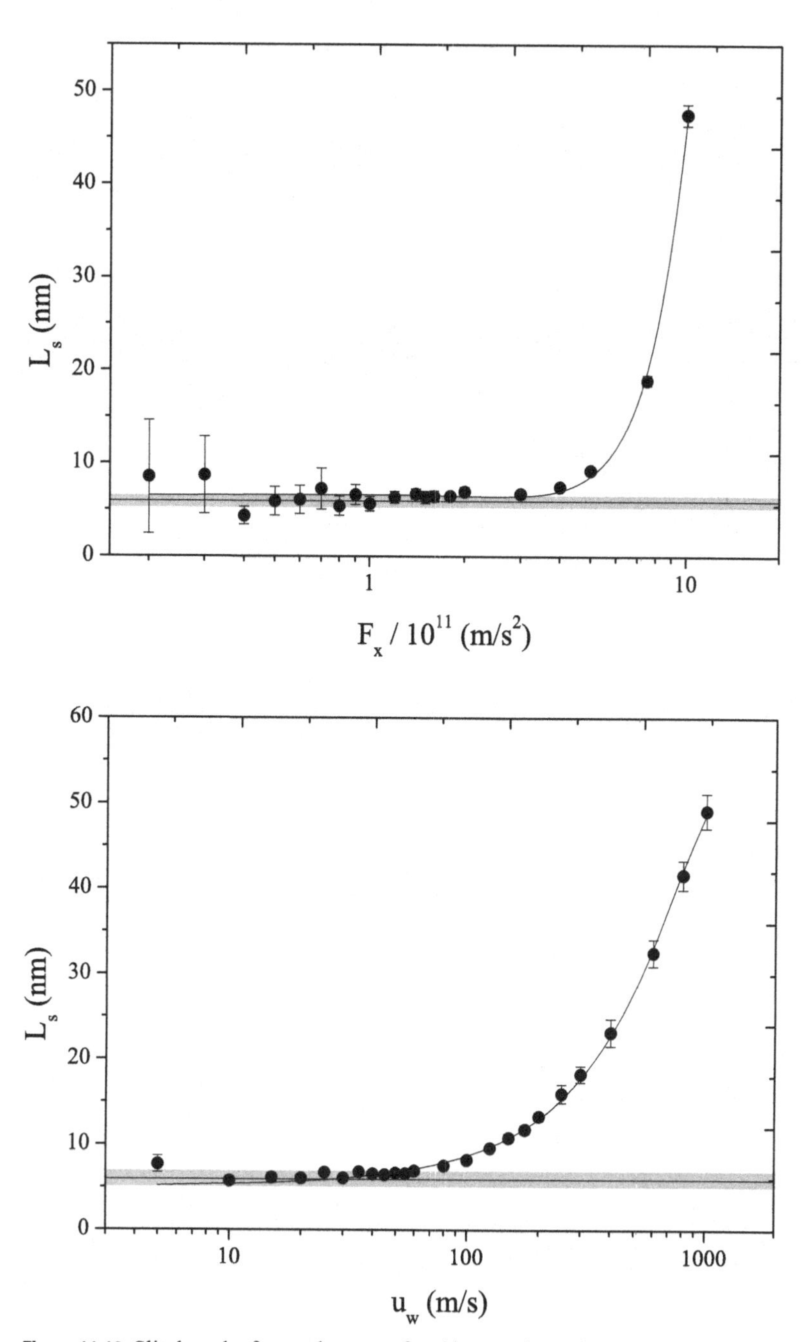

Figure 11.13 Slip lengths for methane confined by graphene slit channels under (a) Poiseuille (where F_x is the magnitude of the external field) and (b) Couette flow (here u_w is the velocity of the upper wall, with lower wall at rest). The data points are from direct NEMD simulation, while the straight line is the equilibrium model prediction (shaded grey area represents error estimates). Reprinted from reference [409] with the permission of AIP Publishing.

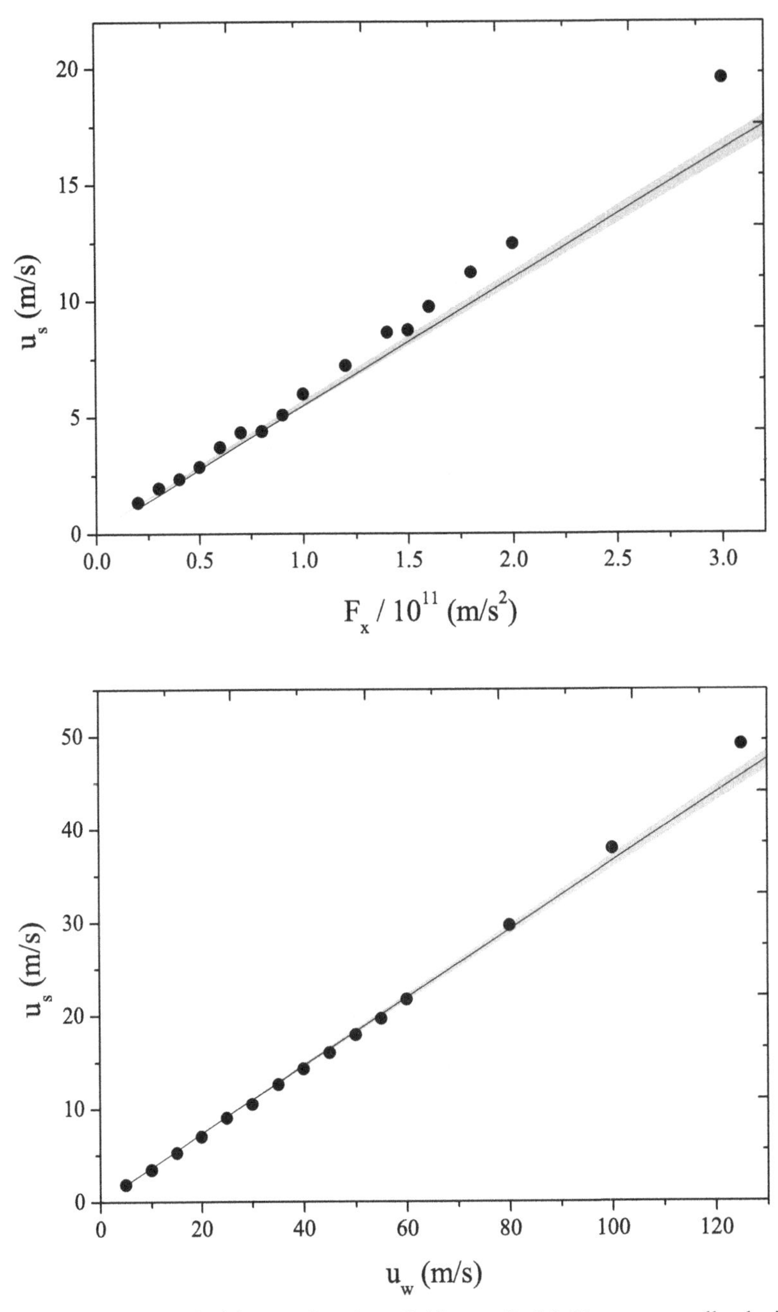

Figure 11.14 Slip velocities as a function of either applied field or upper wall velocity for methane confined by graphene slit channels under (a) Poiseuille and (b) Couette flow. The data points are from direct NEMD simulation, while the straight line is the equilibrium model prediction (shaded grey area represents error estimates). Reprinted from reference [409] with the permission of AIP Publishing.

length is independent of flow geometry in the limit as $\Delta \to 0$, as expected. The model has also been applied to systems of water confined to graphene nanochannels [410] and carbon nanotubes [411, 412] and works extremely well. An important conclusion to emerge from this work is that NEMD is actually an *unreliable* technique when determining the slip length/velocity for high-slip systems, such as the flow of water in carbon nanotubes or graphene. This is because for high-slip systems the effective strain rate in the fluid is quite weak, even if the applied field (in the case of field or pressure-driven flow) or wall velocity (in the case of Couette flow) is very large. This in turn means that very small errors in the estimate of the velocity gradient (of the order of 1–2%) at the wall-fluid interface lead to very large errors in the estimated slip lengths (of the order of 100% or more). This could be one reason why there is such a large discrepancy in the literature for the prediction of slip lengths for water flowing in graphene or carbon nanotube systems. There are also controversies about experimental measurements of water flow in such systems. The various experimental and simulation measurements, along with some theoretical predictions, are displayed in Figure 11.15, which is sourced from Kannam *et al.* [412]. Insofar as simulations are concerned, the equilibrium based

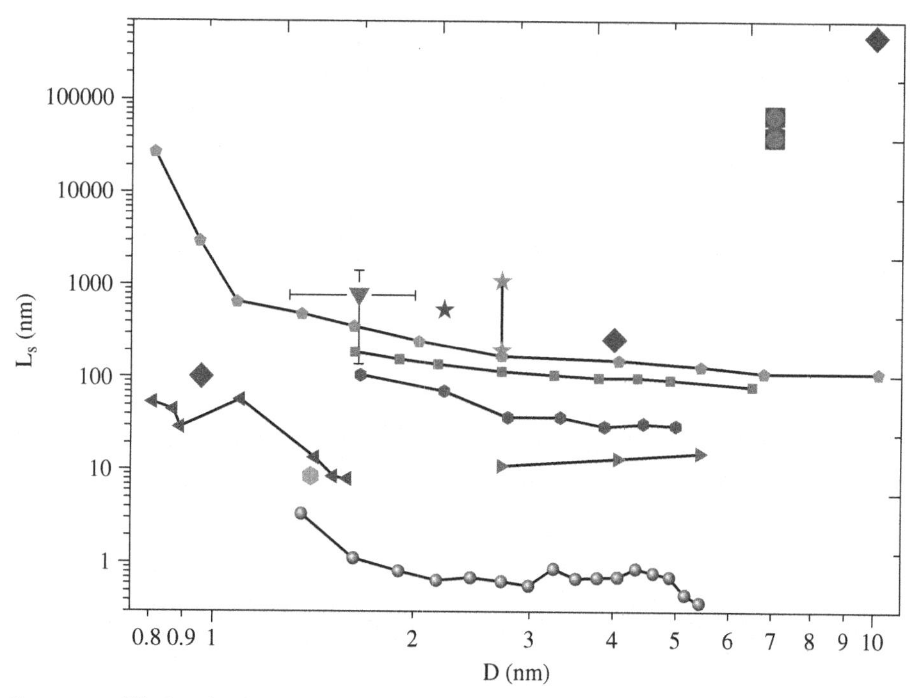

Figure 11.15 Slip length of water in carbon nanotubes of various diameter. The square connected symbols are from the model proposed by Hansen *et al.* [408] and the other symbols are from the literature reviewed and defined in reference [412]. These include results from experimental, theoretical and simulation work. Reprinted from reference [412] with the permission of AIP Publishing.

model is more accurate and computationally less demanding than standard NEMD for high slip systems. The model has the additional advantage of allowing one to simulate systems with different walls, hence different slip lengths at each surface. Greater details of the model for both planar and cylindrical geometry can be obtained in references [408, 411].

Bibliography

[1] B. D. Todd and P. J. Daivis. Homogeneous non-equilibrium molecular dynamics simulations of viscous flow: techniques and applications. *Mol. Simul.*, 33: 189, 2007.

[2] D. J. Evans and G. P. Morriss. *Statistical Mechanics of Nonequilibrium Liquids*. Cambridge University Press, Cambridge, 2nd edition, 2008.

[3] D. A. McQuarrie. *Statistical Mechanics*. Harper Collins, New York, 1976.

[4] S. R. de Groot and P. Mazur. *Non-Equilibrium Thermodynamics*. Dover, New York, 1984.

[5] M. P. Allen and D. J. Tildesley. *Computer Simulation of Liquids*. Clarendon Press, Oxford, 1987.

[6] D. Rapaport. *The Art of Molecular Dynamics Simulation*. Cambridge University Press, Cambridge, 1995.

[7] D. Frenkel and B. Smit. *Understanding Molecular Simulation: From Algorithms to Applications*. Academic Press, San Diego, 2002.

[8] R. J. Sadus. *Molecular Simulation of Fluids: Theory, Algorithms and Object-Orientation*. Elsevier, Amsterdam, 1999.

[9] B. J. Alder and T. E. Wainwright. Phase transition for a hard sphere system. *J. Chem. Phys.*, 27: 1208, 1957.

[10] A. Rahman. Correlations in the motion of atoms in liquid argon. *Phys. Rev.*, 136: A105, 1964.

[11] L. Verlet. Computer "experiments" on classical fluids. I. Thermodynamical properties of Lennard-Jones molecules. *Phys. Rev.*, 159: 98, 1967.

[12] B. J. Alder, D. M. Gass, and T. E. Wainwright. Studies in molecular dynamics. VIII. The transport coefficients for a hard-sphere fluid. *J. Chem. Phys.*, 53: 3813, 1970.

[13] M. S. Green. Markoff random processes and the statistical mechanics of time-dependent phenomena. II. Irreversible processes in fluids. *J. Chem. Phys.*, 22: 398, 1954.

[14] R. Kubo. Statistical-mechanical theory of irreversible processes. 1. General theory and simple applications to magnetic and conduction problems. *J. Phys. Soc. Japan*, 12: 570, 1957.

[15] A. W. Lees and S. F. Edwards. The computer study of transport processes under extreme conditions. *J. Phys. C*, 5: 1921, 1972.

[16] E. M. Gosling, I. R. McDonald, and K. Singer. On the calculation by molecular dynamics of the shear viscosity of a simple fluid. *Mol. Phys.*, 26: 1475, 1973.

[17] W. T. Ashurst and W. G. Hoover. Dense-fluid shear viscosity via nonequilibrium molecular dynamics. *Phys. Rev. A*, 11: 658, 1975.

[18] W. G. Hoover. Atomistic nonequilibrium computer simulations. *Physica*, 118A: 111, 1983.

[19] W. G. Hoover. Nonequilibrium molecular dynamics: the first 25 years. *Physica A*, 194: 450, 1993.

[20] W. G. Hoover, D. J. Evans, R. B. Hickman, A. J. C. Ladd, W. T. Ashurst, and B. Moran. Lennard-Jones triple-point bulk and shear viscosities. Green–Kubo theory, Hamiltonian mechanics, and nonequilibrium molecular dynamics. *Phys. Rev. A*, 22: 1690, 1980.

[21] D. J. Evans and G. P. Morriss. Nonlinear-response theory for steady planar Couette flow. *Phys. Rev. A*, 30(3): 1528, 1984.

[22] G. Ciccotti and G. Jacucci. Direct computation of dynamical response by molecular dynamics: The mobility of a charged Lennard-Jones particle. *Phys. Rev. Lett.*, 35: 789, 1975.

[23] D. J. Evans and G. P. Morriss. Transient-time-correlation functions and the rheology of fluids. *Phys. Rev. A*, 38: 4142, 1988.

[24] F. Müller-Plathe. Reversing the perturbation in nonequilibrium molecular dynamics: An easy way to calculate the shear viscosity of fluids. *Phys. Rev. E*, 59: 4894, 1999.

[25] B. Hafskjold, T. Ikeshoji, and S. Kjelstrup Ratkje. On the molecular mechanism of thermal diffusion in liquids. *Mol. Phys.*, 80: 1389, 1993.

[26] D. Jou, G. Lebon, and J. Casas-Vázquez. *Extended Irreversible Thermodynamics*. Springer, 4th edition, 2010.

[27] H. C. Öttinger. *Beyond Equilibrium Thermodynamics*. John Wiley & Sons, Hoboken, New Jersey, 2005.

[28] R. B. Bird, R. C. Armstrong, and O. Hassager. *Dynamics of Polymeric Liquids, Volume 1 Fluid Mechanics*. John Wiley & Sons, New York, 2nd edition, 1987.

[29] R. B. Bird, C. F. Curtiss, R. C. Armstrong, and O. Hassager. *Dynamics of Polymeric Liquids, Volume 2 Kinetic Theory*. John Wiley & Sons, New York, 2nd edition, 1987.

[30] R. I. Tanner. *Engineering Rheology*. Oxford University Press, 2nd edition, 2000.

[31] R. R. Huilgol and N. Phan-Thien. *Fluid Mechanics of Viscoelasticity*. Elsevier, Amsterdam, 1997.

[32] C. Truesdell and W. Noll. *The Non-Linear Field Theories of Mechanics*. Springer-Verlag, 3rd edition, 2004.

[33] H. J. Juretschke. *Crystal Physics*. W. A. Benjamin Inc., 1974.

[34] P. G. de Gennes and J. Prost. *The Physics of Liquid Crystals*. Oxford University Press, 2nd edition, 1993.

[35] R. F. Snider and K. S. Lewchuk. Irreversible thermodynamics of a fluid system with spin. *J. Chem. Phys.*, 46: 3163, 1967.

[36] D. J. Evans and W. B. Streett. Transport properties of homonuclear diatomics II. Dense fluids. *Mol. Phys.*, 36: 161, 1978.

[37] J. A. McLennan. *Introduction to Nonequilibrium Statistical Mechanics*. Prentice Hall, New Jersey, 1989.

[38] B. C. Eu. *Nonequilibrium Statistical Mechanics: Ensemble Method*. Kluwer Academic, 1998.

[39] R. Zwanzig. *Nonequilibrium Statistical Mechanics*. Oxford University Press, 2001.

[40] D. N. Zubarev, V. G. Morozov, and G. Röpke. *Statistical Mechanics of Nonequilibrium Processes*. Akademie Verlag, 1996.

[41] P. Gaspard. *Chaos, Scattering and Statistical Mechanics*. Cambridge University Press, 1998.

[42] H. Goldstein. *Classical Mechanics*. Addison-Wesley, Reading, MA, 1980.

[43] R. C. Tolman. *The Principles of Statistical Mechanics*. Dover reprinting of the 1938 edition published by Oxford University Press, 1979.

[44] S. R. Williams and D. J. Evans. Time-dependent response theory and nonequilibrium free-energy relations. *Phys. Rev. E*, 78: 021119, 2008.

[45] T. Yamada and K. Kawasaki. Nonlinear effects in shear viscosity of critical mixtures. *Prog. Theor. Phys.*, 38: 1031, 1967.

[46] T. Yamada and K. Kawasaki. Application of mode-coupling theory to nonlinear stress tensor in fluids. *Prog. Theor. Phys.*, 53: 111, 1975.

[47] K. Kawasaki and J. D. Gunton. Theory of nonlinear transport processes: Nonlinear shear viscosity and normal stress effects. *Phys. Rev. A*, 8: 2048, 1973.

[48] W. M. Visscher. Transport processes in solids and linear-response theory. *Phys. Rev. A*, 10: 2461, 1974.

[49] J. W. Dufty and M. J. Lindenfeld. Nonlinear transport in the Boltzmann limit. *J. Stat. Phys.*, 20: 259, 1979.

[50] E. G. D. Cohen. Kinetic theory of non-equilibrium fluids. *Physica A*, 118: 17, 1983.

[51] G. P. Morriss and D. J. Evans. Isothermal response theory. *Mol. Phys.*, 54: 629, 1985.

[52] G. P. Morriss and D. J. Evans. Application of transient correlation-functions to shear-flow far from equilibrium. *Phys. Rev. A*, 35: 792, 1987.

[53] B. D. Todd. Application of transient time correlation functions to nonequilibrium molecular dynamics simulations of elongational flow. *Phys. Rev. E*, 56: 6723–6728, 1997.

[54] J. Petravic and D. J. Evans. Nonlinear response for time-dependent external fields. *Phys. Rev. Lett.*, 78: 1199, 1997.

[55] J. Petravic and D. J. Evans. Nonlinear response for nonautonomous systems. *Phys. Rev. E*, 56: 1207, 1997.

[56] J. Petravic and D. J. Evans. Approach to the non-equilibrium time-periodic state in a "steady" shear flow model. *Mol. Phys.*, 95: 219, 1998.

[57] J. Petravic and D. J. Evans. Nonlinear response theory for time-dependent external fields: Shear flow and color conductivity. *Int. J. Thermophys.*, 19: 1049, 1998.

[58] J. Petravic and D. J. Evans. Time dependent nonlinear response theory. *Trends in Statistical Physics*, 2: 85, 1998.

[59] J. Petravic and D. J. Evans. The Kawasaki distribution function for nonautonomous systems. *Phys. Rev. E*, 58: 2624, 1998.

[60] B. D. Todd. Nonlinear response theory for time-periodic elongational flows. *Phys. Rev. E*, 58: 4587, 1998.

[61] J. P. Hansen and I. R. McDonald. *Theory of Simple Liquids*. Academic Press, New York, 1986.

[62] D. M. Heyes. *The Liquid State: Applications of Molecular Simulations*. Wiley, Chichester, 1997.

[63] P. J. Daivis and D. J. Evans. Comparison of constant pressure and constant volume nonequilibrium simulations of sheared model decane. *J. Chem. Phys.*, 100: 541, 1994.

[64] D. J. Evans, E. G. D. Cohen, and G. P. Morriss. Probability of 2nd law violations in shearing steady-states. *Phys. Rev. Lett.*, 71: 2401, 1993.

[65] E. N. Lorenz. Deterministic nonperiodic flow. *J. Atmos. Sci.*, 20: 130, 1963.

[66] J. D. Weeks, D. Chandler, and H. C. Andersen. Role of repulsive forces in determining the equilibrium structure of simple liquids. *J. Chem. Phys.*, 54: 5237, 1971.

[67] D. J. Evans and D. J. Searles. Equilibrium microstates which generate second law violating steady states. *Phys. Rev. E*, 50: 1645, 1994.

[68] G. Gallavotti and E. G. D. Cohen. Dynamical ensembles in nonequilibrium statistical mechanics. *Phys. Rev. Lett.*, 74: 2694, 1995.

[69] G. M. Wang, E. M. Sevick, E. Mittag, D. J. Searles, and D. J. Evans. Experimental demonstration of violations of the second law of thermodynamics for small systems and short time scales. *Phys. Rev. Lett.*, 89: 050601, 2002.
[70] J. C. Maxwell. Tait's "Thermodynamics" II. *Nature*, 17: 278, 1878.
[71] D. J. Evans and D. J. Searles. The fluctuation theorem. *Advances in Physics*, 51: 1529, 2002.
[72] C. Bustamante, J. Liphardt, and F. Ritort. The nonequilibrium thermodynamics of small systems. *Physics Today*, 58: 43, 2005.
[73] D. J. Evans, D. J. Searles, and S. R. Williams. *Fundamentals of Classical Statistical Thermodynamics: Dissipation, Relaxation and Fluctuation Theorems*. Wiley, 2016.
[74] D. J. Evans, D. J. Searles, and L. Rondoni. Application of the Gallavotti-Cohen fluctuation relation to thermostated steady states near equilibrium. *Phys. Rev. E*, 71: 056120, 2005.
[75] D. J. Evans, D. J. Searles, and S. R. Williams. On the fluctuation theorem for the dissipation function and its connection with response theory. *J. Chem. Phys.*, 128: 014504, 2008.
[76] C. Jarzynski. Nonequilibrium equality for free energy differences. *Phys. Rev. Lett.*, 78: 2690, 1997.
[77] C. Jarzynski. Equilibrium free-energy differences from nonequilibrium measurements: A master-equation approach. *Phys. Rev. E*, 56: 5018, 1997.
[78] G. E. Crooks. Entropy production fluctuation theorem and the nonequilibrium work relation for free energy differences. *Phys. Rev. E*, 60: 2721, 1999.
[79] G. E. Crooks. Nonequilibrium measurements of free energy differences for microscopically reversible Markovian systems. *J. Stat. Phys.*, 90: 1481, 1998.
[80] J. Casas-Vázquez and D. Jou. Temperature in non-equilibrium states: a review of open problems and current proposals. *Rep. Prog. Phys.*, 66: 1937, 2003.
[81] O. G. Jepps, G. Ayton, and D. J. Evans. Microscopic expressions for the thermodynamic temperature. *Phys. Rev. E*, 62: 4757, 2000.
[82] G. Rickayzen and J.G. Powles. Temperature in the classical microcanonical ensemble. *J. Chem. Phys.*, 114: 4333, 2001.
[83] H. H. Rugh. Dynamical approach to temperature. *Phys. Rev. Lett.*, 78: 772, 1997.
[84] A. Baranyai. Temperature of nonequilibrium steady-state systems. *Phys. Rev. E*, 62: 5989, 2000.
[85] J. H. Irving and J. G. Kirkwood. The statistical mechanical theory of transport processes. 4. The equations of hydrodynamics. *J. Chem. Phys.*, 18: 817, 1950.
[86] B. D. Todd, D. J. Evans, and P. J. Daivis. Pressure tensor for inhomogeneous fluids. *Phys. Rev. E*, 52: 1627, 1995.
[87] D. R. J. Monaghan and G. P. Morriss. Microscopic study of steady convective flow in periodic systems. *Phys. Rev. E*, 56: 476, 1997.
[88] B. D. Todd, P. J. Daivis, and D. J. Evans. Heat flux vector in highly inhomogeneous nonequilibrium fluids. *Phys. Rev. E*, 51: 4362, 1995.
[89] P. J. Daivis, K. P. Travis, and B.D. Todd. A technique for the calculation of mass, energy and momentum densities at planes in molecular dynamics simulations. *J. Chem. Phys.*, 104: 9651, 1996.
[90] O.G. Jepps and S. K. Bhatia. Method for determining the shear stress in cylindrical systems. *Phys. Rev. E*, 67: 041206, 2003.
[91] D. M. Heyes, E. R. Smith, D. Dini, and T. A. Zaki. The method of planes pressure tensor for a spherical subvolume. *J. Chem. Phys.*, 140: 054506, 2014.

[92] D. M. Heyes, E. R. Smith, D. Dini, and T. A. Zaki. The equivalence between volume averaging and method of planes definitions of the pressure tensor. *J. Chem. Phys.*, 135: 024512, 2011.

[93] R. J. Hardy. Formulas for determining local properties in molecular-dynamics simulations: Shock waves. *J. Chem. Phys.*, 76: 622, 1982.

[94] J. Cormier, J. M. Rickman, and T. J. Delph. Stress calculation in atomistic simulations of perfect and imperfect solids. *J. Appl. Phys.*, 89: 99, 2001.

[95] R. Hartkamp, T. A. Hunt, and B. D. Todd. A method-of-planes approach for the calculation of position-dependent self-diffusion coefficients in confined fluids. Unpublished.

[96] K. P. Travis, B. D. Todd, and D. J. Evans. Departure from Navier-Stokes hydrodynamics in confined liquids. *Phys. Rev. E*, 55: 4288, 1997.

[97] S. H. Lee and P. T. Cummings. Shear viscosity of model mixtures by nonequilibrium molecular dynamics. I. Argon-krypton mixtures. *J. Chem. Phys.*, 99: 3919, 1993.

[98] S. H. Lee and P. T. Cummings. Effect of three-body forces on the shear viscosity of liquid argon. *J. Chem. Phys.*, 101: 6206, 1994.

[99] G. Marcelli, B. D. Todd, and R. J. Sadus. Analytic dependence of the pressure and energy of an atomic fluid under shear. *Phys. Rev. E*, 63: 021204, 2001.

[100] J. Zhang and B. D. Todd. Pressure tensor and heat flux vector for confined nonequilibrium fluids under the influence of three-body forces. *Phys. Rev. E*, 69: 031111, 2004.

[101] J. A. Barker, R. A. Fisher, and R. O. Watts. Liquid argon: Monte Carlo and molecular dynamics calculations. *Mol. Phys.*, 21: 657, 1971.

[102] B. M. Axilrod and E. Teller. Interaction of the van der Waals' type between three atoms. *J. Chem. Phys.*, 11: 299, 1943.

[103] D. Torii, T. Nakano, and T. Ohara. Contribution of inter- and intramolecular energy transfers to heat conduction in liquids. *J. Chem. Phys.*, 128: 044504, 2008.

[104] J. F. Lutsko. Stress and elastic constants in anisotropic solids: Molecular dynamics techniques. *J. Appl. Phys.*, 64: 1152, 1988.

[105] E. R. Smith, D. M. Heyes, D. Dini, and T. A. Zaki. Control-volume representation of molecular dynamics. *Phys. Rev. E*, 85: 056705, 2012.

[106] P. P. Ewald. The calculation of optical and electrostatic grid potential. *Ann. Phys. (Leipzig)*, 64: 253, 1921.

[107] J. Lekner. Summation of Coulomb fields in computer-simulated disordered systems. *Physica A*, 176: 485, 1991.

[108] J. Lekner. Coulomb forces and potentials in systems with an orthorhombic unit cell. *Molec. Simul.*, 20: 357, 1998.

[109] D. Wolf. Reconstruction of NaCl surfaces from a dipolar solution to the Madelung problem. *Phys. Rev. Lett.*, 68: 3315, 1992.

[110] D. Wolf, S. R. Keblinski, S. R. Phillpot, and J. Eggebrecht. Exact method for the simulation of Coulombic systems by spherically truncated, pairwise r^{-1} summation. *J. Chem. Phys.*, 110: 8254, 1999.

[111] L. Onsager. Electric moments of molecules in liquids. *J. Am. Chem. Soc.*, 58: 1486, 1936.

[112] J. A. Barker and R. O. Watts. Monte-Carlo studies of dielectric properties of water-like models. *Mol. Phys.*, 26: 789, 1973.

[113] D. M. Heyes. Electrostatic potentials and fields in infinite point charge lattices. *J. Chem. Phys.*, 74: 1924, 1981.

[114] D. R. Wheeler, N. G. Fuller, and R. L. Rowley. Non-equilibrium molecular dynamics simulation of the shear viscosity of liquid methanol: Adaption of the Ewald sum to Lees-Edwards boundary conditions. *Mol. Phys.*, 92: 55, 1997.

[115] J. Alejandre, D. J. Tildesley, and G. A. Chapela. Molecular dynamics simulation of the orthobaric densities and surface tension of water. *J. Chem. Phys.*, 102: 4574, 1995.

[116] D. M. Heyes. Pressure tensor of partial-charge and point-dipole lattices with bulk and surface geometries. *Phys. Rev. B*, 49: 755, 1994.

[117] S. Nosé and M. L. Klein. Constant pressure molecular dynamics for molecular systems. *Molec. Phys.*, 50: 1055, 1983.

[118] N. Galamba, C. A. N. de Castro, and J. F. Ely. Thermal conductivity of molten alkali halides from equilibrium molecular dynamics simulations. *J. Chem. Phys.*, 120: 8676, 2004.

[119] J. Petravic. Thermal conductivity of ethanol. *J. Chem. Phys.*, 123: 174503, 2005.

[120] D. E. Parry. Electrostatic potential in surface region of an ionic-crystal. *Surf. Sci.*, 49: 433, 1975.

[121] D. E. Parry. Correction. *Surf. Sci.*, 54: 195, 1976.

[122] D. M. Heyes, M Barber, and J. H. R. Clarke. Molecular-dynamics computer-simulation of surface properties of crystalline potassium-chloride. *Faraday Trans. II*, 73: 1485, 1977.

[123] J. Muscatello and F. Bresme. A comparison of Coulombic interaction methods in non-equilibrium studies of heat transfer in water. *J. Chem. Phys.*, 135: 234111, 2011.

[124] C. J. Fennell and J. D. Gezelter. Is the Ewald summation still necessary? pairwise alternatives to the accepted standard for long-range electrostatics. *J. Chem. Phys.*, 124: 234104, 2006.

[125] D. J. Evans and G. P. Morriss. Non-Newtonian molecular dynamics. *Comput. Phys. Rep.*, 1: 297, 1984.

[126] W. G. Hoover, C. G. Hoover, and J. Petravic. Simulation of two- and three-dimensional dense-fluid shear flows via nonequilibrium molecular dynamics: Comparison of time-and-space-averaged stresses from homogeneous Doll's and Sllod shear algorithms with those from boundary-driven shear. *Phys. Rev. E*, 78: 046701, 2008.

[127] A. J. C. Ladd. Equations of motion for non-equilibrium molecular dynamics simulations of viscous flow in molecular fluids. *Mol. Phys.*, 53: 459, 1984.

[128] P. J. Daivis and B. D. Todd. A simple, direct derivation and proof of the validity of the SLLOD equations of motion for generalised homogeneous flows. *J. Chem. Phys.*, 124: 194103, 2006.

[129] A. M. Kraynik and D. A. Reinelt. Extensional motions of spatially periodic lattices. *Int. J. Multiphase Flow*, 18: 1045, 1992.

[130] B. D. Todd and P. J. Daivis. Nonequilibrium molecular dynamics simulations of planar elongational flow with spatially and temporally periodic boundary conditions. *Phys. Rev. Lett.*, 81: 1118, 1998.

[131] B. D. Todd and P. J. Daivis. A new algorithm for unrestricted duration molecular dynamics simulations of planar elongational flow. *Computer Physics Communications*, 117: 191, 1999.

[132] B. D. Todd and P. J. Daivis. The stability of nonequilibrium molecular dynamics simulations of elongational flows. *J. Chem. Phys.*, 112: 40, 2000.

[133] A. Baranyai and P. T. Cummings. Steady state simulation of planar elongation flow by nonequilibrium molecular dynamics. *J. Chem. Phys.*, 110: 42, 1999.

[134] T. A. Hunt, S. Bernardi, and B. D. Todd. A new algorithm for extended nonequilibrium molecular dynamics simulations of mixed flow. *J. Chem. Phys.*, 133: 154116, 2010.

[135] S. Bernardi, S. J. Brookes, D. J. Searles, and D. J. Evans. Response theory for confined systems. *J. Chem. Phys.*, 137: 074114, 2012.

[136] S. Bernardi and D. J. Searles. Local response in nanopores. *Molec. Simul.*, 42: 463, 2016.

[137] C. Baig, B. J. Edwards, D. J. Keffer, and H. D. Cochran. A proper approach for nonequilibrium molecular dynamics simulations of planar elongational flow. *J. Chem. Phys.*, 122: 114103, 2005.

[138] B. J. Edwards, C. Baig, and D. J. Keffer. An examination of the validity of nonequilibrium molecular-dynamics simulation algorithms for arbitrary steady-state flows. *J. Chem. Phys.*, 123: 114106, 2005.

[139] B. J. Edwards, C. Baig, and D. J. Keffer. A validation of the p-SLLOD equations of motion for homogeneous steady-state flows. *J. Chem. Phys.*, 124: 194104, 2006.

[140] B. J. Edwards and M. Dressler. A reversible problem in non-equilibrium thermodynamics: Hamiltonian evolution equations for non-equilibrium molecular dynamics simulations. *J. Non-Newtonian Fluid Mech.*, 96: 163, 2001.

[141] I. Borzsák, P. T. Cummings, and D. J. Evans. Shear viscosity of a simple fluid over a wide range of strain rates. *Mol. Phys.*, 100: 2735, 2002.

[142] T. A. Hunt and B. D. Todd. On the Arnold cat map and periodic boundary conditions for planar elongational flow. *Mol. Phys.*, 101: 3445, 2003.

[143] B. D. Todd. Cats, maps and nanoflows: Some recent developments in nonequilibrium nanofluidics. *Mol. Simul.*, 31: 411, 2005.

[144] F. Frascoli, D. J. Searles, and B. D. Todd. Chaotic properties of planar elongational flows and planar shear flows: Lyapunov exponents, conjugate-pairing rule and phase space contraction. *Phys. Rev. E*, 73: 046206, 2006.

[145] R. Bhupathiraju, P. T. Cummings, and H. D. Cochran. An efficient parallel algorithm for non-equilibrium molecular dynamics simulations of very large systems in planar Couette flow. *Mol. Phys.*, 88: 1665, 1996.

[146] D. P. Hansen and D. J. Evans. A parallel algorithm for nonequilibrium molecular dynamics simulation of shear flow on distributed memory machines. *Mol. Simul.*, 13: 375, 1994.

[147] B. D. Todd and P. J. Daivis. Elongational viscosities from nonequilibrium molecular dynamics simulations of oscillatory elongational flow. *J. Chem. Phys.*, 107: 1617, 1997.

[148] A. Baranyai and P. T. Cummings. Nonequilibrium molecular dynamics study of shear and shear-free flows in simple fluids. *J. Chem. Phys.*, 103: 10217, 1995.

[149] J. C. Sprott. *Chaos and Time Series Analysis*. Oxford University Press, Oxford, 2003.

[150] A. Katok and B. Hasselblatt. *Introduction to the Modern Theory of Dynamical Systems*. Cambridge University Press, Cambridge, 1995.

[151] F. Frascoli, D. J. Searles, and B. D. Todd. Boundary condition independence of molecular dynamics simulations of planar elongational flow. *Phys. Rev. E*, 75: 066702, 2007.

[152] F. Frascoli, D. J. Searles, and B. D. Todd. Chaotic properties of isokinetic-isobaric atomic systems under planar shear and elongational flows. *Phys. Rev. E*, 77: 056217, 2008.

[153] D. J. Evans, W. G. Hoover, B. H. Failor, B. Moran, and A. J. C. Ladd. Nonequilibrium molecular dynamics via Gauss's principle of least constraint. *Phys. Rev. A*, 28: 1016, 1983.

[154] S. Nosé. A unified formulation of the constant temperature molecular-dynamics methods. *J. Chem. Phys.*, 81: 511, 1984.

[155] S. Nosé. A molecular-dynamics method for simulations in the canonical ensemble. *Mol. Phys.*, 52: 255, 1984.

[156] W. G. Hoover. Canonical dynamics: Equilibrium phase-space distributions. *Phys. Rev. A*, 31: 1695, 1985.

[157] B. D. Butler, G. Ayton, O. G. Jepps, and D. J. Evans. Configurational temperature: Verification of Monte Carlo simulations. *J. Chem. Phys.*, 109: 6519, 1998.

[158] L. Lue and D. J. Evans. Configurational temperature for systems with constraints. *Phys. Rev. E*, 62: 4764, 2000.

[159] J. Delhommelle and D. J. Evans. Configurational temperature thermostat for fluids undergoing shear flow: application to liquid chlorine. *Mol. Phys.*, 99: 1825, 2001.

[160] L. Lue, O. G. Jepps, J. Delhommelle, and D. J. Evans. Configurational thermostats for molecular systems. *Mol. Phys.*, 100: 2387, 2002.

[161] J. Delhommelle and D. J. Evans. Correspondence between configurational temperature and molecular kinetic temperature thermostats. *J. Chem. Phys.*, 117: 6016, 2002.

[162] C. Braga and K. P. Travis. A configurational temperature Nosé-Hoover thermostat. *J. Chem. Phys.*, 123: 134101, 2005.

[163] K. P. Travis and C. Braga. Configurational temperature and pressure molecular dynamics: review of current methodology and applications to the shear flow of a simple fluid. *Mol. Phys.*, 104: 3735, 2006.

[164] K. P. Travis and C. Braga. Configurational temperature control for atomic and molecular systems. *J. Chem. Phys.*, 128: 014111, 2008.

[165] D. J. Evans and B.L. Holian. Shear viscosities away from the melting line - a comparison of equilibrium and non-equilibrium molecular-dynamics. *J. Chem. Phys.*, 78: 5147, 1983.

[166] D. J. Evans and B. L. Holian. The Nosé-Hoover thermostat. *J. Chem. Phys.*, 83: 4069, 1985.

[167] D. J. Evans and S. Sarman. Equivalence of thermostatted nonlinear responses. *Phys. Rev. E*, 48: 65, 1993.

[168] S. Y. Liem, D. Brown, and J. H. R. Clarke. Investigation of the homogeneous-shear nonequilibrium-molecular-dynamics method. *Phys. Rev. A*, 45: 3706, 1992.

[169] P. Padilla and S. Toxvaerd. Simulating shear flow. *J. Chem. Phys.*, 104: 5956, 1996.

[170] P. J. Daivis, B. A. Dalton, and T. Morishita. Effect of kinetic and configurational thermostats on claculations of the first normal stress coefficient in nonequilibrium molecular dynamics simulations. *Phys. Rev. E*, 86: 056707, 2012.

[171] J. Petravic. Time dependence of phase variables in a steady shear flow algorithm. *Phys. Rev. E*, 71: 011202, 2005.

[172] P. J. Daivis and B. D. Todd. Frequency dependent elongational viscosity by nonequilibrium molecular dynamics. *Int. J. Thermophys.*, 19: 1063, 1998.

[173] A. Baranyai and D. J. Evans. New algorithm for constrained molecular-dynamics simulation of liquid benzene and naphthalene. *Molec. Phys.*, 70(1): 53, 1990.

[174] J. N. Bright, D. J. Evans, and D. J. Searles. New observations regarding deterministic, time-reversible thermostats and Gauss's principle of least constraint. *J. Chem. Phys.*, 122: 194106, 2005.

[175] S. Sarman, D. J. Evans, and A. Baranyai. Extremum properties of the Gaussian thermostat. *Physica A*, 208: 191, 1994.

[176] D. J. Evans, E. G. D. Cohen, and G. P. Morriss. Viscosity of a simple fluid from its maximal Lyanpunov exponents. *Phys. Rev. A*, 42: 5990, 1990.

[177] S. Sarman, D. J. Evans, and G. P. Morriss. Conjugate pairing rule and thermal-transport coefficients. *Phys. Rev. A*, 45: 2233–2242, 1992.

[178] F. D. Ditolla and M. Ronchetti. Applicability of Nosé isothermal reversible dynamics. *Phys. Rev. E*, 48: 1726, 1993.

[179] B. L. Holian, A. F. Voter, and R. Ravelo. Thermostatted molecular-dynamics – how to avoid the Toda demon hidden in Nosé-Hoover dynamics. *Phys. Rev. E*, 52: 2338, 1995.

[180] S. Toxvaerd and O. H. Olsen. Canonical molecular-dynamics of molecules with internal degrees of freedom. *Ber. Bunsenges. Phys. Chem.*, 93: 274, 1990.

[181] G. J. Martyna, M. L. Klein, and M. E. Tuckerman. Nosé-Hoover chains: The canonical ensemble via continuous dynamics. *J. Chem. Phys.*, 97: 2635, 1992.

[182] A. C. Branka. Nosé-Hoover chain method for nonequilibrium molecular dynamics simulation. *Phys. Rev. E*, 61: 4769, 2000.

[183] A. C. Branka, M. Kowalik, and K. W. Wojciechowski. Generalization of the Nosé-Hoover approach. *J. Chem. Phys.*, 119: 1929, 2003.

[184] J. J. Erpenbeck. Shear viscosity of the hard-sphere fluid via nonequilibrium molecular-dynamics. *Phys. Rev. Lett.*, 52: 1333, 1984.

[185] D. J. Evans and G. P. Morriss. Shear thickening and turbulence in simple fluids. *Phys. Rev. Lett.*, 56: 2172, 1986.

[186] J. Delhommelle, J. Petravic, and D. J. Evans. Reexamination of string phase and shear thickening in simple fluids. *Phys. Rev. E*, 68: 031201, 2003.

[187] W. Loose and S. Hess. Rheology of dense model fluids via nonequilibrium molecular dynamics – shear thinning and ordering transition. *Rheol. Acta*, 28: 91, 1989.

[188] D. J. Evans, S. T. Cui, H. J. M. Hanley, and G. C. Straty. Conditions for the existence of a reentrant solid-phase in a sheared atomic fluid. *Phys. Rev. A*, 46: 6731, 1992.

[189] J. Delhommelle, J. Petravic, and D. J. Evans. On the effects of assuming flow profiles in nonequilibrium simulations. *J. Chem. Phys.*, 119: 11005, 2003.

[190] J. Delhommelle and D. J. Evans. Comparison of thermostatting mechanisms in NVT and NPT simulations of decane under shear. *J. Chem. Phys.*, 115: 43, 2001.

[191] D. Kusnezov, A. Bulgac, and W. Bauer. Canonical ensembles from chaos. *Ann. Phys.*, 204: 155, 1990.

[192] C. Braga and K. P. Travis. Configurational constant pressure molecular dynamics. *J. Chem. Phys.*, 124: 104102, 2006.

[193] M. E. Tuckerman, C. J. Mundy, S. Balasubramanian, and M. L. Klein. Modified nonequilibrium molecular dynamics for fluid flows with energy conservation. *J. Chem. Phys.*, 106: 5615, 1997.

[194] D. J. Evans and G. P. Morriss. Isothermal-isobaric molecular dynamics. *Chem. Phys.*, 77: 63, 1983.

[195] S. Melchionna, G. Ciccotti, and B. L. Holian. Hoover NPT dynamics for systems varying in shape and size. *Mol. Phys.*, 78: 533, 1993.

[196] S. Bernardi. *Private communication.*

[197] P. J. Daivis, M. L. Matin, and B. D. Todd. Nonlinear shear and elongational rheology of model polymer melts by non-equilibrium molecular dynamics. *J. Non-Newtonian Fluid Mech.*, 111: 1, 2003.

[198] F. Frascoli and B. D. Todd. Molecular dynamics simulation of planar elongational flow at constant pressure and constant temperature. *J. Chem. Phys.*, 126: 044506, 2007.

[199] T. T. Perkins, D. E. Smith, R. G. Larson, and S. Chu. Stretching of a single tethered polymer in a uniform flow. *Science*, 268: 83, 1995.

[200] M. Dobson. Periodic boundary conditions for long-time nonequilibrium molecular dynamics simulations of incompressible flows. *J. Chem. Phys.*, 141: 184103, 2014.

[201] T. A. Hunt. Periodic boundary conditions for the simulation of uniaxial extensional flow of arbitrary duration. *Molec. Simul.*, 42: 347, 2016.

[202] J. G. H. Cifre, S. Hess, and M. Kröger. Linear viscoelastic behavior of unentangled polymer melts via non-equilibrium molecular dynamics. *Macromol. Theory Simul.*, 13: 748, 2004.

[203] H. A. Barnes, J. F. Hutton, and K. Walters. *An Introduction to Rheology*. Elsevier, Amsterdam, 1989.

[204] A. Jain, C. Sasmal, R. Hartkamp, B. D. Todd, and J. R. Prakash. Brownian dynamics simulations of planar mixed flows of polymer solutions at finite concentrations. *Chem. Eng. Sci.*, 121: 245, 2015.

[205] P. M. Adler and H. Brenner. Spatially periodic suspensions of convex particles in linear shear flows. 1. Description and kinematics. *Int. J. Multiphase Flow*, 11: 361, 1985.

[206] P. M. Adler, M. Zuzovsky, and H. Brenner. Spatially periodic suspensions of convex particles in linear shear flows. *Int. J. Multiphase Flow*, 11: 387, 1985.

[207] J. Ge, G. Marcelli, B. D. Todd, and R. J. Sadus. Energy and pressure of fluids under shear at different state points. *Phys. Rev. E*, 64: 021201, 2001.

[208] J. Ge, G. Marcelli, B. D. Todd, and R. J. Sadus. Erratum: Energy and pressure of fluids under shear at different state points. *Phys. Rev. E*, 65: 069901(E), 2002.

[209] J. Ge, B. D. Todd, G. Wu, and R. J. Sadus. Scaling behaviour for the pressure and energy of shearing fluids. *Phys. Rev. E*, 67: 061201, 2003.

[210] B. D. Todd. Power-law exponents for the shear viscosity of non-Newtonian simple fluids. *Phys. Rev. E*, 72: 041204, 2005.

[211] C. Desgranges and J. Delhommelle. Universal scaling law for energy and pressure in a shearing fluid. *Phys. Rev. E*, 79: 052201, 2009.

[212] K. P. Travis, D. J. Searles, and D. J. Evans. Strain rate dependent properties of a simple fluid. *Mol. Phys.*, 95: 195, 1998.

[213] M. Ferrario, G. Ciccotti, B. L. Holian, and J. P. Ryckaert. Shear-rate dependence of the viscosity of the Lennard-Jones liquid at the triple point. *Phys. Rev. A*, 44: 6936, 1991.

[214] P. J. Daivis. Thermodynamic relationships for shearing linear viscoelastic fluids. *J. Non-Newtonian Fluid Mech.*, 152: 120, 2008.

[215] P. J. Daivis and D. J. Evans. Thermal conductivity of a shearing fluid. *Phys. Rev. E*, 48: 1058, 1993.

[216] M. R. Spiegel. *Theory and Problems of Vector Analysis and an Introduction to Tensor Analysis*. McGraw-Hill, Singapore, 1974.

[217] P. J. Daivis, M. L. Matin, and B. D. Todd. Nonlinear shear and elongational rheology of model polymer melts at low strain rates. *J. Non-Newtonian Fluid Mech.*, 147: 35, 2007.

[218] J. Ge, G.-W. Wu, B. D. Todd, and R. J. Sadus. Equilibrium and nonequilibrium molecular dynamics methods for detemining solid-liquid phase coexistence at equilibrium. *J. Chem. Phys.*, 119(21): 11017, 2003.

[219] M. L. Matin, B. D. Todd, and P. J. Daivis. Various aspects of non-equilibrium molecular dynamics simulation of polymer rheology. *Swinburne University Internal Report*, 2003.

[220] H. S. Green. *The Molecular Theory of Fluids*. North-Holland Interscience, New York, 1952.

[221] J. A. Pryde. *The Liquid State*. Hutchinson University Library, London, 1966.

[222] H. J. M. Hanley and D. J. Evans. Equilibrium and non-equilibrium radial distribution functions in mixtures. *Mol. Phys.*, 39: 1039, 1980.

[223] S. Hess. Shear-flow-induced distortion of the pair-correlation function. *Phys. Rev. A*, 22: 2844, 1980.

[224] S. Hess. Similarities and differences in the non-linear flow behavior of simple and molecular liquids. *Physica A*, 118: 79, 1983.

[225] Y. V. Kalyuzhnyi, S. T. Cui, P. T. Cummings, and H. D. Cochran. Distribution functions of a simple fluid under shear: Low shear rates. *Phys. Rev. E*, 60: 1716, 1999.

[226] H. H. Gan and B. C. Eu. Theory of the nonequilibrium structure of dense simple fluids – effects of shearing. *Phys. Rev. A*, 45: 3670, 1992.

[227] H. H. Gan and B. C. Eu. Theory of the nonequilibrium structure of dense simple fluids – effects of shearing. 2. High-shear-rate effects. *Phys. Rev. A*, 46: 6344, 1992.

[228] J. Ge. *The State Point Dependence of Classical Fluids under Shear*. PhD thesis, Swinburne University of Technology, 2004.

[229] C. Desgranges and J. Delhommelle. Rheology of liquid fcc metals: Equilibrium and transient-time correlation-function nonequilibrium molecular dynamics simulations. *Phys. Rev. B*, 78: 184202, 2008.

[230] C. Desgranges and J. Delhommelle. Shear viscosity of liquid copper at experimentally accessible shear rates: Application of the transient-time correlation function formalism. *J. Chem. Phys.*, 128: 084506, 2008.

[231] C. Desgranges and J. Delhommelle. Molecular simulation of transport in nanopores: Application of the transient-time correlation function formalism. *Phys. Rev. E*, 77: 027701, 2008.

[232] C. Desgranges and J. Delhommelle. Estimating the conductivity of a nanoconfined liquid subjected to an experimentally accessible external field. *Mol. Simul.*, 34: 177, 2008.

[233] G. Pan and C. McCabe. Prediction of viscosity for molecular fluids at experimentally accessible shear rates using the transient time correlation function formalism. *J. Chem. Phys.*, 125: 194527, 2006.

[234] R. Hartkamp, S. Bernardi, and B. D. Todd. Transient-time correlation function applied to mixed shear and elongational flows. *J. Chem. Phys.*, 136: 064105, 2012.

[235] D. J. Evans. Homogeneous NEMD algorithm for thermal conductivity - application of non-canonical linear response theory. *Phys. Lett. A*, 91: 457, 1982.

[236] W. W. Wood. Long-time tails of the Green – Kubo integrands for a binary mixture. *J. Stat. Phys.*, 57: 675, 1989.

[237] D. J. Evans and H. J. M. Hanley. Heat induced instability in a model liquid. *Molec. Phys.*, 68: 97, 1989.

[238] D. P. Hansen and D. J. Evans. A generalized heat flow algorithm. *Mol. Phys.*, 81: 767, 1994.

[239] D. J. Evans. Thermal conductivity of the Lennard-Jones fluid. *Phys. Rev. A*, 34: 1449, 1986.

[240] N. Galamba, C. A. N. de Castro, and J. F. Ely. Equilibrium and nonequilibrium molecular dynamics simulations of the thermal conductivity of molten alkali halides. *J. Chem. Phys.*, 126: 204511, 2007.

[241] H. J. V. Tyrrell and K. R. Harris. *Diffusion in Liquids*. Elsevier, 1984.

[242] D. MacGowan and D. J. Evans. Heat and matter transport in binary-liquid mixtures. *Phys. Rev. A*, 34: 2133, 1986.

[243] S. Sarman, D. J. Evans, and P. T. Cummings. Recent developments in non-Newtonian molecular dynamics. *Phys. Rep.*, 305: 1, 1998.

[244] S. Sarman and D. J. Evans. Heat flow and mass diffusion in binary Lennard-Jones mixtures. *Phys. Rev. A*, 45: 2370, 1992.

[245] S. Sarman and D. J. Evans. Heat flow and mass diffusion in binary Lennard-Jones mixtures. II. *Phys. Rev. A*, 46: 1960, 1992.

[246] E. J. Maginn, A. T. Bell, and D. N. Theodorou. *J. Phys. Chem.*, 97: 4173, 1993.

[247] D. R. Wheeler and J. Newman. Molecular dynamics simulations of multicomponent diffusion. 2. Nonequilibrium method. *J. Phys. Chem. B*, 108: 18362, 2004.

[248] D. MacGowan and D. J. Evans. A comparison of NEMD algorithms for thermal conductivity. *Phys. Lett. A*, 117: 414, 1986.

[249] D. MacGowan and D. J. Evans. Addendum to heat and matter transport in binary-liquid mixtures. *Phys. Rev. A*, 36: 948, 1987.

[250] D. J. Evans and P. T. Cummings. Non-equilibrium molecular dynamics algorithm for the calculation of thermal diffusion in simple fluid mixtures. *Molec. Simul.*, 72: 893, 1991.

[251] A. Perronace, J.-M. Simon, B. Rousseau, and G. Ciccotti. Flux expression and NEMD perturbations for models of semi-flexible molecules. *Molec. Phys.*, 99(13): 1139, 2001.

[252] K. Mandadapu, R. E. Jones, and P. Papadopoulos. A homogeneous nonequilibrium molecular dynamics method for calculating the heat transport coefficient of mixtures and alloys. *J. Chem. Phys.*, 133: 034122, 2010.

[253] A. Perronace, C. Leppla, F. Leroy, B. Rousseau, and S. Wiegand. Soret and mass diffusion measurements and molecular dynamics simulations of n-pentane and n-decane mixtures. *J. Chem. Phys.*, 116: 3718, 2002.

[254] J. G. Kirkwood and F. P. Buff. The statistical mechanical theory of solutions. I. *J. Chem. Phys.*, 19: 774, 1951.

[255] N. A. T. Miller, P. J. Daivis, I. K. Snook, and B. D. Todd. Computation of thermodynamic and tranport properties to predict thermophoretic effects in an argon-krypton mixture. *J. Chem. Phys.*, 139: 144504, 2013.

[256] J.-P. Hansen and I. R. McDonald. *Theory of Simple Liquids*. Academic Press, 3rd edition, 2006.

[257] P. Krüger, D. Bedeaux, S. K. Schnell, S. Kjelstrup, T. J. H. Vlugt, and J.-M. Simon. Kirkwood-buff integrals for finite volumes. *J. Phys. Chem. Lett.*, 4: 235, 2013.

[258] J. W. Nichols, S. G. Moore, and D. R. Wheeler. Improved implementation of Kirkwood-Buff solution theory in periodic molecular simulations. *Phys. Rev. E*, 80: 051203, 2009.

[259] S. D. W. Hannam, P. J. Daivis, and G. Bryant. Dynamics of a model colloidal suspension from dilute to freezing. Submitted, 2016.

[260] Y. Zhou and G. H. Miller. Green–Kubo formulas for mutual difusion coefficients in multicomponent systems. *J. Phys. Chem.*, 100: 5516, 1996.

[261] D. J. Evans and S. Murad. Thermal conductivity in molecular fluids. *Molec. Phys.*, 68(6): 1219, 1989.

[262] P. J. Daivis and D. J. Evans. Non-equilibrium molecular dynamics calculation of thermal conductivity of flexible molecules: butane. *Mol. Phys.*, 81: 1289, 1994.

[263] P. J. Daivis and D. J. Evans. Temperature dependence of the thermal conductivity for two models of liquid butane. *Chem. Phys.*, 198: 25, 1995.

[264] G. Marechal and J. P. Ryckaert. Atomic versus molecular description of transport properties in polyatomic fluids: n-butane as an illustration. *Chem. Phys. Lett.*, 101: 548, 1983.

[265] S. Toxvaerd. Molecular dynamics calculation of the equation of state of alkanes. *J. Chem. Phys.*, 93(6): 4290, 1990.

[266] D. Reith, M. Pütz, and F. Müller-Plathe. Deriving effective mesoscale potentials from atomistic simulations. *J. Comput. Chem.*, 24: 1624, 2003.

[267] M. S. Shell. The relative entropy is fundamental to multiscale and inverse thermodynamic problems. *J. Chem. Phys.*, 129: 144108, 2008.

[268] R. Potestio, C. Peter, and K. Kremer. Computer simulations of soft matter: Linking the scales. *Entropy*, 16: 4199, 2014.

[269] G. Raabe and R. J. Sadus. Molecular dynamics simulation of the effect of bond flexibility on the transport properties of water. *J. Chem. Phys.*, 137: 104512, 2012.

[270] D. J. Evans and S. Murad. Singularity free algorithm for molecular dynamics simulation of rigid polyatomics. *Molec. Phys.*, 34(2): 327, 1977.

[271] S. Hess. Rheological properties via nonequilibrium molecular dynamics: From simple towards polymeric liquids. *J. Non-Newtonian Fluid Mech.*, 23: 305, 1987.

[272] H. R. Warner Jr. Kinetic theory and rheology of dilute suspensions of finitely extendible dumbbells. *Ind. Eng. Chem. Fundam.*, 11(3): 379, 1972.

[273] G. S. Grest and K. Kremer. Molecular dynamics simulation for polymers in the presence of a heat bath. *Phys. Rev. A*, 33(5): 3628, 1986.

[274] K. Kremer and G. S. Grest. Dynamics of entangled linear polymer melts – a molecular-dynamics simulation. *J. Chem. Phys.*, 92: 5057, 1990.

[275] I. Snook. *Langevin and Generalised Langevin Approach to the Dynamics of Atomic, Polymeric and Colloidal Systems*. Elsevier, Amsterdam, 2007.

[276] J. K. Johnson, E. A. Müller, and K. E. Gubbins. Equation of state for Lennard-Jones chains. *J. Phys. Chem.*, 98: 6413, 1994.

[277] J.-P. Ryckaert and A. Bellemans. Molecular dynamics of liquid n-butane near its boiling point. *Chem. Phys. Lett.*, 30(1): 123, 1975.

[278] J.-P. Ryckaert, G. Ciccotti, and H. J. C. Berendsen. Numerical integration of the Cartesian equations of motion of a system with constraints: Molecular dynamics of n-alkanes. *J. Comput. Phys.*, 23: 327, 1977.

[279] H. C. Andersen. Rattle: A "velocity" version of the shake algorithm for molecular dynamics calculations. *J. Comput. Phys.*, 52: 24, 1983.

[280] G. J. Martyna, M. E. Tuckerman, D. J. Tobias, and M. L. Klein. Explicit reversible integrators for extended systems dynamics. *Molec. Phys.*, 87(5): 1117, 1996.

[281] S. Balasubramanian, C. J. Mundy, and M. L. Klein. Shear viscosity of polar fluids: Molecular dynamics calculations of water. *J. Chem. Phys.*, 105(24): 11190, 1996.

[282] R. Edberg, D. J. Evans, and G. P. Morriss. Constrained molecular dynamics: Simulations of liquid alkanes with a new algorithm. *J. Chem. Phys.*, 84: 6933, 1986.

[283] A. Baranyai and D. J. Evans. NEMD investigation of the rheology of oblate molecules: shear flow in liquid benzene. *Molec. Phys.*, 71(4): 835, 1990.

[284] G. Ciccotti, M. Ferrario, and J.-P. Ryckaert. Molecular dynamics of rigid systems in cartesian coordinates: A general formulation. *Molec. Phys.*, 47(6): 1253, 1982.

[285] G. P. Morriss and D. J. Evans. A constraint algorithm for the computer simulation of complex molecular liquids. *Comput. Phys. Commun.*, 62: 267, 1991.

[286] R. D. Olmsted and R. F. Snider. Differences in fluid dynamics associated with an atomic versus a molecular description of the same system. *J. Chem. Phys.*, 65: 3407, 1976.

[287] H. Yamakawa. *Modern Theory of Polymer Solutions*. Harper & Row, New York, 1971.

[288] M. P. Allen. Atomic and molecular representations of molecular hydrodynamic variables. *Mol. Phys.*, 52: 705, 1984.

[289] G. Ciccotti and J. P. Ryckaert. Molecular dynamics simulation of rigid molecules. *Computer Physics Reports*, 4: 345, 1986.

[290] R. Edberg, D. J. Evans, and G. P. Morriss. On the nonlinear Born effect. *Mol. Phys.*, 62: 1357, 1987.

[291] S. T. Cui, P. T. Cummings, and H. D. Cochran. The calculation of the viscosity from the autocorrelation function using molecular and atomic stress tensors. *Mol. Phys.*, 88: 1657, 1996.

[292] D. J. Evans. Non-equilibrium molecular dynamics study of the rheological properties of diatomic liquids. *Mol. Phys.*, 42: 1355, 1981.

[293] K. P. Travis, P. J. Daivis, and D. J. Evans. Computer simulation algorithms for molecules undergoing planar Couette flow: A nonequilibrium molecular dynamics study. *J. Chem. Phys.*, 103: 1109, 1995.

[294] K. P. Travis, P. J. Daivis, and D. J. Evans. Thermostats for molecular fluids undergoing shear flow: Application to liquid chlorine. *J. Chem. Phys.*, 103: 10638, 1995.

[295] K. P. Travis, P. J. Daivis, and D. J. Evans. Erratum: Thermostats for molecular fluids undergoing shear flow: Application to liquid chlorine. *J. Chem. Phys.*, 105: 3893, 1996.

[296] A. Baranyai, D. J. Evans, and P. J. Daivis. Isothermal shear-induced heat flow. *Phys. Rev. A*, 46: 7593, 1992.

[297] R. Edberg, G. P. Morriss, and D. J. Evans. Rheology of n-alkanes by nonequilibrium molecular dynamics. *J. Chem. Phys.*, 86: 4555, 1987.

[298] P. T. Cummings and D. J. Evans. Nonequilibrium molecular dynamics approaches to transport properties and non-newtonian fluid rheology. *Ind. Eng. Chem. Res.*, 31: 1237, 1992.

[299] O. Reynolds. On the dilatancy of media composed of rigid particles in contact. With experimental illustrations. *Phil. Mag.*, 20(127): 469, 1885.

[300] D. J. Tildesley and P. A. Madden. Time correlation functions for a model of liquid carbon disulphide. *Molec. Phys.*, 48(1): 129, 1983.

[301] P. Prathiraja, P. J. Daivis, and I. K. Snook. A molecular simulation study of shear viscosity and thermal conductivity of liquid carbon disulphide. *J. Mol. Liq.*, 154: 6, 2010.

[302] M. L. Matin, P. J. Daivis, and B. D. Todd. Comparison of planar Couette flow and planar elongational flow for systems of small freely jointed chain molecules. *J. Chem. Phys.*, 113: 9122, 2000.

[303] M. L. Matin, P. J. Daivis, and B. D. Todd. Erratum: "Comparison of planar Couette flow and planar elongational flow for systems of small freely jointed chain molecules" [J. Chem. Phys. 113, 9122 (2000)]. *J. Chem. Phys.*, 115: 5338, 2001.

[304] M. L. Matin, P. J. Daivis, and B. D. Todd. Cell neighbour list method for planar elongational flow: rheology of a diatomic fluid. *Comput. Phys. Commun.*, 151: 35, 2003.

[305] F. Müller-Plathe. Coarse-graining in polymer simulation: From the atomic to the mesoscopic scale and back. *ChemPhysChem*, 3: 754, 2002.

[306] J. T. Padding and W. J. Briels. Coarse-grained molecular dynamics simulations of polymer melts in transient and steady shear flow. *J. Chem. Phys.*, 118: 10276, 2003.

[307] M. Kröger, W. Loose, and S. Hess. Rheology and structural changes of polymer melts via nonequilibrium molecular dynamics. *J. Rheol.*, 37: 1057, 1993.

[308] J. D. Ferry. *Viscoelastic Properties of Polymers*. Wiley, New York, 1980.

[309] M. Kröger and S. Hess. Rheological evidence for a dynamical crossover in polymer melts via nonequilibrium molecular dynamics. *Phys. Rev. Lett.*, 85: 1128, 2000.

[310] J. T. Bosko, B. D. Todd, and R. J. Sadus. Viscoelastic properties of dendrimers in the melt by nonequilibrium molecular dynamics. *J. Chem. Phys.*, 121: 12050, 2004.

[311] T. A. Hunt and B. D. Todd. A comparison of model linear chain molecules with constrained and flexible bond lengths under planar Couette and extensional flows. *Mol. Simul.*, 35: 1153, 2009.

[312] R. K. Prud'homme and R. B. Bird. The dilational properties of suspensions of gas bubbles in incompressible Newtonian and non-Newtonian fluids. *J. Non-Newtonian Fluid Mech.*, 3: 261, 1977/1978.

[313] S. Sarman, P. J. Daivis, and D. J. Evans. Self-diffusion of rodlike molecules in strong shear fields. *J. Chem. Phys.*, 47: 1784, 1993.

[314] T. A. Hunt. Diffusion of linear polymer melts in shear and extensional flows. *J. Chem. Phys.*, 131: 054904, 2009.

[315] G. G. Stokes. *Mathematical and Physical Papers. Volume 1*. Oxford Press, Oxford, 1880.

[316] C. Clarke and R. Carswell. *Principles of Astrophysical Fluid Dynamics*. Cambridge University Press, Cambridge, 1995.

[317] G. Rubbert and G. Saaris. A general three-dimensional potential-flow method applied to V/STOL aerodynamics. *SAE*, 680304: 945, 1968.

[318] P. Tabeling. *Introduction to Microfluidics*. Oxford University Press, New York, 2005.

[319] H. Bruus. *Theoretical Microfluidics*. Oxford University Press, New York, 2008.

[320] K. P. Travis and K. E. Gubbins. Poiseuille flow of Lennard-Jones fluids in narrow slit pores. *J. Chem. Phys.*, 112: 1984, 2000.

[321] W. E. Alley and B. J. Alder. Generalised transport coefficients for hard spheres. *Phys. Rev. A*, 27: 3158, 1983.

[322] B. D. Todd, J. S. Hansen, and P. J. Daivis. Non-local shear stress for homogeneous fluids. *Phys. Rev. Lett.*, 100: 195901, 2008.

[323] S. Hess. Viscoelasticity of a simple liquid in the pre-freezing regime. *Phys. Lett. A*, 90: 293, 1982.

[324] J. S. Hansen, P. J. Daivis, K. P. Travis, and B. D. Todd. Parameterisation of the nonlocal viscosity kernel for an atomic fluid. *Phys. Rev. E*, 76: 041121, 2007.

[325] D. Bertolini and A. Tani. Generalized hydrodynamics and the acoustic modes of water – theory and simulation results. *Phys. Rev. E*, 51: 1091, 1995.

[326] D. Bertolini and A. Tani. Stress tensor and viscosity of water – molecular-dynamics and generalized hydrodynamics results. *Phys. Rev. E*, 52: 1699, 1995.

[327] R. M. Puscasu, B. D. Todd, P. J. Daivis, and J. S. Hansen. Viscosity kernel of molecular fluids: butane and polymer melts. *Phys. Rev. E*, 82: 011801, 2010.

[328] R. M. Puscasu, B. D. Todd, P. J. Daivis, and J. S. Hansen. Non-local viscosity of polymer melts approaching their glassy state. *J. Chem. Phys.*, 133: 144907, 2010.

[329] K. P. Travis, D. J. Searles, and D. J. Evans. On the wavevector dependent shear viscosity of a simple fluid. *Mol. Phys.*, 97: 415, 1999.

[330] B. D. Todd and D. J. Evans. Temperature profile for Poiseuille flow. *Phys. Rev. E*, 55: 2800, 1997.

[331] P. J. Daivis and J. L. K. Coelho. Generalized Fourier law for heat flow in a fluid with a strong, nonuniform strain rate. *Phys. Rev. E*, 61: 6003, 2000.

[332] P. Cordero and D. Risso. Nonlinear transport laws for low density fluids. *Physica A*, 257: 36, 1998.

[333] M. Criado-Sancho, D. Jou, and J. Casas-Vazquez. Nonequilibrium kinetic temperatures in flowing gases. *Phys. Lett. A*, 350: 339, 2006.

[334] Casas-Vázquez and D. Jou. Nonequilibrium temperature versus local-equilibrium temperature. *Phys. Rev. E*, 49: 1040, 1994.

[335] M. Han and J. S. Lee. Method for calculating the heat and momentum fluxes of inhomogeneous fluids. *Phys. Rev. E*, 70: 061205, 2004.

[336] G. Ayton, O.G. Jepps, and D. J. Evans. On the validity of Fourier's law in systems with spatially varying strain rates. *Mol. Phys.*, 96: 915, 1999.

[337] B. D. Todd and D. J. Evans. The heat flux vector for highly inhomogeneous nonequilibrium fluids in very narrow pores. *J. Chem. Phys.*, 103: 9804, 1995.

[338] H. Hoang and G. Galliero. Shear viscosity of inhomogeneous fluids. *J. Chem. Phys.*, 136: 124902, 2012.

[339] B. A. Dalton, K. S. Glavatskiy, P. J. Daivis, B. D. Todd, and I. K. Snook. Linear and nonlinear density response functions for a simple atomic fluid. *J. Chem. Phys.*, 139: 044510, 2013.

[340] B. A. Dalton, P. J. Daivis, J. S. Hansen, and B. D. Todd. Effects of nanoscale inhomogeneity on shearing fluids. *Phys. Rev. E*, 88: 052143, 2013.

[341] K. S. Glavatskiy, B. A. Dalton, P. J. Daivis, and B. D. Todd. Nonlocal response functions for predicting shear flow of strongly inhomogeneous fluids. I. Sinusoidally driven shear and sinusoidally driven inhomogeneity. *Phys. Rev. E*, 91: 062132, 2015.

[342] B. A. Dalton, K. S. Glavatskiy, P. J. Daivis, and B. D. Todd. Nonlocal response functions for predicting shear flow of strongly inhomogeneous fluids. II. Sinusoidally driven shear and multisinusoidal inhomogeneity. *Phys. Rev. E*, 92: 012108, 2015.

[343] H. Hoang and G. Galliero. Local viscosity of a fluid confined in a narrow pore. *Phys. Rev. E*, 86: 021202, 2012.

[344] H. Hoang and G. Galliero. Local shear viscosity of strongly inhomogeneous dense fluids: from the hard-sphere to the Lennard-Jones fluids. *J. Phys.: Condens. Matter*, 25: 485001, 2013.

[345] B. D. Todd and J. S. Hansen. Nonlocal viscous transport and the effect on fluid stress. *Phys. Rev. E*, 78: 051202, 2008.

[346] B. D. Todd, D. J. Evans, K. P. Travis, and P. J. Daivis. Comment on: Molecular simulation and continuum mechanics study of simple fluids in non-isothermal planar Couette flows. *J. Chem. Phys.*, 111: 10730, 1999.

[347] S. Bernardi, B. D. Todd, and D. J. Searles. Thermostatting highly confined fluids. *J. Chem. Phys.*, 132: 244706, 2010.

[348] S. De Luca, B. D. Todd, J. S. Hansen and P. J. Daivis. A new and effective method for thermostatting confined fluids. *J. Chem. Phys.* 140: 054502, 2014.

[349] K. P. Travis and D. J. Evans. Molecular spin in a fluid undergoing Poiseuille flow. *Phys. Rev. E*, 55: 1566, 1997.

[350] Couette code was developed by S. Bernardi, based on the MD library of J. S. Hansen (http://www.jshansen.dk/resources.html).

[351] A. C. Eringen. *Contributions to Mechanics*. Pergamon, Oxford, 1969.

[352] K. P. Travis, B. D. Todd, and D. J. Evans. Poiseuille flow of molecular fluids. *Physica A*, 240: 315, 1997.

[353] S. Sarman and D. J. Evans. Statistical mechanics of viscous flow in nematic fluids. *J. Chem. Phys.*, 99: 9021, 1993.

[354] M. Kröger. *Models for Polymeric and Anisotropic Liquids*, volume 675 of *Lecture Notes in Physics*. Springer, New York, 2005.

[355] J. Zhang, J. S. Hansen, B. D. Todd, and P. J. Daivis. Structural and dynamical properties for confined polymers undergoing planar Poiseuille flow. *J. Chem. Phys.*, 126: 144907, 2007.

[356] M. Doi. *Introduction to Polymer Physics*. Oxford, New York, 1996.

[357] H. Münstedt, Schmidt M., and E. Wassner. Stick and slip phenomena during extrusion of polyethylene melts as investigated by laser-doppler velocimetry. *J. Rheol.*, 44: 413, 2000.

[358] L. Robert, Y. Demay, and B. Vergnes. Stick-slip flow of high density polyethylene in a transparent slit die investigated by laser doppler velocimetry. *Rheol. Acta*, 43: 89, 2004.

[359] J. S. Hansen, P. J. Daivis, and B. D. Todd. Viscous properties of isotropic fluids composed of linear molecules: Departure from the classical Navier-Stokes theory in nano-confined geometries. *Phys. Rev. E*, 80: 046322, 2009.

[360] J. S. Hansen, H. Bruus, B. D. Todd, and P. J. Daivis. Rotational and spin viscosities of water: Application to nanofluidics. *J. Chem. Phys.*, 133: 144906, 2010.

[361] J. S. Hansen, B. D. Todd, and P. J. Daivis. Dynamical properties of a confined diatomic fluid undergoing zero mean oscillatory flow: Effect of molecular rotation. *Phys. Rev. E*, 77: 066707, 2008.

[362] J. S. Hansen, P. J. Daivis, and B. D. Todd. Molecular spin in nano-confined fluidic flows. *Microfluid. Nanfluid.*, 6: 785, 2009.

[363] J. S. Hansen, J. C. Dyre, P. J. Daivis, B. D. Todd, and H. Bruus. Nanoflow hydrodynamics. *Phys. Rev. E*, 84: 036311, 2011.

[364] D. L. Bonthuis, D. Horinek, L. Bocquet, and R. R. Netz. Electrohydraulic power conversion in planar nanochannels. *Phys. Rev. Lett.*, 103: 144503, 2009.

[365] D. L. Bonthuis, D. Horinek, L. Bocquet, and R. R. Netz. Electrokinetics at aqueous interfaces without mobile charges. *Langmuir*, 26: 12614, 2010.

[366] S. De Luca, B. D. Todd, J. S. Hansen, and P. J. Daivis. Electropumping of water with rotating electric fields. *J. Chem. Phys.*, 138: 154712, 2013.

[367] S. De Luca, B. D. Todd, J. S. Hansen, and P. J. Daivis. Molecular dynamics study of nanoconfined water flow driven by rotating electric fields under realistic experimental conditions. *Langmuir*, 30: 3095, 2014.

[368] A. Menzel. In preparation, 2016.

[369] A. Z. Akcasu and E. Daniels. Fluctuation analysis in simple fluids. *Phys. Rev. A*, 2: 962, 1970.

[370] N. K. Ailawadi, B. J. Berne, and D. Forster. Hydrodynamics and collective angular-momentum fluctuations in molecular fluids. *Phys. Rev. A*, 3: 1462, 1971.

[371] J. P. Boon and S. Yip. *Molecular Hydrodynamics*. McGraw-Hill, New York, 1980.

[372] B. C. Eu. *Generalised Thermodynamics: The Thermodynamics of Irreversible Processes and Generalised Hydrodynamics*. Kluwer, Dordrecht, 2002.

[373] B. L. Holian, W. G. Hoover, B. Moran, and G. K. Straub. Shock-wave structure via nonequilibrium molecular dynamics and Navier-Stokes continuum mechanics. *Phys. Rev. A*, 22: 2798, 1980.

[374] B. L. Holian and P. S. Lomdahl. Plasticity induced by shock waves in nonequilibrium molecular-dynamics simulations. *Science*, 280: 2085, 1998.

[375] E. J. Reed, L. E. Fried, W. D. Henshaw, and C. M. Tarver. Analysis of simulation technique for steady shock waves in materials with analytical equations of state. *Phys. Rev. E*, 74: 056706, 2006.

[376] D. Jou, J. Casas-Vazquez, and G. Lebon. *Extended Irreversible Thermodynamics*. Springer, Heidelberg, 2001.

[377] J. K. G. Dhont. A constitutive relation describing the shear-banding transition. *Phys. Rev. E*, 60: 4534, 1999.

[378] C. Masselon, J.-B. Salmon, and A. Colin. Nonlocal effects in flows of wormlike micellar solutions. *Phys. Rev. Lett.*, 100: 038301, 2008.

[379] R. L. Schiek and E. S. G. Shaqfeh. A nonlocal theory for stress in bound, Brownian suspensions of slender, rigid fibers. *J. Fluid. Mech.*, 296: 271, 1995.

[380] J. Goyon, A. Colin, G. Ovarlez, A. Ajdari, and L. Bocquet. Spatial cooperativity in soft glassy flows. *Nature*, 454: 84, 2008.

[381] E. Akhmatskaya, B. D. Todd, P. J. Daivis, D. J. Evans, K. E. Gubbins, and L. A. Pozhar. A study of viscosity inhomogeneity in porous media. *J. Chem. Phys.*, 106: 4684, 1997.

[382] K. P. Travis. Personal communication.

[383] B. J. Palmer. Transverse-current autocorrelation-function calculations of the shear viscosity for molecular liquids. *Phys. Rev. E*, 49: 359, 1994.

[384] B. Smith, J. S. Hansen, and B. D. Todd. Nonlocal viscosity kernel of mixtures. *Phys. Rev. E*, 85: 022201, 2012.

[385] F. Lado. Numerical Fourier transforms in one, two, and three dimensions for liquid state calculations. *J. Comput. Phys.*, 8: 417, 1971.

[386] R. M. Puscasu, B. D. Todd, P. J. Daivis, and J. S. Hansen. An extended analysis of the viscosity kernel for monatomic and diatomic fluids. *J. Phys: Condens. Matter*, 22: 195105, 2010.

[387] P. J. Cadusch, B. D. Todd, J. Zhang, and P. J. Daivis. A non-local hydrodynamic model for the shear viscosity of confined fluids: analysis of a homogeneous kernel. *J. Phys. A: Math. Theor.*, 41: 035501, 2008.

[388] K. S. Glavatskiy, B. A. Dalton, P. J. Daivis, and B. D. Todd. Non-local viscosity. In preparation.

[389] B. A. Dalton, K. S. Glavatskiy, P. J. Daivis, and B. D. Todd. Non-local density dependent constitutive relations. In preparation.

[390] B. A. Dalton. *The effects of density inhomogeneity and non-locality on nanofluidic flow*. PhD thesis, RMIT University, 2014.

[391] I. Bitsanis, J. J. Magda, M. Tirrell, and H. T. Davis. Molecular dynamics of flow in micropores. *J. Chem. Phys.*, 87: 1733, 1987.

[392] I. Bitsanis, T. K. Vanderlick, M. Tirrell, and H. T. Davis. A tractable molecular theory of flow in strongly inhomogeneous fluids. *J. Chem. Phys.*, 89: 3152, 1988.

[393] C. L. M. H. Navier. Memoire sur les lois du mouvement des fluides. *Mem. Acad. Sci. Inst. Fr.*, 6: 389, 1823.

[394] L. Bocquet and J.-L. Barrat. Hydrodynamic boundary-conditions, correlation-functions, and Kubo relations for confined fluids. *Phys. Rev. E*, 49: 3079, 1994.

[395] J. Petravic and P. Harrowell. On the equilibrium calculation of the friction coefficient for liquid slip against a wall. *J. Chem. Phys.*, 127: 174706, 2007.

[396] J. Petravic and P. Harrowell. On the equilibrium calculation of the friction coefficient for liquid slip against a wall. *J. Chem. Phys.*, 128: 209901, 2008.

[397] S. K. Bhatia and D. Nicholson. Modeling mixture transport at the nanoscale: Departure from existing paradigms. *Phys. Rev. Lett.*, 100: 236103, 2008.

[398] J. Koplik, J. Banavar, and J. Willemsen. Molecular-dynamics of fluid-flow at solid-surfaces. *Phys. Fluids A*, 1: 781, 1989.

[399] F. Brochard and P. G. de Gennes. Shear-dependent slippage at a polymer solid interface. *Langmuir*, 8: 3033, 1992.

[400] Z. Guo, T. S. Zhao, and Y. Shi. Simple kinetic model for fluid flows in the nanometer scale. *Phys. Rev. E*, 71: 035301, 2005.

[401] O. I. Vinogradova. Drainage of a thin liquid-film confined between hydrophobic surfaces. *Langmuir*, 11: 2213, 1995.

[402] C. J. Mundy, S. Balasubramanian, and M. L. Klein. Hydrodynamic boundary conditions for confined fluids via a nonequilibrium molecular dynamics simulation. *J. Chem. Phys.*, 105: 3211, 1996.

[403] S. Heidenreich, P. Ilg, and S. Hess. Boundary conditions for fluids with internal orientational degrees of freedom: Apparent velocity slip associated with the molecular alignment. *Phys. Rev. E*, 75: 066302, 2007.

[404] V. P. Sokhan and N. Quirke. Slip coefficient in nanoscale pore flow. *Phys. Rev. E*, 78: 015301, 2008.

[405] C. Denniston and M. O. Robbins. General continuum boundary conditions for miscible binary fluids from molecular dynamics simulations. *J. Chem. Phys.*, 125: 214102, 2006.

[406] M. Cieplak, J. Koplik, and J. Banavar. Boundary conditions at a fluid-solid interface. *Phys. Rev. Lett.*, 86: 803, 2001.

[407] K. Huang and I. Szlufarska. Green–Kubo relation for friction at liquid-solid surfaces. *Phys. Rev. E*, 89: 032118, 2014.

[408] J. S. Hansen, B. D. Todd, and P. J. Daivis. Prediction of fluid velocity slip at solid surfaces. *Phys. Rev. E*, 84: 016313, 2011.

[409] S. K. Kannam, B. D. Todd, J. S. Hansen, and P. J. Daivis. Slip flow in graphene nanochannels. *J. Chem. Phys.*, 135. 144701, 2011.

[410] S. K. Kannam, B. D. Todd, J. S. Hansen, and P. J. Daivis. Slip length of water on graphene: Limitations of non-equilibrium molecular dynamics simulations. *J. Chem. Phys.*, 136: 024705, 2012.

[411] S. K. Kannam, B. D. Todd, J. S. Hansen, and P. J. Daivis. Interfacial slip friction at a fluid-solid cylindrical boundary. *J. Chem. Phys.*, 136: 244704, 2012.

[412] S. K. Kannam, B. D. Todd, J. S. Hansen, and P. J. Daivis. How fast does water flow in carbon nanotubes? *J. Chem. Phys.*, 138: 094701, 2013.

[413] J. S. Hansen, P. J. Daivis, J. Dyre, B. D. Todd, and H. Bruus. Generalized extended Navier-Stokes theory. *J. Chem. Phys.* 138: 034503, 2013.

[414] J. S. Hansen. Generalized extended Navier-Stokes theory: Multiscale spin relaxation in molecular fluids. *Phys. Rev. E.* 88: 032101, 2013.

[415] J. S. Hansen, J. C. Dyer, P. J. Daivis, B. D. Todd, and H. Bruus. Continuum nanofluidics. *Langmuir* 31:13275, 2015.

Index

Printed in the United States
by Baker & Taylor Publisher Services